Progress in Mathematics
Volume 136

Series Editors

H. Bass
J. Oesterlé
A. Weinstein

Algebraic Topology: New Trends in Localization and Periodicity

Barcelona Conference on Algebraic Topology,
Sant Feliu de Guíxols, Spain, June 1–7, 1994

Carles Broto
Carles Casacuberta
Guido Mislin
Editors

Birkhäuser Verlag
Basel · Boston · Berlin

Editors:

Carles Broto and Carles Casacuberta
Departament de Matemàtiques
Universitat Autònoma de Barcelona
E–08193 Bellaterra
Spain

Guido Mislin
Mathematik
ETH Zentrum
8092 Zürich
Switzerland

A CIP catalogue record for this book is available from the Library of Congress, Washington D.C., USA

Deutsche Bibliothek Cataloging-in-Publication Data

Algebraic topology: new trends in localization and periodicity /
Barcelona Conference on Algebraic Topology, Sant Feliu de
Guixols, Spain, June 1–7, 1994. Carles Broto . . . ed. – Basel ;
Boston ; Berlin : Birkhäuser, 1996
   (Progress in mathematics ; Vol. 136)
   ISBN-13: 978-3-0348-9869-0   e-ISBN-13: 978-3-0348-9018-2
   DOI: 10.1007/978-3-0348-9018-2
NE: Broto, Carles [Hrsg.]; Barcelona Conference on Algebraic Topology
   <1994>; GT
AMS-Classification 55P60.

© 1996 Birkhäuser Verlag, P.O. Box 133, CH-4010 Basel, Switzerland
Softcover reprint of the hardcover1st edition 1996

Printed on acid-free paper produced of chlorine-free pulp. TCF ∞

9 8 7 6 5 4 3 2 1

Table of contents

Foreword

The 1994 Barcelona Conference on Algebraic Topology (BCAT) was held from June 1 to 7 in Sant Feliu de Guíxols, a small town on the Northern Catalan coast. As in the previous meetings of the same series, the conference was organized by the Centre de Recerca Matemàtica (CRM), a mathematical research institute sponsored by the Institut d'Estudis Catalans. The 1994 BCAT was a part of the activities which took place during a semester devoted to algebraic topology at the CRM. The very stimulating atmosphere created by semester visitors and conference participants is reflected in these proceedings. Some of the articles were completed —or even born— in Barcelona during that period.

Articles were submitted between July and November 1994. We are deeply indebted to the authors for their prompt and generous response to our invitation to send us their most recent work. Thus, the volume reports on current progress in algebraic topology, focusing on advances in localization and periodicity in homotopy theory, as a central but not exclusive topic.

All articles have been thoroughly refereed. We take this opportunity to thank the referees once more for their help in the selection of the articles, as well as for their valuable contribution in the improvement of several of them.

The conference organizers were Jaume Aguadé, Manuel Castellet, and Guido Mislin. They join us in thanking the secretaries Consol Roca, Maria Julià, and Sylvia Hoemke for their help before, during, and after the conference. We extend our thanks to the Direcció General de Recerca de la Generalitat de Catalunya and the Dirección General de Investigación Científica y Técnica for their partial financial support. Finally, thanks are due to Birkhäuser Verlag for their interest in the publication of this volume.

Carles Broto
Carles Casacuberta
Guido Mislin

Bellaterra, April 1995

List of participants

J. Aguadé

J. Amorós

M. Arkowitz

D. Arlettaz

G. Arone

M. Bendersky

D. Blanc

A. K. Bousfield

C. Broto

M. Brunetti

R. Budney

J. Cabeza

H. E. A. Campbell

C. Casacuberta

M. Castellet

A. Cavicchioli

B. Cenkl

W. Chachólski

D. Christensen

F. R. Cohen

J. A. Crespo

M. Crossley

J. Cruickshank

W. Dreckmann

E. Dror Farjoun

W. G. Dwyer

C. Elvira

M. I. Gálvez

H. H. Glover

M. Golasiński

M. Grandis

V. Halperin

M. Hartl

A. Hatcher

F. Hegenbarth

J. R. Hubbuck

N. H. V. Hu'ng

J. R. Hunton

P. Igodt

I. M. James

A. Jeanneret

R. Kane

M. Kankaanrinta

T. Kashiwabara

L. Langsetmo

I. J. Leary

K. Lesh

R. Levi

I. Llerena

L. Lomonaco

M. Mahowald

J. Martino

C. A. McGibbon

H. R. Miller

G. Mislin

J. M. Møller

A. Murillo

J. L. Navarro

J. Neisendorfer

F. Neumann

M. Neusel

A. Nofech

D. Notbohm

E. Ossa

A. Osse

G. Peschke

M. Pfenniger

S. B. Priddy

D. Randall

D. C. Ravenel

J. L. Rodríguez

Y. Rudyak

J. W. Rutter

M. Santos

L. Saumell

D. Scevenels

B. Schuster

R. Shank

B. Shipley

L. Smith

S. Smith

D. Tamaki

M. Tanabe

P. Turner

V. V. Vershinin

A. Viruel

Y. Xia

K. Xu

S. Zarati

P. Zhang

BCAT

BARCELONA
CONFERENCE
ON ALGEBRAIC
TOPOLOGY
■ 1 9 9 4

1 M. Castellet
2 W. Dwyer
3 M. Neusel
4 J. Cruickshank
5 Y. Rudyak
6 R. Kane
7 A. Jeanneret
8 M. Mahowald
9 D. Blanc
10 M. Crossley
11 W. Dreckmann
12 I. Leary
13 G. Peschke
14 A. Murillo

15 P. Turner
16 P. Igodt
17 F. Neumann
18 H. Glover
19 F. Cohen
20 R. Shank
21 M. Hartl
22 F. Hegenbarth
23 R. Levi
24 S. Priddy
25 K. Xu
26 J. Martino
27 A. Hatcher
28 E. Campbell

29 Y. Xia
30 H. Miller
31 D. Arlettaz
32 L. Langsetmo
33 J. Rutter
34 P. Zhang
35 W. Chacholski
36 J. Navarro
37 D. Notbohm
38 D. Christensen
39 G. Arone
40 M. Pfenniger
41 J. Hubbuck
42 M. Golasinski

43 J. Hunton
44 B. Schuster
45 S. Smith
46 D. Tamaki
47 D. Scevenels
48 C. Casacuberta
49 D. Ravenel
50 L. Saumell
51 J. Møller
52 V. Halperin
53 M. Bendersky
54 J. Neisendorfer
55 D. Randall
56 B. Cenkl

57 L. Smith
58 I. James
59 G. Mislin
60 J. Aguadé
61 C. Broto
62 A. Cavicchioli
63 A. Osse
64 C. Elvira
65 C. McGibbon
66 J. Rodriguez
67 S. Zarati
68 E. Dror Farjoun
69 I. Gálvez
70 K. Lesh

71 I. Llerena
72 E. Ossa
73 M. Grandis
74 M. Kankaanrinta
75 B. Shipley
76 L. Lomonaco
77 M. Brunetti
78 J. Crespo
79 A. Viruel
80 M. Tanabe
81 T. Kashiwabara
82 N. Hung
83 V. Vershinin
84 M. Santos

Titles of talks

(in chronological order)

A. K. Bousfield, K-theoretic localizations and periodizations of spaces

D. C. Ravenel, Telescopes in stable homotopy theory

C. Casacuberta, Homotopy idempotent functors and fundamental group

L. Langsetmo, The K-theory of loop spaces: examples, applications, and generalizations

J. M. Møller, Completely reducible p-compact groups

D. Notbohm, On the classification of p-compact groups

R. Levi, On homological rate of growth and a conjecture of Cohen

F. R. Cohen, On applications of combinatorial group theory to topology

Y. Xia, Periodicity in the mapping class group and its cohomology

B. Shipley, On the convergence of second quadrant homology spectral sequences

R. Kane, Elementary abelian p-subgroups of Lie groups

W. G. Dwyer, The centralizer diagram for BG

M. Mahowald, Goodwillie towers and the v_n-localization of odd spheres

S. Zarati, Tor and Ext dimension of an unstable H^*V-A-module

M. Bendersky, The structure of the v_1-periodic homotopy groups of Lie groups

D. Blanc, A model category for periodic homotopy

E. Dror Farjoun, On the cellular structure of K-acyclic spaces

W. Chachólski, Closed classes and localization functors

G. Peschke, Local global principles and genus in group theory and homotopy theory

J. Hubbuck, Intrinsic structure of the homology of an elementary abelian 2-group

Progress in Mathematics, Vol. 136
© 1996 Birkhäuser Verlag Basel/Switzerland

On the Nilpotency of Subgroups of Self-Homotopy Equivalences

Martin Arkowitz and Gregory Lupton*

1 Introduction

If X is a topological space, we denote by $\mathcal{E}(X)$ the set of homotopy classes of self-homotopy equivalences of X. Then $\mathcal{E}(X)$ is a group with group operation given by composition of homotopy classes. The group $\mathcal{E}(X)$ is a natural object in homotopy theory and has been studied extensively —see [**Ar**] for a survey of known results and applications of $\mathcal{E}(X)$. In this paper we continue our investigation of $\mathcal{E}_{\#}(X)$, the subgroup of $\mathcal{E}(X)$ consisting of homotopy classes which induce the identity on homotopy groups, and, to a lesser extent, of $\mathcal{E}_{*\#}(X)$, the subgroup of $\mathcal{E}_{\#}(X)$ consisting of homotopy classes which also induce the identity on homology groups (see §2 for precise definitions), which was begun in [**A-L**]. These groups are nilpotent and we focus primarily on the nilpotency class of $\mathcal{E}_{\#}(X)$. The determination of this nilpotency class appears in the list of problems on $\mathcal{E}(X)$ in [**Ka, Problem 10**]. For rational spaces we obtain both general results on the nilpotency class and a complete determination of the nilpotency class in specific cases. This leads to a lower bound for the nilpotency class of the groups $\mathcal{E}_{\#}(X)$ for certain finite complexes X by using derationalization techniques.

We now describe the organization of the paper and our results. In §2, after fixing notation, we indicate the relationship between $\mathcal{E}_{\#}(X)$ and $\mathcal{E}_{\#}(X_{\mathbb{Q}})$ when X is a finite complex with rationalization $X_{\mathbb{Q}}$. We study $\mathcal{E}_{\#}(X_{\mathbb{Q}})$ by means of the Sullivan minimal model $\mathcal{M}$ of X. This involves a purely algebraic investigation of the group $\mathcal{E}_{\#}(\mathcal{M})$, the analogue of $\mathcal{E}_{\#}(X_{\mathbb{Q}})$ in the homotopy category of rational minimal algebras. Similar considerations hold for the group $\mathcal{E}_{*\#}(X)$ and its minimal model analogue $\mathcal{E}_{\#}^{*}(\mathcal{M})$. We begin our investigation in §3 by considering the automorphisms $\mathrm{Aut}_{\#}(\mathcal{M})$ of a rational minimal algebra $\mathcal{M}$ which induce the identity on indecomposables. If $\varphi \in \mathrm{Aut}_{\#}(\mathcal{M})$, then we denote by $l(\varphi)$ the length of the shortest decomposable term obtained by applying φ to each generator of $\mathcal{M}$. Our main result in §3 is $l((\varphi, \psi)) \geq l(\varphi) + l(\psi) - 1$, where (φ, ψ) is the commutator of $\varphi, \psi \in \mathrm{Aut}_{\#}(\mathcal{M})$ (Proposition 3.2). This yields an upper bound on $\mathrm{nil}\,\mathcal{E}_{\#}(\mathcal{M})$ which depends on the degrees of the generators of $\mathcal{M}$. In §4 we consider elements in $\mathrm{Aut}_{\#}(\mathcal{M})$ which are the identity on all but one generator of $\mathcal{M}$ and obtain formulas for commutators of these automorphisms (Lemma 4.3 and Proposition 4.4). These formulas are used in

---

*)   The first named author would like to thank the Centre de Recerca Matemàtica of Barcelona for its hospitality during the time that this paper was written.

the calculations in the last two sections. In Section 5 we study the relationship of nil $\mathcal{E}_\#(\mathcal{M})$ to the cup-length $c(\mathcal{M})$ of $H^*(\mathcal{M})$ and to the e_0-invariant $e_0(\mathcal{M})$ of $\mathcal{M}$, the maximum length of all non-exact cocycles of $\mathcal{M}$. For formal minimal algebras $\mathcal{M}$, we show that nil $\mathcal{E}_\#(\mathcal{M}) < c(\mathcal{M})$ (Theorem 5.2) and for formal or coformal minimal algebras $\mathcal{M}$, we show that nil $\mathcal{E}_\#(\mathcal{M}) < e_0(\mathcal{M})$ (Corollary 5.3 and Theorem 5.5). In §6 we consider two stage algebras $\mathcal{M}$, a class of minimal algebras which includes the minimal models of homogeneous spaces, and obtain several results on $\mathcal{E}_\#^*(\mathcal{M})$. In the last two sections we compute $\mathcal{E}_\#(\mathcal{M})$ for specific two stage algebras $\mathcal{M}$. In §7 we determine nil $\mathcal{E}_\#(\mathcal{M})$ for many minimal algebras $\mathcal{M}$ with trivial differential. These include the minimal models of (1) the Lie groups $SU(n)$ (2) the Lie groups $Sp(n)$ (3) the classifying spaces $BU(n)$ and $BSp(n)$. We conclude the paper in Section 8 by calculating nil $\mathcal{E}_\#(\mathcal{M})$ for $\mathcal{M}$ the minimal model of the homogeneous spaces $U(n)/(U(p) \times U(q))$.

The approach of this paper can be dualized to a study of $\mathcal{E}_*(X)$, the subgroup of $\mathcal{E}(X)$ of homotopy classes which induce the identity on homology groups. This can be done by studying $\mathcal{E}_*(\mathcal{L})$, where $\mathcal{L}$ is the Quillen Lie algebra minimal model of X. However, the examples of two stage minimal Lie algebras are Quillen minimal models of two-cones, i.e., of the cofibre of a map between wedges of spheres. These appear to be less interesting than the examples of two stage minimal algebras such as those considered in Sections 7 and 8.

We conclude the introduction by emphasizing two points. First of all, although we use the techniques of rational homotopy theory and most of our results are stated for rational minimal algebras, our main interest is in $\mathcal{E}_\#(X)$ for a CW-complex X, most especially for a finite, 1-connected complex X. To underscore this point, we have, whenever applicable, included an interpretation of our results for $\mathcal{E}_\#(X)$. Frequently this gives a lower bound for nil $\mathcal{E}_\#(X)$ for certain finite complexes X. Although the lower bounds are most likely very crude (see Examples 7.4(1)), they are to our knowledge the first general results for this invariant. Secondly, our work here and in [A-L] should be thought of as a method for the study of $\mathcal{E}_\#(X)$, $\mathcal{E}_*(X)$ and $\mathcal{E}_{*\#}(X)$. The examples we have worked out in Sections 7 and 8 are to be regarded as illustrations of this method. It is clear that many other spaces —for instance, different Lie groups and homogeneous spaces— could be chosen and similar calculations could be made.

2 Minimal Algebras and Rationalization

We begin with some algebraic preliminaries, using standard conventions (see [G-M], [H-S] and [A-L]). If for each positive integer i, V^i is a vector space over the rationals $\mathbb{Q}$, then we call the collection $V = \{V^i\}$ a graded vector space. If $v_1, \ldots, v_k, \ldots$, is a basis of V, we write $V = \langle v_1, \ldots, v_k, \ldots \rangle$.

Now let $\mathcal{A}$ denote a differential graded commutative algebra over $\mathbb{Q}$ (DG algebra, for short). By $x \in \mathcal{A}$ is meant $x \in \mathcal{A}^p$, for some p. We then write $|x| = p$ for the degree of x. If $x \in \mathcal{A}$ is a cocycle, then $[x] \in H^*(\mathcal{A})$ denotes the cohomology class of x. By a map $\varphi: \mathcal{A} \to \mathcal{B}$ of DG algebras we mean a DG algebra homomorphism. A map φ induces a map $\varphi^*: H^*(\mathcal{A}) \to H^*(\mathcal{B})$ of cohomology algebras. The identity map of a DG algebra will always be denoted by ι. If $\mathcal{A}$ is the free commutative graded algebra generated by a graded vector space V, then we write $\mathcal{A} = \Lambda(V)$. If $V = \langle v_1, \dots, v_k, \dots \rangle$, then $\mathcal{A} = \Lambda(V) = \Lambda(v_1, \dots, v_k, \dots)$. If V has a finite basis, we call $\mathcal{A} = \Lambda(V)$ finitely-generated. A DG algebra $\mathcal{M}$ with differential d is a *minimal algebra* if (1) $\mathcal{M} = \Lambda(V)$ for some graded vector space V and (2) there exists a basis $v_1, \dots, v_k, \dots$ of V with $|v_1| \leq \cdots \leq |v_k| \leq \cdots$ such that $d(v_i) \in \Lambda(v_1, \dots, v_{i-1})$, for all i. In addition, we always assume that $\mathcal{M}$ has finite type (i.e., $H^i(\mathcal{M})$ is finite dimensional for each i) and that $\mathcal{M}$ is 1-connected (i.e., $\mathcal{M}^0 = \mathbb{Q}$ and $\mathcal{M}^1 = 0$). The latter hypothesis can be weakened in many of our results, but we assume it to avoid complications.

For maps $\varphi, \psi \colon \mathcal{M} \to \mathcal{N}$ of minimal algebras we use the notion of homotopy given in [**H-S, p. 240**] which we now describe. Suppose $\mathcal{M} = \Lambda(V)$ with differential d and define a DG algebra $\mathcal{M}^I = \Lambda(V \oplus \overline{V} \oplus \widehat{V})$ with differential also called d as follows: $\widehat{V}$ is an isomorphic copy of V and $\overline{V}$ is the desuspension of V (i.e., $\overline{V}^p = V^{p+1}$). Furthermore, the differential d of $\mathcal{M}^I$ agrees with the differential on $\mathcal{M}$, $d(\overline{v}) = \hat{v}$ and $d(\hat{v}) = 0$, for $\overline{v} \in \overline{V}$ and $\hat{v} \in \widehat{V}$. In addition, there is a degree -1 derivation $i \colon \mathcal{M}^I \to \mathcal{M}^I$ defined on generators by $i(v) = \overline{v}$, $i(\overline{v}) = 0$ and $i(\hat{v}) = 0$. We then obtain a degree 0 derivation $\gamma \colon \mathcal{M}^I \to \mathcal{M}^I$ defined by setting $\gamma = di + id = [d, i]$ (the bracket of graded derivations). Finally, we have a map $\alpha \colon \mathcal{M}^I \to \mathcal{M}^I$ defined by

$$\alpha = \sum_{n=0}^{\infty} \frac{1}{n!} \gamma^n.$$

A *homotopy* from $\mathcal{M}$ to $\mathcal{N}$ is a map $H \colon \mathcal{M}^I \to \mathcal{N}$. Then H is a homotopy from the map φ to the map ψ if $H|_{\mathcal{M}} = \varphi$ and $H\alpha|_{\mathcal{M}} = \psi$. We say that H begins at φ and ends at ψ, and we write $\varphi \simeq \psi$.

For a minimal algebra $\mathcal{M}$, the group of homotopy classes of homotopy equivalences from $\mathcal{M}$ to $\mathcal{M}$ is denoted $\mathcal{E}(\mathcal{M})$. In this paper we study the following subgroups of $\mathcal{E}(\mathcal{M})$: (1) $\mathcal{E}_\#(\mathcal{M})$ the subgroup of homotopy classes which induce the identity on the vector space $Q^i(\mathcal{M})$ of indecomposables of $\mathcal{M}$ for all i and (2) $\mathcal{E}_\#^*(\mathcal{M})$, the subgroup of $\mathcal{E}_\#(\mathcal{M})$ of homotopy classes which also induce the identity on cohomology $H^*(\mathcal{M})$. We also denote by $\mathcal{E}_{\#N}(\mathcal{M})$ the subgroup of $\mathcal{E}(\mathcal{M})$ of homotopy classes which induce the identity on $Q^i(\mathcal{M})$, for $i \leq N$.

For a group G, we denote the commutator $x^{-1}y^{-1}xy$ of elements $x, y \in G$ by (x, y). For n elements $x_1, x_2, \dots, x_{n-1}, x_n \in G$, the n-fold commutator (or commutator of length n) is $(x_1, (x_2, \dots, (x_{n-1}, x_n) \dots))$, denoted $(x_1, x_2, \dots, x_{n-1}, x_n)$. Recall that G is nilpotent if all n-fold commutators in G are trivial for all sufficiently large n. In this case one defines the nilpotency class $\operatorname{nil} G$ to be the integer k such that there is a non-trivial k-fold commutator and all $(k+1)$-fold commutators are trivial. We regard the trivial group as the group having nilpotency class 0.

All topological spaces will be based and have the based homotopy type of a CW-complex of finite type. Furthermore, all spaces that we consider will be 1-connected, though it is possible to weaken this condition. All maps and homotopies are to preserve base point. A space Y is called rational if $\pi_i(Y)$ is a vector space over $\mathbb{Q}$ for all i. For a space X, let $X_{\mathbb{Q}}$ denote the rationalization of X and for a homotopy class α let $\alpha_{\mathbb{Q}}$ denote the rationalization of α [H-M-R]. The Sullivan minimal model of X is denoted $\mathcal{M}_X$ [G-M].

For a space X we let $\mathcal{E}(X)$ denote the group of homotopy classes of homotopy equivalences of X to itself, as stated in the introduction. Now let X have the homotopy type of a finite complex of dimension N or of the rationalization of such a finite complex. We then define $\mathcal{E}_\#(X)$ to be the kernel of the natural homomorphism

$$\mathcal{E}(X) \longrightarrow \sum_{i \leq N} \operatorname{Aut} \pi_i(X).$$

Then $\mathcal{E}_\#(X)$ is a nilpotent group [D-Z] and so its rationalization (or localization at the empty set) $(\mathcal{E}_\#(X))_{\mathbb{Q}}$ exists [H-M-R]. Then Maruyama [Ma] has proved the natural map $e \colon \mathcal{E}_\#(X) \to \mathcal{E}_\#(X_{\mathbb{Q}})$ defined by $e(\alpha) = \alpha_{\mathbb{Q}}$ is, up to isomorphism, the rationalization homomorphism. Thus we have the following result.

2.1 PROPOSITION. *Let X have the homotopy type of a finite complex of dimension N and let $\mathcal{M}$ be the minimal model of X. Then $\mathcal{E}_\#(\mathcal{M})$ is a nilpotent group and*

$$\operatorname{nil} \mathcal{E}_\#(\mathcal{M}) \leq \operatorname{nil} \mathcal{E}_\#(X).$$

Furthermore, if $\mathcal{M} = \Lambda(v_1, \ldots, v_k)$ with $|v_1| \leq \cdots \leq |v_k| \leq N$ and

$$\operatorname{nil} \mathcal{E}_\#(\mathcal{M}) = r,$$

then all $(r+1)$-fold commutators in $\mathcal{E}_\#(X)$ are of finite order.

Proof. Since $\mathcal{E}_\#(X_{\mathbb{Q}})$ is the rationalization of $\mathcal{E}_\#(X)$, it follows that $\mathcal{E}_\#(X_{\mathbb{Q}})$ is nilpotent and $\operatorname{nil} \mathcal{E}_\#(X_{\mathbb{Q}}) \leq \operatorname{nil} \mathcal{E}_\#(X)$ [H-M-R, p. 7]. Now the indecomposables $Q^i(\mathcal{M})$ of the minimal model $\mathcal{M}$ correspond to the rational homotopy groups of X [G-M, p. 136], and so $\mathcal{E}_{\#N}(\mathcal{M})$ is anti-isomorphic to $\mathcal{E}_\#(X_{\mathbb{Q}})$ (see [A-L, Rem. 2.3]). Since $\mathcal{E}_\#(\mathcal{M}) \subseteq \mathcal{E}_{\#N}(\mathcal{M})$, the group $\mathcal{E}_\#(\mathcal{M})$ is nilpotent and its nilpotency class is less than or equal to that of $\mathcal{E}_\#(X)$. For the second assertion, note that under the hypothesis, $\mathcal{E}_\#(\mathcal{M}) = \mathcal{E}_{\#N}(\mathcal{M})$, and so $(r+1)$-fold commutators in $\mathcal{E}_\#(X)$ are in the kernel of the rationalization homomorphism. Thus they have finite order. □

2.2 Remark. One defines the group $\mathcal{E}_{*\#}(X)$ for a finite complex X of dimension N (or its rationalization) to be the kernel of the natural homomorphism

$$\mathcal{E}(X) \longrightarrow \left(\sum_{i \leq N} \operatorname{Aut} \pi_i(X) \right) \times \left(\sum_{i \leq N} \operatorname{Aut} H_i(X) \right).$$

Since $\mathcal{E}_{*\#}(X) \subseteq \mathcal{E}_\#(X)$, $\mathcal{E}_{*\#}(X)$ is a nilpotent group. One can then show that the natural map $e' \colon \mathcal{E}_{*\#}(X) \to \mathcal{E}_{*\#}(X_{\mathbb{Q}})$ is the rationalization homomorphism (see [A-L, Prop.2.1 and Rem.2.2]). It then follows that Proposition 2.1 holds with $\mathcal{E}_{*\#}(X)$ replacing $\mathcal{E}_\#(X)$ and $\mathcal{E}_\#^*(\mathcal{M})$ replacing $\mathcal{E}_\#(\mathcal{M})$.

3 Commutators in $\mathrm{Aut}_{\#}(\mathcal{M})$

In this section we study commutators of certain automorphisms of a minimal algebra and obtain some simple but useful results. Let $\mathcal{M}$ be a minimal algebra with differential d. We fix a basis of algebra generators of $\mathcal{M}$ and write $\mathcal{M} = \Lambda(v_1, \dots, v_k, \dots)$, where the degree $|v_j| = n_j$ and $n_1 \leq \cdots \leq n_k \leq \cdots$. Let $\mathrm{Aut}_{\#}(\mathcal{M})$ be the group of automorphisms $\varphi \colon \mathcal{M} \to \mathcal{M}$ such that φ induces the identity on the vector space of indecomposables. Then for each j,

$$\varphi(v_j) = v_j + \chi_j \,,$$

where χ_j is decomposable. The relation of homotopy for maps of minimal algebras (see §2) induces an equivalence relation on $\mathrm{Aut}_{\#}(\mathcal{M})$. The set of homotopy classes of $\mathrm{Aut}_{\#}(\mathcal{M})$ is the group $\mathcal{E}_{\#}(\mathcal{M})$ defined in §2.

Now let $\chi \in \mathcal{M} = \Lambda(v_1, \dots, v_k, \dots)$ and $\varphi \in \mathrm{Aut}_{\#}(\mathcal{M})$. We write χ as a polynomial in $v_1, \dots, v_k, \dots$.

3.1 DEFINITIONS. *If $\chi \neq 0$, the word length $w(\chi)$ is defined to be the length of the shortest monomial in χ. If $\chi = 0$, set $w(\chi) = \infty$. Define the (perturbation) length $l(\varphi)$ to be $\min\{w(\varphi(v_j) - v_j)\}$ over all generators v_j.*

Note that $\varphi = \iota$, the identity map of $\mathcal{M}$, if and only if $l(\varphi) = \infty$. Note too that $l(\varphi) \geq 2$ for every $\varphi \in \mathrm{Aut}_{\#}(\mathcal{M})$. Finally, it is not difficult to show that $l(\varphi^{-1}) = l(\varphi)$ for every $\varphi \in \mathrm{Aut}_{\#}(\mathcal{M})$. We next consider the length of a commutator in $\mathrm{Aut}_{\#}(\mathcal{M})$.

3.2 PROPOSITION. *If $\varphi, \psi \in \mathrm{Aut}_{\#}(\mathcal{M})$ and $(\varphi, \psi) \in \mathrm{Aut}_{\#}(\mathcal{M})$ is the commutator of φ and ψ, then*

$$l((\varphi, \psi)) \geq l(\varphi) + l(\psi) - 1 \,.$$

Proof. Suppose $l(\varphi) = r$ and $l(\psi) = s$ and for each $j = 1, \dots, k, \dots$ write

$$\varphi(v_j) = v_j + \chi_j^r + \chi_j^{r+1} + \cdots + \chi_j^{N_j} \qquad \text{and}$$
$$\psi(v_j) = v_j + \xi_j^s + \xi_j^{s+1} + \cdots + \xi_j^{M_j} \,,$$

where each χ_j^l and ξ_j^l is either zero or a homogeneous polynomial of length l in $v_1, \dots, v_k, \dots$. Since $l(\varphi) = r$, we have, for any j and l, a congruence modulo terms of length $\geq l - 1 + r$,

$$\varphi(\xi_j^l) \equiv \xi_j^l \,. \tag{1}$$

Similarly, since $l(\psi^{-1}) = s$, modulo terms of length $\geq l - 1 + s$,

$$\psi^{-1}(\chi_j^l) \equiv \chi_j^l \,, \tag{2}$$

for any j and l. Furthermore, modulo terms of length $\geq r + s - 1$,

$$\varphi^{-1}(v_j + \chi_j^r + \cdots + \chi_j^{r+s-2}) \equiv v_j \qquad \text{and} \tag{3}$$
$$\psi^{-1}(v_j + \xi_j^s + \cdots + \xi_j^{r+s-2}) \equiv v_j \,. \tag{4}$$

Now we calculate $(\varphi, \psi)(v_j) = (\varphi^{-1}\psi^{-1}\varphi\psi)(v_j)$ modulo terms of length $\geq r + s - 1$. Using (1), we have

$$(\varphi, \psi)(v_j) \equiv \varphi^{-1}\psi^{-1}\varphi(v_j + \xi_j^s + \cdots + \xi_j^{r+s-2})$$
$$\equiv \varphi^{-1}\psi^{-1}(v_j + \chi_j^r + \cdots + \chi_j^{r+s-2} + \xi_j^s + \cdots + \xi_j^{r+s-2}).$$

It then follows from (4) and (2) that

$$(\varphi, \psi)(v_j) \equiv \varphi^{-1}(\psi^{-1}(v_j + \xi_j^s + \cdots + \xi_j^{r+s-2}) + \psi^{-1}(\chi_j^r + \cdots + \chi_j^{r+s-2}))$$
$$\equiv \varphi^{-1}(v_j + \chi_j^r + \cdots + \chi_j^{r+s-2}).$$

Therefore by (3),

$$(\varphi, \psi)(v_j) \equiv v_j,$$

modulo terms of length $\geq r + s - 1$. It then follows that $l(\varphi, \psi) \geq r + s - 1$.
□

The next result follows by induction.

3.3 COROLLARY. *If $\varphi_i \in \mathrm{Aut}_\#(\mathcal{M})$, $i = 1, \ldots, r$ and $(\varphi_1, \varphi_2, \ldots, \varphi_r)$ is the r-fold commutator, then*

$$l((\varphi_1, \varphi_2, \ldots, \varphi_r)) \geq l(\varphi_1) + l(\varphi_2) + \cdots + l(\varphi_r) - (r - 1).$$

In particular, if φ is an r-fold commutator of elements of $\mathrm{Aut}_\#(\mathcal{M})$, then $l(\varphi) \geq r + 1$.

In the case that $\mathcal{M}$ is finitely-generated, then we have as an immediate consequence of Corollary 3.3 that $\mathcal{E}_\#(\mathcal{M})$ is a nilpotent group (see Proposition 2.1). A crude estimate of $\mathrm{nil}\,\mathcal{E}_\#(\mathcal{M})$ can now easily be made.

3.4 PROPOSITION. *If $\mathcal{M} = \Lambda(v_1, \ldots, v_k)$ is finitely-generated and $n_k < (r + 1)n_1$, then $\mathrm{nil}\,\mathcal{E}_\#(\mathcal{M}) \leq r - 1$.*

Proof. If $\varphi \in \mathrm{Aut}_\#(\mathcal{M})$ is an r-fold commutator of elements of $\mathrm{Aut}_\#(\mathcal{M})$, then $l(\varphi) \geq r + 1$. If $\varphi \neq \iota$, then for some $i = 1, \ldots, k$,

$$n_i = |v_i| = |\varphi(v_i) - v_i| \geq (r + 1)n_1.$$

Since this contradicts the hypothesis, $\varphi = \iota$. Thus $\mathrm{nil}\,\mathcal{E}_\#(\mathcal{M}) \leq r - 1$.   □

In the case all the n_i are odd, the following sharpening of the above conclusions is evident.

3.5 PROPOSITION. *Let $\mathcal{M} = \Lambda(v_1, \ldots, v_k, \ldots)$ be a minimal algebra with all $n_i = |v_i|$ odd. If φ is an r-fold commutator of elements of $\mathrm{Aut}_\#(\mathcal{M})$, then $l(\varphi) \geq 2r + 1$. If $\mathcal{M} = \Lambda(v_1, \ldots, v_k)$ is finitely-generated and $n_k < (2r + 1)n_1$, then $\mathrm{nil}\,\mathcal{E}_\#(\mathcal{M}) \leq r - 1$.*

Of course Propositions 3.4 and 3.5 can be restated for rational spaces. We explicitly give such a result.

3.6 PROPOSITION. *Let X be the rationalization of an $(n-1)$-connected finite complex of dimension N, $n \geq 2$. If $N < (r+1)n$, then $\mathrm{nil}\,\mathcal{E}_{\#}(X) \leq r-1$. In particular, if $N < 3n$, then $\mathcal{E}_{\#}(X)$ is abelian.*

Proof. This is an immediate consequence of the fact that $\mathcal{E}_{\#}(X) \approx \mathcal{E}_{\#}(X^{(N)})$, where $X^{(N)}$ is the N-th Postnikov section of X (see [**A-C, p. 21**]). $\qquad\square$

4 Elementary Automorphisms

In this section we consider a class of automorphisms in $\mathrm{Aut}_{\#}(\mathcal{M})$ which plays an important role in this paper.

4.1 DEFINITION. *If $\mathcal{M} = \Lambda(v_1, \dots, v_k, \dots)$ is a minimal algebra with $|v_1| \leq \cdots \leq |v_k| \leq \cdots$, then for each $j = 1, \dots, k, \dots$ and decomposable element $\chi \in \mathcal{M}$ with $|v_j| = |\chi|$, define an algebra automorphism $\varphi_\chi^j \colon \mathcal{M} \to \mathcal{M}$ by*

$$\varphi_\chi^j(v_i) = \begin{cases} v_i, & \text{if } i \neq j \\ v_j + \chi, & \text{if } i = j. \end{cases}$$

In the definition, we allow $\chi = 0$, in which case $\varphi_\chi^j = \iota$. If $\varphi_\chi^j \in \mathrm{Aut}_{\#}(\mathcal{M})$, then we call φ_χ^j an *elementary automorphism*. Note that if φ_χ^j is an elementary automorphism, then χ is a cocycle. However, we show next that the converse of this is not true.

4.2 Example. Suppose that $\mathcal{M} = \Lambda(v_1, \dots, v_k, \dots)$ with differential d and that χ is a decomposable cocycle of degree n_j. Then $\varphi_\chi^j \colon \mathcal{M} \to \mathcal{M}$ as defined above is an algebra automorphism, but it does not necessarily commute with d. For, if $l > j$ and $d(v_l) = v_j \xi + \zeta$, where ξ and ζ are polynomials in $v_1, \dots, v_{j-1}, v_{j+1}, \dots$, then $d\varphi_\chi^j(v_l) = d(v_l) = v_j \xi + \zeta$ and $\varphi_\chi^j d(v_l) = \varphi_\chi^j(v_j \xi + \zeta) = v_j \xi + \chi \xi + \zeta$. Thus if $\chi\xi \neq 0$, then φ_χ^j does not commute with d. It is clear that one can construct many concrete examples where this holds.

We remark that the composition of two elementary automorphisms is not necessarily elementary, but that the inverse of an elementary automorphism is an elementary automorphism, in fact, $(\varphi_\chi^j)^{-1} = \varphi_{-\chi}^j$. We consider next the commutator of two elementary automorphisms.

4.3 LEMMA. *If φ_χ^j and φ_ξ^l are elementary automorphisms and $j > l$, then*

$$(\varphi_\chi^j, \varphi_\xi^l) = \varphi_\zeta^j,$$

where $\zeta = \varphi_{-\xi}^l(\chi) - \chi$. Consequently, the commutator of two elementary automorphisms is an elementary automorphism.

Proof. The first assertion follows by direct calculation. For the second assertion, note that $(\varphi_\chi^j, \varphi_\xi^j) = \iota$ and that $(\varphi_\xi^l, \varphi_\chi^j) = (\varphi_\chi^j, \varphi_\xi^l)^{-1} = \varphi_{-\zeta}^j$. $\qquad\square$

We next obtain a formula for the s-fold commutator of elementary automorphisms in a special case. The formula is not the most general result of this type which can be proved, but it is sufficient for our purposes in §§7 and 8. As is usual, a circumflex over a symbol that appears in a list indicates that symbol is to be omitted.

4.4 PROPOSITION. *Let* $\mathcal{M} = \Lambda(v_1,\ldots,v_k,\ldots)$ *with* $|v_1| \leq \cdots \leq |v_k| \leq \cdots$ *and let* $(i_1,\ldots,i_s)$ *be a sequence of positive integers with* $i = \max(i_1,\ldots,i_s)$. *Suppose for each* $j = 1,\ldots,s$, *there is a* $\varphi_{\chi_j}^{i_j} \in \mathrm{Aut}_{\#}(\mathcal{M})$ *such that* $|v_{i_j}|$ *is odd and* χ_j *is a monomial in* $v_1,\ldots,v_k,\ldots$. *Define a monomial* $\alpha \in \mathcal{M}$ *as follows: List all the* v_r *which occur as factors in* $\chi_1,\chi_2,\ldots,\chi_s$ *(allowing repetition). If each of* $v_{i_1},\ldots,\hat{v}_i,\ldots,v_{i_s}$ *does not appear in the list, set* $\alpha = 0$. *Otherwise, delete one occurence of each of* $v_{i_1},\ldots,\hat{v}_i,\ldots,v_{i_s}$ *from the list and set* α *equal to the product of the remaining members. Then the* s-*fold commutator*

$$\left(\varphi_{\chi_s}^{i_s},\ldots,\varphi_{\chi_2}^{i_2},\varphi_{\chi_1}^{i_1}\right)$$

is either ι *or* $\varphi_{\pm\alpha}^i$.

Proof. The result is proved by induction on s using Lemma 4.3. The proof is straightforward and hence omitted. $\square$

5 The e_0 Invariant

In this section we relate the nilpotency of $\mathcal{E}_{\#}(\mathcal{M})$ to some well-known numerical invariants of a minimal model, namely the cup-length $c(\mathcal{M})$ —that is, the length of the longest non-trivial product in $H^*(\mathcal{M})$— and Toomer's e_0 invariant of $\mathcal{M}$ defined below. In each of our results, we require an additional hypothesis on the minimal model, such as formality or coformality.

We begin with a brief review of some terminology and notation. Suppose that the minimal algebra $\mathcal{M}$ with differential d is *formal*. Then $\mathcal{M} = \Lambda(V)$ is a *bigraded model* —see **[H-S]** for a full discussion. This means, in particular, that we can write $V = \bigoplus_{j\geq 0} V_j$, and if this lower grading is extended to the algebra $\Lambda(V)$ in the natural way, then we have $d : V_j \to \left(\Lambda(V)\right)_{j-1}$ for each $j \geq 1$. Furthermore, $d(V_0) = 0$ and, transferring the lower grading to cohomology, we have $H_i^*(\mathcal{M}) = 0$ for $i \geq 1$ and $H^*(\mathcal{M}) \cong H_0^*(\mathcal{M})$ as algebras.

Now consider $\varphi \in \mathrm{Aut}_{\#}(\mathcal{M})$ with $\mathcal{M}$ formal. We use the extra structure of the bigraded model to sharpen the notion of length introduced in Section 3. On each basis element v_j of V, we write

$$\varphi(v_j) = v_j + \chi_{(j,0)} + \chi_{(j,1)} + \cdots + \chi_{(j,N_j)},$$

where $\chi_{(j,k)}$ is that part of the decomposable element of lower degree k. Thus we define the *lower degree zero (perturbation) length* of φ by

$$\ell_0(\varphi) = min\{w\big(\chi_{(j,0)}\big)\}$$

over all j.

5.1 LEMMA. *Suppose* $\mathcal{M}$ *is formal and* $c(\mathcal{M}) = N$. *If* $\varphi \in \mathrm{Aut}_{\#}(\mathcal{M})$ *with* $\ell_0(\varphi) \geq N + 1$ *and if* $\varphi = \iota$ *on generators of degree* $\leq n$, *then* φ *is homotopic to some* $\psi \in \mathrm{Aut}_{\#}(\mathcal{M})$ *such that* $\ell_0(\psi) \geq N + 1$ *and* $\psi = \iota$ *on generators of degree* $\leq n + 1$.

Proof. Let v_j be a generator of degree $n + 1$ and write $\varphi(v_j)$ as above. Then $w(\chi_{(j,0)}) \geq N + 1$ by assumption. Since $\chi_{(j,0)}$ is an element of $(\Lambda(V))_0$, $\chi_{(j,0)}$

is a cocycle. Furthermore the cohomology class that it represents is a sum of products of cohomology classes of length at least $N+1$. Hence $\chi_{(j,0)}$ is a coboundary. On the other hand, $\varphi = \iota$ on generators of degree $\leq n$, and it follows that $\chi_{(j,1)} + \chi_{(j,2)} + \cdots + \chi_{(j,N_j)}$ is a cocycle. Since this latter term represents a cohomology class in $\bigoplus_{i \geq 1} H_i^*(\mathcal{M})$ which is zero, we have that $\chi_{(j,0)} + \chi_{(j,1)} + \cdots + \chi_{(j,N_j)}$ is a coboundary. So for each generator v_j of degree $n+1$, write $d(\eta_j) = \chi_{(j,0)} + \chi_{(j,1)} + \cdots + \chi_{(j,N_j)}$. Since d is zero on V_0, we can assume that $\eta_j \in \bigoplus_{i \geq 1} \left(\Lambda(V) \right)_i$. Now define a homotopy $H : \Lambda(V, \overline{V}, \widehat{V}) \to \Lambda(V)$ starting at φ by

$$H(\bar{v}_j) = \begin{cases} -\eta_j \text{ if } |v_j| = n+1 \\ 0 \text{ otherwise} \end{cases}$$

$$H(\hat{v}_j) = \begin{cases} -(\chi_{(j,0)} + \chi_{(j,1)} + \cdots + \chi_{(j,N_j)}) \text{ if } |v_j| = n+1 \\ 0 \text{ otherwise.} \end{cases}$$

By construction, H ends at a map $\psi = H\alpha$ that equals ι on generators v_t of degree $\leq n+1$.

We claim that on each generator v_t of degree $\geq n+2$, ψ has the following form:

$$\psi(v_t) = v_t + \chi_{(t,0)} + \xi_{(t,1)} + \xi_{(t,2)} + \cdots + \xi_{(t,M_t)},$$

where $\chi_{(t,0)}$ is the lower degree zero term of $\varphi(v_t)$ and each $\xi_{(t,k)}$ is a term of lower degree k. For if v_t is any generator, then

$$\psi(v_t) = H\alpha(v_t) = H(v_t + \hat{v}_t + id(v_t) + \frac{\gamma^2}{2}(v_t) + \cdots)$$
$$= \varphi(v_t) + H(\hat{v}_t) + H(\zeta_t),$$

for some $\zeta_t \in \left(\overline{V} \right)$, the ideal of $\Lambda(V, \overline{V}, \widehat{V})$ generated by $\overline{V}$ since $\gamma(\overline{V}) = 0$. In addition, by construction, we have $H(\overline{V}) \subseteq \bigoplus_{i \geq 1}(\Lambda(V))_i$ and for a generator v_t of degree $\geq n+2$, $H(\hat{v}_t) = 0$. Therefore $\psi(v_t) = \varphi(v_t) + H(\zeta_t)$ and the claim follows. $\square$

5.2 THEOREM. *Let $\mathcal{M}$ be formal with cup-length $c(\mathcal{M})$. Then*

$$\text{nil } \mathcal{E}_\#(\mathcal{M}) \leq c(\mathcal{M}) - 1.$$

Proof. If $c(\mathcal{M}) = N$, then we must show that N-fold commutators of equivalences in $\mathcal{E}_\#(\mathcal{M})$ are trivial. Suppose that $\varphi \in \text{Aut}_\#(\mathcal{M})$ is an N-fold commutator. By Corollary 3.3, $\ell(\varphi) \geq N+1$, and hence $\ell_0(\varphi) \geq N+1$. Now an induction argument, using Lemma 5.1 for the inductive step, shows that φ is homotopic to a map φ_n, which equals ι on generators of degree $\leq n$, for all n. But it is known that this implies that φ is homotopic to ι (see, e.g., **[D-R, Lem. 2.4]**). $\square$

Notice that in case $\mathcal{M}$ is formal, Theorem 5.2 implies Proposition 3.6, but Theorem 5.2 gives a sharper bound on nil $\mathcal{E}_\#(\mathcal{M})$ in general.

We now elaborate on Theorem 5.2. Recall from [**F-H, p. 3**] the definition of the e_0 invariant: For a minimal algebra $\mathcal{M}$, this is the largest value of $w(\chi)$ as χ ranges over all representative cocycles of non-zero classes in $H^*(\mathcal{M})$. It is well-known that there are inequalities $c(\mathcal{M}) \leq e_0(\mathcal{M}) \leq \mathrm{cat}_0(\mathcal{M})$, where $\mathrm{cat}_0(\mathcal{M})$ denotes the category of $\mathcal{M}$. In case $\mathcal{M}$ is formal, these three invariants coincide [**F-H, Cor. 4.10 and Rem. 9.3(4)**] and so from Theorem 5.2 we have the following:

5.3 COROLLARY. *If $\mathcal{M}$ is formal, then*

$$\mathrm{nil}\, \mathcal{E}_{\#}(\mathcal{M}) \leq c(\mathcal{M}) - 1 = e_0(\mathcal{M}) - 1 = \mathrm{cat}_0(\mathcal{M}) - 1.$$

If $\mathcal{M}$ is not formal, then the above relation between $\mathrm{nil}\,\mathcal{E}_{\#}(\mathcal{M})$ and the cup-length $c(\mathcal{M})$ does not necessarily hold. Of course, this is not unexpected since it is only in the formal case that $\mathcal{M}$ is determined by $H^*(\mathcal{M})$. In general, the inequalities of Corollary 5.3 do not hold. However, we will see that the inequality $\mathrm{nil}\,\mathcal{E}_{\#}(\mathcal{M}) \leq e_0(\mathcal{M}) - 1$ of Corollary 5.3 does hold under different hypotheses on $\mathcal{M}$.

We say that a minimal algebra $\mathcal{M} = \Lambda(V)$ has *homogeneous length differential* (of length L), if there is some integer $L \geq 2$ such that on each basis element $v_t \in V$, $d(v_t)$ is zero or a homogeneous length L polynomial in $\Lambda(V)$. If d has homogeneous length 2, then $\mathcal{M}$ is called *coformal* [**Ta, p. 65**].

With this terminology, we show the following:

5.4 LEMMA. *Suppose $\mathcal{M}$ has homogeneous length differential and $e_0(\mathcal{M}) = N$. If $\varphi \in \mathrm{Aut}_{\#}(\mathcal{M})$ with $\ell(\varphi) \geq N + 1$ and if $\varphi = \iota$ on generators of degree $\leq n$, then φ is homotopic to $\psi \in \mathrm{Aut}_{\#}(\mathcal{M})$ such that $\ell(\psi) \geq N + 1$ and $\psi = \iota$ on generators of degree $\leq n + 1$.*

Proof. The proof is analogous to that of Lemma 5.1. Let v_j be a generator of degree $n + 1$. We have

$$\varphi(v_j) = v_j + \chi_j,$$

with $w(\chi_j) \geq N + 1$ by assumption. Now $\varphi = \iota$ on generators of degree $\leq n$, so χ_j is a cocycle. Since $e_0(\mathcal{M}) = N$, χ_j is a coboundary. Thus for each generator v_j of degree $n + 1$, write $d(\eta_j) = \chi_j$. Since d is of homogeneous length $L \geq 2$, we can assume without loss of generality that $w(\eta_j) \geq N + 1 - (L - 1)$. Now define a homotopy $H : \Lambda(V, \overline{V}, \widehat{V}) \to \Lambda(V)$ starting at φ by setting

$$H(\bar{v}_j) = \begin{cases} -\eta_j & \text{if } |v_j| = n + 1 \\ 0 & \text{otherwise} \end{cases} \quad \text{and} \quad H(\hat{v}_j) = \begin{cases} -\chi_j & \text{if } |v_j| = n + 1 \\ 0 & \text{otherwise.} \end{cases}$$

Notice that wherever H is non-zero on the ideal $(\overline{V})$ of $\Lambda(V, \overline{V}, \widehat{V})$, it extends length by at least $N + 1 - L$.

By construction, H ends at a map $\psi = H\alpha$ that equals ι on generators v_t of degree $\leq n+1$. We claim that on each generator v_t with $|v_t| \geq n+2$, ψ has the form $\psi(v_t) = v_t + \xi_t$, with $w(\xi_t) \geq N+1$. To see this, note that

$$\alpha(v_t) = v_t + \hat{v}_t + id(v_t) + \frac{\gamma^2}{2}(v_t) + \cdots$$

and that $\gamma^2(v_t) = \gamma id(v_t)$. Since d is of length L, $w(id(v_t)) \geq L$ and hence $w\left(id(v_t) + \frac{\gamma^2}{2}(v_t) + \cdots\right) \geq L$. Furthermore, since $id(v_t) + \frac{\gamma^2}{2}(v_t) + \cdots \in (\overline{V})$, it follows that $w\left(H(id(v_t) + \frac{\gamma^2}{2}(v_t) + \cdots)\right) \geq N+1$. In particular, for generators v_t with $|v_t| \geq n+2$, we have $H(\hat{v}_t) = 0$ by construction, and so

$$\psi(v_t) = H\alpha(v_t) = H(v_t + \hat{v}_t + id(v_t) + \frac{\gamma^2}{2}(v_t) + \cdots)$$

$$= \varphi(v_t) + H(id(v_t) + \frac{\gamma^2}{2}(v_t) + \cdots),$$

$$= v_t + \chi_t + H(id(v_t) + \frac{\gamma^2}{2}(v_t) + \cdots).$$

Hence $w\left(\psi(v_t) - v_t\right) = w\left(\chi_t + H(id(v_t) + \frac{\gamma^2}{2}(v_t) + \cdots)\right) \geq N+1$ and the claim follows. □

5.5 THEOREM. *Let $\mathcal{M}$ have a homogeneous length differential. Then*

$$\mathrm{nil}\,\mathcal{E}_{\#}(\mathcal{M}) \leq e_0(\mathcal{M}) - 1.$$

In particular, this inequality holds if $\mathcal{M}$ is coformal.

Proof. This follows from Lemma 5.4 in the same way that Theorem 5.2 follows from Lemma 5.1. □

We finish this section with some examples. The first example shows the inequalities in the above theorems cannot be replaced by equality, indeed that $\mathrm{nil}\,\mathcal{E}_{\#}(\mathcal{M})$ can be less than $e_0(\mathcal{M})$ by an arbitrarily large amount. The second example shows that $\mathrm{nil}\,\mathcal{E}_{\#}(\mathcal{M})$ can equal $e_0(\mathcal{M}) - 1$.

5.6 Examples. (1) We observe that there are many examples of minimal models $\mathcal{M}$ that have $\mathcal{E}_{\#}(\mathcal{M}) = \{\iota\}$ in conjunction with a large cup-length. For example, if $\mathcal{M}$ is the minimal model of either $\mathbb{C}P^n$ or a product of n even-dimensional spheres all of the same dimension, then $c(\mathcal{M}) = n$ and yet $\mathcal{E}_{\#}(\mathcal{M}) = \{\iota\}$. Notice that in either case, $\mathcal{M}$ is both formal and has homogeneous length differential. Indeed, the product of spheres is both formal and coformal.

(2) Suppose $Y = S^3 \vee S^5 \vee \cdots \vee S^{2n+3}$ is an $(n+1)$-fold wedge of spheres and $X = \mathbb{C}P^n \times Y$. Then $\mathcal{M}_X = \Lambda(x,y) \otimes \mathcal{M}_Y$, where $|x| = 2$ and $|y| = 2n+1$. Now Y is formal and has bigraded model $\mathcal{M}_Y = \Lambda(V)$, where $V = \bigoplus_{i \geq 0} V_i$ with $V_0 = \langle u_1, u_2, \ldots, u_{n+1} \rangle$ and $|u_j| = 2j+1$. It is straightforward to construct

maps $\varphi \in \text{Aut}_\#(\mathcal{M}_X)$, for $j = 2, \dots, n+1$, with the following properties: For each j, $\varphi_j(x) = x$, $\varphi_j(y) = y$ and

$$\varphi_j(u_i) = \begin{cases} u_j + xu_{j-1} & \text{if } i = j \\ u_i & \text{if } i \neq j. \end{cases}$$

The construction of these φ_j's proceeds inductively over the generators in $V_1, V_2, \dots$ of $\mathcal{M}_X$. Now consider the n-fold commutator

$$\varphi = (\varphi_{n+1}, \varphi_n, \dots, \varphi_3, \varphi_2).$$

We have $\varphi(u_{n+1}) = u_{n+1} + (-1)^{n+1}x^n u_1$, and so φ is not homotopic to ι since it does not induce ι on cohomology. Hence $\text{nil}\,\mathcal{E}_\#(\mathcal{M}_X) \geq n$. On the other hand, $c(\mathcal{M}_X) = e_0(\mathcal{M}_X) = n+1$. Therefore by Corollary 5.3, $\text{nil}\,\mathcal{E}_\#(\mathcal{M}_X) = c(\mathcal{M}_X) - 1 = e_0(\mathcal{M}_X) - 1$.

6 Two Stage Algebras

In this section we consider the automorphisms of a special class of minimal algebras, the two stage algebras. This class of algebras contains the minimal models of many interesting examples, including the minimal models of homogeneous spaces (see [A-L]). The results of this section are for $\mathcal{E}_\#^*(\mathcal{M})$ for a two stage algebra.

6.1 DEFINITION. *A two stage algebra is a minimal algebra $\mathcal{M}$ such that we can write $\mathcal{M} = \Lambda(V_0 \oplus V_1)$ for graded vector spaces V_0 and V_1 with $d(V_0) = 0$, $d|_{V_1} : V_1 \to \Lambda(V_0)$ and $d|_{V_1}$ injective.*

In all cases of interest to us, a two-stage minimal algebra $\mathcal{M}$ will be finitely generated, although in general this need not be the case. For this section, then, we assume that our minimal algebras are finitely generated and write them as such.

We shall use the obstruction theory for homomorphisms defined on a two stage algebra which was introduced in [A-L]. In particular, if $\varphi, \psi : \mathcal{M} \to \mathcal{M}$ are maps of a two stage algebra $\mathcal{M}$ such that $\varphi|_{V_0} = \psi|_{V_0}$, then there is an obstruction homomorphism $\mathcal{O}(\varphi, \psi) : V_1 \to H^*(\mathcal{M})$. We have proved that if $\mathcal{O}(\varphi, \psi) = 0$, then φ and ψ are homotopic [A-L, **Prop. 3.3**].

Our first result relates $\text{nil}\,\mathcal{E}_\#^*(\mathcal{M})$ and $e_0(\mathcal{M})$ and is analogous to Theorems 5.2 and 5.5. We write $V_0 = \langle v_1, \dots, v_s \rangle$ and $V_1 = \langle v_{s+1}, \dots, v_r \rangle$, where $|v_1| \leq \dots \leq |v_s|$ and $|v_{s+1}| \leq \dots \leq |v_r|$, but we do not assume that $|v_s| \leq |v_{s+1}|$.

6.2 THEOREM. *Let $\mathcal{M} = \Lambda(V_0 \oplus V_1)$ be two stage. Then*

$$\text{nil}\,\mathcal{E}_\#^*(\mathcal{M}) \leq e_0(\mathcal{M}) - 1.$$

Proof. The proof is simpler than that of Theorem 5.2 or Theorem 5.5 in that it does not require an inductive argument. Denote $e_0(\mathcal{M})$ by N. First we observe that if $\varphi \in \text{Aut}_\#(\mathcal{M})$ has $l(\varphi) \geq N+1$ and if $\varphi = \iota$ on generators in V_0, then φ is homotopic to ι. To see this, note that on each generator $v_j \in V_1$, $j = s+1, \ldots, r$,

$$\varphi(v_j) = v_j + \chi_j \,,$$

where χ_j is a cocycle and $w(\chi_j) \geq N+1$ by assumption. Thus each $\chi_j = d(\eta_j)$ for some $\eta_j \in \mathcal{M}$. Now define a homotopy $H \colon \Lambda(V \oplus \overline{V} \oplus \widehat{V}) \to \Lambda(V)$ starting at φ. For $j = s+1, \ldots, r$ set $H(\bar{v}_j) = -\eta_j$ and $H(\hat{v}_j) = -\chi_j$ and for $j = 1, \ldots, s$, set $H(\bar{v}_j) = 0 = H(\hat{v}_j)$. One easily sees that H ends at ι. Note that the generators of V_1 do not appear in the differential of any other generator.

Now consider an N-fold commutator of equivalences

$$f = (f_1, \ldots, f_{N-1}, f_N),$$

where $f_j \in \mathcal{E}_\#^*(\mathcal{M})$. We must show that this commutator is trivial. Since each f_j induces the identity on cohomology, we can choose a representative φ_j in each homotopy class f_j, with $\varphi_j|_{V_0} = \iota|_{V_0}$ [**A-L, Cor. 3.4**]. Then f is represented by the N-fold commutator of automorphisms $(\varphi_1, \ldots, \varphi_{N-1}, \varphi_N)$. Clearly this latter commutator is ι on V_0 and has length $\geq N+1$ by Corollary 3.3. It follows that this commutator is homotopically trivial. $\qquad\square$

Next let $\mathcal{M}$ be any minimal algebra and consider the set S of elementary automorphisms in $\mathrm{Aut}_\#(\mathcal{M})$. Then S generates a subgroup of $\mathrm{Aut}_\#(\mathcal{M})$ denoted $\mathrm{EAut}_\#(\mathcal{M})$. By taking homotopy classes we obtain a subgroup $E\mathcal{E}_\#(\mathcal{M})$ of $\mathcal{E}_\#(\mathcal{M})$ consisting of all homotopy classes which have a representative in $\mathrm{EAut}_\#(\mathcal{M})$. In the case when $\mathcal{M}$ is a two stage algebra, $\mathcal{M} = \Lambda(V_0 \oplus V_1)$, we consider the *restricted elementary automorphisms* $\varphi \in \mathrm{Aut}_\#(\mathcal{M})$ to be those elementary automorphisms with $\varphi|_{V_0} = \iota|_{V_0}$. These generate a subgroup $\mathrm{REAut}_\#(\mathcal{M})$ of $\mathrm{EAut}_\#(\mathcal{M})$ and so we obtain a subgroup $RE\mathcal{E}_\#(\mathcal{M})$ of $E\mathcal{E}_\#(\mathcal{M})$. It is often easier to study $E\mathcal{E}_\#(\mathcal{M})$ or its subgroups than $\mathcal{E}_\#(\mathcal{M})$ (see §7 and §8), though in general $E\mathcal{E}_\#(\mathcal{M})$ does not equal $\mathcal{E}_\#(\mathcal{M})$.

6.3 PROPOSITION. *If $\mathcal{M}$ is a two stage algebra, then*

$$\mathcal{E}_\#^*(\mathcal{M}) \subseteq RE\mathcal{E}_\#(\mathcal{M}) \,.$$

Proof. Let $\mathcal{M} = \Lambda(V_0 \oplus V_1)$ with differential d, $V_0 = \langle v_1, \ldots, v_s \rangle$ and $V_1 = \langle v_{s+1}, \ldots, v_r \rangle$. Suppose $\psi \in \mathrm{Aut}_\#(\mathcal{M})$ and $\psi^* = \iota \colon H^*(\mathcal{M}) \to H^*(\mathcal{M})$. By Corollary 3.4 of [**A-L**], ψ is homotopic to an automorphism φ such that $\varphi|_{V_0} = \iota|_{V_0}$. Thus for each $j = s+1, \ldots, r$,

$$\varphi(v_j) = v_j + \chi_j \,,$$

for some decomposable cocycle χ_j. Then $\varphi_{\chi_j}^j$ is an elementary automorphism, where $j = s+1, \ldots, r$, since $d\varphi_{\chi_j}^j = \varphi_{\chi_j}^j d$. Furthermore,

$$\varphi = \varphi_{\chi_r}^r \cdots \varphi_{\chi_{s+1}}^{s+1} \,.$$

This shows $\mathcal{E}_\#^*(\mathcal{M}) \subseteq RE\mathcal{E}_\#(\mathcal{M})$. $\qquad\square$

To prove a corollary of Proposition 6.3 we need the following purely group-theoretic result. The proof is straightforward and hence omitted. The result will be used again in §8.

6.4 PROPOSITION. *Let G be a group and S a generating subset of G such that $S = S^{-1}$. If all n-fold commutators of elements of S are trivial, then* $\operatorname{nil} G \leq n - 1$.

The following corollary is now an immediate consequence of Propositions 6.3 and 6.4.

6.5 COROLLARY. *If $\mathcal{M}$ is a two stage algebra and s-fold commutators of restricted elementary automorphisms of $\mathcal{M}$ are homotopically trivial, then* $\operatorname{nil} \mathcal{E}_{\#}^*(\mathcal{M}) \leq s - 1$.

The additional hypothesis of formality allows us to conclude that $\mathcal{E}_{\#}^*(\mathcal{M})$ is abelian.

6.6 PROPOSITION. *If $\mathcal{M}$ is a formal two-stage algebra, then $\mathcal{E}_{\#}^*(\mathcal{M})$ is abelian.*

Proof. Let $\mathcal{M} = \Lambda(V_0 \oplus V_1)$ with $V_0 = \langle v_1, \dots, v_s \rangle$ and $V_1 = \langle v_{s+1}, \dots, v_r \rangle$. We first prove that if $\varphi_\chi^j \in \operatorname{Aut}_\#(\mathcal{M})$ is a restricted elementary automorphism, $j = s+1, \dots, r$, then φ_χ^j is homotopic to $\varphi_{\chi'}^j$, where $\chi' \in \Lambda(V_0)$. Write $\chi = \chi' + \chi''$ with $\chi' \in \Lambda(V_0)$ and such that each monomial in χ'' has some v_i as a factor, $i = s+1, \dots, r$. Then

$$d(v_j) = \varphi_\chi^j d(v_j) = d\varphi_\chi^j(v_j) = d(v_j) + d(\chi') + d(\chi'').$$

Since $d(\chi') = 0$, $d(\chi'') = 0$ and so χ'' is a cocycle of $\mathcal{M}$. But $\mathcal{M}$ is formal, and so χ'' is a coboundary (cf. the discussion in the proof of Lemma 5.3 of [A-L]). We now consider the restricted elementary automorphisms $\varphi_{\chi'}^j$ and compute the obstruction $\mathcal{O}(\varphi_\chi^j, \varphi_{\chi'}^j)$. From the definition (cf. [A-L, §3]),

$$\mathcal{O}(\varphi_\chi^j, \varphi_{\chi'}^j)(v_j) = [\varphi_\chi^j(v_j) - \varphi_{\chi'}^j(v_j)] = [\chi''] = 0$$

in $H^*(\mathcal{M})$. Thus $\mathcal{O}(\varphi_\chi^j, \varphi_{\chi'}^j) = 0$, and so φ_χ^j and $\varphi_{\chi'}^j$ are homotopic. Thus any element of $RE\mathcal{E}_\#(\mathcal{M})$ can be represented by a restricted elementary automorphism $\varphi_{\chi'}^j$ with $j = s+1, \dots, r$ and $\chi' \in \Lambda(V_0)$. If $\varphi_{\xi'}^l$ is another such restricted elementary automorphism, $l \in \{s+1, \dots, r\}$ and $\xi' \in \Lambda(V_0)$, then by Corollary 6.5 it suffices to show $(\varphi_{\chi'}^j, \varphi_{\xi'}^l) = \iota$. Without loss of generality assume $j > l$ and apply Lemma 4.3 to get $(\varphi_{\chi'}^j, \varphi_{\xi'}^l) = \varphi_\zeta^j$, where $\zeta = \varphi_{-\xi'}^l(\chi') - \chi'$. Since $\chi' \in \Lambda(V_0)$ and $l = s+1, \dots, r$, $\zeta = 0$. Thus $(\varphi_{\chi'}^j, \varphi_{\xi'}^l) = \iota$. $\square$

6.7 Remark. Let X be a finite complex (of dimension N) which is a formal space whose minimal model $\mathcal{M}$ is a finitely-generated, two-stage algebra. Then all commutators in $\mathcal{E}_{*\#}(X)$ are of finite order. To see this, suppose without loss of generality that $f \in \mathcal{E}_{\#N}^*(\mathcal{M})$ is represented by $\phi : \mathcal{M} \to \mathcal{M}$ with $\phi|_{V_0} = \iota|_{V_0}$ [A-L, Cor. 3.4] and that $w_k \in V_1$ has degree $> N$. Then $\phi(w_k) - w_k$ is a cocycle of degree $> N$. Hence it is a coboundary and so decomposable. Therefore $f \in \mathcal{E}_{\#}^*(\mathcal{M})$ and we have $\mathcal{E}_{\#N}^*(\mathcal{M}) = \mathcal{E}_{\#}^*(\mathcal{M})$. By Proposition 6.6 above, $\mathcal{E}_{\#N}^*(\mathcal{M})$ is abelian. It now follows from Proposition 2.1 for $\mathcal{E}_{*\#}(X)$ (see

Remark 2.2) that all commutators in $\mathcal{E}_{*\#}(X)$ have finite order. For example, if X is a product of spheres, then all commutators in $\mathcal{E}_{*\#}(X)$ are of finite order even though $\mathcal{E}_{*\#}(X)$ may be an infinite group (see [**A-L**, §7]). More generally, this remark applies to any formal, homogeneous space.

7 Rational H-Spaces

In this section we consider finitely-generated minimal algebras of the form $\mathcal{M} = \Lambda(v_1, \dots, v_k)$ with differential $d = 0$. These are just the minimal models of rational H-spaces. Note that they can also be thought of as two stage algebras with $V_1 = 0$. Since $H^*(\mathcal{M}) = \mathcal{M} = \Lambda(v_1, \dots, v_k)$, it easily follows that two maps $\varphi, \psi \colon \mathcal{M} \to \mathcal{M}$ are homotopic if and only if they are equal. Thus to calculate $\mathcal{E}_\#(\mathcal{M})$ in this case it suffices to determine $\mathrm{Aut}_\#(\mathcal{M})$. We first consider the case where all the generators of $\mathcal{M}$ are odd.

7.1 PROPOSITION. *Let* $\mathcal{M} = \Lambda(v_1, \dots, v_k)$ *be a minimal algebra with* $d = 0$, $|v_i| = n_i$ *odd and* $n_1 \leq \cdots \leq n_k$. *Let* $(t_1, \dots, t_r)$ *be a sequence of odd integers such that* $3 \leq t_1 < t_2 < \cdots < t_r$. *Assume that for each* t_s, $s = 1, \dots, r$, *there is a choice of* i_s *from* $1, \dots, k$ *with* $n_1 + \cdots + n_{t_s} = n_{i_s}$ *such that* $t_{s+1} < i_s$ *for* $s = 1, \dots, r - 1$. *Then* $\mathrm{nil}\, \mathcal{E}_\#(\mathcal{M}) \geq r$.

Proof. We exhibit a non-trivial r-fold commutator of elementary automorphisms. Consider

$$\varphi = \left(\varphi^{i_r}_{v_{i_{r-1}} v_{t_{r-1}+1} \cdots v_{t_r}}, \dots, \varphi^{i_2}_{v_{i_1} v_{t_1+1} \cdots v_{t_2}}, \varphi^{i_1}_{v_1 \cdots v_{t_1}} \right).$$

By hypothesis these elementary automorphisms exist. By Proposition 4.4,

$$\varphi = \varphi^{i_r}_{\pm v_1 v_2 \cdots v_{t_r}}.$$

Since this elementary automorphism is non-trivial, the result follows. □

Although the hypotheses appear cumbersome, they are easily checked when the generators are equally spaced, as in the following proposition.

7.2 PROPOSITION. *Let* $\mathcal{M} = \Lambda(v_1, \dots, v_k)$ *with* $d = 0$ *and* $|v_i| = n_i$. *Suppose the degrees of the generators form an arithmetic progression,* $n_j = a + (j - 1)u$, *for* $a \geq 3$ *an odd integer and* $u \geq 2$ *even, with* a *and* u *relatively prime. If* $ra + \frac{ru(ru+1)}{2} \leq k - 1$, *then* $\mathrm{nil}\, \mathcal{E}_\#(\mathcal{M}) \geq r$. *If* $ra + \frac{ru(ru+1)}{2} > k - 1$, *then* $\mathrm{nil}\, \mathcal{E}_\#(\mathcal{M}) < r$.

Proof. We first prove the first assertion as a corollary of Proposition 7.1. Set $t_1 = u + 1$, $t_2 = 2u + 1, \dots, t_k = ku + 1$. Clearly for any l, $n_1 + \cdots + n_l = la + \frac{l(l-1)}{2} u$. Thus for each $s = 1, \dots, r$,

$$n_1 + \cdots + n_{su+1} = (su + 1)a + \frac{(su+1)su}{2} u$$

$$= a + \left[sa + \frac{(su+1)su}{2} \right] u.$$

But $sa+\frac{(su+1)su}{2} \leq ra+\frac{(ru+1)ru}{2} \leq k-1$ by hypothesis. Thus $n_1 + \cdots + n_{su+1} = n_{i_s}$ for

$$i_s = sa + \frac{(su+1)su}{2} + 1 \leq k.$$

Next we verify that the remaining hypothesis of Proposition 7.1 holds. Now $t_{s+1} = (s+1)u + 1$ and so it follows that $t_{s+1} < i_s$ for $s = 1, \ldots, r-1$.

Now we prove the second assertion of Proposition 7.2. First note that for $\psi \in \mathrm{Aut}_\#(\mathcal{M})$ written $\psi(v_j) = v_j + \chi_j$, we have that any monomial occuring in χ_j must have length at least $u+1$. This follows since u and a are relatively prime. Consequently $l(\psi) \geq u+1$. Now let φ be an r-fold commutator of elements of $\mathrm{Aut}_\#(\mathcal{M})$. By Corollary 3.3,

$$l(\varphi) \geq r(u+1) - (r-1) = ru + 1.$$

From this it follows that if $\varphi(v_j) = v_j + \xi_j$ with $\xi_j \neq 0$, $j = 1, \ldots, k$, then

$$|\xi_j| \geq n_1 + \cdots + n_{ru+1} = (ru+1)a + \frac{(ru+1)ru}{2}u,$$

i.e.,

$$a + (j-1)u \geq a + \left[ra + \frac{(ru+1)ru}{2}\right]u$$

or

$$j - 1 \geq ra + \frac{(ru+1)ru}{2}.$$

But $j - 1 \leq k - 1 < ra + \frac{(ru+1)ru}{2}$ by hypothesis. This contradiction implies $\xi_j = 0$. Thus $\varphi = \iota$ and so $\mathrm{nil}\,\mathcal{E}_\#(\mathcal{M}) < r$. $\qquad\square$

Next we briefly consider equally spaced even dimensional generators.

7.3 PROPOSITION. *Let u be a fixed even integer ≥ 2 and consider $\mathcal{M} = \Lambda(v_1, \ldots, v_k)$ with $|v_i| = iu$, $i = 1, \ldots, k$, and $d = 0$. Then $\mathrm{nil}\,\mathcal{E}_\#(\mathcal{M}) = k - 1$.*

Proof. If φ is a k-fold commutator of elements of $\mathrm{Aut}_\#(\mathcal{M})$, then by Corollary 3.3, $l(\varphi) \geq k+1$. For dimension reasons it now follows that $\varphi = \iota$, and so $\mathrm{nil}\,\mathcal{E}_\#(\mathcal{M}) \leq k-1$. However, the commutator of elementary automorphisms

$$(\varphi^k_{v_{k-1}v_1}, \ldots, \varphi^3_{v_2 v_1}, \varphi^2_{v_1 v_1})$$

equals $\varphi^k_{(-1)^k v_1^k}$ by Lemma 4.3. Since this is non-trivial, $\mathrm{nil}\,\mathcal{E}_\#(\mathcal{M}) = k-1$. $\qquad\square$

We now use the preceding results to obtain information on $\mathrm{nil}\,\mathcal{E}_\#(X)$ for certain CW-complexes X:

7.4 Examples. (1) Let $n \geq 2$ and consider $\mathcal{M} = \Lambda(v_1, \ldots, v_{n-1})$ with $|v_i| = 2i + 1$ and $d = 0$. Then $\mathcal{M}$ is the minimal model of the special unitary group

$SU(n)$ since $\mathcal{M} = H^*(SU(n); \mathbb{Q}) = \Lambda(v_1, \dots, v_{n-1})$ [**Bo**, §9]. Let r be the unique integer such that

$$2r^2 + 4r + 2 \leq n < 2(r+1)^2 + 4(r+1) + 2 .$$

Then by Proposition 7.2, nil $\mathcal{E}_\#(\mathcal{M}) = r$. Consequently, nil $\mathcal{E}_\#(SU(n)_\mathbb{Q}) = r$, nil $\mathcal{E}_\#(SU(n)) \geq r$ and all $(r+1)$-fold commutators in $\mathcal{E}_\#(SU(n))$ are of finite order. Maruyama has proved that $\mathcal{E}_\#(SU(n))/\text{torsion}$ is abelian for $n \leq 17$ and that $\mathcal{E}_\#(SU(4))$ is not abelian [**Ma**]. We observe that our results yield $\mathcal{E}_\#(SU(n)_\mathbb{Q})$ is abelian for $n \leq 17$ and non-abelian for $n > 17$. The fact that $\mathcal{E}_\#(SU(4))$ is not abelian shows that our lower bound for $\mathcal{E}_\#(X)$ need not be sharp.

(2) Let $n \geq 1$ and consider $\mathcal{M} = \Lambda(v_1, \dots, v_n)$ with $|v_i| = 4i - 1$ and $d = 0$. Then $\mathcal{M}$ is the minimal model of the symplectic group $Sp(n)$. Let r be the unique integer such that

$$8r^2 + 5r + 1 \leq n < 8(r+1)^2 + 5(r+1) + 1 .$$

Then by Proposition 7.2, nil $\mathcal{E}_\#(\mathcal{M}) = r$. Consequently, nil $\mathcal{E}_\#(Sp(n)_\mathbb{Q}) = r$, nil $\mathcal{E}_\#(Sp(n)) \geq r$ and all $(r+1)$-fold commutators in $\mathcal{E}_\#(Sp(n))$ are of finite order.

(3) If $u = 2$ in Proposition 7.3, then $\mathcal{M}$ is the minimal model of the classifying space $BU(k)$. If $u = 4$ in Proposition 7.3, then $\mathcal{M}$ is the minimal model of the classifying space $BSp(k)$. In either case we obtain that nil $\mathcal{E}_\#(\mathcal{M}) = k - 1$.

8 The Homogeneous Spaces $U(n)/(U(p) \times U(q))$

Let X be $U(n)/(U(p) \times U(q))$ with $1 \leq p \leq q$ and $p + q \leq n$ and let $\mathcal{M}$ be the minimal model of X. The purpose of this section is to prove the following theorem which completely determines nil $\mathcal{E}_\#(\mathcal{M})$.

8.1 THEOREM. *If $\mathcal{M}$ is the minimal model of $U(n)/(U(p) \times U(q))$ and $r = n - (p + q) \geq 0$, then*

$$\text{nil}\,\mathcal{E}_\#(\mathcal{M}) = \begin{cases} r - 1, & \text{if } r \leq pq + 1 \\ l + pq, & \text{if } r > pq + 1 , \end{cases}$$

where l is the largest integer such that

$$2l(p + q + l + 1) + pq + 1 \leq r .$$

For the cases $r = 0$ or 1, we interpret nil $\mathcal{E}_\#(\mathcal{M}) = -1$ or 0 to mean that $\mathcal{E}_\#(\mathcal{M})$ is the trivial group.

We begin with a description of the two stage algebra $\mathcal{M} = \Lambda(V_0 \oplus V_1)$ which we recall from [**A-L**, §6]. We have

$$V_0 = \langle u_1, \dots, u_p, w_1, \dots, w_r \rangle \quad \text{and} \quad V_1 = \langle v_1, \dots, v_p \rangle ,$$

where $|u_i| = 2i$, $|w_j| = 2(p + q + j) - 1$, $|v_i| = 2(q + i) - 1$, $d|_{V_0} = 0$ and $d|_{V_1}: V_1 \rightarrow \Lambda(u_1, \dots, u_p)$. If $\widetilde{M} = \Lambda(u_1, \dots, u_p, v_1, \dots, v_p)$ with differential induced from d, then $\widetilde{M}$ is the minimal model of $U(n)/(U(p) \times U(q))$ with $r = 0$, i.e., of the Grassmannian manifold $G(p, p+q) = U(p+q)/(U(p) \times U(q))$. Hence $[u_1], \dots, [u_p]$ are generators of $H^*(\widetilde{M})$, $H^j(\widetilde{M}) = 0$ for $j > 2pq$ and $[u_1]^{pq} \neq 0$.

The proof of Theorem 8.1 is achieved by a sequence of lemmas and propositions.

8.2 LEMMA. *If $\varphi \in \mathrm{Aut}_\#(\mathcal{M})$, then φ is homotopic to $\sigma \in \mathrm{Aut}_\#(\mathcal{M})$ with $\sigma|_{\widetilde{M}} = \iota$.*

Proof. Let $\widetilde{M}$ be as above. Then by restriction φ induces $\widetilde{\varphi} \in \mathrm{Aut}_\#(\widetilde{M})$. By **[A-L, Prop. 6.4(i)]**, $\widetilde{\varphi}$ is homotopic to the identity ι of $\widetilde{M}$. But then one easily extends the homotopy to a homotopy between φ and some $\sigma \in \mathrm{Aut}_\#(\mathcal{M})$ such that $\sigma|_{\widetilde{M}} = \iota$. $\qquad\square$

8.3 LEMMA. *If $\sigma \in \mathrm{Aut}_\#(\mathcal{M})$ is such that $\sigma|_{\widetilde{M}} = \iota$, then there is $\psi \in \mathrm{Aut}_\#(\mathcal{M})$ homotopic to σ with $\psi|_{\widetilde{M}} = \iota$ and $\psi(w_j) = w_j + \chi_j$ for $j = 1, \dots, r$, where χ_j is a decomposable element of $\Lambda(V_0)$.*

Proof. Write $\sigma(w_j) = w_j + \alpha_j + \beta_j$, where $\alpha_j \in \Lambda(V_0)$ and each monomial of β_j has some v_i as a factor. Then β_j is a cocycle and so $[\beta_j] \in H^*(\mathcal{M})$. But $H^*(\mathcal{M})$ is generated by $[u_1], \dots, [u_p], [w_1], \dots, [w_r]$. Thus $\beta_j = \gamma_j + d(\xi_j)$ for some $\gamma_j \in \Lambda(V_0)$ and $\xi_j \in \mathcal{M}$. Setting $\chi_j = \alpha_j + \gamma_j \in \Lambda(V_0)$, we have

$$\sigma(w_j) = w_j + \chi_j + d(\xi_j).$$

Now define ψ by $\psi(w_j) = w_j + \chi_j$ for each $j = 1, \dots, r$ and $\psi = \iota$ on the other generators. Then $\psi \in \mathrm{Aut}_\#(\mathcal{M})$ and ψ is homotopic to σ. A homotopy is defined by $H|_{\mathcal{M}} = \sigma$, $H(\bar{w}_j) = -\xi_j$, $H(\hat{w}_j) = -d(\xi_j)$ and $H = 0$ on all other generators. Then H starts at σ and ends at ψ. $\qquad\square$

Lemmas 8.2 and 8.3 yield Theorem 8.1 in the cases $r = 0$ and $r = 1$. We assume $r \geq 2$ for the rest of this section.

We next define *special elementary automorphisms* ψ^j of $\mathcal{M}$ to be elementary automorphisms such that $\psi^j_\chi(w_j) = w_j + \chi$ and $\psi^j_\chi = \iota$ on the other generators, where $j = 1, \dots, r$ and χ is a decomposable cocycle and a monomial in $\Lambda(V_0)$. These are not the same as the restricted elementary automorphisms of §6. Note that given any such χ, namely, a decomposable, monomial cocycle in $\Lambda(V_0)$ with $|\chi| = |w_j|$, then ψ^j_χ defined as above is a special elementary automorphism, i.e., $d\psi^j_\chi = \psi^j_\chi d$ (cf. Example 4.2).

By putting Lemmas 8.2 and 8.3 together we obtain

8.4 PROPOSITION. *Suppose that $\varphi \in \mathrm{Aut}_\#(\mathcal{M})$. Then φ is homotopic to some $\psi \in \mathrm{Aut}_\#(\mathcal{M})$ such that $\psi(u_i) = u_i$, $\psi(v_i) = v_i$ and $\psi(w_j) = w_j + \chi_j$ for $\chi_j \in \Lambda(V_0)$. Furthermore, ψ is a product of special elementary automorphisms.*

Now let $S \subseteq \mathcal{E}_{\#}(\mathcal{M})$ be the set of homotopy classes of special elementary automorphisms. Then $S = S^{-1}$ and by Proposition 8.4, S generates the group $\mathcal{E}_{\#}(\mathcal{M})$. Therefore Proposition 6.4 implies the following corollary.

8.5 COROLLARY. *If all $(t+1)$-fold commutators of special elementary automorphisms of $\mathcal{M}$ are homotopically trivial, then $\operatorname{nil}\mathcal{E}_{\#}(\mathcal{M}) \leq t$.*

This corollary will give an upper bound on $\mathcal{E}_{\#}(\mathcal{M})$. We obtain a lower bound by exhibiting a non-trivial commutator.

8.6 PROPOSITION. *If $2 \leq s \leq r$ and $s - 1 \leq pq$, then $\operatorname{nil}\mathcal{E}_{\#}(\mathcal{M}) \geq s - 1$.*

Proof. Consider the $(s-1)$-fold commutator

$$\left(\psi^s_{u_1 w_{s-1}}, \dots, \psi^3_{u_1 w_2}, \psi^2_{u_1 w_1} \right).$$

By Lemma 4.3 this commutator is ψ^s_χ, where $\chi = (-1)^s u_1^{s-1} w_1$. But in cohomology,

$$(\psi^s_\chi)^*[w_s] = [w_s] + (-1)^s[u_1]^{s-1}[w_1].$$

Since $s - 1 \leq pq$, $[u_1]^{s-1} \neq 0$. Therefore $[u_1]^{s-1}[w_1] \neq 0$ and so $(\psi^s_\chi)^* \neq \iota$. Thus ψ^s_χ is not homotopic to ι, and so $\operatorname{nil}\mathcal{E}_{\#}(\mathcal{M}) \geq s - 1$. □

As a consequence of this, we prove the following proposition.

8.7 PROPOSITION. *If $r - 1 \leq pq$, then $\operatorname{nil}\mathcal{E}_{\#}(\mathcal{M}) = r - 1$.*

Proof. Let ψ be a commutator of r special elementary automorphisms. By Proposition 8.6 and Corollary 8.5, it suffices to show that ψ is homotopic to ι. By Proposition 4.4, ψ is a special elementary automorphism of the form ψ^j_χ, where $\chi \in \Lambda(V_0)$ is a monomial and, by Corollary 3.3, $l(\psi^j_\chi) \geq r + 1$. Suppose $\chi \neq 0$ and write

$$\chi = a u_{i_1} \cdots u_{i_k} w_{j_1} \cdots w_{j_l},$$

where $1 \leq i_1 \leq \cdots \leq i_k \leq p$, $1 \leq j_1 < \cdots < j_l \leq r$, $l \geq 1$, $k + l \geq r + 1$ and $a \neq 0 \in \mathbb{Q}$. Then

$$\begin{aligned} 2(p+q+j) - 1 = |\chi| = |u_{i_1} \cdots u_{i_k} w_{j_1} \cdots w_{j_l}| &\geq 2k + |w_{j_l}| \\ &\geq 2k + 2(p+q+l) - 1 = 2(p+q+(l+k)) - 1 \\ &\geq 2(p+q+r+1) - 1. \end{aligned}$$

Thus $j \geq r + 1$ which is impossible. Hence $\chi = 0$ and so $\psi^j_\chi = \iota$. □

This proves Theorem 8.1 in the case $r \leq pq + 1$. We now deal with the case $r > pq + 1$ and complete the proof with the following result.

8.8 PROPOSITION. *If $r > pq + 1$ and l is the largest integer such that $2l(p + q + l + 1) + pq + 1 \leq r$, then $\operatorname{nil} \mathcal{E}_\#(\mathcal{M}) = l + pq$.*

Proof. We first show the existence of a non-trivial $(l + pq)$-fold commutator which will be a commutator of $l + pq$ special elementary automorphisms. Consider

$$\left(\psi^{pq+1}_{u_1 w_{pq}}, \ldots, \psi^3_{u_1 w_2}, \psi^2_{u_1 w_1} \right)$$

which by Proposition 4.4 equals $\psi^{pq+1}_{\pm u_1^{pq} w_1}$. Next consider the sequence of l special elementary automorphisms

$$\psi^{f(1)}_{w_2 w_3 w_{pq+1}}, \psi^{f(2)}_{w_4 w_5 w_{f(1)}}, \ldots, \psi^{f(i)}_{w_{2i} w_{2i+1} w_{f(i-1)}}, \ldots, \psi^{f(l)}_{w_{2l} w_{2l+1} w_{f(l-1)}}, \qquad (*)$$

for integers $f(1), \ldots, f(l)$, where $2(p + q + f(1)) - 1 = |w_2| + |w_3| + |w_{pq+1}|$ and $2(p + q + f(i)) - 1 = |w_{2i}| + |w_{2i+1}| + |w_{f(i-1)}|$, $i = 2, \ldots, l$. This yields $f(1) = 2p + 2q + pq + 5$ and the recursion formula

$$f(i) = 2p + 2q + 4i + f(i - 1).$$

Therefore,

$$f(l) = 2l(p + q + l + 1) + pq + 1.$$

To ensure that the l special elementary automorphisms in $(*)$ exist and are non-trivial, we must have (1) $f(l) \leq r$ (2) $2i + 1 < f(i - 1)$ for $i = 2, \ldots, l$ and (3) $3 < pq + 1$. But (1) follows by hypothesis and (2) is a consequence of $1 \leq p$. Furthermore, (3) holds except in the cases $p = 1$, $q = 1$ and $p = 1$, $q = 2$. In each of these last two cases, the existence of a non-trivial $(l + pq)$-fold commutator is easily verified directly. We now sketch the proof of this in the case $p = 1$, $q = 1$ — the other case is verified similarly. Consider the $(l+1)$-fold commutator

$$\psi = \left(\psi^{g(l)}_{w_{2l} w_{2l+1} w_{g(l-1)}}, \ldots, \psi^{g(3)}_{w_6 w_7 w_{g(2)}}, \psi^{g(2)}_{w_2 w_5 w_{g(1)}}, \psi^{g(1)}_{w_2 w_3 w_4}, \psi^2_{u_1 w_1} \right),$$

with $g(1) = 12$ and $g(i) = 2i^2 + 6i + 2$ for $i \geq 2$. Clearly $2i + 1 < g(i - 1)$ and $g(l) \leq r$. It follows from Proposition 4.4 that $\psi = \psi^{g(l)}_{\pm u_1 w_1 w_2 \cdots w_{2l+1}}$. Thus ψ is not homotopic to the identity since it does not induce the identity on $H^*(\mathcal{M})$. We now return to the main part of the proof. By Proposition 4.4, for the $(l + pq)$-fold commutator as follows we have

$$\left(\psi^{f(l)}_{w_{2l} w_{2l+1} w_{f(l-1)}}, \ldots, \psi^{f(1)}_{w_2 w_3 w_{pq+1}}, \psi^{pq+1}_{u_1 w_{pq}}, \ldots, \psi^2_{u_1 w_1} \right)$$

$$= \left(\psi^{f(l)}_{w_{2l} w_{2l+1} w_{f(l-1)}}, \ldots, \psi^{f(1)}_{w_2 w_3 w_{pq+1}}, \psi^{pq+1}_{\pm u_1^{pq} w_1} \right) = \psi^{f(l)}_{\pm u_1^{pq} w_1 w_2 \cdots w_{2l+1}}.$$

Then $\psi^{f(l)}_{\pm u_1^{pq} w_1 w_2 \cdots w_{2l+1}}$ is not homotopic to ι since it does not induce the identity on $H^*(\mathcal{M})$.

We next show that every s-fold commutator of special elementary auto-morphisms, $s = l + pq + 1$, is trivial when $r > pq + 1$. By Corollary 8.5, this suffices to prove the proposition. Let

$$\psi = (\psi^{i_s}_{\chi_s}, \dots , \psi^{i_1}_{\chi_1})$$

and let $i = \max(i_1, \dots , i_s)$. Then $\psi = \psi^i_{\pm\alpha}$, where α is as described in Proposi-tion 4.4. We assume that $\alpha \neq 0$. Each χ_j either contains some u_a as a factor or else contains no factor from $u_1, \dots , u_p$. For parity reasons, in the former case χ_j contains some w_b as factor and in the latter case χ_j contains at least three distinct w_b's. If more than pq of the χ_j contain some u_a as factor, then the monomial α can be written as $\alpha = \beta\gamma$, where β is a product of more than pq of the u_a's and γ is some product of w_b's. Since in this case β is a coboundary — see the comments on $\mathcal{M}$ preceding Lemma 8.2 — and γ is a cocycle, then α is a coboundary. Thus $\psi = \psi^i_{\pm\alpha}$ is homotopic to ι. Therefore we assume there are t of the χ_j which contain some u_a as factor and that $t \leq pq$. Then $\pm\alpha$ is a product of at least t u_a's and k distinct w_b's, for some number k. But

$$k \geq t + 3(s - t) - (s - 1) = 2(s - t) + 1 .$$

Thus

$$|\alpha| \geq 2t + |w_1| + \cdots + |w_{2(s-t)+1}|$$

and so

$$|\alpha| \geq 2t + 2(2(s - t) + 1)p + 2(2(s - t) + 1)q + (2(s - t) + 1)^2$$

with $t \leq pq$, since $|w_1| + \cdots + |w_N| = 2Np + 2Nq + N^2$. Now the expression on the right side of the inequality is smallest when $t = pq$. Since $s = l + pq + 1$, we obtain

$$|\alpha| \geq 2pq + 2(2l + 3)p + 2(2l + 3)q + (2l + 3)^2 .$$

Thus

$$|\alpha| \geq 2\{p + q + 2(l + 1)(p + q + l + 2) + pq + 1\} - 1 .$$

But by hypothesis, this last expression is $> 2(p + q + r) - 1 = |w_r|$. Hence if $\alpha \neq 0$, we have $\psi = \psi^i_{\pm\alpha}$ with $|\alpha| > |w_r| \geq |w_i|$. Therefore, $\alpha = 0$ and so $\psi = \iota$.
□

8.9 Remarks. (1) If $X = U(n)/(U(p) \times U(q))$, where $r = n - (p + q) \geq 3$ and $pq \geq 2$, then $\mathcal{E}_\#(X_\mathbb{Q})$ is not abelian.

(2) If $X = U(p+q+pq+1)/(U(p) \times U(q))$, then nil $\mathcal{E}_\#(X_\mathbb{Q}) = pq$. In particular, nil $\mathcal{E}_\#((U(2n + 2)/(U(1) \times U(n)))_\mathbb{Q}) = n$.

Finally, we state the obvious consequence of Theorem 8.1.

8.10 Remark. If $X = U(n)/(U(p) \times U(q))$, then

$$\mathrm{nil}\, \mathcal{E}_\#(X) \geq \begin{cases} r - 1, & \text{if } r \leq pq + 1 \\ l + pq, & \text{if } r > pq + 1, \end{cases}$$

where l is the largest integer such that $2l(p+q+l+1)+pq+1 \leq r$. In addition, all commutators of $\mathcal{E}_\#(X)$ of length $\geq$ nil $\mathcal{E}_\#(X_\mathbb{Q})$ are of finite order.

References

[Ar]   Arkowitz, M., *The Group of Self-Homotopy Equivalences — A Survey*, Springer-Verlag **1425** (1990), 170–203.

[A-C]   Arkowitz, M. and Curjel, C., *Groups of Homotopy Classes*, Lecture Notes in Math., Springer-Verlag **4** (1964).

[A-L]   Arkowitz, M. and Lupton, G., *On Finiteness of Subgroups of Self-Homotopy Equivalences*, Proceedings of the Čech Memorial Conference (to appear).

[Bo]   Borel, A., *Sur la Cohomologie des Espaces Fibrés principaux et des Espaces Homogènes de Groupes de Lie Compacts*, Ann. of Math. **57** (1953), 115–207.

[D-R]   Douglas, R. and Renner, L., *Uniqueness of Product and Coproduct Decompositions in Rational Homotopy Theory*, Trans. Amer. Math. Soc. **264** (1981), 165–180.

[D-Z]   Dror, E. and Zabrodsky, A., *Unipotency and Nilpotency in Homotopy Equivalences*, Topology **18** (1979), 187–197.

[F-H]   Félix, Y. and Halperin, S., *Rational L-S Category and its Applications*, Trans. Amer. Math. Soc. **372** (1982), 1–37.

[G-M]   Griffiths, P. and Morgan, J., *Rational Homotopy and Differential Forms*, Progress in Math. Birkhäuser **15** (1981).

[H-S]   Halperin, S. and Stasheff, J., *Obstructions to Homotopy Equivalences*, Adv. in Math. **32** (1979), 233–279.

[H-M-R] Hilton, P., Mislin, G. and Roitberg, J., *Localization of Nilpotent Groups and Spaces*, Notas de Matemática, North Holland **15** (1975).

[Ka]   Kahn, D.W., *Some Research Problems on Homotopy Self-Equivalences*, Lecture Notes in Math., Springer-Verlag **1425** (1990), 204–207.

[Ma]   Maruyama, K.I., *Localization of a Certain Subgroup of Self-Homotopy Equivalences*, Pac. J. of Math. **136** (1989), 293–301.

[Ta]   Tanré, D., *Homotopie Rationnelle: Modèles de Chen, Quillen, Sullivan*, Lecture Notes in Math., Springer-Verlag **1025** (1983).

Department of Mathematics, Dartmouth College, Hanover, NH 03755, USA
Martin.Arkowitz@dartmouth.edu
Department of Mathematics, Cleveland State University, Cleveland, OH 44115, USA
Lupton@math.csuohio.edu

Progress in Mathematics, Vol. 136
© 1996 Birkhäuser Verlag Basel/Switzerland

Linear group homology properties of the inclusion of a ring of integers into a number field

DOMINIQUE ARLETTAZ AND PIOTR ZELEWSKI*

1 Introduction and statement of the main results

Let F be a number field, O the ring of algebraic integers in F, and let θ denote the inclusion $O \hookrightarrow F$. The localization exact sequence in algebraic K-theory splits into short exact sequences

$$0 \longrightarrow K_n O \xrightarrow{\theta_\sharp} K_n F \longrightarrow \bigoplus_m K_{n-1}(O/m) \longrightarrow 0$$

for all positive integers n, where $\theta_\sharp$ is the homomorphism induced by θ in K-theory and where m runs over the set of maximal ideals of O (see Section 5 of [Q1], Theorem 8 of [Q2] and Théorème 1 of [S2]); in particular, $\theta_\sharp$ is always injective. On the other hand, G. Banaszak investigated the subgroup of divisible elements in $K_n F$ and explained the important role of these elements in relation with the Lichtenbaum-Quillen conjecture and étale K-theory (see [B1], [B2], [BG], [BZ]). If n is odd, $K_n F$ is a finitely generated abelian group and has therefore no non-trivial divisible elements. If n is even, $K_n F$ is a large torsion group but all its divisible elements belong to the image of $\theta_\sharp$ because $\bigoplus_m K_{n-1}(O/m)$ is a direct sum of finite cyclic groups and hence contains no non-trivial divisible elements.

In this note, we consider similar questions about the homomorphism

$$\theta_* : H_n(SL(O); \mathbb{Z}) \longrightarrow H_n(SL(F); \mathbb{Z})$$

induced by θ on the integral homology of the infinite special linear group ($n \geq 0$). Our main objective is to answer the following two questions.

QUESTION 1.1. For which dimensions n is the homomorphism θ_* injective?

QUESTION 1.2. For which dimensions n does the image of θ_* contain all divisible elements of $H_n(SL(F); \mathbb{Z})$?

Both questions concern the structure of the groups $H_n(SL(F); \mathbb{Z})$. Remember that $H_n(SL(O); \mathbb{Z})$ is a finitely generated abelian group for all $n \geq 0$, while $H_n(SL(F); \mathbb{Z})$ is not finitely generated. However, it was shown in Section 2 of [A1] that $H_n(SL(F); \mathbb{Z})$ is the direct sum of a free abelian group of finite (and known) rank and a torsion group. Some information about this torsion

---

*) The second author wishes to thank the Swiss National Science Foundation for financial support while this research was being carried out.

group is given in [AZ], where we prove that for any $n \geq 0$, it contains finitely many non-trivial divisible homology classes. Since $SL(F) = \varinjlim_S SL(O_S)$ for any exhaustion of the set of maximal ideals in O by finite subsets S (here O_S denotes the ring where the elements of the ideals in S are inverted), the image of the induced homomorphism $H_n(SL(O_S); \mathbb{Z}) \to H_n(SL(F); \mathbb{Z})$ for a certain finite set S of maximal ideals contains the whole subgroup of classes divisible in $H_n(SL(F); \mathbb{Z})$.

Now, let us formulate the main results of the paper.

DEFINITION 1.3. For a prime p, let d_p denote the smallest positive integer n for which $K_n F$ contains non-trivial p-torsion divisible elements (observe that d_p is even and that it depends on F); if there are no non-trivial p-torsion divisible elements in $K_n F$ for all $n \geq 1$, we say that $d_p = \infty$. For instance if $F = \mathbb{Q}$ and p an odd prime $< 125'000$, then according to Theorem 3 of [B2] and Corollary 4 of [BG], $d_p = \infty$ if and only if p is regular, and if p is irregular, d_p is the smallest even integer $2i$, with i odd, such that p divides the numerator of $\dfrac{B_{i+1}}{i+1}$, where B_{i+1} is the $(i+1)$-st Bernoulli number.

THEOREM 1.4. *For any prime number p, the homomorphism*

$$\theta_* : H_n(SL(O); \mathbb{Z}_{(p)}) \longrightarrow H_n(SL(F); \mathbb{Z}_{(p)})$$

is injective for all integers n such that $2 \leq n \leq \min(2p - 2, d_p + 1)$.

THEOREM 1.5. *Let p be a prime number and n an integer satisfying $2 \leq n \leq \min(2p - 2, 2d_p - 1)$. Then all p-torsion divisible homology classes in $H_n(SL(F); \mathbb{Z})$ belong to the image of the homomorphism*

$$\theta_* : H_n(SL(O); \mathbb{Z}) \to H_n(SL(F); \mathbb{Z}).$$

Sections 2 and 3 are devoted to the proof of Theorems 1.4. and 1.5. respectively and to the discussion of related problems.

2 The injectivity of θ_*

The purpose of this section is to investigate Question 1.1. Write again θ for the map $BSL(O)^+ \to BSL(F)^+$, induced by the inclusion $O \hookrightarrow F$, between the spaces obtained by performing the plus construction on the classifying spaces of the respective special linear groups (remember that these spaces have the same homology as the associated special linear groups). Our argument is based on the comparison of these spaces with the corresponding products of Eilenberg-MacLane spaces.

LEMMA 2.1. *Let p be an odd prime number, n an integer ≥ 2, and consider the product map*

$$\eta = \prod_{j=2}^{n} \eta_j : \prod_{j=2}^{n} K(K_j O, j) \longrightarrow \prod_{j=2}^{n} K(K_j F, j),$$

where $\eta_j : K(K_jO, j) \to K(K_jF, j)$ is induced by the map θ (for $2 \le j \le n$). If $d_p = \infty$, the induced homomorphism

$$\eta_* : H_n(\prod_{j=2}^{n} K(K_jO, j); \mathbb{Z}_{(p)}) \longrightarrow H_n(\prod_{j=2}^{n} K(K_jF, j); \mathbb{Z}_{(p)})$$

is split injective for any $n \ge 2$; if $d_p < \infty$, it is split injective for $2 \le n \le d_p - 1$ or $n = d_p + 1$, and injective for $n = d_p$.

Proof. For all positive integers j, the inclusion $\theta : O \hookrightarrow F$ induces an injection $K_jO \to K_jF$ as mentioned in the introduction. Moreover, Banaszak proved in [B2], Corollary 1 that $K_j(O; \mathbb{Z}_{(p)}) \to K_j(F; \mathbb{Z}_{(p)})$ is a split injection for $j \le d_p - 1$ (or for all j's if $d_p = \infty$) if p is odd. If $d_p = \infty$, η_* is obviously split injective for all $n \ge 2$. If $d_p < \infty$, the assertion is trivial for $n = 2$, and for $3 \le n \le d_p + 1$, we conclude by Künneth formula that

$$\eta_* : H_n(\prod_{j=2}^{n} K(K_jO, j); \mathbb{Z}_{(p)}) \cong K_n(O; \mathbb{Z}_{(p)}) \oplus H_n(\prod_{j=2}^{n-2} K(K_jO, j); \mathbb{Z}_{(p)}) \longrightarrow$$

$$K_n(F; \mathbb{Z}_{(p)}) \oplus H_n(\prod_{j=2}^{n-2} K(K_jF, j); \mathbb{Z}_{(p)}) \cong H_n(\prod_{j=2}^{n} K(K_jF, j); \mathbb{Z}_{(p)})$$

is an injection on the first factor (which splits if n is even $\le d_p - 1$ and which is an isomorphism if n is odd) and a split injection on the second factor since $n - 2 \le d_p - 1$. $\qquad\square$

DEFINITION 2.2.   Let $M_1 := 1$, and for an integer $h \ge 2$ let M_h be the product of all primes $p \le \frac{h}{2} + 1$.

DEFINITION 2.3.   Let $L_1 := 1$, and for an integer $k \ge 2$ let L_k denote the product of all primes p for which there exists a sequence of non-negative integers $(a_1, a_2, a_3, \dots)$ satisfying:
(a) $a_1 \equiv 0 \pmod{2p - 2}$, $a_i \equiv 0$ or $1 \pmod{2p - 2}$ for $i \ge 2$,
(b) $a_i \ge p a_{i+1}$ for $i \ge 1$,
(c) $\sum_{i=1}^{\infty} a_i = k$.

Notice that L_k divides M_h if $k \le h$. These integers occur in the computation of the stable homology groups of Eilenberg-MacLane spaces (see [C], Théorème 2):

PROPOSITION 2.4.   *For any abelian group G and any pair of integers i and j with $j < i < 2j$, one has $L_{i-j} H_i(K(G, j); \mathbb{Z}) = 0$. Consequently, if $n \ge 1$, the integer M_n fulfills $M_n H_i(K(G, j); \mathbb{Z}) = 0$ for all i and j with $j < i < 2j$ and $i - j \le n$.*

DEFINITION 2.5.   Let $R_j := \prod_{k=1}^{j} L_k$ for $j \ge 2$. For example, $R_2 = 2$, $R_3 = 4$, $R_4 = 24$, $R_5 = 144$, $R_6 = 288, \dots$ Then, define $\overline{R}_i := \prod_{j=2}^{i} R_j$ for $i \ge 2$. It turns out that a prime number p divides $\overline{R}_i$ if and only if $p \le \frac{i}{2} + 1$.

If X is a CW-complex and n a positive integer, let us write $X[n]$ for the n-th Postnikov section of X (i.e., $X[n]$ is a CW-complex with $\pi_i X[n] = 0$ for $i > n$ and $\pi_i X \cong \pi_i X[n]$ for $i \leq n$).

PROPOSITION 2.6.. *(a) If X is an $(m-1)$-connected infinite loop space (with $m \geq 2$) and n an integer $\geq m+1$, then there exist maps*

$$X[n] \xrightarrow{\varphi} \prod_{j=m}^{n} K(\pi_j X, j) \xrightarrow{\psi} X[n]$$

such that the composition is homotopic to the $\overline{R}_{n-m+1}$-th power map.
(b) If $f : X \to Y$ is an infinite loop map between $(m-1)$-connected infinite loop spaces and n an integer $\geq m+1$, then there is commutative diagram

$$
\begin{array}{ccccc}
X[n] & \xrightarrow{\varphi} & \prod\limits_{j=m}^{n} K(\pi_j X, j) & \xrightarrow{\psi} & X[n] \\
\downarrow{\scriptstyle f} & & \downarrow{\scriptstyle \xi} & & \downarrow{\scriptstyle f} \\
Y[n] & \xrightarrow{\varphi} & \prod\limits_{j=m}^{n} K(\pi_j Y, j) & \xrightarrow{\psi} & Y[n],
\end{array}
$$

where $\xi_* : \pi_j X \to \pi_j Y$ is exactly the homomorphism induced by f for $m \leq j \leq n$.
(c) Moreover, if η denotes the product map $\prod_{j=m}^{n} \eta_j$, where η_j is the map $K(\pi_j X, j) \to K(\pi_j Y, j)$ induced by f (for $m \leq j \leq n$), and if we set $\zeta = \xi - \eta$, then, for any prime number p, the induced homomorphism

$$\zeta_* : H_n\Big(\prod_{j=m}^{n} K(\pi_j X, j); \mathbb{Z}_{(p)}\Big) \longrightarrow H_n\Big(\prod_{j=m}^{n} K(\pi_j Y, j); \mathbb{Z}_{(p)}\Big)$$

is trivial assuming that $n \leq m + 2p - 3$.

Proof. The Postnikov k-invariants of X satisfy $R_{n-m+1} k^{n+1}(X) = 0$ in $H^{n+1}(X[n-1]; \pi_n X)$ for all $n \geq m+1$, since X is an infinite loop space (see for example [A3], Remark 1.6). Thus, Assertions (a) and (b) follow from the argument explained in Section 1 of [A3]. Because of the property of ξ_* stated in (b), the restriction of $\zeta = \xi - \eta$ to $K(\pi_j X, j)$ is actually an infinite loop map

$$\zeta_j : K(\pi_j X, j) \longrightarrow Z_j := \prod_{i=j+1}^{n} K(\pi_i Y, i) \hookrightarrow \prod_{i=m}^{n} K(\pi_i Y, i)$$

for any j such that $m \leq j \leq n-1$. Now, fix a prime p. It follows from [C] that for $j+1 \leq i \leq j+2p-3$, the homology groups with coefficients localized at p $H_i(K(\pi_j X, j); \mathbb{Z}_{(p)})$ contain only sums of products of homology classes represented by integral multiples of cycles of the form $\gamma_s(x)$ if j is even, respectively $x\gamma_s(y)$ if j is odd, where $x \in H_j(K(\pi_j X, j); \mathbb{Z}_{(p)})$ and $\deg(y) = j+1$ ($\gamma_s(-)$ denotes the s-th divided power); notice that $1 \le s \le p - 1$ since $i \le j + 2p - 3$. The ring homomorphism $(\zeta_j)_* : H_*(K(\pi_j X, j); \mathbb{Z}_{(p)}) \to H_*(Z_j; \mathbb{Z}_{(p)})$ maps any class x of degree j onto 0 because $H_j(Z_j; \mathbb{Z}_{(p)}) = 0$. If j is even, this implies that $s!(\zeta_j)_*(\gamma_s(x)) = (\zeta_j)_*(x^s) = ((\zeta_j)_*(x))^s = 0$, and consequently that $(\zeta_j)_*(\gamma_s(x)) = 0$ since p does not divide $s!$. Similarly if j is odd, it is obvious that $(\zeta_j)_*(x\gamma_s(y)) = 0$. The vanishing of the homomorphism

$$\zeta_* : H_n\big(\prod_{j=m}^n K(\pi_j X, j); \mathbb{Z}_{(p)}\big) \longrightarrow H_n\big(\prod_{j=m}^n K(\pi_j Y, j); \mathbb{Z}_{(p)}\big)$$

follows now from Künneth formula. $\qquad\square$

PROOF OF THEOREM 1.4. If $p = 2$, the statement of the theorem concerns only the dimension $n = 2$ where $\theta_\sharp : K_2 O \to K_2 F$ is injective. Thus, we may assume that p is an odd prime. According to the previous proposition for the map $\theta : BSL(O)^+ \to BSL(F)^+$ (with $m = 2$), we obtain a commutative diagram in homology localized at p.

$$
\begin{array}{ccccc}
H_n(BSL(O)^+; \mathbb{Z}_{(p)}) & \xrightarrow{\varphi_*} & H_n(\prod_{j=2}^n K(K_j O, j); \mathbb{Z}_{(p)}) & \xrightarrow{\psi_*} & H_n(BSL(O)^+; \mathbb{Z}_{(p)}) \\
\downarrow{\scriptstyle\theta_*} & & \downarrow{\scriptstyle\xi_*} & & \downarrow{\scriptstyle\theta_*} \\
H_n(BSL(F)^+; \mathbb{Z}_{(p)}) & \xrightarrow{\varphi_*} & H_n(\prod_{j=2}^n K(K_j F, j); \mathbb{Z}_{(p)}) & \xrightarrow{\psi_*} & H_n(BSL(F)^+; \mathbb{Z}_{(p)}),
\end{array}
$$

such that both horizontal compositions are multiplication by the p-primary part $(\overline{R}_{n-1})_p$ of $\overline{R}_{n-1}$. Since $n \le 2p - 2$, it turns out by Proposition 2.6 (c) that ξ_* is exactly the homomorphism η_* introduced in Lemma 2.1; hence, Lemma 2.1 and the hypothesis $n \le d_p + 1$ (if $d_p < \infty$) imply that ξ_* is injective. Consequently, if x belongs to the kernel of $\theta_* : H_n(BSL(O)^+; \mathbb{Z}_{(p)}) \to H_n(BSL(F)^+; \mathbb{Z}_{(p)})$, then $\varphi_*(x) = 0$ and $(\overline{R}_{n-1})_p x = \psi_*\varphi_*(x) = 0$. The conclusion follows from the fact that p does not divide $\overline{R}_{n-1}$ because $n \le 2p - 2$.

REMARK 2.7. The homomorphism $\theta_* : H_n(SL(O); \mathbb{Z}) \to H_n(SL(F); \mathbb{Z})$ is not injective for all n. Consider the case $O = \mathbb{Z}$, $F = \mathbb{Q}$, $n = 4(i-1)$ with i even, p a properly irregular prime with $p > i$ and such that p divides the numerator of $\dfrac{B_i}{i}$. There is a p-torsion element y in $H_{2(i-1)}(SL(\mathbb{Z}); \mathbb{Z})$ (see [S1], p. 290 and [A2], Section 2), and $\theta_*(y)$ is divisible in $H_{2(i-1)}(SL(\mathbb{Q}); \mathbb{Z})$ (by [B2], Theorem 3, and [AZ], Corollary 2.5). If w is a generator of the cyclic direct summand of $H_{2(i-1)}(SL(\mathbb{Z}); \mathbb{Z})$ containing y, then the Pontryagin product yw is non-trivial in $H_n(SL(\mathbb{Z}); \mathbb{Z})$, according to Corollary 2.3 of [A2], and belongs to the kernel of θ_* since $\theta_*(yw) = \theta_*(y)\theta_*(w) = 0$ because of the divisibility of $\theta_*(y)$. (We would like to thank C. Ausoni for this example.)

3 Divisible homology classes

Partial answers to Question 1.2 are given by the first two propositions while Theorem 1.5 is proved at the end of this section. If S is a set of maximal ideals of O, let $O_{(S)}$ denote the localization of O at S (i.e., the ring where the elements which are not in the ideals of S are inverted) and $\theta_{(S)}$ the inclusion $O_{(S)} \hookrightarrow F$.

PROPOSITION 3.1. *Let n be an integer ≥ 2 and x any divisible homology class in $H_n(SL(F); \mathbb{Z})$. Then, for all finite sets S of maximal ideals of O, x belongs to the image of the homomorphism*

$$(\theta_{(S)})_* : H_n(SL(O_{(S)}); \mathbb{Z}) \longrightarrow H_n(SL(F); \mathbb{Z})$$

induced by $\theta_{(S)}$. Moreover, there is a divisible element $y \in H_n(SL(O_{(S)}); \mathbb{Z})$ such that $(\theta_{(S)})_*(y) = x$.

Proof. There is a fibration

$$\prod_{m \in S} BSL(O/m)^+ \longrightarrow BSL(O_{(S)})^+ \xrightarrow{\theta_{(S)}} BSL(F)^+$$

(see [Q2], Theorem 4 or [Q1], Section 7, Proposition 3.2). The associated Serre spectral sequence

$$E^2_{s,t} \cong H_s(BSL(F)^+; H_t(\prod_{m \in S} BSL(O/m)^+; \mathbb{Z}) \Longrightarrow H_{s+t}(BSL(O_{(S)})^+; \mathbb{Z})$$

has the property that $H_t(\prod_{m \in S} BSL(O/m)^+; \mathbb{Z})$ is finite whenever $t \geq 1$, because the set S is finite and the O/m's are finite fields. This shows that $E^r_{s,t}$ is a group of finite exponent, and consequently has no divisible elements except 0, for any $r \geq 2$, $s \geq 0$, $t \geq 1$. If x is divisible in $H_n(SL(F); \mathbb{Z}) \cong E^2_{n,0}$, then $d^r(x)$ must also be divisible in $E^r_{n-r,r-1}$: thus, $d^r(x) = 0$ for all $r \geq 2$ and $x \in E^\infty_{n,0}$, in other words, x belongs to the image of $(\theta_{(S)})_*$.

Another way to prove this assertion is to observe that all homotopy groups of the fibre of the above fibration are finite and to deduce, using Serre class theory, that the homomorphism $(\theta_{(S)})_*$ has finite kernel and finite cokernel. Since its cokernel is finite, x is actually divisible in the image of $(\theta_{(S)})_*$ and the argument of Lemma 11 of [B1] enables us to conclude that there exists a divisible element y in $H_n(SL(O_{(S)})^+; \mathbb{Z})$ such that $(\theta_{(S)})_*(y) = x$. □

PROPOSITION 3.2. *Let p be a prime and n an integer with $2 \leq n \leq 2p - 2$. If a p-torsion divisible element x of $H_n(SL(F); \mathbb{Z})$ belongs to the image of the Hurewicz homomorphism $h_n : K_n F \to H_n(SL(F); \mathbb{Z})$, then*
(a) there is a p-torsion divisible element $z \in K_n F$ such that $h_n(z) = x$,
(b) x belongs to the image of $\theta_* : H_n(SL(O); \mathbb{Z}) \to H_n(SL(F); \mathbb{Z})$.

Proof. Assertion (a) is obvious since h_n is split injective on p-torsion for large primes p, i.e., if $p \geq \frac{n}{2} + 1$ (see Corollary 2.4 of [AZ]). Assertion (b) follows from the fact, mentioned in the introduction, that any divisible element z of $K_n F$ belongs to the image of $\theta_\sharp : K_n O \to K_n F$. □

LEMMA 3.3. *Let p be an odd prime such that $d_p < \infty$, n an integer satisfying $2 \leq n \leq 2d_p - 1$, and set $\mu = 1$ if $n < 4p - 4$ and $\mu = p$ if $n \geq 4p - 4$. If x is a p-torsion divisible element in $H_n(\prod_{j=2}^{n} K(K_j F, j); \mathbb{Z})$, then μx is the image of a p-torsion class under the homomorphism*

$$\eta_* : H_n(\prod_{j=2}^{n} K(K_j O, j); \mathbb{Z}) \longrightarrow H_n(\prod_{j=2}^{n} K(K_j F, j); \mathbb{Z})$$

induced by the map η introduced in the statement of Lemma 2.1.

Proof. Let us write

$$H_n(\prod_{j=2}^{n} K(K_j F, j); \mathbb{Z}_{(p)}) = H_n(Y \times \prod_{j=[\frac{n}{2}]+1}^{n} K(K_j F, j); \mathbb{Z}_{(p)})$$

where $Y = \prod_{j=2}^{[\frac{n}{2}]} K(K_j F, j)$ and $[\frac{n}{2}]$ is the integral part of $\frac{n}{2}$. If j is odd, $K_j F \cong K_j O$, and if j is even $\leq [\frac{n}{2}]$, $K_j(F; \mathbb{Z}_{(p)}) \cong K_j(O; \mathbb{Z}_{(p)}) \oplus (\bigoplus_m K_{j-1}(O/m; \mathbb{Z}_{(p)}))$ again by Corollary 1 of [B2], because $[\frac{n}{2}] \leq d_p - 1$. Therefore, for $2 \leq j \leq [\frac{n}{2}]$, $K_j(F; \mathbb{Z}_{(p)})$ is a direct sum of a free $\mathbb{Z}_{(p)}$-module of finite rank with an infinite direct sum of finite cyclic p-groups. On the other hand, since $\mu H_i(K(K_j F, j); \mathbb{Z}_{(p)}) = 0$ for $[\frac{n}{2}] + 1 \leq j < i \leq n$ according to Proposition 2.4 (notice that μ is the p-primary part of $M_{[\frac{n}{2}]}$), we get

$$\mu H_i(\prod_{j=[\frac{n}{2}]+1}^{n} K(K_j F, j); \mathbb{Z}_{(p)}) \cong \mu K_i(F; \mathbb{Z}_{(p)})$$

for $[\frac{n}{2}] + 1 \leq i \leq n$ and deduce from Künneth formula that

$$\mu H_n(\prod_{j=2}^{n} K(K_j F, j); \mathbb{Z}_{(p)}) \cong \mu \left(\bigoplus_{i=[\frac{n}{2}]+1}^{n} K_i(F; \mathbb{Z}_{(p)}) \otimes H_{n-i}(Y; \mathbb{Z}_{(p)}) \right)$$
$$\oplus \mu \left(\bigoplus_{i=[\frac{n}{2}]+1}^{n-3} \mathrm{Tor}(K_i(F; \mathbb{Z}_{(p)}), H_{n-i-1}(Y; \mathbb{Z}_{(p)})) \right).$$

In that formula the groups $H_{n-i}(Y; \mathbb{Z}_{(p)})$, respectively $H_{n-i-1}(Y; \mathbb{Z}_{(p)})$, are direct sums of free $\mathbb{Z}_{(p)}$-modules of finite rank with infinite direct sums of finite cyclic p-groups. It then follows from the fact that $\otimes$ and Tor commute with direct sums that

$$\mu H_n(\prod_{j=2}^{n} K(K_j F, j); \mathbb{Z}_{(p)}) \cong \mu \left((\bigoplus_{i=[\frac{n}{2}]+1}^{n} K_i(F; \mathbb{Z}_{(p)}) \otimes A_i) \oplus G \right),$$

where the groups A_i are free $\mathbb{Z}_{(p)}$-modules of finite rank and G is an infinite direct sum of groups of finite exponent; consequently, only the factors $K_i(F; \mathbb{Z}_{(p)}) \otimes A_i$ (for $[\frac{n}{2}] + 1 \leq i \leq n$) may contain divisible elements. However, we explained in the introduction that any p-torsion divisible element of $K_i F$ is the image of a p-torsion element of $K_i O$ under the induced homomorphism $K_i O \to K_i F$ and we obtain the desired assertion. $\qquad \square$

PROOF OF THEOREM 1.5. We may assume that $d_p < \infty$ since otherwise $H_n(SL(F);\mathbb{Z})$ contains no non-trivial divisible elements by Theorem 3.1 of [AZ]. If $p = 2$, n must be 2 and the theorem is a trivial consequence of the corresponding statement for the homomorphism $\theta_\sharp : K_2O \to K_2F$. If p is odd, look again at the diagram given by Proposition 2.6 for the map $\theta : BSL(O)^+ \to BSL(F)^+$ (with $m = 2$)

$$
\begin{array}{ccccc}
H_n(BSL(O)^+;\mathbb{Z}) & \xrightarrow{\varphi_*} & H_n(\prod_{j=2}^n K(K_jO,j);\mathbb{Z}) & \xrightarrow{\psi_*} & H_n(BSL(O)^+;\mathbb{Z}) \\
\downarrow{\scriptstyle \theta_*} & & \downarrow{\scriptstyle \xi_*} & & \downarrow{\scriptstyle \theta_*} \\
H_n(BSL(F)^+;\mathbb{Z}) & \xrightarrow{\varphi_*} & H_n(\prod_{j=2}^n K(K_jF,j);\mathbb{Z}) & \xrightarrow{\psi_*} & H_n(BSL(F)^+;\mathbb{Z}),
\end{array}
$$

in which both horizontal compositions are multiplication by $\overline{R}_{n-1}$. If x is a p-torsion divisible homology class in $H_n(BSL(F)^+;\mathbb{Z})$, then $\varphi_*(x)$ is divisible in $H_n(\prod_{j=2}^n K(K_jF,j);\mathbb{Z})$, and Lemma 3.3 (in which $\mu = 1$ because of the condition $n \le 2p - 2$) implies that there is a p-torsion element $w \in H_n(\prod_{j=2}^n K(K_jO,j);\mathbb{Z})$ such that $\eta_*(w) = \varphi_*(x)$. It follows from Proposition 2.6 (c) that $\xi_*(w) = \eta_*(w) = \varphi_*(x)$. Therefore, $\overline{R}_{n-1}\,x = \psi_*(\phi_*(x))$ belongs to the image of θ_*. The proof is then complete because p does not divide $\overline{R}_{n-1}$ since $n \le 2p - 2$.

References

[A1] D. Arlettaz: On the homology of the special linear group over a number field, *Comment. Math. Helv.* **61** (1986), 556–564.

[A2] D. Arlettaz: Torsion classes in the cohomology of congruence subgroups, *Math. Proc. Cambridge Philos. Soc.* **105** (1989), 241–248.

[A3] D. Arlettaz: Exponents for extraordinary homology groups, *Comment. Math. Helv.* **68** (1993), 653–672.

[AZ] D. Arlettaz and P. Zelewski: Divisible homology classes in the special linear group of a number field, preprint.

[B1] G. Banaszak: Algebraic K-theory of number fields and rings of integers and the Stickelberger ideal, *Ann. of Math.* **135** (1992), 325–360.

[B2] G. Banaszak: Generalization of the Moore exact sequence and the wild kernel for higher K-groups, *Compositio Math.* **86** (1993), 281–305.

[BG] G. Banaszak and W. Gajda: Euler systems for higher K-theory of number fields, preprint.

[BZ] G. Banaszak and P. Zelewski: Continuous K-theory, preprint.

[C] H. Cartan: Algèbres d'Eilenberg-MacLane et homotopie, exposé 11, *Séminaire H. Cartan Ecole Norm. Sup.* (1954/1955).

[Q1] D. Quillen: Higher algebraic K-theory I, in Higher K-theories, *Lecture Notes in Math.* **341** (Springer 1973), 85–147.

[Q2] D. Quillen: Higher K-theory for categories with exact sequences, in New Developments in Topology, *London Math. Soc. Lecture Note Ser.* **11** (Cambridge University Press 1974), 95–103.

[S1] C. Soulé: K-théorie des anneaux d'entiers de corps de nombres et cohomologie étale, *Invent. Math.* **55** (1979), 251–295.

[S2] C. Soulé: Groupes de Chow et K-théorie des variétés sur un corps fini, *Math. Ann.* **268** (1984), 317–345.

Dominique Arlettaz, Institut de mathématiques,
Université de Lausanne, CH–1015 Lausanne, Switzerland
e-mail: dominique.arlettaz@ima.unil.ch

Piotr Zelewski, Department of Mathematics and Statistics,
McMaster University, Hamilton, Ontario L8S 4K1, Canada
e-mail: piotr@icarus.math.mcmaster.ca

Progress in Mathematics, Vol. 136
© 1996 Birkhäuser Verlag Basel/Switzerland

Unstable localization and periodicity

A.K. BOUSFIELD*

Introduction

In the 1980's, remarkable advances were made by Ravenel, Hopkins, Devinatz, and Smith toward a global understanding of stable homotopy theory, showing that some major features arise "chromatically" from an interplay of periodic phenomena arranged in a hierarchy (see [20], [21], [28]). We would like very much to achieve a similar understanding in unstable homotopy theory and shall describe some progress in that direction. In particular, we shall explain and extend some results of our papers [4], [11], and some closely related results of Dror Farjoun and Smith [17], [18], [19].

Periodic phenomena in stable homotopy theory are quite effectively exposed by localizations with respect to various periodic homology theories such as the Morava K-theories [6], [27]. This approach remains promising in unstable homotopy theory, but a different sort of localization, called the W-*nullification* or W-*periodization* for a chosen space W, now seems more fundamental and effective. It simply trivializes the $[W, -]_*$-homotopy of spaces in a universal way. In Section 1 of this article, we recall the general theory of nullifications, including some crucial properties which have only recently been discovered. In Section 2, we introduce a corresponding theory of nullifications for spectra which we apply to determine nullifications of Eilenberg-MacLane spaces and other infinite loop spaces.

In Section 3, we begin to classify spaces according to the nullification functors which they produce, and prove a classification theorem for finite suspension complexes similar to the Hopkins-Smith classification theorem for finite spectra. In Section 4, we study the arithmetic nullifications, which act very much like classical localizations and completions of spaces. We apply them to determine arbitrary nullifications of Postnikov spaces and to extend the classification results of Section 3 beyond finite suspension complexes.

In Section 5, we present an unstable chromatic tower providing successive approximations to a space, incorporating higher and higher types of periodicity. In Section 6, we introduce a sequence of monochromatic homotopy categories containing the successive fibres of chromatic towers. Using work of Kuhn [22] and others, we show that the nth stable monochromatic homotopy category embeds as a categorical retract of its unstable counterpart. Finally, in Section 7, we apply some of the preceding work to obtain general results on E_*-acyclicity and E_*-equivalences of spaces for various spectra E.

---

*)   The author was partially supported by the National Science Foundation.

For simplicity, we work primarily in the pointed homotopy category Ho_* of CW-complexes and use the natural free and pointed function complexes, $\mathrm{map}(X, Y)$ and $\mathrm{map}_*(X, Y)$, in Ho_*.

1 Nullifications of spaces

For spaces $W, Y \in Ho_*$, we say that Y is W-*null* or W-*periodic* if $W \to *$ induces an equivalence $Y \simeq \mathrm{map}(W, Y)$. When Y is connected, this just means that $\mathrm{map}_*(W, Y) \simeq *$ or equivalently that $[\Sigma^i W, Y] = *$ for each $i \geq 0$. A W-*nullification* or W-*periodization* of X consists of a map $\alpha : X \to X'$ such that X' is W-null and

$$\mathrm{map}(\alpha, Y) : \mathrm{map}(X', Y) \simeq \mathrm{map}(X, Y)$$

for each W-null space Y. By [4, Cor. 7.2], [11], or [17], we have

THEOREM 1.1. *For each $W, X \in Ho_*$, there exists a W-nullification of X.*

This is unique up to equivalence and will be denoted by $\alpha : X \to P_W X$. Roughly speaking, $P_W X$ may be constructed from X by repeatedly attaching mapping cones to trivialize maps coming in from W and its suspensions, continuing to an appropriate transfinite colimit. Among the best known examples are

EXAMPLE 1.2. If $W = S^{n+1}$, then $P_W X$ is the nth Postnikov section of X.

EXAMPLE 1.3. If $W = S^1 \cup_p e^2$, then $P_W X$ is the Anderson localization [2], [14] of X away from p. This is equivalent to the standard localization $X[1/p]$ when X is simply connected.

The W-nullification is actually a special case of the very general f-*localization* introduced in [4, Cor. 7.2] and [17] for a map f of spaces, and many results on W-nullifications can at least partially be generalized to f-localizations.

As seen from Example 1.2, the W-nullification need not preserve fibrations. However, by [11, 4.1] and [18], it mixes with the ΣW-nullification to give

THEOREM 1.4. *For a space $W \in Ho_*$ and a fibre sequence $F \to X \to B$ of pointed spaces with B connected, there is a natural fibre sequence $P_W F \to \overline{X} \to P_{\Sigma W} B$ together with a natural P_W-equivalence $X \to \overline{X}$ where $\overline{X}$ is ΣW-null.*

We may obtain $\overline{X}$ as the orbit space of $P_W F$ under the principal action by $P_W \Omega B$. The following case, first noted by Dror Farjoun, is particularly useful.

COROLLARY 1.5. *For $W \in Ho_*$, P_W preserves each fibre sequence $F \to X \to B$ of pointed spaces such that B is W-null and connected.*

In the natural Postnikov tower

$$P_W X \leftarrow P_{\Sigma W} X \leftarrow P_{\Sigma^2 W} X \leftarrow \ldots,$$

we long suspected that the higher fibres might be Eilenberg-MacLane spaces
for many choices of W beyond the classical spheres. We very much wanted
to prove such a result because we knew that it would imply strong fibration
theorems for nullification functors and allow us to bring some important parts
of stable localization and periodicity theory into the unstable realm. In 1991,
we finally succeeded by using a version of

KEY LEMMA 1.6. *For a connected space V and a connected ΣV-null H-space Y, the inclusion $V \subset SP^\infty V$ induces an equivalence $\mathrm{map}_*(SP^\infty V, Y) \simeq \mathrm{map}_*(V, Y)$.*

Proof. This follows by [11, Cor. 6.9] since $\mathrm{map}_*(V, Y)$ is homotopically discrete
with $\pi_1 Y$ acting trivially on $[V, Y]$.

A space is called a *GEM* when it is equivalent to a product of Eilenberg-MacLane spaces $K(G_n, n)$ for a sequence of abelian groups $\{G_n\}_{n \geq 1}$. For the
connected ΣV-null H-space Y, the Key Lemma shows that each map $V \to Y$
has a canonical factorization through the GEM

$$SP^\infty V \simeq \prod_{n=1}^{\infty} K(H_n V, n).$$

This easily implies

THEOREM 1.7. *For a space W and a connected H-space X, if $P_W X \simeq *$ then $P_{\Sigma W} X$ is a GEM.*

Proof. $P_{\Sigma W} X$ is a connected H-space since nullification functors preserve finite
products. Moreover,

$$\mathrm{map}_*(\Sigma P_{\Sigma W} X, P_{\Sigma W} X) \simeq \mathrm{map}_*(P_{\Sigma W} X, \Omega P_{\Sigma W} X) \simeq *$$

since $P_W(P_{\Sigma W} X) \simeq P_W X \simeq *$ and $\Omega P_{\Sigma W} X$ is W-null. Hence, by the Key
Lemma, $P_{\Sigma W} X$ is a retract of the GEM $SP^\infty P_{\Sigma W} X$.

This immediately generalizes to f-localizations, and a relative version is
given by Dror Farjoun and Smith [19]. For p prime and $n \geq 1$, we say that
a space $W \in Ho_*$ satisfies the *n-supported p-torsion condition* when $\widetilde{H}_* W$ is
$(n-1)$-connected p-torsion with $H_n(W; Z/p) \neq 0$. We now recover the following
result of [11, 7.2] and [19].

THEOREM 1.8. *For connected spaces W, $X \in Ho_*$ and $i \geq 1$, the fibre F of the
Postnikov map $P_{\Sigma^{i+1} W} X \to P_{\Sigma^i W} X$ is a GEM. Moreover, when W satisfies
the n-supported p-torsion condition, $F \simeq K(G, n+i)$ for some p-torsion abelian
group G.*

Proof. The space F is 1-connected and $\Sigma^{i+1} W$-null with $P_{\Sigma^i W} F \simeq *$ by Corollary 1.5. Thus F is an H-space by [19, 2.1], and is a GEM by Theorem 1.7. The
last statement follows as in [11, 7.6], where the obvious H-space ΩF is used
instead of F.

The fibre of the lowest Postnikov map $P_{\Sigma W} X \to P_W X$ can be much more
complicated: it equals X when X is acyclic and $W = X$, but it remains a GEM
when X is a connected H-space. Theorem 1.8 combines with Theorem 1.4 to
give the strong fibration theorem of [11, Thm. 8.1] and [19].

THEOREM 1.9. *For a connected space $W \in Ho_*$ and a fibre sequence $F \to X \to B$ of pointed spaces with B connected, the fibre E of the map*

$$P_{\Sigma W} F \to \mathrm{fib}(P_{\Sigma W} X \to P_{\Sigma W} B)$$

is a GEM. Moreover, when W satisfies the n-supported p-torsion condition, then $E \simeq K(G, n)$ for some p-torsion abelian group G.

Thus $P_{\Sigma W}$ preserves fibre sequences up to a "small abelian error term" E.

2 Nullifications of spectra

We now introduce nullifications of spectra and show that they have almost the same basic properties as nullifications of spaces, but with easier proofs. By virtue of Theorem 2.10 below, they determine the unstable nullifications of Eilenberg-MacLane spaces and other infinite loop spaces.

We work in the homotopy category Ho^s of CW-spectra [1] and call a spectrum E *connective* when $\pi_i E = 0$ for $i < 0$. For spectra W and Y, we let $F^c(W, Y)$ denote the connective cover of the function spectrum $F(W, Y)$, and say that Y is W-*null* or W-*periodic* when $F^c(W, Y) \simeq 0$. This means that $[W, Y]_i \cong 0$ for each $i \geq 0$. A W-*nullification* or W-*periodization* of a spectrum X consists of a map $\alpha : X \to X'$ of spectra such that X' is W-null and

$$F^c(\alpha, Y) : F^c(X', Y) \simeq F^c(X, Y)$$

for each W-null spectrum Y.

THEOREM 2.1. *For each W, $X \in Ho^s$, there exists a W-nullification of X.*
Proof. We may view Ho^s as the associated homotopy category of the closed simplicial model category of spectra in [12, 2.4] and apply [4, Cor. 7.2] to give W-nullifications in Ho^s.

The W-nullification in Ho^s is unique up to equivalence and will be denoted by $\alpha : X \to P_W X$. It is a special case of the f-localization which exists in Ho^s for each map f of spectra. The W-nullification mixes with the ΣW-nullification to give

THEOREM 2.2. *For $W \in Ho^s$ and a cofibre sequence $F \to X \to B$ of spectra, there is a natural cofibre sequence $P_W F \to \overline{X} \to P_{\Sigma W} B$ together with a natural P_W-equivalence $X \to \overline{X}$ where $\overline{X}$ is ΣW-null.*
Proof. Use the cofibre sequence of $P_W(\Sigma^{-1} B) \to P_W F$.

COROLLARY 2.3. *For $W \in Ho^s$, P_W preserves each cofibre sequence $F \to X \to B$ of spectra such that B is W-null.*

To obtain stronger (co)fibration results, we let H be the spectrum of integral homology and use

KEY LEMMA 2.4. *For spectra V, $Y \in Ho^s$, if $F^c(\Sigma V, Y) \simeq 0$, then the Hurewicz map $V \to H \wedge V$ induces an equivalence $F^c(H \wedge V, Y) \simeq F^c(V, Y)$.*
Proof. This follows since the cofibre of the unit map $S \to H$ is 1-connected.

A spectrum X is called a *stable GEM* if it is equivalent to a wedge (and thus a product) of Eilenberg-MacLane spectra $\{\Sigma^n HG_n\}_{n \in Z}$. This happens if and only if X admits a module structure over the ring spectrum H (i.e. a map $H \wedge X \to X$ in Ho^s satisfying the associativity and unit conditions). As in 1.7, the Key Lemma implies

THEOREM 2.5. *For spectra W, $X \in Ho^s$, if $P_W X \simeq 0$, then $P_{\Sigma W} X$ is a stable GEM with a canonical H-module structure.*

This immediately generalizes to f-localizations. For p prime and $n \in Z$, we say that a spectrum W satisfies the *n-supported p-torsion condition* when $H_* W$ is $(n-1)$-connected p-torsion with $H_n(W; Z/p) \neq 0$. As in 1.8, we deduce

THEOREM 2.6. *For spectra W, $X \in Ho^s$ and $i \in Z$, the fibre F of the Postnikov map $P_{\Sigma^{i+1}W} X \to P_{\Sigma^i W} X$ is a stable GEM with a canonical H-module structure. Moreover, when W satisfies the n-supported p-torsion condition, then $F \simeq \Sigma^{n+i} HG$ for some p-torsion abelian group G.*

This combines with Theorem 2.2 to give a strong fibration theorem

THEOREM 2.7. *For $W \in Ho^s$ and a cofibre sequence $F \to X \to B$ of spectra, the fibre E of the map*

$$P_W F \to \mathrm{fib}(P_W X \to P_W B)$$

is a stable GEM. Morever, when W satisfies the n-supported p-torsion condtion, then $E \simeq \Sigma^{n-1} HG$ for some p-torsion abelian group G.

Since smash products with connective spectra preserve P_W-equivalences in Ho^s, we have

PROPOSITION 2.8. *For $W \in Ho^s$, if A is a connective ring spectrum and M is an A-module spectrum, then $\alpha : M \to P_W M$ is a map of A-module spectra.*

COROLLARY 2.9. *For $W \in Ho^s$, if M is a stable GEM, then so is $P_W M$. Moreover, if $M = \Sigma^n HG$, then $\pi_i P_W M = 0$ unless $i = n$, $n + 1$.*
Proof. This follows since $P_W M$ is an H-module spectrum by 2.8, and since each H-module map out of $\Sigma^n HG$ has a retract of the required form.

To relate the stable and unstable nullifications, we use the adjoint functors $\Sigma^\infty : Ho_* \to Ho^s$ and $\Omega^\infty : Ho^s \to Ho_*$.

THEOREM 2.10. *For a space $W \in Ho_*$ and a spectrum $E \in Ho^s$, the natural map*

$$P_W \Omega^\infty E \to \Omega^\infty P_{\Sigma^\infty W} E$$

is an equivalence.

Proof. Let $Ho^{sc} \subset Ho^s$ be the full subcategory of connective spectra. The proof in [8, Thm. 1.1] is easily adapted to show the existence of an idempotent functor $T : Ho^{cs} \to Ho^{cs}$ and $\eta : Id \to T$, such that for any $X \in Ho^{cs}$ the map $\Omega^{\infty}\eta : \Omega^{\infty}X \to \Omega^{\infty}TX$ is a W-nullification. Moreover, (T, η) must be equivalent to the idempotent functor $(P_{\Sigma^{\infty}W}, \alpha)$ on Ho^{cs} since a connective spectrum X is $\Sigma^{\infty}W$-null if and only if $\Omega^{\infty}X$ is W-null. Thus, $\Omega^{\infty}\alpha : \Omega^{\infty}E \to \Omega^{\infty}P_{\Sigma^{\infty}W}E$ is a W-nullification for all connective E, and hence for all E.

We may now destabilize the preceding corollary to give

COROLLARY 2.11. *For spaces $W, Y \in Ho_*$, if Y is a GEM, then so is $P_W Y$. Moreover, if $Y = K(G, n)$, then $\pi_i P_W Y = 0$ unless $i = n, n + 1$.*

2.12. TRIVIALIZATIONS OF SPECTRA. In [5, 1.7] for $W \in Ho^s$, we introduced $[W, -]_*$-*trivializations* of spectra. These may be defined in the same way as W-nullifications, using $F(W, -)$ instead of $F^c(W, -)$. From the present standpoint, $[W, -]_*$-trivializations of spectra are just $\Sigma^* W$-nullifications, where $\Sigma^* W$ is the wedge of $\{\Sigma^n W\}_{n \in Z}$. They always preserve cofibre sequences of spectra since $\Sigma(\Sigma^* W) \simeq \Sigma^* W$.

3 Nullity classes

We can now begin to classify spaces and spectra according to the nullification functors which they produce. For spaces $X, Y \in Ho_*$, we say that X *kills* Y when the following equivalent conditions hold:

(i) each X-null space is Y-null;
(ii) $Y \to *$ is an X-nullification;
(iii) $P_X Y \simeq *$.

We say that X and Y have the *same nullity* when they kill each other and thus produce equivalent nullifications. The resulting *nullity classes* or *P-classes* $\langle X \rangle$ have a partial ordering, where $\langle X \rangle \geq \langle Y \rangle$ means that X kills Y, and have operations

$$\bigvee_{\alpha} \langle X_{\alpha} \rangle = \langle \bigvee_{\alpha} X_{\alpha} \rangle \qquad \langle X \rangle \wedge \langle Y \rangle = \langle X \wedge Y \rangle$$

with the expected properties as explained more fully in [11, Sect. 9] and [18]. However, we warn that the inequality "$\geq$" may be defined oppositely.

The above notions extend immediately to spectra, and we write $\langle E \rangle^s$ for the nullity class of $E \in Ho^s$. By Theorem 2.10 we have

PROPOSITION 3.1. *For a space $W \in Ho_*$ and connective spectrum $X \in Ho^s$, the condition $\langle W \rangle \geq \langle \Omega^{\infty}X \rangle$ is equivalent to $\langle \Sigma^{\infty}W \rangle^s \geq \langle X \rangle^s$. Thus $\langle W \rangle \geq \langle \Omega^{\infty}\Sigma^{\infty}W \rangle$ and $\langle \Sigma^{\infty}\Omega^{\infty}X \rangle^s \geq \langle X \rangle^s$.*

COROLLARY 3.2. *Let V, $W \in Ho_*$ be connected spaces.*

(i) *If $\langle V \rangle \geq \langle W \rangle$, then $\langle SP^\infty V \rangle \geq \langle SP^\infty W \rangle$.*
(ii) *$\langle V \rangle \geq \langle SP^\infty V \rangle$.*
(iii) *If V is a GEM, then $\langle V \rangle = \langle SP^\infty V \rangle$.*

Proof. This follows by 3.1 since $SP^\infty V \simeq \Omega^\infty(H \wedge \Sigma^\infty V)$ and since a GEM is a homotopy retract of its infinite symmetric product.

THEOREM 3.3. *For a connected space $W \in Ho_*$ and $k \geq 1$,*
$$\langle \Sigma W \rangle = \langle \Sigma^k W \rangle \vee \langle SP^\infty \Sigma W \rangle.$$

Proof. By Theorem 1.8 for $i \geq 1$, $P_{\Sigma^{i+1}W}(\Sigma^i W)$ is a GEM killed by $\Sigma^i W$. Hence, by 3.2, it is also killed by $SP^\infty \Sigma^i W$, and we have
$$\langle \Sigma^i W \rangle \leq \langle \Sigma^{i+1} W \rangle \vee \langle SP^\infty \Sigma^i W \rangle.$$

This inductively implies
$$\langle \Sigma W \rangle \leq \langle \Sigma^k W \rangle \vee \langle SP^\infty \Sigma W \rangle,$$

and the opposite inequality is evident.

This theorem enables us to partially destabilize the Hopkins-Smith classification of finite CW-spectra [20], [21], [28]. Over a finite prime p and for $n \geq 0$, let $K(n)$ denote the nth Morava K-spectrum, where $K(0) = HQ$. The *p-type* of a space X is the smallest integer n such that $\widetilde{K}(n)_* X \neq 0$, or is ∞ when $\widetilde{K}(n)_* X = 0$ for all n. It is denoted by type$_p X$. We shall see in Corollary 7.2 that type$_p X = \infty$

if and only if $\widetilde{H}_*(X; Z_{(p)}) = 0$. By Mitchell [26] or Hopkins-Smith [21], for each positive integer n, there exists a finite p-torsion complex of p-type n. We say that two spaces $X, Y \in Ho_*$ have the *same stabilized nullity* if $\langle X \rangle \geq \langle \Sigma^i Y \rangle$ and $\langle Y \rangle \geq \langle \Sigma^j X \rangle$ for some $i, j \geq 0$. The resulting *stabilized nullity classes* $\{X\}$ are partially ordered with finite wedge and smash operations. As noted by Dror Farjoun in the p-local case, the Hopkins-Smith classification shows

THEOREM 3.4. *For finite connected complexes V, $W \in Ho_*$, the condition $\{V\} = \{W\}$ holds if and only if type$_p V = $ type$_p W$ for each prime p.*
Proof. Taking suspensions, we can assume that V and W are 1-connected. Given that type$_p V = $ type$_p W$ for each prime p, we apply the "thick subcategory theorem" as in [11, 9.14] to deduce that $\{V_{(p)}\} = \{W_{(p)}\}$ for each p. When $\widetilde{H}_*(V; Q) = 0$ this implies that $\{V\} = \{W\}$ by wedge decomposition. When $\widetilde{H}_*(V; Q) \neq 0$, the p-types of V and W are all 0, and hence $\{V_{(p)}\} = \{S^1_{(p)}\} = \{W_{(p)}\}$ for each p. Thus $\{V\} \geq \{M\} \leq \{W\}$ for each finite complex M with $\widetilde{H}_*(M; Q) = 0$. Using a cofibre sequence $B \to \Sigma^i V \to M$ where B is a wedge of spheres and M is as above, we deduce that $\{V\} = \{S^1\} = \{W\}$.

The *Z/p-connectivity* of a space X is the largest integer n such that $\widetilde{H}_n(X; Z/p) = 0$, or is ∞ when $\widetilde{H}_*(X; Z/p) = 0$. It is denoted by conn$_p X$ and is a nullity class invariant since it may be expressed as a cohomological connectivity. As shown p-locally in [11, 9.15], the Hopkins-Smith classification destabilizes to

THEOREM 3.5. *For finite connected complexes V, $W \in Ho_*$, the condition $\langle \Sigma V \rangle = \langle \Sigma W \rangle$ is equivalent to the joint conditions:*

(i) $\mathrm{type}_p \Sigma V = \mathrm{type}_p \Sigma W$ *for each prime p;*
(ii) $\mathrm{conn}_p \Sigma V = \mathrm{conn}_p \Sigma W$ *for each prime p.*

Proof. This follows from Theorems 3.3 and 3.4 since condition (ii) implies $\langle SP^\infty \Sigma V \rangle = \langle SP^\infty \Sigma W \rangle$.

The preceding results 3.2-3.5 have the expected versions for spectra, culminating in

THEOREM 3.6. *For finite CW spectra X, $Y \in Ho^s$, the condition $\langle X \rangle^s = \langle Y \rangle^s$ is equivalent to the joint conditions:*

(i) $\mathrm{type}_p X = \mathrm{type}_p Y$ *for each prime p;*
(ii) $\mathrm{conn}_p X = \mathrm{conn}_p Y$ *for each prime p.*

3.7. RELATED CLASSIFICATIONS OF SPECTRA. For a spectrum $X \in Ho^s$, we let $\langle X \rangle^t$ be the class of all spectra Y such that $[Y, -]_*$ and $[X, -]_*$ have the same trivial spectra, and thus give the same trivialization functors (2.12). As in [5], we let $\langle X \rangle$ be the class of all spectra Y such that the homology theories X_* and Y_* have the same acyclic spectra, and thus give the same localization functors. In general $\langle X \rangle^s \subset \langle X \rangle^t \subset \langle X \rangle$, and for a finite CW spectrum X the class $\langle X \rangle^t = \langle X \rangle$ is determined by the Hopkins-Smith invariants $\{\mathrm{type}_p X\}_p$.

4 The arithmetic nullifications

When W is a wedge of 1-connected Moore spaces, the W-nullification acts very much like a classical localization or completion functor, transforming homotopy groups in an elementary arithmetic way. We shall describe these arithmetic nullifications quite explicitly, and then apply them to determine arbitrary nullifications of Postnikov spaces and to extend our nullity classification results.

For a sequence $\{G_i\}_{i \geq 2}$ of abelian groups, let $M(G_i, i)$ be the Moore space with $H_i M(G_i, i) = G_i$ and take the wedge

$$MG(n) = M(G_2, 2) \vee \cdots \vee M(G_n, n).$$

Let J be the set of all primes p such that $p : G_i \cong G_i$ for $2 \leq i \leq n$. By [11, Sect. 5], we have

THEOREM 4.1. *For a space $Y \in Ho_*$ and $m > n$, there is a natural isomorphism*

$$\pi_m P_{MG(n)} Y \cong \pi_m Y \otimes Z_{(J)}$$

when $G_2, \ldots, G_n$ are all torsion, and there is a splittable natural short exact sequence

$$0 \to \prod_{p \in J} \mathrm{Ext}(Z_{p^\infty}, \pi_m Y) \to \pi_m P_{MG(n)} Y \to \prod_{p \in J} \mathrm{Hom}(Z_{p^\infty}, \pi_{m-1} Y) \to 0$$

when $G_2, \ldots, G_n$ are not all torsion.

The required Ext-p-completion is discussed in [13] and is given by

$$\text{Ext}(Z_{p^\infty}, N) \cong \lim_n N/p^n N$$

when the p torsion elements of N are of bounded order. To extend Theorem 4.1, we need another algebraic notion. For abelian groups B and X, we call X B-null or B-reduced when $\text{Hom}(B, X) = 0$. Each abelian group A has a maximal B-null quotient group $A//B$ as in [11, 5.1]. For instance, when B is p-torsion with $B/pB \neq 0$, then $A//B$ is the quotient of A by its p-torsion subgroup. We shall need

LEMMA 4.2. *If A is J-local for a set J of primes, then so is $A//B$ for all B. Proof.* This follows when $Q \otimes B \neq 0$ since each Q-null quotient of A is J-local.

Now let

$$MG = M(G_2, 2) \vee M(G_3, 3) \vee \ldots$$

be an infinite wedge of Moore spaces. Let $\overline{G}_{n+1} = G_{n+1}$ when $G_2, G_3 \ldots, G_{n+1}$ are all torsion or when $G_{n+1} \otimes Q \neq 0$, and let $\overline{G}_{n+1} = G_{n+1} \oplus Q$ otherwise.

THEOREM 4.3. *For a space $Y \in Ho_*$, there is a natural isomorphism*

$$\pi_{n+1} P_{MG} Y \cong \begin{cases} \pi_{n+1} Y & \text{for } n < 1 \\ (\pi_{n+1} P_{MG(n)} Y)//\overline{G}_{n+1} & \text{for } n \geq 1. \end{cases}$$

Proof. This follows since

$$\pi_{n+1} P_{MG} Y \cong \pi_{n+1} P_{MG(n+1)} Y \cong \pi_{n+1} P_{MG(n) \vee N} Y \cong \pi_{n+1} P_N P_{MG(n)} Y$$

for $N = M(\overline{G}_{n+1}, n+1)$ by [11, Sect. 5] and Lemma 4.2.

In this theorem, we could replace $\overline{G}_{n+1}$ by $G_2 \oplus \cdots \oplus G_{n+1}$, but not by G_{n+1} as seen from

EXAMPLE 4.4. For $G_2 = Z[1/p]$, $G_3 = Z/p$, and $Y = K(\bigoplus_j Z/p^j, 3)$, we have $\pi_3 P_{MG} Y = 0$ while $(\pi_3 P_{MG(2)} Y)//G_3 \neq 0$.

Theorems 4.1 and 4.3 combine to express $\pi_* P_{MG} Y$ algebraically in terms of $\pi_* Y$, and hence

COROLLARY 4.5. *A space $Y \in Ho_*$ is killed by MG if and only if Y is 1-connected and $K(\pi_n Y, n)$ is killed by MG for each $n \geq 2$.*

For a 1-connected space W, we let MHW denote the associated wedge $M(\widetilde{H}_* W)$ of Moore spaces, and we note that MHW kills W since it successively kills the homology groups of W. Thus, there is a natural map $P_W Y \to P_{MHW} Y$ for $Y \in Ho_*$.

THEOREM 4.6. *For a 1-connected space W and connected Postnikov space $Y \in Ho_*$, the map $P_W Y \to P_{MHW} Y$ is an equivalence.*

Proof. The homotopy fibre F of this map is a 1-connected Postnikov space with $P_{MHW}F \simeq *$ by Corollary 1.5. Hence for $n \geq 2$, $\langle MHW \rangle \geq \langle K(\pi_n F, n) \rangle$ by Corollary 4.5, and $\langle W \rangle \geq \langle SP^\infty W \rangle = \langle SP^\infty MHW \rangle \geq \langle K(\pi_n F, n) \rangle$ by Corollary 3.2. Hence, W kills F and $Y \to P_{MHW}Y$ is a W-nullification.

By this theorem, the W-nullification always acts arithmetically on Postnikov spaces. Of course, it acts arithmetically on arbitrary spaces when $\langle W \rangle = \langle MHW \rangle$.

THEOREM 4.7. *If $W \in Ho_*$ is 1-connected with $\langle \Sigma^r W \rangle = \langle \Sigma^r MHW \rangle$ for some $r \geq 0$, then $\langle W \rangle = \langle MHW \rangle$.*
Proof. We may assume $\langle \Sigma W \rangle = \langle \Sigma MHW \rangle$ and must show $P_W Y \simeq P_{MHW}Y$ for each connected $Y \in Ho_*$. Since MHW is a suspension and since $P_W Y$ is ΣMHW-null, the fibre F of $P_W Y \to P_{MHW}Y$ is a GEM by Theorem 1.8. Since MHW kills the GEM F, it kills $K(\pi_n F, n)$ for $n \geq 2$. As in the proof of 4.6, this implies that W kills $K(\pi_n F, n)$ for $n \geq 2$. Thus W kills the GEM F, and $P_W Y \to P_{MHW}Y$ is a P_W-equivalence. Since $P_{MHW}Y$ is W-null, we have $P_W Y \simeq P_{MHW}Y$ as required.

We can now supplement Theorem 3.5 with a nullity classification theorem for some nonsuspension spaces.

THEOREM 4.8. *Let V, $W \in Ho_*$ be finite 1-connected complexes such that $\text{type}_p V$ and $\text{type}_p W$ belong to $\{0, 1, \infty\}$ for all p. Then the condition $\langle V \rangle = \langle W \rangle$ is equivalent to the joint conditions:*

   (i) $\text{type}_p V = \text{type}_p W$ *for all p;*
   (ii) $\text{conn}_p V = \text{conn}_p W$ *for all p.*

Proof. We have $\langle \Sigma V \rangle = \langle \Sigma MHV \rangle = \langle \Sigma MHW \rangle = \langle \Sigma W \rangle$ by Theorem 3.5, and thus have $\langle V \rangle = \langle W \rangle$ by Theorem 4.7.

The nullity classes covered by this theorem have canonical representatives of the form $\bigvee_{p \in J} M(Z/p, n_p)$ or $S^n \vee \bigvee_{p \in J} M(Z/p, n_p)$ where J is a finite set of primes and n_p, $n \geq 2$ are integers with $n_p < n$ for each $p \in J$. For instance, when W is a finite 1-connected complex with $H_*(W; Q) \neq 0$, its nullity class has a canonical representative of the second sort, where n is the smallest integer with $\tilde{H}_n(W; Q) \neq 0$.

We do not know whether the assumption that the p-types belong to $\{0, 1, \infty\}$ is actually required in Theorem 4.8. However, the 1-connectivity is required. For instance, a homology n-sphere M^n and S^n satisfy the joint conditions, but $\langle M^n \rangle \neq \langle S^n \rangle$ when $P^{n-1} M^n$ is noncontractible.

5 Chromatic towers

In [11] we introduced an unstable chromatic tower $\{P_{v_n} X\}_{n \geq 0}$ providing successive approximations to a space X, incorporating higher and higher types of periodicity. We now present a simple version of this tower and compare it with a stable chromatic tower of Ravenel [29] and others. We work over a fixed prime p.

5.1 AN UNSTABLE CHROMATIC TOWER. For each $n \geq 0$, we choose a finite p-torsion complex $\overline{V}_n$ of p-type $n+1$ such that $\mathrm{conn}_p \overline{V}_n$ is minimal. For instance, we may choose $\overline{V}_0 = S^1 \cup_p e^2$ and choose $\overline{V}_1$ for p odd to be the cofibre of the Adams map

$$A : \Sigma^{2p-2}(S^2 \cup_p e^3) \to S^2 \cup_p e^2$$

constructed in [15]. In general, $\mathrm{conn}_p \overline{V}_n$ must be at least n by [11, 9.16], and actually equals n in the only known cases above. By Theorem 3.5 the nullity class $\langle \Sigma \overline{V}_n \rangle$ is well-defined and satisfies

$$\langle \Sigma \overline{V}_n \rangle = \langle \Sigma \overline{V}_n \vee \Sigma \overline{V}_{n+1} \rangle > \langle \Sigma \overline{V}_{n+1} \rangle.$$

The $\Sigma \overline{V}_n$-nullification of a space $X \in Ho_*$ is called the v_n-periodization and is denoted by $P_n X$. There is a natural unstable chromatic tower

$$P_0 X \leftarrow P_1 X \leftarrow P_2 X \leftarrow \cdots ,$$

which has the obvious convergence property, $\mathrm{holim}\, P_n X \simeq X$, since $\pi_i P_n X \cong \pi_i X$ for $i < \mathrm{conn}_p \overline{V}_n + 2$. The v_n-periodization $P_n X$ may perhaps "become active" at a higher dimension than the more sophisticated $P_{v_n} X$ of [11, 10.2], but there is essential agreement since $\pi_i P_n X \cong \pi_i P_{v_n} X$ for $i > \mathrm{conn}_p \overline{V}_n + 2$. To explain the chromatic properties of our tower, we recall

5.2. THE v_n-PERIODIC HOMOTOPY GROUPS. For a finite p-torsion complex $W \in Ho_*$ of p-type n, a v_n-map is a map $\omega : \Sigma^d W \to W$ with $d > 0$ such that $\widetilde{K}(n)_* \omega$ is an isomorphism and $\widetilde{K}(m)_* \omega = 0$ for all $m \neq n$. For instance, the above Adams map is a v_1-map. The Hopkins-Smith "periodicity theorem" [21] ensures that each finite p-torsion complex of p-type n has a v_n-map after sufficient suspension, and that such a v_n-map is unique up to stable iteration. For each $n \geq 1$, we choose a finite p-torsion complex V_{n-1} of p-type n having a v_n-map ω. Then for a space $Y \in Ho_*$, we obtain the v_n-periodic homotopy groups

$$v_n^{-1} \pi_*(Y; V_{n-1}) = Z[\omega, \omega^{-1}] \otimes_{Z[\omega]} \pi_*(Y; V_{n-1})$$

by inverting the action of ω on $\pi_*(Y; V_{n-1})$. These depend on V_{n-1}, but not on the choice of ω. By [11, Thm. 11.5], we have

THEOREM 5.3. For a space $X \in Ho_*$ and $n \geq 0$, the v_n-periodization $X \to P_n X$ induces

$$v_m^{-1} \pi_*(P_n X; V_{m-1}) \cong \begin{cases} v_m^{-1} \pi_*(X; V_{m-1}) & \text{for } 1 \leq m \leq n \\ 0 & \text{for } m > n. \end{cases}$$

Thus we may regard the spaces $\{P_n X\}_{n \geq 0}$ as successive approximations to X capturing higher and higher types of periodicity at the prime p. To isolate the nth type of periodicity, we simply take the fibre $\widetilde{P}_n X$ of the tower map $P_n X \to P_{n-1} X$.

COROLLARY 5.4. *For a space $X \in Ho_*$ and $n \geq 1$, $\widetilde{P}_n X$ is an n-connected p-torsion space with*

$$v_m^{-1}\pi_*(\widetilde{P}_n X; V_{m-1}) \cong \begin{cases} v_n^{-1}\pi_*(X; V_{n-1}) & \text{for } m = n \\ 0 & \text{for } m \neq n. \end{cases}$$

Since the cofibre of $\omega : \Sigma^d V_{n-1} \to V_{n-1}$ has p-type $n + 1$, we find

$$v_n^{-1}\pi_t(X; V_{n-1}) \cong \pi_t(P_n X; V_{n-1}) \cong \pi_t(\widetilde{P}_n X; V_{n-1})$$

for $n \geq 1$ and $t \geq 2$, so that the v_n-periodic homotopy groups of X are exposed as ordinary homotopy groups of $P_n X$ and $\widetilde{P}_n X$. Our unstable chromatic tower is closely related to

5.5. A STABLE CHROMATIC TOWER. For a spectrum E, we obtain the *stable chromatic tower* $\{L_n' E\}_{n\geq 0}$ of Ravenel [29] and others by letting $L_n' E$ be the $[W_n, -]_*$-trivialization (2.12) of E for a p-torsion finite CW-spectrum W_n of p-type $n + 1$. This tower must be distinguished from Ravenel's original chromatic tower $\{L_n E\}_{n\geq 0}$ in view of his refutation of the telescope conjecture. The fibre of the tower map $L_n' E \to L_{n-1}' E$ is denoted by $M_n' E$, and the tower $\{L_n' E\}_{n\geq 0}$ sorts the v_n-periodic homotopy groups of a spectrum E in the same way as the tower $\{P_n X\}_{n\geq 0}$ sorts the v_n-periodic homotopy groups of a space X. The chromatic tower of a spectrum E and that of $\Omega^\infty E$ are related by a natural map

$$\{P_n \Omega^\infty E\}_{n\geq 0} \to \{\Omega^\infty L_n' E\}_{n\geq 0}.$$

THEOREM 5.6. *There are induced isomorphisms $\pi_i P_n \Omega^\infty E \cong \pi_i \Omega^\infty L_n' E$ and $\pi_i \widetilde{P}_n \Omega^\infty E \cong \pi_i \Omega^\infty M_n' E$ for $i > \text{conn}_p \overline{V}_n + 2$.*
Proof. For $W_n = \Sigma^\infty(\Sigma \overline{V}_n)$, the $[W_n, -]_*$-trivialization of E is given by the homotopy colimit of the $\Sigma^{-k} W_n$-nullifications of E as $k \to \infty$. Hence by Theorems 2.6 and 2.10,

$$\pi_i P_n \Omega^\infty E \cong \pi_i P_{W_n} E \cong \pi_i L_n' E$$

for $i > \text{conn}_p \overline{V}_n + 2$.

6 The monochromatic homotopy categories

Working over a fixed prime p for $n \geq 1$, we let $\widetilde{P}_n Ho_* \subset Ho_*$ and $M_n' Ho^s \subset Ho^s$ be the full subcategories whose objects are equivalent to the nth chromatic layers $\widetilde{P}_n X$ and $M_n' E$ of spaces $X \in Ho_*$ and spectra $E \in Ho^s$. We call $\widetilde{P}_n Ho_*$ and $M_n' Ho^s$ *monochromatic homotopy categories* and now show that $M_n' Ho^s$ embeds faithfully as a categorical retract of its unstable counterpart $\widetilde{P}_n Ho_*$.

THEOREM 6.1. *For $n \geq 1$, the functor $\widetilde{P}_n \Omega^\infty : M_n' Ho^s \to \widetilde{P}_n Ho_*$ has a left inverse Φ_n.*

The required functor Φ_n is given by the following theorem which extends results of Kuhn [22], Davis-Mahowald [16], and the author [10].

THEOREM 6.2. *For $n \geq 1$, there exists a functor $\Phi_n : Ho_* \to M'_n Ho^s$ such that:*

(i) *there is a natural equivalence $\Phi_n \Omega^\infty E \simeq M'_n E$ for $E \in Ho^s$;*
(ii) *the functor Φ_n preserves fibre sequences and homotopy direct limits of directed systems of pointed spaces;*
(iii) *if $V_{n-1} \in Ho_*$ is a finite p-torsion complex of p-type n with a v_n-map, then $v_n^{-1} \pi_*(X; V_{n-1}) \cong [V_{n-1} \Phi_n X]_*$ and $\Omega^\infty F(\Sigma^2 V_{n-1}, \Phi_n X) \simeq \mathrm{map}_*(\Sigma^2 V_{n-1}, P_n X)$ for $X \in Ho_*$;*
(iv) *if $f : X \to Y$ is a map in Ho_* with $f_* : v_n^{-1} \pi_*(X; V_{n-1}) \cong v_n^{-1} \pi_*(Y; V_{n-1})$, then $\Phi_n f : \Phi_n X \simeq \Phi_n Y$.*

This will be proved in 6.8. To avoid some technical difficulties, we shall construct Φ_n as the composite of functors $\widehat{\Phi}_n : Ho_* \to \widehat{M'_n} Ho^s$ and $\Gamma_n : \widehat{M'_n} Ho^s \simeq M'_n Ho^s$ where $\widehat{M'_n} Ho^s$ is a different form of the nth monochromatic stable homotopy category. We first explain

6.3. THE FUNCTORS Γ_n AND $\widehat{\Gamma}_n$.

For $n \geq 0$ let W_n be a p-torsion finite CW-spectrum of p-type $n+1$. As in [5], we say that a spectrum C is $[W_n, -]_*$-*colocal* if each $[W_n, -]_*$-equivalence of spectra is a $[C, -]_*$-equivalence, and say that a map of spectra $X' \to X$ is a $[W_n, -]_*$-*colocalization* if it is a $[W_n, -]_*$-equivalence with X' $[W_n, -]_*$-colocal. Each spectrum X has a natural $[W_n, -]_*$-colocalization given by the fibre of the $[W_n, -]_*$-trivialization $X \to L'_n X$, and we let $\Gamma_n Ho^s \subset Ho^s$ denote the full subcategory of $[W_n, -]_*$-colocal spectra. In addition, each spectrum X has a natural W_{n*}-localization $X \to \widehat{\Gamma}_n X$ as in [6], and we let $\widehat{\Gamma}_n Ho^s \subset Ho^s$ denote the full subcategory of W_{n*}-local spectra. The functor $\widehat{\Gamma}_n : Ho^s \to Ho^s$ is right adjoint to $\Gamma_n : Ho^s \to Ho^s$ since there are natural equivalences $\widehat{\Gamma}_n X \simeq F(\Gamma_n S, X)$ and $\Gamma_n X \simeq X \wedge \Gamma_n S$ by [5, p. 375]. In unpublished work [7, 2.7], we noted

THEOREM 6.4. *For $n \geq 0$, there are adjoint equivalences of categories $\Gamma_n : \widehat{\Gamma}_n Ho^s \simeq \Gamma_n Ho^s : \widehat{\Gamma}_n$.*
Proof. For $X \in \widehat{\Gamma}_n Ho^s$, the map $\Gamma_n X \to X$ is a W_{n*}-localization since its cofibre $L'_n X$ is W_{n*}-acyclic. Hence the adjunction unit $X \to \Gamma_n \widehat{\Gamma}_n X$ is an equivalence, and the adjunction counit is likewise.

For $n = 0$, this theorem gives a correspondence between spectra with p-torsion homotopy groups and those with Ext-p-complete homotopy groups. To identify spectra in $\Gamma_n Ho^s$ and $\widehat{\Gamma}_n Ho^s$ for $n \geq 0$, we need

LEMMA 6.5. *A spectrum E belongs to $\Gamma_n Ho^s$ if and only if $\pi_* E$ is p-torsion and $v_i^{-1} \pi_*(E; W_{i-1}) = 0$ for each $i \leq n$.*
Proof. These conditions hold if and only if $L'_n E \simeq 0$.

LEMMA 6.6. *If p acts nilpotently on a spectrum E and if a v_i-map of W_{i-1} acts nilpotently on $F(W_{i-1}, E)$ for each $i \leq n$, then E belongs to both $\Gamma_n Ho^s$ and $\widehat{\Gamma}_n Ho^s$.*
Proof. For a W_{n*}-acyclic spectrum A, $F(A, E)$ is trivial since it is both $[W_n, -]_*$-trivial and $[W_n, -]_*$-colocal by Lemma 6.5. Hence E is W_{n*}-local.

Finally we need

6.7. THE EQUIVALENT CATEGORIES $M'_n Ho^s$ AND $\widehat{M'_n} Ho^s$. For $n \geq 1$, the nth chromatic layer of a spectrum E is now given by $M'_n E = \Gamma_{n-1} L'_n E$, and the nth monochromatic homotopy category $M'_n Ho^s$ consists of the $[W_n, -]_*$-trivial $[W_{n-1}, -]_*$-colocal spectra. Similarly, we let $\widehat{M'_n} E = \widehat{\Gamma}_{n-1} L'_n E$ and the homotopy category $\widehat{M'_n} Ho^s$ consists of the $[W_n, -]_*$-trivial W_{n-1*}-local spectra. These would be the $K(n)_*$-local spectra if the telescope conjecture were valid. We may view $\widehat{M'_n} Ho^s$ and $M'_n Ho^s$ as alternative forms of the nth stable monochromatic homotopy category since there are adjoint equivalences

$$\Gamma_n : \widehat{M'_n} Ho^s \simeq M'_n Ho^s : \widehat{\Gamma}_n$$

by Theorem 6.4. In both $\widehat{M'_n} Ho^s$ and $M'_n Ho^s$, each $[W_{n-1}, -]_*$-equivalence is a homotopy equivalence. However, $\widehat{M'_n} Ho^s$ is closed under homotopy inverse limits, while $M'_n Ho^s$ is closed under homotopy direct limits.

6.8. PROOF OF THEOREM 6.2. In [22], Kuhn constructed a functor $\phi_n : Ho_* \rightarrow Ho^s$ for $n \geq 1$ such that $\phi_n \Omega^\infty : Ho^s \rightarrow Ho^s$ is the $K(n)_*$-localization. His work may now be adapted to give a functor $\widehat{\Phi}_n : Ho_* \rightarrow Ho^s$ for $n \geq 1$ such that $\widehat{\Phi}_n \Omega^\infty : Ho^s \rightarrow Ho^s$ is $\widehat{M'_n}$, and the resulting functor $\Phi_n = \Gamma_n \widehat{\Phi}_n$ has the required properties. In more detail, choose a sequence $C_1 \rightarrow C_2 \rightarrow C_3 \rightarrow \cdots$ of finite p-torsion spectra of p-type $n-1$ with homotopy direct limit $\Gamma_{n-1} S$, by starting with $C_0 = 0$ and successively attaching finite sets of "$\Sigma^i W_{n-1}$-cells" to give a sequence of complexes C_k over S with

$$\mathrm{colim}_k [W_{n-1}, C_k]_* \cong [W_{n-1}, S]_*.$$

By the Hopkins-Smith "periodicity theorem", the complexes C_k for $k \geq 1$ can successively be equipped with v_n-maps $\omega_k : \Sigma^{d_k} C_k \rightarrow C_k$ such that each ω_{k+1} is compatible with a power of ω_k. As in [22], for $X \in Ho_*$, there are associated "function spectra" $\phi(C_k, P_n X)$ of C_k into $P_n X$. Each $\phi(C_k, P_n X)$ belongs to $\widehat{M'_n} Ho^s$ by Lemma 6.6, and there are natural equivalences

$$\phi(C_k, P_n \Omega^\infty E) \simeq \phi(C_k, \Omega^\infty L'_n E) \simeq F(C_k, L'_n E)$$

for $E \in Ho^s$. We construct the spectrum $\widehat{\Phi}_n X$ as the homotopy inverse limit of the tower $\{\phi(C_k, P_n X)\}_{k \geq 1}$, working in the underlying categories of spaces and spectra as in [10]. Each $\widehat{\Phi}_n X$ belongs to $\widehat{M'_n} Ho^s$, and there are natural equivalences

$$\widehat{\Phi}_n(\Omega^\infty E) \simeq F(\Gamma_{n-1} S, L'_n E) \simeq \widehat{M'_n} E$$

for $E \in Ho^s$. By [10] and 6.7, the functor $\Phi_n = \Gamma_n \widehat{\Phi}_n$ has the required properties.

7 E_*-acyclicity and E_*-equivalences of spaces

We shall apply some of the preceding work to obtain general results on E_*-acyclicity and E_*-equivalences of spaces for various spectra E. For p prime and $E \neq 0$, the E_*-acyclicity of $K(Z/p, n)$ implies that of $K(Z/p, n+1)$, and we define the p-transition $\mathrm{tran}_p E$ of E to be the largest integer n such that $\widetilde{E}_* K(Z/p, n) \neq 0$, or to be ∞ when $\widetilde{E}_* K(Z/p, n) \neq 0$ for all n. For instance, $\mathrm{tran}_p HZ/p = \infty$ and $\mathrm{tran}_p K(n) = n$ by [30]. In [9], we proved

THEOREM 7.1. *Each E_*-equivalence of spaces is an $H_i(-; Z/p)$-equivalence for $i \leq \mathrm{tran}_p E$. The condition $\mathrm{tran}_p E = 0$ holds if and only if $E \simeq E[1/p]$.*

COROLLARY 7.2. *If E is a p-local spectrum with $\mathrm{tran}_p E = 0$ or ∞, then the E_*-equivalences of spaces are the same as the $H_*(-; G)$-equivalences for $G = Z/p$, $Z_{(p)}$, or Q*

Thus, for an infinite wedge $E = \bigvee_{i=0}^{\infty} K(n_i)$ of Morava K-spectra with $n_i < n_j$ for $i < j$, the E_*-equivalences of spaces are the same as the $H_*(-; Z_{(p)})$-equivalences when $n_0 = 0$ and as the $H_*(-; Z/p)$-equivalences when $n_0 > 0$. In view of this corollary, we are primarily interested in p-local spectra E with extraordinary p-transitions $\mathrm{tran}_p E = n$ where $0 < n < \infty$. In general, if a loop space ΩX is E_*-acyclic, then so is the space X, but the converse will obviously fail when E has an extraordinary p-transition. We now show that such failures are quite limited.

THEOREM 7.3. *If a simply connected H-space X is E_*-acyclic for a spectrum E, then $(\Omega X)_E$ is an E_*-local GEM and $B(\Omega X)_E$ is an E_*-acyclic GEM, where $B(\Omega X)_E$ denotes the classifying space of the E_*-localized loop space.*
Proof. As in the proof of Theorem 1.7, we have

$$\mathrm{map}_*\big(\Sigma B(\Omega X)_E, B(\Omega X)_E\big) \simeq \mathrm{map}_*\big(B(\Omega X)_E, (\Omega X)_E\big) \simeq *$$

because $B(\Omega X)_E$ is E_*-acyclic and $(\Omega X)_E$ is E_*-local. Thus, by the Key Lemma 1.6, $B(\Omega X)_E$ is a retract of $SP^\infty B(\Omega X)_E$ and is therefore a GEM.

This may also be deduced from the f-generalization of Theorem 1.7 and is closely related to results of [19]. It implies

THEOREM 7.4. *Let E be a p-local spectrum with $\mathrm{tran}_p E = n$ where $0 < n < \infty$. If X is an E_*-acyclic $(n+1)$-connected H-space and $\pi_{n+2} X$ is torsion, then ΩX is also E_*-acyclic.*
Proof. By Theorem 7.3, $K\big(\pi_i(\Omega X)_E, i\big)$ is E_*-local for all i, and either $\pi_* E$ or $\pi_*(\Omega X)_E$ is torsion. Thus by Lemma 7.5 below, $\pi_{n+1}(\Omega X)_E$ is torsion-free and $\pi_i(\Omega X)_E = 0$ for $i \geq n+2$. Hence the map $\Omega X \to (\Omega X)_E$ is nullhomotopic, and ΩX is E_*-acyclic.

It is straightforward to show

LEMMA 7.5. *For E as above, an Eilenberg-MacLane space $K(G,i)$ is E_*-acyclic if G is torsion and $i \geq n+1$, or if $\pi_* E$ is torsion and $i \geq n+2$.*

We now investigate E_*-equivalences in the full subcategory $Ho_{n+2} \subset Ho_*$ of $(n+2)$-connected spaces, letting $X\langle k \rangle$ denote the k-connected cover of a space X.

PROPOSITION 7.6. *For E as above and for a map $g : X \to Y$ in the Ho_{n+2}, the following are equivalent:*

(i) *g is an E_*-equivalence;*
(ii) *g is an $(EZ/p)_*$-equivalence and an $(EQ)_*$-equivalence;*
(iii) *$g\langle k \rangle : X\langle k \rangle \to Y\langle k \rangle$ is an E_*-equivalence for all k.*

Proof. We have (i) $\Leftrightarrow$ (ii) since E is p-local, and obtain (ii) $\Leftrightarrow$ (iii) since the maps $X\langle k \rangle \to X\langle k-1 \rangle$ are $(EZ/p)_*$-equivalences by Lemma 7.5.

We let $\widetilde{\Omega} : Ho_{n+2} \to Ho_{n+2}$ denote the $(n+2)$-connected loop functor $\widetilde{\Omega}X = (\Omega X)\langle n+2 \rangle$, and we say that a map $g : X \to Y$ in Ho_{n+2} is a *durable E_*-equivalence* when $\widetilde{\Omega}^m g : \widetilde{\Omega}^m X \to \widetilde{\Omega}^m Y$ is an E_*-equivalence for all $m \geq 0$.

THEOREM 7.7. *Let E be a p-local spectrum with $\mathrm{tran}_p E = n$ where $0 < n < \infty$. A map $g : X \to Y$ in Ho_{n+2} is a durable E_*-equivalence under each of the following conditions:*

(i) *g, $\widetilde{\Omega}g$, and $\widetilde{\Omega}^2 g$ are E_*-equivalences;*
(ii) *g and $\widetilde{\Omega}g$ are E_*-equivalences, and the fibre of g is an H-space.*

Proof. Use Theorem 7.4 to show that $\widetilde{\Omega}^m(\mathrm{fib}\, g)$ is E_*-acyclic for all m.

When $E = K(1)$ or KZ/p, we have convenient homotopical criteria for durability.

THEOREM 7.8. *A map $g : X \to Y$ in Ho_3 is a durable $K(1)_*$-equivalance (or $K_*(-;Z/p)$-equivalence) if and only if it satisfies the following equivalent conditions:*

(i) *$g_* : v_1^{-1}\pi_*(X;Z/p) \cong v_1^{-1}\pi_*(Y;Z/p)$;*
(ii) *$\Phi_1 g : \Phi_1 X \simeq \Phi_1 Y$;*
(iii) *$\widetilde{P}_1 g_* : \pi_j \widetilde{P}_1 X \cong \pi_j \widetilde{P}_1 Y$ for $j > \mathrm{conn}\, \overline{V}_1 + 2$.*

Proof. This is proved for (i) in [11, Thm. 14.7] using work of Thompson to verify the "only if" part. The equivalences (i) $\Leftrightarrow$ (ii) $\Leftrightarrow$ (iii) follow by Corollary 5.4 and Theorem 6.2.

This is an unstable version of the result, proved in [6] using work of Mahowald and Miller, that a map of spectra is a $K_*(-;Z/p)$-equivalence if and only if it is a $v_1^{-1}\pi_*(-;Z/p)$-equivalence. Theorems 7.7 and 7.8 provide tools for studying the $K_*(-;Z/p)$-homology of iterated loop spaces in some previously inaccessible cases. For instance, in [11] we used the "if" part of 7.8 to deduce that the Snaith map

$$s : \Omega_0^{2n+1} S^{2n+1} \to QRP^{2n}$$

is a $K_*(-; Z/2)$-equivalence from Mahowald's result that it is a $v_1^{-1}\pi_*(-; Z/p)$-equivalence.

This confirmed an old conjecture of Miller-Snaith [25] and allowed us to determine $K_*(\Omega_0^{2n+1} S^{2n+1}; Z/2)$ from their computation of $K_*(QRP^{2n}; Z/2)$. Lisa Langsetmo has likewise determined $K_*(\Omega^j S^{2n+1}; Z/p)$ for all $j < 2n$ using $v_1^{-1}\pi_*(-; Z/p)$-equivalences of Mahowald-Thompson [24], and we are currently obtaining similar results with S^{2n+1} replaced by an H-space X such that $K^*(X; Z_p^\wedge)$ is a p-adic exterior algebra.

Theorem 7.8 cannot easily be generalized to durable $K(n)_*$-equivalences, although the implications (i) $\Leftrightarrow$ (ii) $\Leftrightarrow$ (iii) clearly remain valid for $v_n^{-1}\pi_*(-; V_{n-1})$, Φ_n, $\widetilde{P}_n$, and $\overline{V}_n$. To see the difficulty for $n = 2$, note that $U\langle 4 \rangle \in Ho_4$ has $v_2^{-1}\pi_*(U\langle 4 \rangle; V_1) = 0$, but also has $\widetilde{K}(2)_*U\langle 4 \rangle \neq 0$ by Theorem 7.4, because $\widetilde{K}(2)_*BSU \neq 0$ since the map $BU \to CP^\infty$ is not a $K(2)_*$-equivalence by [30, p. 709] and [31, p. 394]. One might reasonably try to generalize Theorem 7.8 to durable $K(n)_*$-equivalences by strengthening condition (i) to "$g_* : v_i^{-1}\pi_*(X; V_{i-1}) \cong v_i^{-1}\pi_*(Y; V_{i-1})$ for $1 \leq i \leq n$" and similarly strengthening conditions (ii) and (iii). These homotopy conditions do indeed imply that a map $g : X \to Y$ in Ho_{n+2} is a durable $K(n)_*$-equivalence by [11, Sect. 13], and the converse would also follow if we knew that the $K(n)_*$-equivalences of spaces were the same as the $(L_n' SZ/p)_*$-equivalences. Such a generalization of Theorem 7.8 would become quite plausible if each $K(n)_*$-equivalence of spaces were shown to be a $K(i)_*$-equivalence for $1 \leq i \leq n$, without the usual finiteness assumptions.

References

1. J.F. Adams, *Stable homotopy and generalized homology*, University of Chicago Press, 1974.

2. D.W. Anderson, *Localizing CW-complexes*, Illinois J. Math. **16** (1972), 519–525.

3. A.K. Bousfield, *The localization of spaces with respect to homology*, Topology **14** (1975), 133–150.

4. _____, *Constructions of factorization systems in categories*, J. Pure Appl. Algebra **9** (1977), 207–220.

5. _____, *The Boolean algebra of spectra*, Comment. Math. Helv. **54** (1979), 368–377.

6. _____, *The localization of spectra with respect to homology*, Topology **18** (1979), 257–281.

7. _____, *Cohomological localizations of spaces and spectra*, unpublished preprint (1979).

8. _____, *K-localizations and K-equivalences of infinite loop spaces*, Proc. London Math. Soc. **44** (1982), 291–311.

9. _____, *On homology equivalences and homological localizations of spaces*, Amer. J. Math. **104** (1982), 1025–1042.

10. _____, *Uniqueness of infinite deloopings for K-theoretic spaces*, Pacific J. Math. **129** (1987), 1–31.

11. _____, *Localization and periodicity in unstable homotopy theory*, J. Amer. Math. Soc. **7** (1994), 831–873.

12. A.K. Bousfield and E.M. Friedlander, *Homotopy theory of Γ-spaces, spectra, and bisimplicial sets*, Lecture Notes in Math, vol. 658, Springer-Verlag, 1978, pp. 80–130.

13. A.K. Bousfield and D.M. Kan, *Homotopy limits, completions and localizations*, Lecture Notes in Math., vol. 304, Springer-Verlag, 1972.

14. C. Casacuberta, *Anderson localization from a modern point of view*, Contemp. Math. (to appear).

15. F. Cohen and J. Neisendorfer, *A note on desuspending the Adams map*, Math. Proc. Camb. Philos. Soc. **99** (1986), 59–64.

16. D.M. Davis and M. Mahowald, *v_1-localizations of finite torsion spectra and spherically resolved spaces*, Topology **32** (1993), 543–550.

17. E. Dror Farjoun, *Homotopy localization and v_1-periodic spaces*, Lecture Notes in Math., vol. 1509, Springer-Verlag, 1992, pp. 104–113.

18. E. Dror Farjoun, *Localizations, fibrations and conic structures* (to appear).

19. E. Dror Farjoun and J.H. Smith, *Homotopy localization nearly preserves fibrations*, Topology (to appear).

20. M.J. Hopkins, *Global methods in homotopy theory*, London Math. Soc. Lecture Note Ser., vol. 117, Cambridge Univ. Press, 1987, pp. 73–96.

21. M.J. Hopkins and J.H. Smith, *Nilpotence and stable homotopy II*, Ann. of Math. (to appear).

22. N.J. Kuhn, *Morava K-theories and infinite loop spaces*, Lecture Notes in Math., vol. 1370, Springer-Verlag, 1989, pp. 243–257.

23. L. Langsetmo, *The K-theory localization of an odd sphere and applications*, Topology **32** (1993), 577–585.

24. M. Mahowald and R. Thompson, *The K-theory localization of an unstable sphere*, Topology **31** (1992), 133–141.

25. H. Miller and V. Snaith, *On the K-theory of the Kahn-Priddy map*, J. London Math. Soc. **20** (1979), 339–342.

26. S.A. Mitchell, *Finite complexes with $A(n)$-free cohomology*, Topology **24** (1985), 227–246.

27. D.C. Ravenel, *Localization with respect to certain periodic homology theories*, Amer. J. Math. **106** (1984), 351–414.

28. _____, *Nilpotence and periodicity in stable homotopy theory*, Ann. of Math. Stud., no. 128, Princeton Univ. Press, 1992.

29. _____, *Life after the telescope conjecture*, in: P.G. Goerss and J.F. Jardine, eds., Algebraic K-Theory and Algebraic Topology, Kluwer Academic Publishers, 1993, 205–222.

30. D.C. Ravenel and W.S. Wilson, *The Morava K-theories of Eilenberg-MacLane spaces and the Conner-Floyd conjecture*, Amer. J. Math. **102** (1980), 691–748.

31. R. Switzer, *Algebraic topology-homotopy and homology*, Springer-Verlag, 1975.

University of Illinois at Chicago
Department of Mathematics, Statistics and Computer Science (M/C 249)
851 South Morgan Street, Chicago, IL 60607 USA

Progress in Mathematics, Vol. 136
© 1996 Birkhäuser Verlag Basel/Switzerland

On the canonical $GL_2(\mathbb{F}_2)$-module structure of $K(n)^*(B\mathbb{Z}/2 \times B\mathbb{Z}/2)$

MAURIZIO BRUNETTI*

ABSTRACT. Let $K(n)^*(-)$ be the n-th Morava K-theory at 2. We discuss the existence in $K(n)^*(B\mathbb{Z}/2 \times B\mathbb{Z}/2)$ of a basis whose elements are just permuted by the canonical action of $GL_2(\mathbb{F}_2) \cong Aut(\mathbb{Z}/2 \times \mathbb{Z}/2)$. We exhibit such a basis when $n = 3$ and prove that it does not exist when $n = 2$.

Introduction

Let p be any prime, A a finite abelian group, and $GL(A)$ the group of automorphisms of A. It is well known that for any cohomology theory h^* an automorphism $\phi : A \to A$ determines a well-defined map

$$h^*(B\phi) : h^*(BA) \longrightarrow h^*(BA),$$

giving $h^*(BA)$ a canonical $GL(A)$-module structure. Let now $K(n)^*(-)$ be the n-th Morava K-theory at p. We say that $K(n)^*(BA)$ is a $GL(A)$-permutation module if it has a basis as a $K(n)^*$-module Ω which is a $GL(A)$-set with reference to the action we just described; in this case $GL(A)$ would act on Ω like a permutation representation. The existence of such Ω would be very helpful to study the behaviour of the spectral sequence

$$H^s(BW; K(n)^t(BA)) \Longrightarrow K(n)^{s+t}(BG)$$

related to the extension

$$0 \longrightarrow A \longrightarrow G \longrightarrow W \longrightarrow 1,$$

especially when there exists additionally a subset $\Omega' \subset \Omega$ such that the action of $\pi_1(BW)$ is transitive on Ω' and trivial on its complement.
It was proved in [B] that the following formula

$$[k]_F x \equiv kx \pmod{([p^n]_F x)},$$

holds when p is an odd prime and

$$k^{p-1} \equiv 1 \pmod{p^n}.$$

---

*AMS 1991 Subject Classification 55N20, 55N22.
The author was supported by the CNR, Consiglio Nazionale delle Ricerche, Italy.

Therefore, when $A = C_{p^r}$, the cyclic group of order p^r, and p is an odd prime, it is easy to verify that $K(n)^*(BA)$ is not a $GL(A)$-permutation module in our sense.

In this paper we always assume on $K(n)^*(BA)$ the above module structure, even if not specified. We analyse here the case $p = 2$ and $A = \mathbb{Z}/2 \times \mathbb{Z}/2$, and prove that $K(2)^*(B\mathbb{Z}/2 \times B\mathbb{Z}/2)$ is not a permutation module by studying in detail its $GL_2(\mathbb{F}_2)$-module structure thanks to the following formula concerning the formal group law $F_{K(n)}$ for Morava K-theories at 2:

$$F_{K(n)}(x, y) \equiv x + y + \nu_n x^{2^{n-1}} y^{2^{n-1}} \pmod{(x,y)^{2^{n+1}}}.$$

We also prove that when $n = 3$ a basis Ω which is a $GL_2(\mathbb{F}_2)$-set exists, and we list its elements in the third section.

I would like to thank Alan Robinson, my research supervisor, Nicholas J. Kuhn who introduced me to the problem discussed here, and the CNR for their financial support.

1. The formal group law $F_{K(n)}$ for $p = 2$

We recall that the Morava K-theories at 2 have coefficients

$$K(n)^*(\{pt\}) = \mathbb{F}_2[\nu_n, \nu_n^{-1}]$$

with $\deg \nu_n = -2(2^n - 1)$. It is known that there exist two non-equivalent products

$$K(n) \wedge K(n) \longrightarrow K(n)$$

which give to $K(n)$ a structure of a ring spectrum [W]. We shall not specify which multiplication we are considering, since in both cases

1. $K(n)^*(\mathbb{CP}^\infty) = K(n)^*[[x]]$ with $\deg x = 2$

2. $[2]_{F_{K(n)}} = \nu_n x^{2^n}$.

We also recall that

$$K(n)^*(B\mathbb{Z}/2 \times B\mathbb{Z}/2) \cong K(n)^*[x, y]/(x^{2^n}, y^{2^n})$$

This result can be achieved by studying the Gysin sequence related to the fibration

$$S^1 \to B\mathbb{Z}/2 \longrightarrow \mathbb{CP}^\infty$$

which gives short exact sequences

$$0 \longrightarrow K(n)^{*-2}(\mathbb{CP}^\infty) \xrightarrow{\cdot [2]x} K(n)^*(\mathbb{CP}^\infty) \longrightarrow K(n)^*(B\mathbb{Z}/2) \longrightarrow 0,$$

and then by using a Künneth isomorphism

$$K(n)^*(X \times Y) \cong K(n)^*(X) \otimes_{K(n)^*} K(n)^*(Y)$$

which holds in K(n)-cohomology for any topological spaces X and Y.

Let $B_n(x,y)$ and $C_n(x,y)$ be the following two polynomials

$$B_n(x,y) = (x+y)^n - x^n - y^n$$

$$C_n = \begin{cases} B_n/p & \text{if } n = p^k \text{ for some prime } p \\ B_n & \text{otherwise.} \end{cases}$$

We use the following Comparison Lemma due to Lazard [L] to calculate

$$F_{K(n)}(x,y)\,(\text{mod}\,(x,y)^{2^{n+1}}).$$

LEMMA 1.1 *Let F and G be two formal group laws over a ring R such that*

$$F \equiv G\,(\text{mod}\,(x,y)^m).$$

Then

$$F \equiv G + a\,C_m\,(\text{mod}\,(x,y)^{m+1})$$

for some $a \in R$.   □

PROPOSITION 1.2 *The formal group law of Morava K-theories at 2 satisfies the following relation:*

$$F_{K(n)}(x,y) \equiv x + y + \nu_n x^{2^{n-1}} y^{2^{n-1}}\,(\text{mod}\,(x,y)^{2^{n+1}})$$

Proof. Since $\deg \nu_n = -2(2^n - 1)$, we have necessarily

$$F_{K(n)}(x,y) \equiv x + y\,(\text{mod}\,(x,y)^{2^n}),$$

and applying lemma 1.1

$$F_{K(n)}(x,y) \equiv x + y + a\,C_{2^n}\,(\text{mod}\,(x,y)^{2^n+1}),$$

where a is in $K(n)^*$, and $C_{2^n} = x^{2^{n-1}} y^{2^{n-1}}\,(\text{mod}\,2)$, since induction gives

$$(x+y)^{2^n} - x^{2^n} - y^{2^n} = 2\,x^{2^{n-1}} y^{2^{n-1}}\,(\text{mod}\,4).$$

We notice now that

$$x +_{F_{K(n)}} x = \nu_n x^{2^n},$$

therefore $a = \nu_n$, and proposition follows, since in $F_{K(n)}(x,y)$ there are no monomials whose degree in x and y is between $2^n + 1$ and 2^{n+1}.   □

2. $K(2)^*(B\mathbb{Z}/2 \times B\mathbb{Z}/2)$ is not a $GL_2(\mathbb{F}_2)$-permutation module

From now on A will denote the group $\mathbb{Z}/2 \times \mathbb{Z}/2$. We fix two generators of $GL_2(\mathbb{F}_2) \cong S_3$

$$\sigma = \begin{pmatrix} 0 & 1 \\ 1 & 0 \end{pmatrix} \quad \tau = \begin{pmatrix} 1 & 0 \\ 1 & 1 \end{pmatrix}$$

Both matrices have order 2. We shall use the symbols α and β to denote respectively the maps

$$K(n)^*(\sigma) : K(n)^*(BA) \longrightarrow K(n)^*(BA)$$

and

$$K(n)^*(\tau) : K(n)^*(BA) \longrightarrow K(n)^*(BA).$$

LEMMA 2.1 *The maps α and β act as follows on $K(n)^*[x,y]/(x^{2^n}, y^{2^n})$:*

$$x \xrightarrow{\beta} x, \quad y \xrightarrow{\beta} x + y + \nu_n x^{2^{n-1}} y^{2^{n-1}}, \quad and \quad x \xrightarrow{\alpha} y.$$

Proof. Let s and p_i denote respectively the ordinary sum in $\mathbb{Z}/2$ and the projection on the i-th coordinate. We immediately verify that

$$B(p_2) \circ B(\tau) = Bs, \quad and \quad B(p_1) \circ B(\tau) = B(p_1),$$

therefore the map β acts as we claimed since

$$K(n)^*(Bs) : x \in K(n)^*(BA) \longmapsto F_{K(n)}(x,y) \in K(n)^*(BA).$$

The reader will find easily the right equalities of maps to prove that α interchanges x and y in $K(n)^*(BA)$.   □

Consider now the case $n = 2$. For any topological space X, it is possible to split $K(2)^*(X)$ in three summands as follows

$$K(2)^*(X) \cong K(2)^{*\equiv 2 \,(\mathrm{mod}\ 6)}(X) \oplus K(2)^{*\equiv 4 \,(\mathrm{mod}\ 6)}(X) \oplus K(2)^{*\equiv 0 \,(\mathrm{mod}\ 6)}(X).$$

We take $X = BA$, and study these three summands separately.

PROPOSITION 2.2 $K(n)^{*\equiv 2(\mathrm{mod}\ 6)}(BA)$ *is not a $GL(A)$-permutation module.*

Proof In $K(2)^2(BA)$ there is a basis formed by monomials, namely

$$x\,, y\,, \nu_2 x^3 y\,, \nu_2 x^2 y^2, \nu_2 x y^3.$$

Elements in other dimensions are just obtained multiplying by a power of ν_2 the elements of this five dimensional $\mathbb{F}_2$-vector space. Suppose that a basis Ω which is a $GL(A)$-set exists. The following elementary properties have to be satisfied:

1. If a is in Ω, then all elements of its orbit are in Ω.

2. Given an orbit in Ω, the sum of its elements has to be a non-zero fixed point.

3. The map

$$\Phi : \{\text{Orbits in } \Omega\} \longrightarrow \{\text{Fixed points}\} \subset K(2)^2(BA)$$

which maps any orbit to the sum of its elements has to be injective.

Since the cardinality of Ω is five, it has to contain at least two full orbits, but computation shows that among the $2^5 - 1$ non-zero elements in $K(2)^2(BA)$, only $\nu_2 x^2 y^2$ is a fixed point, therefore we get a contradiction.   □

PROPOSITION 2.3 $K(2)^{* \equiv 4 \,(\text{mod } 6)}(BA)$ is not a $GL(A)$-permutation module.

Proof The argument is the same as above. Among the $2^5 - 1$ non-zero elements of $K(2)^4(BA)$ generated by the five monomials

$$x^2 , y^2 , xy , \nu_2 x^2 y^3 , \nu_2 x^3 y^2,$$

there is only one fixed point, namely

$$x^2 + y^2 + xy + \nu_2(x^3 y^2 + x^2 y^3).$$

Hence there are no bases in $K(2)^4(BA)$ which are $GL(A)$-sets.   □

PROPOSITION 2.4 In $K(2)^6(BA)$ there are exactly four bases on which the group $GL(A)$ acts by permuting elements. Therefore $K(2)^{* \equiv 0 \,(\text{mod } 6)}(BA)$ is a $GL(A)$-permutation module.

Proof $K(2)^6(BA)$ is generated as an $\mathbb{F}_2$-vector space by the following monomials

$$x^3 , y^3 , x^2 y, xy^2 , \nu_2 x^3 y^3.$$

The subspace of fixed points has dimension 2; its three non zero elements are

$$\nu_2 x^3 y^3, x^2 y + xy^2, x^2 y + xy^2 + \nu_2 x^3 y^3.$$

By the argument used in the proof of proposition 2.2, only the elements of the following four orbits can form a basis which is a $GL(A)$-set

$\alpha \; \circlearrowleft \; x^3 + y^3 + \nu_2 x^3 y^3 \; \xrightarrow{\beta} \; y^3 + x^2 y + xy^2 + \nu_2 x^3 y^3 \; \xrightarrow{\alpha} \; x^3 + x^2 y + xy^2 + \nu_2 x^3 y^3 \; \circlearrowright \; \beta$

$\beta \; \circlearrowleft \; x^3 + \nu_2 x^3 y^3 \; \xrightarrow{\alpha} \; y^3 + \nu_2 x^3 y^3 \; \xrightarrow{\beta} \; x^3 + y^3 + x^2 y + xy^2 + \nu_2 x^3 y^3 \; \circlearrowright \; \alpha$

$$x^3 + y^3 + x^2 y + \nu_2 x^3 y^3 \; \underset{\beta}{\overset{\alpha}{\rightleftarrows}} \; x^3 + y^3 + xy^2 + \nu_2 x^3 y^3$$

$$x^3 + y^3 + x^2 y \; \underset{\beta}{\overset{\alpha}{\rightleftarrows}} \; x^3 + y^3 + xy^2$$

For any choice of one orbit of length 2 and one orbit of length 3, we have a basis, in fact the determinants of those matrices relating the four sets of vectors with the 'canonical' basis containing monomials are non-zero. Such bases are obviously $GL(A)$-sets.   □

As Bjorn Schuster pointed out, for any n the summand

$$K(n)^{*\equiv 0(\mathrm{mod}\,2(2^n-1))}$$

is always a $GL_2(\mathbb{F}_2)$-permutation module; in fact $GL_2(\mathbb{F}_2)$ fixes $\nu_n x^{2^n-1} y^{2^n-1}$, and the isomorphism Φ of $\mathbb{Z}/2$-vector spaces

$$K(n)^{2(2^n-1)}(BA) \longrightarrow \frac{K(n)^{2(2^n-1)}(BA)}{(\nu_n x^{2^n-1} y^{2^n-1})} \oplus\ <\nu_n x^{2^n-1} y^{2^n-1}>$$

is actually a splitting of modules. Furthermore there is an obvious isomorphism Θ of $\mathbb{Z}/2$-vector spaces

$$\frac{K(n)^{2(2^n-1)}(BA)}{(\nu_n x^{2^n-1} y^{2^n-1})} \longrightarrow H^{2^n-1}(BA; \mathbb{Z}/2).$$

which makes Θ an isomorphism of $GL_2(\mathbb{F}_2)$-permutation modules after equipping $H^{2^n-1}(BA; \mathbb{Z}/2)$ with its canonical module structure.

3.   The $n=3$ case

The third Morava K-theory at 2 is $2(2^3-1)$-periodic, therefore we split $K(3)^*(BA)$ in seven different summands, each related to a different even number modulo 14, and having a basis of nine monomials. Calculations show that in every summand there are two independent fixed points, and there are at least two orbits, one of length 3 and one of 6, whose elements form a basis. To convince the reader that a basis of sixtythree elements giving $K(3)^*(\mathbb{Z}/2 \times B\mathbb{Z}/2)$ a structure of a $GL_2(\mathbb{F}_2)$-permutation module really exists, we list now fourteen orbits. Elements of the $k-th$ couple of consecutive orbits belong to $K(3)^{2k}(\mathbb{Z}/2 \times B\mathbb{Z}/2)$, and their independence can be easily proved just verifying that the determinant of the 9×9 matrix relating them with the basis of monomials is non-zero. In the drawn orbits maps α and β act alternately, therefore it is sufficient to name only one map in every orbit.

$$\alpha \left(\begin{array}{ccc} \nu_3 x^3 y^5 & \longleftarrow \nu_3(x^3 y^5 + x^7 y + x^4 y^4) \longleftarrow & \nu_3(x^5 y^3 + xy^7 + x^4 y^4) \\ \nu_3 x^5 y^3 & \longrightarrow \nu_3(x^5 y^3 + x^6 y^2 + x^7 y) \longrightarrow & \nu_3(x^3 y^5 + x^2 y^6 + xy^7) \end{array} \right)$$

$$x \xrightarrow{\ \alpha\ } y \longrightarrow x + y + \nu_3 x^4 y^4$$

$$\alpha \left(\begin{array}{ccc} \nu_3 x^6 y^3 & \longleftarrow \nu_3(x^7 y^2 + x^6 y^3) \longleftarrow & \nu_3(x^2 y^7 + x^3 y^6) \\ \nu_3 x^3 y^6 & \longrightarrow \nu_3(x^7 y^2 + x^5 y^4 + x^3 y^6) \longrightarrow & \nu_3(x^2 y^7 + x^4 y^5 + x^6 y^3) \end{array} \right)$$

$$xy \xrightarrow{\ \beta\ } x^2 + xy + \nu_3 x^5 y^4 \longrightarrow y^2 + xy + \nu_3 x^4 y^5$$

$$\begin{array}{ccc} xy^2 + \nu_3 x^3 y^7 & \longrightarrow & x^3 + xy^2 + \nu_3(x^3 y^7 + x^4 y^6 + x^5 y^5 + x^6 y^4 + x^7 y^3) \\ x^2 y + \nu_3 x^7 y^3 & & y^3 + x^2 y + \nu_3(x^7 y^3 + x^6 y^4 + x^5 y^5 + x^4 y^6 + x^3 y^7) \\ \beta \quad x^3 + x^2 y + \nu_3(x^6 y^4 + x^7 y^3) & \longrightarrow & y^3 + xy^2 + \nu_3(x^4 y^6 + x^3 y^7) \end{array}$$

$$\nu_3 x^5 y^5 \longrightarrow \nu_3(x^5 y^5 + x^6 y^4) \xrightarrow{\ \alpha\ } \nu_3(x^5 y^5 + x^4 y^6)$$

$$x^4 + x^3y + x^2y^2 + xy^3 + \nu_3(x^7y^4 + x^5y^6) \longrightarrow y^4 + xy^3 + x^2y^2 + x^3y + \nu_3(x^4y^7 + x^6y^5)$$

$$xy^3 \xrightarrow{\alpha} x^3y \longrightarrow x^4 + x^3y + \nu_3 x^7 y^4 \longrightarrow y^4 + xy^3 + \nu_3 x^4 y^7$$

$$x^4 + x^2y^2 \xrightarrow{\alpha} y^4 + x^2y^2 \longrightarrow x^4 + y^4 + x^2y^2$$

$$x^3y^2 \xrightarrow{\alpha} x^2y^3 \longrightarrow x^5 + x^4y + x^3y^2 + x^2y^3 + \nu_3 x^6 y^6$$

$$x^5 + x^3y^2 \longleftarrow y^5 + x^2y^3 \longleftarrow y^5 + xy^4 + x^2y^3 + x^3y^2 + \nu_3 x^6 y^6 \longleftarrow \quad \alpha$$

$$\nu_3 x^7 y^5 \xrightarrow{\alpha} \nu_3 x^5 y^7 \longrightarrow \nu_3(x^5y^7 + x^6y^6 + x^7y^5)$$

$$\alpha \quad xy^5 \longrightarrow xy^5 + x^2y^4 + x^5y + x^6 \longrightarrow y^6 + xy^5 + x^4y^2 + x^5y$$

$$x^5y \longleftarrow x^6 + x^5y \longleftarrow y^6 + xy^5 \longleftarrow$$

$$x^3y^3 \longrightarrow x^6 + x^5y + x^4y^2 + x^3y^3 + \nu_3 x^7 y^6 \xrightarrow{\alpha} y^6 + xy^5 + x^2y^4 + x^3y^3 + \nu_3 x^6 y^7$$

$$\alpha \quad x^2y^5 \longrightarrow x^7 + x^2y^5 + x^3y^4 + x^6y \longrightarrow y^7 + xy^6 + x^4y^3 + x^5y^2$$

$$x^5y^2 \longleftarrow x^7 + x^5y^2 \longleftarrow y^7 + x^2y^5 \longleftarrow$$

$$x^3y^4 + x^4y^3 + \nu_3 x^7 y^7 \qquad\qquad x^3y^4 + x^4y^3 + xy^6 + x^2y^5 + \nu_3 x^7 y^7$$

$$x^3y^4 + x^4y^3 + x^6y + x^5y^2 + \nu_3 x^7 y^7 \quad \alpha$$

The number of elements in $K(n)^2(B\mathbb{Z}/2 \times B\mathbb{Z}/2)$ is 2^{2^n+1}, therefore calculations by hand become virtually impossible for $n > 3$. However we noticed that the sum of the orbit (of length 3) containing the monomial

$$\nu_n x^k y^{2^n-k}$$

with $k \neq 2^{n-1}$, is non-zero if and only if neither k or $2^n - k$ are a non-negative power of 2. As n increases, the number of independent fixed points grows very rapidly, therefore we are tempted to believe that for $n > 3$ it is always possible to find the sufficient number of orbits containing independent elements to form a basis.

References

[B]   M. Brunetti, 'A family of $2(p-1)$-sparse cohomology theories and some actions on $h^*(BC_{p^n})$', to appear on *Math. Proc. Camb. Phil. Soc.*

[L]   M. Lazard, 'Lois de groupes et analyseurs', *Ann. Sci. École Norm. Sup.* **72** (1955), 299–400.

[W]   U. Würgler, 'Commutative ring-spectra of characteristic 2', *Comment. Math. Helv.* **61** (1986), 33–45.

Dipartimento di Matematica e applicazioni, Università di Napoli
via Cintia, I-80126 Napoli, Italy

Progress in Mathematics, Vol. 136
© 1996 Birkhäuser Verlag Basel/Switzerland

Manifolds with Highly Connected Universal Covers

ALBERTO CAVICCHIOLI AND FRIEDRICH HEGENBARTH

ABSTRACT. We apply the homology theory with local coefficients to study closed even-dimensional manifolds with highly connected universal covering spaces. Then we obtain simple algebraic characterizations of aspherical manifolds and discuss some properties regarding the minimality of their Euler characteristics.

1 Introduction

In this paper we study closed connected oriented $(2n)$-manifolds X satisfying the property $\Pi_i(X) = 0$ for $2 \leq i \leq n-1$, i. e. the universal covering space $\tilde{X}$ of X is $(n-1)$-connected. If $\Lambda = \mathbb{Z}[\Pi_1(X)]$ is the integral group ring of $\Pi_1(X)$, this is equivalent to $H_j(X; \Lambda) = 0$, for $1 \leq j \leq n-1$, because there are isomorphisms $H_*(X; \Lambda) \cong H_*(\tilde{X}; \mathbb{Z})$. Examples are given by aspherical $(2n)$-manifolds and manifolds with universal covering space $\mathbb{S}^k \times \mathbb{R}^{2n-k}$ for some $k \geq n$. The case $\mathbb{S}^3 \times \mathbb{R}$ for closed 4-manifolds was treated in [14]. Here the author determined first the homotopy type of this class of 4-manifolds and then characterized them, up to TOP homeomorphism, by making use of topological surgery (see [8], [22]).

To study $(2n)$-manifolds X with highly connected universal covers $\tilde{X}$ we use the (co)homology theory with local coefficients (see for example [1], [11], [21], [22]). Because $\tilde{X}$ will not in general be compact, we can not apply the simple Poincaré duality over $\mathbb{Z}$ and the usual universal coefficient theorem to control the homology groups $H_i(\tilde{X}; \mathbb{Z})$ for $i \geq n + 1$. However, these groups depend of course on the n-skeleton $X^{(n)}$ of X. Here we always assume that X is provided with a CW-structure, up to homotopy. To describe these groups we take coefficients for (co)homology in the right Λ-module $L = \mathrm{Hom}_{\mathbb{Z}}(\Lambda, \mathbb{Z})$. Our main results are algebraic characterizations of aspherical $(2n)$-manifolds among those having highly connected universal covering spaces. These facts are suggested, in some sense, by the topological classification of closed 4-manifolds X with $\Pi_3(X) = 0$ exposed in section 4. Then we show that the homotopy type of the above-mentioned class of $(2n)$-manifolds can be described by their Euler

---

1991 *Mathematics Subject Classification.* 55 N 25, 55 P 15, 55 P 20, 57 N 13.

Key words and phrases. Homology with local coefficients, highly connected universal covers, manifolds, aspherical manifolds, Euler characteristic, homotopy type.

*) Work performed under the auspicies of the C.N.R. of Italy and partially supported by Ministero per la Ricerca Scientifica e Tecnologica of Italy within the projects "Geometria Reale e Complessa" and "Topologia".

characteristics. Some properties about the minimality of the Euler characteristic complete the paper. These results are related to analogous ones presented in [4], [5], [10], [13], [15], [18] and [19]. Finally we note that a general treatment of duality is given in [11], [21], [22] for Poincaré complexes, both for the orientable case and for the non orientable case.

2   General remarks on Poincaré duality

In this section we recall some basic results about the Poincaré duality with local coefficients. As general references we refer to [1], [11], [21] and [22]. Let X^m be a closed connected oriented TOP (PL or DIFF) m-manifold and let $\widetilde{X}$ be the universal covering space of X. Let us consider a CW-complex structure on X, up to homotopy. Let $X^{(k)}$ denote the k-skeleton of X. The universal covering space $\widetilde{X}$ admits a CW-complex structure satisfying the property $\widetilde{X}^{(k)} = \widetilde{X^{(k)}}$. Obviously, $\Pi_1(X)$ acts on $\widetilde{X}^{(k)}$ by permuting the k-cells of $\widetilde{X}^{(k)}$ which lie over a fixed k-cell of $X^{(k)}$. There is an induced free right Λ-module structure on the chain complex $C_*(\widetilde{X}) = H_*(\widetilde{X}^{(*)}, \widetilde{X}^{(*-1)})$, where Λ is the integral group ring of $\Pi_1(X)$.

Let B be a right Λ-module. Then the homology of X with coefficients in B is defined by $H_*(X; \bar{B}) = H_*(C_*(\widetilde{X}) \otimes_\Lambda \bar{B})$. To define $\otimes_\Lambda$ the right Λ-structure of B is turned into the left one, written $\bar{B}$, by setting $\lambda b = b\bar{\lambda}$ for any $\lambda \in \Lambda$ and $b \in B$. Here $\bar{}: \Lambda \to \Lambda$ denotes the canonical anti-involution of Λ obtained first by sending any $g \in \Pi_1(X)$ to its inverse g^{-1} and then extending this correspondence by linearity.

Similarly, the cohomology of X with coefficients in B is defined by setting $H^*(X; B) = H^*(\text{Hom}_\Lambda(C_*(\widetilde{X}), B))$. Note that $\text{Hom}_\Lambda(C_*(\widetilde{X}), B)$ carries a right Λ-module structure by setting $(\varphi \lambda)(c) = \bar{\lambda}\varphi(c)$ for any $\varphi \in \text{Hom}_\Lambda(C_*(\widetilde{X}), B)$, $\lambda \in \Lambda$ and $c \in C_*(\widetilde{X})$.

Now the Poincaré duality with local coefficients can be stated in the following way.

THEOREM 1.1. *Let X^m be a closed connected oriented TOP (PL or DIFF) m-manifold. Then the map*

$$[X] \cap : H^*(X; B) \to H_{m-*}(X; \bar{B})$$

is an isomorphism for any right Λ-module B. Here $[X] \in H_m(X; \mathbb{Z})$ denotes the fundamental class of X.

In particular, if $B = \mathbb{Z}$, then we obtain the usual integral (co)homology of X. If $B = \Lambda$, then $H_*(X; \Lambda) \cong H_*(\widetilde{X}; \mathbb{Z})$ and $H^*(X; \Lambda) \cong H_c^*(\widetilde{X}; \mathbb{Z})$, where H_c^* denotes the cohomology of $\widetilde{X}$ with compact supports.

Let $L = \text{Hom}_\mathbb{Z}(\Lambda, \mathbb{Z})$ be the right Λ-module defined by $(\varphi\lambda)(\mu) = \varphi(\lambda\mu)$ for any $\varphi \in L$ and $\lambda, \mu \in \Lambda$ (see for example [3]). The natural isomorphism

$$\text{Hom}_\Lambda(C_i(\widetilde{X}), \text{Hom}_\mathbb{Z}(\Lambda, \mathbb{Z})) \xrightarrow[\cong]{} \text{Hom}_\mathbb{Z}(C_i(\widetilde{X}) \otimes_\Lambda \Lambda, \mathbb{Z}) = \text{Hom}_\mathbb{Z}(C_i(\widetilde{X}), \mathbb{Z})$$

establishes a canonical identification $H^i(X;L) \cong H^i(\widetilde{X};\mathbb{Z})$. Then the Poincaré duality gives an isomorphism $H^i(X;L) \underset{\cong}{\to} H_{m-i}(X;\bar{L})$, hence the composition map $H^i(\widetilde{X};\mathbb{Z}) \to H_{m-i}(X;\bar{L})$ is an isomorphism for any $i \geq 0$.

Summarizing we have the following result.

THEOREM 1.2. *Let X^m be a closed connected oriented TOP (PL or DIFF) m-manifold and $\widetilde{X}$ its universal covering space. If Λ is the group ring of $\Pi_1(X)$ and $L = \mathrm{Hom}_{\mathbb{Z}}(\Lambda, \mathbb{Z})$, then there are isomorphisms $H_*(X;\Lambda) \cong H_*(\widetilde{X};\mathbb{Z})$ and $H^*(X;L) \cong H^*(\widetilde{X};\mathbb{Z})$. Furthermore, the composition map*

$$H^i(\widetilde{X};\mathbb{Z}) \to H_{m-i}(X;\bar{L})$$

gives the Poincaré duality isomorphism with local coefficients for any $i \geq 0$.

3 Manifolds with highly connected universal covers

Let X^{2n} be a closed connected oriented $(2n)$-dimensional manifold such that $\Pi_i(X) = 0$ for $2 \leq i \leq n-1$. As remarked above, this is equivalent to $H_j(X;\Lambda) \cong H_j(\widetilde{X};\mathbb{Z}) = 0$ for $1 \leq j \leq n-1$. To control the groups $H_q(\widetilde{X};\mathbb{Z})$ for $q \geq n+1$ we use the right Λ-module $L = \mathrm{Hom}_{\mathbb{Z}}(\Lambda, \mathbb{Z})$. More precisely, we have the following results.

LEMMA 3.1. *With the above notation, suppose $H_j(X;\Lambda) \cong H_j(X;\bar{L}) \cong 0$ for $1 \leq j \leq n-1$. Then we have $H_q(\widetilde{X};\mathbb{Z}) \cong 0$, for $n+1 \leq q < 2n$, and $H_n(\widetilde{X};\mathbb{Z})$ is $\mathbb{Z}$-free. If $\Pi_1(X)$ is infinite, then $H_{2n}(\widetilde{X};\mathbb{Z}) \cong 0$ and $\widetilde{X}$ is homotopy equivalent to a wedge of n-spheres. If $\Pi_1(X)$ is finite, then $H_{2n}(\widetilde{X};\mathbb{Z}) \cong \mathbb{Z}$ and $\widetilde{X}$ is homotopy equivalent to $\vee_r \mathbb{S}^n \cup_\alpha e^{2n}$, where α is the attaching map of an $(2n)$-cell.*

Proof. We apply the usual universal coefficient theorem to

$$H_j(X;\bar{L}) \cong H^{2n-j}(\widetilde{X};\mathbb{Z}) \cong 0$$

for $1 \leq j \leq n-1$. Then we obtain

$$\mathrm{Hom}_{\mathbb{Z}}(H_{2n-j}(\widetilde{X}),\mathbb{Z}) \cong \mathrm{Ext}_{\mathbb{Z}}(H_{2n-j-1}(\widetilde{X}),\mathbb{Z}) \cong 0$$

for $1 \leq j \leq n-1$ (see for example [2], [17]). We can assume that $H_n(\widetilde{X})$ is at most countable because the fundamental group $\Pi_1(X)$ is finitely presented. Therefore, $\mathrm{Ext}_{\mathbb{Z}}(H_n(\widetilde{X}),\mathbb{Z}) \cong 0$ implies that $H_n(\widetilde{X})$ is $\mathbb{Z}$-free as claimed (see [17], p. 109). □

LEMMA 3.2. *Suppose $H_j(X;\Lambda) \cong 0$ for $1 \leq j \leq n-1$. If $\Pi_{n+1}(X) \cong 0$, then $H_{n+1}(X;\Lambda) \cong 0$.*

Proof. The proof follows by a direct calculation using the following commutative diagram with exact rows

$$
\begin{array}{ccccc}
\Pi_{n+2}\left(X^{(n+2)}, X^{(n+1)}\right) & \longrightarrow & \Pi_{n+1}\left(X^{(n+1)}\right) & \longrightarrow & \Pi_{n+1}\left(X^{(n+2)}\right) \\
\Big\downarrow{\scriptstyle\partial_{n+2}} & & \Big\downarrow{\scriptstyle i_*} & & \\
\Pi_{n+1}\left(X^{(n+1)}, X^{(n)}\right) & = = = & \Pi_{n+1}\left(X^{(n+1)}, X^{(n)}\right) & & \\
\Big\downarrow{\scriptstyle\partial_{n+1}} & & \Big\downarrow{\scriptstyle k_*} & & \\
\Pi_n\left(X^{(n)}, X^{(n-1)}\right) & \xleftarrow{\ \ j_*\ \ } & \Pi_n(X^{(n)}) & \longleftarrow & \Pi_n(X^{(n-1)}),
\end{array}
$$

where $\Pi_{n+1}\left(X^{(n+2)}\right) \cong \Pi_{n+1}(X) \cong 0$ and $\Pi_n\left(X^{(n-1)}\right) \cong 0$ because $\widetilde{X}^{(n-1)}$ is contractible. This implies that

$$
H_{n+1}(\widetilde{X}) = \frac{\operatorname{Ker}\partial_{n+1}}{\operatorname{Im}\partial_{n+2}} \cong \frac{\operatorname{Ker}k_*}{\operatorname{Im}i_*} \cong 0
$$

as claimed. $\qquad\square$

Finally we describe the cohomology groups of X with coefficients in B as follows.

LEMMA 3.3. *If* $H_j(X;\Lambda) \cong H_j(X;\bar{L}) \cong 0$ *for* $1 \leq j \leq n-1$, *then we have*

$$
H^q(X;B) \cong
\begin{cases}
H^q\left(B\Pi_1(X); B\right) & if & q \leq n-1 \\
H^n\left(B\Pi_1(X); B\right) \oplus \operatorname{Hom}_\Lambda(H_n(\widetilde{X}), B) & if & q = n \\
H^q\left(B\Pi_1(X); B\right) \oplus \operatorname{Ext}_\Lambda^{q-n}(H_n(\widetilde{X}), B) & if & n < q < 2n,
\end{cases}
$$

where B *is an arbitrary right* Λ-*module and* $B\Pi_1(X)$ *denotes the Eilenberg-MacLane space* $K(\Pi_1(X), 1)$.

Proof. We apply the spectral sequence (see [9])

$$
\operatorname{Ext}_\Lambda^p(H_q(\widetilde{X}), B) \Longrightarrow H^{p+q}(X;B).
$$

Using Lemma 3.1 we see that the spectral sequence collapses, so we get the results of the statement. $\qquad\square$

4 Four-manifolds with vanishing third homotopy group

In order to study $(2n)$-manifolds with highly connected universal covers, we first consider the case of closed TOP 4-manifolds X^4 (possibly non orientable) satisfying the property $\Pi_3(X) = 0$. Then we generalize this condition in the next section to obtain algebraic characterizations of aspherical $(2n)$-manifolds.

THEOREM 4.1. *Let X^4 be a closed connected TOP 4-manifold such that $\Pi_3(X) = 0$. If $\Pi_1(X)$ is finite, then X is TOP homeomorphic to either $\mathbb{S}^4$, $\mathbb{C}P^2$, $*\mathbb{C}P^2$ (the Chern manifold), $\mathbb{R}P^4$ or the unique non-smoothable homotopy $\mathbb{R}P^4$. If $\Pi_1(X)$ is infinite, then X is aspherical, i. e. $X \simeq K(\Pi_1(X), 1)$.*

Proof. Let us first assume that $\widetilde{X}$ is compact, i. e. $\Pi_1(X)$ is a finite group. Then $\widetilde{X}^{(3)} = \widetilde{X} \backslash \overset{\circ}{D}{}^4$ is homotopy equivalent to a wedge $\vee_{i=1}^r \mathbb{S}_i^2$. Because $\oplus_r \Pi_3(\mathbb{S}_i^2)$ injects into $\Pi_3(\widetilde{X}^{(3)})$, the exact homotopy sequence

$$\Pi_4(\widetilde{X}, \widetilde{X}^{(3)}) \cong H_4(\widetilde{X}, \widetilde{X}^{(3)}) \cong \mathbb{Z} \to \Pi_3(\widetilde{X}^{(3)}) \to \Pi_3(X) \cong 0$$

implies that $r \leq 1$, hence there are only $\widetilde{X} = \mathbb{S}^4$, $\mathbb{C}P^2$ or $*\mathbb{C}P^2$, up to homotopy equivalence. By Freedman's theorem, these are all (see [7], [8]). Therefore the only possibilities for X are finite quotients of $\mathbb{S}^4$, $\mathbb{C}P^2$ and $*\mathbb{C}P^2$. Again by [7], [8] and [20], X must be TOP homeomorphic to either $\mathbb{S}^4$, $\mathbb{C}P^2$, $*\mathbb{C}P^2$, $\mathbb{R}P^4$ or the unique non-smoothable homotopy $\mathbb{R}P^4$. Surgery theory establishes the existence of such a homotopy $\mathbb{R}P^4$, while Ruberman gave an explicit construction in [20]. We recall that Freedman proved that every smooth fake $\mathbb{R}P^4$ is homeomorphic to the standard $\mathbb{R}P^4$ (see [7] and [8]). We also remark that this non-smoothable $\mathbb{R}P^4$ can not be written as $*\mathbb{R}P^4$ since the latter is formally $\mathbb{R}P^4$ by [8], p. 167.

Let us assume now that $\widetilde{X}$ is not compact, i. e. $\Pi_1(X)$ is infinite. Then we have $H_q(\widetilde{X}; \mathbb{Z}) \cong 0$ for $q \neq 0, 2$ by Lemmas 3.1 and 3.2. Thus we have only to consider the group $H_2(\widetilde{X}) \cong \Pi_2(\widetilde{X}) \cong \Pi_2(X)$. Suppose $\varphi \colon \mathbb{S}^2 \to \widetilde{X}$ represents an element of $H_2(\widetilde{X})$. Because $\Pi_3(X) = 0$, the composition

$$\mathbb{S}^3 \xrightarrow{\quad \eta \quad} \mathbb{S}^2 \xrightarrow{\quad \varphi \quad} \widetilde{X}$$

is homotopic to zero, η being the Hopf map. Thus φ extends to a map

$$\Phi \colon \mathbb{S}^2 \cup_\eta D^4 = \mathbb{C}P^2 \to \widetilde{X}.$$

Observe that $H_3(\widetilde{X}) \cong H_4(\widetilde{X}) \cong 0$ yield $H^4(\widetilde{X}) \cong 0$ by using the universal coefficient sequence

$$0 \to \mathrm{Ext}_{\mathbb{Z}}(H_3(\widetilde{X}), \mathbb{Z}) \to H^4(\widetilde{X}) \to \mathrm{Hom}_{\mathbb{Z}}(H_4(\widetilde{X}), \mathbb{Z}) \to 0.$$

Let us consider now the induced homomorphism $\Phi^* \colon H^*(\widetilde{X}) \to H^*(\mathbb{C}P^2)$. We are going to prove that $\Phi^q = 0$ for any $q > 0$. Let $x \in H^2(\widetilde{X})$ with $\Phi^2(x) = u \in H^2(\mathbb{C}P^2)$. Since $H^4(\widetilde{X}) \cong 0$, it follows that $x^2 = 0$, hence $u^2 = (\Phi^2(x))^2 = 0$. This gives $u = 0$, i. e. $\Phi^q = 0$ for any $q > 0$. As a consequence, we obtain $\varphi \simeq 0$, hence $H_2(\widetilde{X}) \cong 0$. Thus X is an aspherical 4-manifold as requested. $\qquad\square$

Remark. If the Borel conjecture would hold in dimension four (compare for example with [6] for higher dimensions), Theorem 4.1 yields a complete list of all homeomorphy types of closed 4-manifolds with $\Pi_3 = 0$. Recall that the conjecture is true for aspherical 4-manifolds whenever the fundamental group is poly-(cyclic or finite) (see [8], Theorem 11.5)

5 Algebraic characterizations of aspherical manifolds

In this section we consider $(2n)$-manifolds X with highly connected universal covers and infinite fundamental group. These manifolds are in some sense generalizations of closed 4-manifolds with $\Pi_3 = 0$. Indeed, assuming $\Pi_1(X)$ infinite yields $H_i(\widetilde{X}) \cong 0$, except for $i = 0, n$. We can generalize the condition $\Pi_3 = 0$ (see section 4) in at least two ways.

PROPOSITION 5.1. *Let X be a closed connected oriented $(2n)$-manifold such that $H_i(\widetilde{X}) \cong 0$ for $i \neq 0, n$. Suppose that $H^{n+1}(\widetilde{X}) \cong 0$ and $\Pi_{n+1}(X) \cong 0$. Then X is aspherical, i. e. X is homotopy equivalent to $K(\Pi_1(X), 1)$.*

Proof. The hypothesis $H^{n+1}(\widetilde{X}) \cong 0$ implies that $H_n(\widetilde{X})$ is $\mathbb{Z}$-free, hence $\widetilde{X}$ is homotopy equivalent to a wedge of n-spheres. Indeed, $H_n(\widetilde{X})$ is at most numerable because $\Pi_1(X)$ is finitely presented. Thus $\mathrm{Ext}_{\mathbb{Z}}(H_n(\widetilde{X}), \mathbb{Z}) \cong 0$ implies that $H_n(\widetilde{X})$ is $\mathbb{Z}$-free by the Stein-Serre theorem (see for example [2], [3], [16]). Let $\varphi_\alpha \colon \mathbb{S}_\alpha^n \to \widetilde{X}$, $\alpha \in J$, present a base of $H_n(\widetilde{X}) \cong \Pi_n(X)$. Obviously, the wedge map

$$\Phi = \vee_\alpha \varphi_\alpha \colon \vee_{\alpha \in J} \mathbb{S}_\alpha^n \to \widetilde{X}$$

is a homotopy equivalence. Because the direct sum $\oplus_{\alpha \in J} \Pi_{n+1}(\mathbb{S}^n)$ injects into $\Pi_{n+1}(\vee_{\alpha \in J} \mathbb{S}_\alpha^n) \cong \Pi_{n+1}(X) \cong 0$, we must have $J = \emptyset$, i. e. $H_n(\widetilde{X}) \cong 0$. Thus $\widetilde{X}$ is contractible or equivalently $X \simeq K(\Pi_1(X), 1)$ as claimed. $\qquad\square$

We observe that the above proof works for each r such that $\Pi_{n+r}(\mathbb{S}^n) \neq 0$.

Using Lemma 3.1 and the facts that $\Pi_{2n-1}(\mathbb{S}^n) \neq 0$ and $\Pi_{n+1}(\mathbb{S}^n) \neq 0$ for all $n \geq 2$ we obtain the following algebraic characterizations of aspherical $(2n)$-manifolds.

COROLLARY 5.2. *Let X^{2n} be a closed connected oriented $(2n)$-manifold with infinite fundamental group. Assume $H_i(X; \Lambda) \cong H_i(X; \bar{L}) \cong 0$ for $1 \leq i \leq n-1$, where $\Lambda = \mathbb{Z}[\Pi_1(X)]$ and $L = \mathrm{Hom}_{\mathbb{Z}}(\Lambda, \mathbb{Z})$. If one of the following conditions*

   (1) $\Pi_{n+1}(X) \cong 0$ *or*
   (2) $\Pi_{2n-1}(X) \cong 0$

holds, then the manifold X^{2n} is a $K(\Pi_1(X), 1)$-space.

Recall that in the proof of Proposition 5.1 we have used the fact that $H_n(\widetilde{X})$ is $\mathbb{Z}$-free. This is insured in Corollary 5.2 by the hypothesis

$$H_{n-1}(X; \bar{L}) \cong H^{n+1}(\widetilde{X}) \cong 0,$$

hence $\mathrm{Ext}_{\mathbb{Z}}(H_n(\widetilde{X}), \mathbb{Z}) \cong 0$. Thus $H_n(\widetilde{X})$ is actually $\mathbb{Z}$-free by the Stein-Serre theorem. However we can avoid this condition (as we did in the 4-dimensional case, see section 4) when n is even, i. e. when the homotopy group $\Pi_{2n-1}(\mathbb{S}^n)$ is detected by the Hopf invariant. More precisely, we have the following result which represents the second generalization of the condition named at the beginning of the section.

PROPOSITION 5.3. *Let n be even and let $H_i(\widetilde{X}) \cong 0$ for $i \neq 0, n$. Assume further that $\Pi_{2n-1}(X) \cong 0$. Then X is homotopy equivalent to $K(\Pi_1(X), 1)$.*

Proof. Given $x \in H_n(\tilde{X}) \cong \Pi_n(\tilde{X})$, let $\varphi\colon \mathbb{S}^n \to \tilde{X}$ be a representative for x. Since the composition

$$\mathbb{S}^{2n-1} \xrightarrow{\ \eta\ } \mathbb{S}^n \xrightarrow{\ \varphi\ } \tilde{X}$$

(η the Whitehead square) is homotopic to zero, the map φ extends to

$$\Phi\colon \mathbb{S}^n \cup_\eta D^{2n} \to \tilde{X}.$$

The same argument as in the proof of Theorem 4.1 implies that $x = [\varphi] = 0$. Here we use the fact that $u^2 \neq 0$ in $H^{2n}\left(\mathbb{S}^n \cup_\eta D^{2n}\right)$, where u denotes a generator of $H^n\left(\mathbb{S}^n \cup_\eta D^{2n}\right)$. $\qquad\square$

At the end of the section we remark that if $\Pi_1(X)$ is finite, then the universal covering space $\tilde{X}$ is TOP homeomorphic to $\mathbb{S}^{2n}$ (assuming the homological hypothesis of the previous corollaries). Thus X is either $\mathbb{S}^{2n}$ or a homotopy real projective $(2n)$-space, including of course the non orientable case (see [21]).

6 Remarks on the Euler characteristic

In this section we study the relation between the asphericity of a manifold and its Euler characteristic. Similar discussions were also treated in different cases in [4], [5], [10], [13], [15], [16], [18] and [19].

Assume that X^{2n} satisfies $H_i(X;\Lambda) \cong H_i(X;\bar{L}) \cong 0$ for $1 \leq i \leq n-1$, so Lemma 3.1 applies. Furthermore, suppose $H_n(X;\Lambda)$ is Λ-projective. Let $c\colon X \to B\Pi_1 = K(\Pi_1,1)$ be the classifying map for the covering $\tilde{X} \to X$. By our hypothesis, the complex $C_*(\widetilde{B\Pi_1}, \tilde{X})$ is a free Λ-complex whose homology is concentrated in dimension $n+1$, i. e. $H_i(B\Pi_1, X; \Lambda) \cong 0$ for any $i \neq n+1$. Here we assume $\Pi_1(X)$ infinite as usual. Furthermore, the homology group $H_{n+1}(B\Pi_1, X; \Lambda) \cong H_n(X;\Lambda)$ is Λ-projective. Applying the spectral sequence (see [9])

$$\mathrm{Tor}_p^\Lambda\left(H_q(B\Pi_1, X; \Lambda), \mathbb{Z}\right) \Longrightarrow H_{p+q}(B\Pi_1, X; \mathbb{Z}),$$

we obtain $H_*(B\Pi_1, X; \Lambda) \otimes_\Lambda \mathbb{Z} \cong H_*(B\Pi_1, X; \mathbb{Z})$. This implies that

$$H_{2n}(X;\mathbb{Z}) \cong \mathbb{Z} \xrightarrow[\cong]{} H_{2n}(B\Pi_1;\mathbb{Z})$$

and $H_i(B\Pi_1;\mathbb{Z}) \cong 0$ for $i > 2n$. Therefore, the Euler characteristic $\chi(B\Pi_1)$ is well-defined and the relation

$$\chi(B\Pi_1) - \chi(X) = (-1)^{n+1}\,\mathrm{rank}(H_n(\tilde{X}) \otimes_\Lambda \mathbb{Z})$$

holds. Thus we have the following result.

PROPOSITION 6.1. *Consider the class of closed connected oriented manifolds X^{2n}, of dimension $2n = 4k$, such that $\Pi_1(X) = \Pi_1$ is an infinite group, $H_i(X;\Lambda) \cong H_i(X;\bar{L}) \cong 0$ for $1 \leq i \leq 2k-1$ and $H_{2k}(X;\Lambda)$ is Λ-projective. Then $\chi(B\Pi_1) \leq \chi(X)$ for all such X and $\chi(B\Pi_1) = \chi(X)$ if and only if $B\Pi_1$ belongs to this class. In the case $2n = 4k+2$, $\chi(B\Pi_1) \geq \chi(X)$ and $\chi(B\Pi_1) = \chi(X)$ if and only if $B\Pi_1$ is in the above-defined class.*

Applying Proposition 6.1 requires the determination of conditions under which $H_n(X; \Lambda)$ is Λ-projective. This seems not to be an easy matter.

References

1. G. A. Anderson, Surgery with coefficients, Lect. Notes in Math. 591, Springer Verlag, Berlin-Heidelberg-New York, 1977.
2. H. Cartan, S. Eilenberg, Homological Algebra, Princeton Univ. Press, Princeton, New Jersey, 1956.
3. P. M. Cohn, Algebra, J. Wiley and Sons, Chichester-New York-Toronto, 1989.
4. B. Eckmann, *Nilpotent group action and Euler characteristic*, Algebraic Topology Barcelona 1986 (J. Aguadé, R. Kane, eds.), Lect. Notes in Math. 1298, Springer Verlag, Berlin-Heidelberg-New York, 1987, pp. 120–123.
5. B. Eckmann, *Amenable groups and Euler characteristic*, Comment. Math. Helvetici **67** (1992), 383–393.
6. F. T. Farrell, L. E. Jones, *A topological analogue of Mostow's rigidity theorem*, J. Amer. Math. Soc. **2** (1989), 257–370.
7. M. H. Freedman, *The topology of four-dimensional manifolds*, J. Differential Geom. **17** (1982), 357–453.
8. M. H. Freedman, F. Quinn, Topology of 4-Manifolds, Princeton Univ. Press, Princeton, New Jersey, 1990.
9. R. Godement, Topologie algébrique et théorie des faisceaux, Publ. de l'Inst. Math. de Strasbourg XII, Hermann, Paris, 1958.
10. J. C. Hausmann, S. Weinberger, *Caracteristiques d'Euler et groupes fondamentaux des varietés de dimension* 4, Comment. Math. Helvetici **60** (1985), 139–144.
11. J. C. Hausmann, P. Vogel, *Poincaré spaces*, to appear.
12. J. A. Hillman, 2-Knots and their Groups, Australian Math. Soc. Lect. Ser. 5, Cambridge Univ. Press, Cambridge, 1989.
13. J. A. Hillman, *Elementary amenable groups and 4-manifolds with Euler characteristic* 0, J. Austral. Math. Soc. **50** (1991), 160–170.
14. J. A. Hillman, *On 4-manifolds with universal covering space* $\mathbb{S}^2 \times \mathbb{R}^2$ *or* $\mathbb{S}^3 \times \mathbb{R}$, Topology its Appl. **52** (1993), 23–42.
15. J. A. Hillman, *Minimal 4-manifolds for groups of cohomological dimension* 2, Proc. Roy Soc. Edinburgh, to appear.
16. J. A. Hillman, *Free products and 4-dimensional connected sums*, to appear.
17. P. J. Hilton, U. Stammbach, A course in Homological Algebra, Springer Verlag, Berlin-Heidelberg-New York, 1971.
18. G. Mislin, *Finitely dominated nilpotent spaces*, Ann. Math. **103** (1976), 547–556.
19. S. Rosset, *A vanishing theorem for Euler characteristic*, Math. Z. **185** (1984), 211–215.
20. D. Ruberman, *Invariant knots of free involutions of* $\mathbb{S}^4$, Topology its Appl. **18** (1984), 217–224.
21. C. T. C. Wall, *Poincaré complexes, I*, Ann. of Math. **86** (1967), 213–245.
22. C. T. C. Wall, Surgery on compact manifolds, Academic Press, London-New York, 1970.

Alberto Cavicchioli, Dipartimento di Matematica, Università di Modena, Via Campi 213/B, 41100 Modena (Italia)

Friedrich Hegenbarth, Dipartimento di Matematica, Università di Milano, Via C. Saldini 50, 20133 Milano (Italia)

Progress in Mathematics, Vol. 136
© 1996 Birkhäuser Verlag Basel/Switzerland

Hochschild and cyclic homology of an almost commutative cochain algebra associated to a nilmanifold

BOHUMIL CENKL AND MICHELINE VIGUÉ-POIRRIER

A nilmanifold, as defined by Malcev [Ma], is a compact manifold N which is the space of cosets of a simply connected Lie group by discrete uniform subgroup G. Thus the manifold N can be identified with the Eilenberg-MacLane space $K(G, 1)$, where $G = \pi_1(N)$ is a finitely generated torsion free nilpotent group.

The cohomology of a nilmanifold N is isomorphic to the cohomology of a Lie algebra associated to N, when the coefficients are rational or even tame, [C-P 4]. In the computation of the integral cohomology of N in [C-P 1], [C-P 2], the authors associate with G a free associative graded cochain algebra $P(G)$ over $\mathbb{Z}$ whose cohomology is isomorphic to $H^*(G, \mathbb{Z}) = H^*(N, \mathbb{Z})$. In order to make the computation effective they consider a finite differential graded cochain algebra $M(G)$ over $\mathbb{Z}$ whose differential and structure product depend only in the group structure of G. Then they get the following isomorphisms:

$$H^*(M(G), \mathbb{Z}) \simeq H^*(P(G), \mathbb{Z}) \simeq H^*(G, \mathbb{Z}) \simeq H^*(N, \mathbb{Z}).$$

In an attempt to understand the cohomology of the free loop space on a nil-manifold, as a special case of a non simply connected space, the first author computed the Hochschild homology of $P(G)$, [Ce]. The cyclic homology of $P(G)$ is computed in [C-V]. The present paper is devoted to further study of these homologies. More precisely, on $P(G)$ and $M(G)$ we define a second gradation called a filtering degree (a norm in [Ce]). This filtering degree is used to define on $P(G)$ and $M(G)$ filtrations which are compatible with the differentials and product structures. The resulting graded ring associated to $M(G)$ has a structure of a commutative differential graded algebra. Therefore $M(G)$, when associative, is an example of an almost commutative differential graded algebra (see [Ka], [K-V]).Using the techniques of [C-P 1] it can be proved that $M(G)$ is associative for the Heisenberg group, for example. Recent study of the homology of non commutative deformations of commutative algebras can be found in [Br], [Fg-T], [K-V], [Ta].

In this paper we associate to the data of an almost commutative differential cochain algebra $(M(G), D)$ over $\mathbb{Q}$ a filtered complex whose homology is the

---

1991 *Mathematics Subject Classification*. Primary 57T15, 16A61; Secondary 18G35, 55N25.

Key words and phrases. Hochschild homology, nilmanifolds, almost commutative algebra.

Hochschild homology of $M(G)$ (or of $P(G)$). The E^0 term of the corresponding spectral sequence is exactly the Hochschild homology of $M(H)$ (or $P(H)$).
Here H is the abelian group with k generators, when k is the dimension of
the nilmanifold N. Thus we recover the main result of [Ce]. In addition we get
more information about the complex involved. That complex can be compared
to the result of Brylinsky concerning Poisson manifolds [Br]. Since $M(H)$ is a
free commutative graded algebra, E^0 is isomorphic to the algebra of differential
forms $\Omega^*_{M(H)}$ on $M(H)$. Here $d^0 = 0$. The differential d^1 on $E^1 = E^0 = \Omega^*_{M(H)}$
(see Theorem 3.6) is the sum of two differentials δ and δ'. δ is exactly the
internal differential on the algebra of differential forms on the commutative
differential graded algebra $(E^1 M(G))$, which is the E^1-term of the spectral
sequence associated to the filtration of $(M(G), D)$. The second differential δ'
is the graded version of Brylinski's differential on $\Omega^*_{M(H)}$, where $M(H)$ has
a structure of Poisson graded algebra determined by the almost commutative
structure of $M(G)$, [Br]. Moreover, $(M(G), D)$ and $(P(G), d)$ are filtered algebras, so that their Hochschild complexes $\mathcal{C}_*(P(G))$, and $\mathcal{C}_*(M(G))$ are filtered
complexes. The resulting spectral sequences $\tilde{E}^r(\mathcal{C}_*(P(G))), \tilde{E}^r(\mathcal{C}_*(M(G)))$ and
the spectral sequence of the Theorem 3.6 are isomorphic at the E^1-levels.

    If $G = G_3$ is the Heisenberg group of matrices

$$\begin{pmatrix} 1 & a_1 & a_3/k \\ 0 & 1 & a_2 \\ 0 & 0 & 1 \end{pmatrix},$$

where k is a fixed positive integer, and $(a_1, a_2, a_3) \in \mathbb{Z}^3$, then there is an
algorithm for the computation of the Hochschild homology of $M(G_3)$ (Theorem
4.1). The three spectral sequences mentioned above collapse at the E^2-level.
Thus the Hochschild homology of G_3 can be computed explicitly (Theorem 4.2).
Some classical results on cyclic homology can be extended to negative cyclic
homology of differential cochain algebras. Furthermore, there is a bounded
exact sequences connecting the Hochschild and negative cyclic homologies of
$(M(G), D)$ and $(P(G), d)$ (Theorem 5.2). In the case of the group G_3 there is
an explicit result (Theorem 5.3).

    As mentioned earlier, the algebras $P(G)$ and $M(G)$ were originally in-
troduced for the computation of the integral cohomology of the nilmanifold
$K(G, 1)$. These algebras make it possible to bypass the difficulties with iterated fibrations. Both of these issues were addressed from different perspective
in the independent works of Huebschmann, Kadeishvili, Lambe and Stasheff,
[H-K], [L-S]. In addition, we would like to point out a related work of Lambe
[L]. Furthermore, the perturbation techniques and the chain homotopies relating $P(G)$ and $M(G)$ can be viewed in the framework of A_∞-structures as
introduced by Stasheff in the sixties. What makes the algebra $P(G)$ so special
(in contrast with the general cochain algebra) is the fact that it is free associative. This makes the computation of the Hochschild and cyclic homologies
possible and the investigation of the cohomology of the free loop space feasible.

    The paper is organized as follows. In §1 some basic properties of $P(G)$ and
$M(G)$ are recalled. The definitions of the Hochschild and cyclic homologies of

differential graded algebras are reviewed in §2. The filtered complex that is used to compute the Hochschild homology of $M(G)$ is constructed in §3. The explicit computations for the group G_3 are done in §4. Finally, in §5, the negative cyclic homologies of $(M(G), D)$ and $(P(G), d)$ are computed.

1 The polynomial cochain algebra $(P(G), d)$ and the finite cochain algebra $(M(G), D)$, [C-P 1], [C-P 2], [C-P 3], [C-V], [Ce].

Let G be a finitely generated torsion free nilpotent group and let $\{g_1, \dots, g_k\}$ be a Malcev basis, [Ma]. An element $g \in G$ can be written in the form $g = g_1^{x_1} \cdots g_k^{x_k}, x_i \in \mathbb{Z}$, and the product of the elements $g = g_1^{x_1} \cdots g_k^{x_k}, g' = g_1^{y_1} \cdots g_k^{y_k}$,

$$g \cdot g' = g_1^{\rho_1(x,y)} \cdots g_k^{\rho_k(x,y)},$$

with $\rho_j(x, y) = x_j + y_j + \tau_j(x, y)$ and where $\tau_j(x, y)$ is a polynomial in $\mathbb{Q}[x_1, \dots, x_{j-1}, y_1, \dots, y_{j-1}]$ of degree ≥ 2.

A canonical basis $\{g_1, \dots, g_k\}$ gives a bijection between G^n and the set of $k \times n$ matrices over $\mathbb{Z}$, extending the map that associates to $g = g_1^{x_1} \cdots g_k^{x_k}$ the $k \times 1$ matrix

$$\begin{pmatrix} x_1 \\ \cdot \\ \cdot \\ \cdot \\ x_k \end{pmatrix}$$

Under this correspondence, $C^n(G)$ is the $\mathbb{Z}$-module of functions from the set of $k \times n$ matrices over $\mathbb{Z}$ with values in $\mathbb{Z}$. Let $P(G) = \oplus_{n \geq 0} P^n(G)$ be those rational polynomial functions that vanish on degenerate chains. Let $P^0(G) = C^0(G) = \mathbb{Z}$. For any $a : \{1, \dots, k\} \to \mathbb{N}$, denote by $m(a)$ the polynomial function over $\mathbb{Z}^k$ defined by

$$\langle m(a), (x_1, \dots, x_k) \rangle = \prod_{1 \leq r \leq k} \binom{x_r}{a_r},$$

where $\binom{x_r}{a_r}$ is the binomial coefficient

$$\frac{x_r(x_r - 1) \cdots (x_r - a_r + 1)}{a_r!}.$$

By convention $\binom{x_r}{0} = 1$. Denote $\binom{x}{1} = x$. It can be proved, [St], that $P^1(G)$ is a free $\mathbb{Z}$-module with basis $m(a)$ for all $a : \{1, \dots, k\} \to \mathbb{N}$ with $\sum a_i > 0$. $P(G) = T(P^1(G))$ is a free associative algebra over $P^1(G)$, and $P(G)$ is a differential subalgebra of $C^*(G)$. An element α of $P^n(G) = P^1(G)^{\otimes n}$ has degree $|\alpha| = n$. The differential d on $P^1(G)$ can be extended to $P(G)$ by the formula

$$d(u \otimes v) = du \otimes v + (-1)^{|u|} u \otimes dv.$$

If $u \in P^1(G), u = \prod_{r=1}^{k} \binom{x_r}{a_r}$, then

$$du = \prod_{r=1}^{k} \binom{y_r}{a_r} + \prod_{r=1}^{k} \binom{z_r}{a_r} - \prod_{r=1}^{k} \binom{\rho_r(y,z)}{a_r}.$$

DEFINITION. $(P(G), d)$ is called the *polynomial cochain algebra* associated with G.

Next we define on $P(G)$ a second gradation, called a filtering degree and denote it by $\| \cdot \|$ (this is called a norm in [C-P 3]). Let $G \supset G_1 \supset G_2 \supset \cdots \supset G_{s+1} = 1$ be the shortest central series such that

(i) G_i/G_{i+1} is a free abelian group,
(ii) the commutator $[G_i, G_j] \subset G_{i+j}$ for all i, j.

Let $\{g_1, \dots, g_k\}$ be a Malcev canonical basis for the group G. Let

$$\{1, \dots, i_1\}, \dots, \{i_{j-1} + 1, \dots, i_j\}, \dots, \{i_{s-1} + 1, \dots, i_s = k\}$$

be a partition of $\{1, \dots, k\}$ into s subsets such that $\{g_{i_{j-1}+1}, \dots, g_{i_j}\}$ is a basis of the abelian group G_j/G_{j+1}.

Let i be an integer that belongs to $\{1, \dots, k\}$. If $i \in \{i_{j-1} + 1, \dots, i_j\}$, set

$$\|x_i\| = j - 1,$$

$$\left\| \binom{x_i}{a} \right\| = a\|x_i\| = a(j-1) \qquad \text{for } a \in \mathbb{N}^*$$

$$\left\| p\binom{x_i}{a} \right\| = \left\| \binom{x_i}{a} \right\| \qquad \text{for } p \in \mathbb{Z}^*,$$

$$\|u + v\| = \max(\|u\|, \|v\|) \quad \text{for } u, v \text{ in } P(G),$$
$$\|u \otimes v\| = \|u\| + \|v\| \quad \text{for } u, v \text{ in } P(G),$$
$$\|\lambda\| = 0 \quad \text{if } \lambda \in T^0(P(G)) = \mathbb{Z}.$$

LEMMA 1.1. *Let $\rho_j(x, y) = x_j + y_j + \tau_j(x, y)$. Then*

$$\|\rho_j(x, y)\| = \|x_j\|,$$
$$\|\tau_j(x, y)\| \leq \|x_j\| - 1.$$

Define an increasing filtration F^i on $P(G)$, with $F^{-1} = 0$, by setting

$$F^i P(G) = \{u \in P(G) \mid \|u\| \leq i\}.$$

Then

LEMMA 1.2. ([C-P 3], LEMMA 3). *The above filtration satisfies*
(i) $F^i P(G) \cdot F^j P(G) \subset F^{i+j} P(G)$,
(ii) $d(F^i P(G)) \subset F^i P(G)$,
(iii) $d = d_0 + d'$, where $d'(F^i P(G)) \subset F^{i-1} P(G)$, and d_0 is the extension to $P(G)$ of the differential defined on $P^1(G)$ by

$$d_0 \left(\prod_{i=1}^{k} \binom{x_i}{a_i} \right) = \prod_{i=1}^{k} \binom{y_i}{a_i} + \prod_{i=1}^{k} \binom{z_i}{a_i} - \prod_{i=1}^{k} \binom{z_i + y_i}{a_i}.$$

Next we introduce the differential cochain algebra $(M(G), D)$ as defined in [C-P 1], [C-P 2]. As a $\mathbb{Z}$-module, $M(G)$ is the integral cohomology of the k-torus. Thus $M(G) = M^{\otimes k}, M = H^*(S^1, \mathbb{Z}) = \mathbb{Z} \oplus \mathbb{Z}e$, where e is a generator of $H^1(S^1, \mathbb{Z})$. As a $\mathbb{Z}$-graded module, $M(G)$ is generated by $e_k^{u_k} \otimes \cdots \otimes e_1^{u_1}$, where $u_i = 0$ or 1. The degree of such an element is

$$| e_k^{u_k} \otimes \cdots \otimes e_1^{u_1} | = \sum u_i.$$

The differential D and the product law on $M(G)$ are expressed in terms of the group structure of G (see [C-P 1] and [C-P 2]). Using the perturbation techniques, one constructs degree preserving maps I, P and a homotopy H so that $PI =$ id and $IP = dH + Hd +$ id,

$$(P(G), d)$$

$$I \uparrow \quad \downarrow P$$

$$(M(G), D).$$

The product on $M(G)$ is defined by

$$a \cdot b = P(I(a)I(b))$$

for $a, b \in M(G)$. It follows that P is onto and that P^* is an isomorphism of $\mathbb{Z}$-graded modules. The formula $DP = Pd$ defines D on $M(G)$.

Let $\| \cdot \|$ be a filtering degree on $M(G)$ defined by

$$\|e_i\| = j - 1 \qquad \text{for } i \in \{i_{j-1} + 1, \ldots, i_j\},$$

and

$$\|e_k^{u_k} \otimes \cdots \otimes e_1^{u_1}\| = \sum_{l=1}^{k} u_l \|e_l\|,$$

$$\|\lambda e_k^{u_k} \otimes \cdots \otimes e_1^{u_1}\| = \|e_k^{u_k} \otimes \cdots \otimes e_1^{u_1}\| \text{ when } \lambda \in \mathbb{Z}^*,$$

$$\|u + v\| = \max (\|u\|, \|v\|).$$

Furthermore, let $F^i \subset F^{i+1}$ be an increasing filtration;

$$F^{-1} = 0, \quad F^i M(G) = \{u \in M(G), \|u\| \leq i\}.$$

It has to be pointed out that in general the product on $M(G)$ is not associative. But there is a decomposition of the product $a \cdot b$ into an associative and non associative parts. Namely

$$a \cdot b = A(a, b) + B(a, b),$$

where $A(A(a, b), c) = A(a, A(b, c))$. It follows that

LEMMA 1.3.

(i) $\|A(a,b)\| \leq \|a\| + \|b\|$,
    $\|B(a,b)\| \leq \|A(a,b)\| - 1$ for $a, b \in M(G)$,
(ii) $D(F^i M(G)) \subset F^i M(G)$,
(iii) $P(F^i P(G)) = F^i M(G)$,
(iv) $I(F^i M(G)) \subset F^i P(G)$ for all i.

From now on we associate to a differential graded cochain algebra over $\mathbb{Z}$, $(A^* = \underset{n \geq 0}{\oplus} A^n, d), d(A^n) \subset A^{n+1}$, a negatively graded differential algebra (A_*, d) with $A_{-n} = A^n$, and $d(A_{-n}) \subset A_{-n-1}$. If (A^*, d) is a filtered algebra with an increasing filtration $\mathbb{Z} \subset F^0 A^* \subset F^1 A^* \subset \cdots$ such that

$$A^* = \bigcup F^p A^*, \quad d(F^p A^*) \subset F^p A^*, \quad F^p A^* \cdot F^q A^* \subset F^{p+q} A^*,$$

then (A_*, d) is endowed with the same filtration having the same properties. Then we define the associated graded algebra

$$\mathrm{gr} A_* = \underset{n \in \mathbb{N}}{\oplus} (\mathrm{gr} A_*)_n, \quad \text{where} \quad (\mathrm{gr} A_*)_n = \underset{p}{\oplus} (F^p A_n / F^{p-1} A_n).$$

The differential $\mathrm{gr} d$ on $\mathrm{gr} A_*$ is defined as follows: $\mathrm{gr} d(\bar{a}) = \overline{da}$, where a belongs to $F^p A$, and $\bar{a}$ is its projection in $F^p A / F^{p-1} A$, $\overline{da}$ also belongs to $F^p A / F^{p-1} A$.

From Lemmas 1.1 and 1.2 it follows that $(\mathrm{gr} P(G)_*, \mathrm{gr} d)$ can be identified with $(P(H)_*, d_0)$, where $P(H)_{-n} = P(H)^n$ and $(P(H) = \underset{n \geq 0}{\oplus} P(H)^n, d_0)$ is just the polynomial cochain algebra for the free abelian group H on k generators.

The key point is the following result, which follows from the construction of P and D in [C-P 1], [C-P 2] and from the filtration in [C-P 3].

THEOREM 1.4. Let G be the fundamental group of a k-dimensional nilmanifold and let $H = \mathbb{Z}^k$ be the free abelian group on k generators. Let $P(G)$ (resp. $P(H)$) be the polynomial cochain algebra associated with G (resp. H). Let $(M(G), D)$ (resp. $(M(H), D_H)$) be the finite dimensional algebra associated with G (resp. H). Then

(a) $M(H)$ is the exterior algebra on k generators $e_1, \ldots, e_k$ with $|e_i| = 1$ for $i \in \{1, \ldots, k\}$, and the differential D_H is zero.
(b) If P_H, I_H are the maps between $(P(H), d)$ and $(M(H), 0)$, as constructed in [C-P 2], then the maps (P, I) between $P(G)$ and $M(G)$, and the differential D, can be constructed so that
(i) $\mathrm{gr}(M(G)) = M(H)$ as graded algebras,
(ii) $\mathrm{gr} D = 0$,
(iii) $\mathrm{gr}(P) = P_H$,
(iv) $\mathrm{gr}(I) = I_H$.

REMARK. The differential and the product on $M(G)$ can be looked at as deformations of such operations on $M(H)$.

COROLLARY 1.5. *Let $(M(G), D)$ be the filtered differential graded algebra as defined above. Then*

(1) $u \cdot v - (-1)^{|u| \cdot |v|} v \cdot u \in F^{i+j-1} M(G)$ *if* $u \in F^i M(G), v \in F^j M(G)$,

(2) $D(F^i M(G)) \subset F^{i-1} M(G)$.

Thus $(M(G), D)$ is a non commutative non associative filtered differential graded algebra whose associated graded algebra is associative and commutative in the graded sense. Therefore there is a graded Poisson bracket on $\mathrm{gr} M(G)$. It is defined as follows. If $f \in F^p M(G)/F^{p-1} M(G)$ (resp. $g \in F^q M(G)/F^{q-1} M(G)$) and $a \in F^p M(G)$ (resp. $b \in F^q M(G)$) with image f (resp. g) in F^p/F^{p-1} (resp. F^q/F^{q-1}), then $[a, b] = a \cdot b - (-1)^{|a||b|} b \cdot a$ is an element of $F^{p+q-1} M(G)$. The image of $[a, b]$ in F^{p+q-1}/F^{p+q-2} is independent of the choice of a and b. Then the Poisson bracket of f and g is

$$\{f, g\} = \text{image of} \ [a, b] \ \text{in} \ F^{p+q-1} M(G)/F^{p+q-2} M(G).$$

It extends as a bilinear map $\mathrm{gr} M(G) \times \mathrm{gr} M(G) \to \mathrm{gr} M(G)$. It follows that $(\mathrm{gr} M(G), \{\ ,\})$ is a Lie graded algebra. Furhermore, $\{\ ,\}$ is a derivation of degree zero, in each variable. $(M(G), \{\ ,\})$ is called a graded Poisson algebra ([Ka], [K-V]).

2 Hochschild and cyclic homologies of a flat differential graded algebra over $\mathbb{Z}$, [Go], [Vi].

Let (A^*, d) be an associative differential graded algebra over $\mathbb{Z}$, which is flat as a $\mathbb{Z}$-module. Suppose that $A^* = \underset{n \geq 0}{\oplus} A^n, A^0 = \mathbb{Z}, d(A^n) \subset A^{n+1}$. Consider the negatively graded chain algebra $A_* = \underset{n \geq 0}{\oplus} A_{-n}$, with $A_{-n} = A^n$ and $d : A_{-n} \to A_{-n-1}$. Recall that with (A_*, d) there is associated the total Hochschild complex

$$\mathcal{C}_{-n} = \underset{p-q=-n}{\oplus} C_{p,-q}$$

$$C_{p,-q} = A_{-i_0} \otimes \overline{A}_{-i_1} \otimes \cdots \otimes \overline{A}_{-i_p}, \quad \sum i_k = q, \quad \overline{A} = A/\mathbb{Z}.$$

Let $b + d$ be the total differenctial on $\mathcal{C}_* = \underset{n}{\oplus} \mathcal{C}_{-n}$, (see [Vi]). It follows that $\mathcal{C}_*$ is uniquely graded in negative degrees.

DEFINITION. $H_*(\mathcal{C}_*, b + d)$ is called the *Hochschild homology* of (A_*, d) and it is denoted by $HH_*(A_*, d)$.

For each $n \geq 0, HH_n(A^*, d) = HH_{-n}(A_*, d) = H_{-n}(\mathcal{C}_*, b + d)$,   and $HH_*(A^*, d)$
$= \underset{n \geq 0}{\oplus} HH_n(A^*, d)$ is called the Hochschild homology of the cochain algebra (A^*, d). Since A^* is flat over $\mathbb{Z}$, we also have

$$HH_n(A^*, d) = \mathrm{Tor}_{-n}^{A_* \otimes A_*^{op}}(A_*, A_*) \quad \text{for all} \ \ n \geq 0,$$

where (A_*^{op}, d^{op}) is the opposite differential algebra of (A_*, d). The structure of an $(A_* \otimes A_*^{op})$-module on A_* is given by the multiplication $m(a \otimes a') = aa'$, [Vi]. Following Connes, we define the operator $B : C_{p,-q} \to C_{p+1,-q}$ (see [Go, III] or [Vi] for the exact formula). Let

$$\mathcal{B}_{-n} = \bigoplus_{p \geq 0} C_{-n+2p}, \qquad \mathcal{B}_* = \bigoplus_n \mathcal{B}_{-n},$$

and let the differential $_B b$ on $\mathcal{B}_*$ be defined by setting

$$_B b(a_n, a_{n-2}, \cdots) = ((b+d)a_n, (b+d)a_{n-2} + Ba_n, (b+d)a_{n-4} + Ba_{n-2}, \cdots)$$

$$\text{for} \quad a_i \in C_{-i}, i \geq 0.$$

In fact, for a given integer $n \geq 0$, the sum $\bigoplus_p C_{-n+2p}$ is finite, and $\mathcal{B}_*$ is uniquely graded in negative degree.

DEFINITION. $H_*(\mathcal{B}_*, _B b)$ is called the *negative cyclic homology of* (A_*, d) and it is denoted by $HC_*^-(A_*, d)$, [Jo]. $HC_*^-(A^*, d) = \bigoplus_{n \geq 0} HC_n^-(A^*, d)$, where $HC_n^-(A^*, d) = HC_{-n}^-(A_*, d)$, is called the *negative cyclic homology of the cochain algebra* (A^*, d).

It follows that $HC_*^-(\mathbb{Z}) = \mathbb{Z}[u]$, where $\mathbb{Z}[u]$ is the polynomial algebra on u of degree $|u| = -2$, and $HC_*^-(A_*, d)$ has a structure of $\mathbb{Z}[u]$-graded module. Thus $HC_*^-(A^*, d)$ is a $\mathbb{Z}[v]$-graded module with $|v| = +2$.

PROPOSITION 2.1. *Let* (A^*, d) *be a differential graded algebra which is flat over* $\mathbb{Z}$, *and such that* $A^0 = \mathbb{Z}, d(A^n) \subset A^{n+1}, A^* = \bigoplus_{n \geq 0} A^n$. *Then*

(1) $HH_0(A^*, d) \simeq HC_0^-(A^*, d)$, *and*
(2) *there is a long exact sequence connecting* $HH_n(A^*, d)$ *and* $HC_n^-(A^*, d)$, *where* S^- *stands for a multiplication by* v:

$$0 \to HC_1^-(A^*, d) \to HH_1(A^*, d) \to HC_0^-(A^*, d) \xrightarrow{S^-} HC_2^-(A^*) \to HH_2(A^*) \to$$

$$\cdots \to HC_n^-(A^*) \xrightarrow{S^-} HC_{n+2}^-(A^*) \to HH_{n+2}(A^*) \to HC_{n+1}^-(A^*) \xrightarrow{S^-} \cdots.$$

Proof. There is an exact sequence of complexes

$$0 \to \mathcal{B}_*(A_*, _B b)_2 \xrightarrow{j} \mathcal{B}_*(A_*, _B b) \xrightarrow{q} (C_*(A_*), b+d) \to 0,$$

where $j(a_n, a_{n-2}, \dots) = (0, a_n, a_{n-2}, \dots)$.

For $a_i \in C_{-i}$, $i \geq 0$, $q(a_n, a_{n-2}, \dots) = a_n$. In degree 0, $\mathcal{B}_0 = C_0$, and q induces an isomorphism. The long exact sequence is just the long exact sequence in homology associated with the short exact sequence of complexes defined above. $\square$

Now assume that (A, d) is a filtered associative differential graded algebra over $\mathbb{Z}$ with either $A_n = 0, \quad n < 0$, or $A_n = 0, \quad n > 0$ and $A_0 = \mathbb{Z}$, where d is a differential of degree -1, and the filtration satisfies

$$F^{-1} = 0 \subset \mathbb{Z} \subset F^0 \subset \cdots \subset F^n \subset F^{n+1} \subset \cdots.$$

The Hochschild complex $C_*(A)$ can be filtered as follows:

$$F^p C_{n,*}(A) = \sum_{k_0 + k_1 + \cdots + k_n \leq p} F^{k_0} \otimes F^{k_1} \overline{A} \otimes \cdots \otimes F^{k_n} \overline{A}.$$

Thus $(C_*(A), b + d)$ is a filtered complex. We have

LEMMA 2.2. [BR]. *Let $(C_*(A), b + d)$ be the total Hochschild complex associated with (A, d) with the filtration described above. Then the resulting spectral sequence $\tilde{E}^r(C_*)$ satisfies:*

(i) *$\tilde{E}^1_{p,q} = HH_{p+q}((\mathrm{gr}A)_p)$,*
    where $HH_{p+q}((\mathrm{gr}A)_p)$ is the homogeneous part of degree p of the graded $\mathbb{Z}$-module $HH_{p+q}(\mathrm{gr}A)$.
(ii) *$\tilde{E}^r_{p,q} \Longrightarrow HH_*(A, d)$.*

LEMMA 2.3. [Go, III]. *Let (A, d) and (B, d) be two filtered differential graded algebras which are flat over $\mathbb{Z}$ and let $f : (A, d) \to (B, d)$ be a morphism of filtered complexes such that $(\mathrm{gr}f)_* : H_*(\mathrm{gr}A, \mathrm{grd}) \to H_*(\mathrm{gr}B, \mathrm{grd})$ is an isomorphism. Then the map f induces morphisms $f^r : \tilde{E}^r_{p,q}(C_*(A)) \to \tilde{E}^r_{p,q}(C_*(B))$ between the two spectral sequences associated to the filtrations of the Hochschild complexes, and the maps f^r are isomorphisms for $r \geq 1$.*

Proof. The map $f^1 : \tilde{E}^1_{p,q}(C_*(A)) = HH_{p+q}(\mathrm{gr}A) \to \tilde{E}^1_{p,q}(C_*(B))$ is the map between the Hochschild homology groups, induced by $\mathrm{gr}f : (\mathrm{gr}A, \mathrm{grd}) \to (\mathrm{gr}B, \mathrm{grd})$. Since $\mathrm{gr}A$ and $\mathrm{gr}B$ are flat over $\mathbb{Z}$, and $\mathrm{gr}f$ induces an isomorphism on homology, it is known, [Go, III], that the map induced by $\mathrm{gr}f$ on Hochschild homology is an isomorphism. Therefore f^r, $r \geq 1$, is an isomorphism. $\square$

As a consequence of Theorem 1.4 and Lemma 2.3 we have

PROPOSITION 2.4. *Let G be a finitely generated torsion free nilpotent group, and let $(P(G), d), (M(G), D)$ be the filtered differential graded algebras as defined in Part 1. Assume that $M(G)$ is an associative differential graded algebra. Then the two spectral sequences associated with the filtrations of the Hochschild complexes of $(P(G), d)$ and $(M(G), D)$ are isomorphic on the E^1-levels.*

COROLLARY 2.5.

$$HH_*(P(G), d) \simeq HH_*(M(G), D),$$
$$HC^-_*(P(G), d) \simeq HC^-_*(M(G), D)$$

are isomorphisms of graded $\mathbb{Z}$-modules.

REMARK. As it was pointed out earlier, $M(G)$ is associative for the Heisenberg group. That example will be studied in details in Part 4.

From now on it is assumed that $M(G)$ is associative, almost commutative differential graded algebra (i.e. its associated graded ring is commutative). We will compute the Hochschild and cyclic homologies of $M(G)$. These homologies were computed earlier for $(P(G), d)$. The method of computation is very close to the method of Kassel. The computation for the enveloping algebras can be found in [Ka], and for the quotients of enveloping algebras in [K-V]. The structure of the Lie algebra on the graded algebra associated to an almost commutative algebra, as defined at the end of Part 1, plays an important role in the computation of the differential d^1 on $\tilde{E}^1_{p,q} = HH_{p+q}(\mathrm{gr} A)$, [Br], [Ka], [K-V].

3 Hochschild homology of $(M(G), D)$

In this paragraph we construct a "small" filtered complex whose homology is $HH_*(M(G), D)$. Recall that $M(G)_* = \bigoplus_{n \geq 0} M(G)_{-n}$, where $M(G)_{-n} = M(G)^n$. Since $M(G)$ is flat over $\mathbb{Z}$,

$$HH_*(M(G)_*, D) = \mathrm{Tor}_*^{M(G)_* \otimes M(G)_*^{op}}(M(G)_*, M(G)_*).$$

From [A-H] it follows that

$$\mathrm{Tor}_*^{M(G)_* \otimes M(G)_*^{op}}(M(G)_*, M(G)_*) = H_*(M(G)_* \otimes_{M(G)_* \otimes M(G)_*^{op}} P'),$$

where $P' \to (M(G)_*, D)$ is a semifree resolution of the $(M(G)_* \otimes M(G)_*^{op})$-differential module $(M(G)_*, D)$.

First, we consider the problem for an abelian group H on k generators. In this case $M(H) = \wedge(e_1, \dots, e_k)$, and $M(H)_*$ is the exterior algebra of the free $\mathbb{Z}$-free module $\bigoplus_{i=1}^{k} \mathbb{Z} e_i$, where $|e_i| = -1$. The differential D on $M(H)$ is zero.

Next, we consider the polynomial algebra $\mathbb{Z}[\bar{e}_1, \dots, \bar{e}_k]$ on k generators $\bar{e}_i$ with $|\bar{e}_i| = 0$. Let $P'_0 = M(H)_* \otimes \mathbb{Z}[\bar{e}_1, \dots, \bar{e}_k] \otimes M(H)_*$ be the tensor product, of graded $\mathbb{Z}$-algebras. Let $b'_0 : P'_0 \to P'_0$,

$$b'_0(a \otimes \bar{x}_1 \cdots \bar{x}_n \otimes a') = (-1)^{|a|}(ax_i \otimes \bar{x}_1 \cdots \overset{\wedge}{\bar{x}_i} \cdots \bar{x}_n \otimes a' - a \otimes \bar{x}_1 \cdots \overset{\wedge}{\bar{x}_i} \cdots \bar{x}_n \otimes x_i a'),$$

where $a, a' \in M(H), \bar{x}_i \in \{\bar{e}_1, \dots, \bar{e}_k\}$, and x_i is the corresponding element of $\{e_1, \dots, e_k\}$. ($\overset{\wedge}{\bar{x}_i}$ means that the variable $\bar{x}_i$ is omitted.)

PROPOSITION 3.1. *Let* $\phi'_0 : (P'_0, b'_0) \to M(H)_*$ *be the algebra map defined by* $\phi'_0(a \otimes a') = aa'$ *for* $a, a' \in M(H), \phi'_0(a \otimes \xi \otimes a') = 0$ *for* $\xi \in \mathbb{Z}^+[\bar{e}_i] = \mathbb{Z}[\bar{e}_i]/\mathbb{Z}$. *Then*

(1) P'_0 *is a* $(M(H) \otimes M(H))$-*free module,*
(2) $\phi'_0 b'_0 = 0$ *and* $(\phi'_0)_* : H_*(P'_0, b'_0) \to M(H)$ *is an isomorphism.*

Proof. The statement is proved by induction on the number of generators of H. If $H = \mathbb{Z}e$, we have $\phi_0' : P_0' = (\wedge e \otimes \mathbb{Z}[\bar{e}] \otimes \wedge e, b_0') \to \wedge e$. Let

$$\psi : \wedge e \otimes \mathbb{Z}[\bar{e}] \otimes \wedge e \to \wedge e \otimes \mathbb{Z}[\bar{e}] \otimes \wedge e,$$

be defined by setting:

$$\psi(e \otimes 1 \otimes 1) = e \otimes 1 \otimes 1,$$

$$\psi(1 \otimes \bar{e} \otimes 1) = 1 \otimes \bar{e} \otimes 1,$$

$$\psi(1 \otimes 1 \otimes e) = -1 \otimes 1 \otimes e + e \otimes 1 \otimes 1.$$

ψ can be extended to an automorphism of graded algebra (in fact $\psi^{-1} = \psi$). The differential d on $\wedge e \otimes \mathbb{Z}[\bar{e}] \otimes \wedge e$ is defined by the formula $d = \psi b_0' \psi^{-1}$. It follows that

$$d(e \otimes 1 \otimes 1) = 0, \quad d(1 \otimes 1 \otimes e) = 0, \quad d(1 \otimes \bar{e} \otimes 1) = 1 \otimes 1 \otimes e.$$

Thus $H_*(\wedge e \otimes \mathbb{Z}[\bar{e}] \otimes \wedge e, d) = \wedge e \otimes H_*(\wedge e \otimes \mathbb{Z}[\bar{e}], d)$. If we denote $K_* = \wedge e \otimes \mathbb{Z}[\bar{e}]$, then $K_0 = \mathbb{Z}[\bar{e}], K_{-1} = e\mathbb{Z}[\bar{e}], K_p = 0$ for $p \neq 0, -1; d(\bar{e}^n) = ne\bar{e}^{n-1}, d(e\bar{e}^n) = 0, n \in \mathbb{N}$. This implies that $H_0(\wedge e \otimes \mathbb{Z}[\bar{e}], d) = \mathbb{Z}, H_n(\wedge e \otimes \mathbb{Z}[\bar{e}], d) = 0, n \neq 0$. Thus, there is an isomorphism $\psi_* : H_*(P_0', b_0') \to H_*(\wedge e \otimes \mathbb{Z}[\bar{e}] \otimes \wedge e, d) = \wedge e$, and $\phi_{0*}' \psi_*^{-1} = \mathrm{id}$. $\square$

Now consider $P_0 = M(H)_* \otimes_{M(H)_* \otimes M(H)_*^{op}}, P_0' \simeq M(H)_* \otimes \mathbb{Z}[\bar{e}_1, \dots, \bar{e}_k]$. The induced differential on P_0 is zero. And from the general theory it follows that $HH_*(M(H)_*) = M(H)_* \otimes \mathbb{Z}[\bar{e}_1, \dots, \bar{e}_k]$. Recall that with a commutative differential graded algebra A over a unitary ring there is associated an algebra of commutative differential forms Ω_A^*, [K-V]. In particular, if $A = \wedge(e_1, \dots, e_k)$ with $|e_i| = -1$ for $1 \leq i \leq k$, then $\Omega_A^* = \wedge(e_1, \dots, e_k) \otimes \mathbb{Z}[\bar{e}_1, \dots \bar{e}_k]$ with $|\bar{e}_i| = 0$ for $1 \leq i \leq k$. Let d be the derivation on Ω_A^* defined as an extension of the map of degree $+1$ given by $\quad de_i = \bar{e}_i, \quad d\bar{e}_i = 0 \quad$ for all i. Then we get

COROLLARY 3.2. $HH_*(M(H)_*) = \Omega_{M(H)_*}^* = \Omega_{\wedge(e_1,\dots,e_k)}^*$, *where* $\wedge(e_1, \dots, e_k)$ *is the exterior algebra of the free $\mathbb{Z}$-module* $\overset{k}{\underset{i=1}{\oplus}} \mathbb{Z}e_i, |e_i| = -1$.

When G is not abelian, then the differential D on $M(G)$ is not zero, and the situation is more complicated. In that case the computation will be over a field of characteristic zero. We will use the techniques of rational homotopy theory as well as an approach of Koszul. In fact from now on, until the end of this paper we work with $M(G) \otimes_{\mathbb{Z}} \mathbb{Q}$, but we shall use the notation $M(G)$ for simplification. In fact all the tensor products are over $\mathbb{Q}$. Furthermore, we assume that $M(G)$ is associative.

From Proposition 1.4 it follows that $(M(G), D)$ is a filtered differential graded algebra with $D = D^1 + D'$, where $D' = D - D^1$ maps $F^i M(G)$ into $F^{i-2} M(G)$. Considering the spectral sequence associated with the filtration on $(M(G)_*, D)$, we obtain: $(E^0 \simeq M(H), D^0 = 0), \quad E^1 \simeq M(H) = \wedge(e_1, \dots, e_k)$ and we see that the differential induced by D on E^1 is just the image of D^1.

Recall a well known result:

PROPOSITION 3.3. [G-H-V]. *Let $(\wedge Z, \partial)$ be a free commutative differential graded algebra over $\mathbb{Q}$, with ∂ of degree -1, such that, either $Z = \bigoplus\limits_{n \geq 0} Z_n$, or $Z = \bigoplus\limits_{n < 0} Z_n$. Then on the commutative graded algebra $\wedge Z \otimes \wedge \overline{Z} \otimes \wedge Z$ (where $\overline{Z} = \bigoplus\limits_{n \geq 0} \overline{Z}_n, \overline{Z}_n = Z_{n-1}$), there exists a differential Δ_0' and there is a morphism $\phi_0' : \wedge Z \otimes \wedge \overline{Z} \otimes \wedge Z \to \wedge Z$ such that*

(1) $\phi_0'(a \otimes a') = aa', \quad \phi_0'(a \otimes \xi \otimes a') = 0; \quad a, a' \in \wedge Z, \xi \in \wedge^+ \overline{Z},$

(2) ϕ_0' *commutes with the differentials and ϕ_0' induces an isomorphism in homology,*

(3) $\Delta_0'(z \otimes 1 \otimes 1) = \partial z, \quad \Delta_0'(1 \otimes 1 \otimes z) = 1 \otimes 1 \otimes \partial z,$
$\Delta_0'(1 \otimes \overline{z} \otimes 1) = z \otimes 1 \otimes 1 - 1 \otimes 1 \otimes z + \Delta_0''(1 \otimes \overline{z} \otimes 1), \quad$ *where*
$\Delta_0''(1 \otimes \overline{z} \otimes 1) \in \wedge(Z_{<n}) \otimes \overline{Z}_{\leq n} \otimes \wedge(Z_{<n}) \quad$ *if* $\quad z \in Z_n,$
$\Delta_0''(1 \otimes \overline{z} \otimes 1) \in \wedge^+(Z_{<n} \oplus Z_{<n}) \otimes \overline{Z}_{\leq n} \quad$ *if* $\quad \partial z \in \wedge^+ Z \cdot \wedge^+ Z$
and $\quad Z_{<n} = \bigoplus\limits_{|p| < |n|} Z_p.$

Denote by τ_p the quotient map

$$F^p M(G) \to F^p M(G) / F^{p-1} M(G) \simeq \wedge(e_i)_p,$$

where $\wedge(e_i)_p$ is the sub-vector space generated by elements of filtering degree p. Denote by η_p the map from $\wedge(e_i)_p$ to $F^p M(G)$, defined by

$$\eta_p(e_k^{u_k} e_{k-1}^{u_{k-1}} \cdots e_1^{u_1}) = e_k^{u_k} \bullet e_{k-1}^{u_{k-1}} \bullet \cdots \bullet e_1^{u_1},$$

where $u_i = 0$ or 1. The product on the left is the product of the exterior algebra, and the product $\bullet$ on $M(G)$ is the product defined in Part 1. We have $\tau_p \circ \eta_p = \mathrm{Id}_{\wedge(e_i)_p}, \eta_p \circ \tau_p(x) = x \bmod F^{p-1} M(G)$, for $x \in F^p M(G)$. Now consider the tensor product of graded algebras

$$P' = M(G)_* \otimes \mathbb{Q}[\overline{e}_1, \ldots, \overline{e}_k] \otimes M(G)_*.$$

We define on $\mathbb{Q}[\overline{e}_i]$ a filtration as of a graded algebra by setting $\|\overline{e}_i\| = \|e_i\|$. Thus P', endowed with the product of the three filtrations, is a filtered graded algebra. We extend η to a map

$$\eta' : \wedge(e_i) \otimes \mathbb{Q}[\overline{e}_i] \otimes \wedge(e_i) \to M(G) \otimes \mathbb{Q}[\overline{e}_i] \otimes M(G),$$

$$\eta'(a \otimes \xi \otimes a') = \eta(a) \otimes \xi \otimes \eta(a').$$

We also define a map β of degree $+1$:

$$\beta : M(G)_* \to M(G)_* \otimes \mathbb{Q}[\overline{e}_1, \ldots, \overline{e}_k] \otimes M(G)_*,$$

$$\beta(e_{i_1} \bullet \cdots \bullet e_{i_p}) = \sum_{h=1}^{p} (-1)^{h-1} e_{i_1} \bullet \cdots \bullet e_{i_{h-1}} \otimes \overline{e}_{i_h} \otimes e_{i_{h+1}} \bullet \cdots \bullet e_{i_p}$$

for $e_{i_j} \in \{e_1, \ldots, e_k\}, \quad i_1 \geq i_2 \geq \cdots \geq i_p$. Recall that such elements form a basis of $M(G)$ over $\mathbb{Q}$. The commutator of two elements $a, a' \in M(G)$ is:

$$[a, a'] = a \bullet a' - (-1)^{|a||a'|} a' \bullet a.$$

If $a \in F^p M(G), \quad b \in F^q M(G), \quad [a, a'] \in F^{p+q-1} M(G)$. We will prove the following

THEOREM 3.4. *Let $(M(G), D)$ be the filtered differential algebra associated with G, and $P' = M(G)_* \otimes \mathbb{Q}[\bar{e}_i] \otimes M(G)_*$ the filtered graded algebra as above. Then*

(1) *on the differential $(M(G) \otimes M(G)^{op})$-module P' there exists a differential Δ', such that*

$$\Delta'(1 \otimes \bar{e}_1^{\alpha_1} \bar{e}_2^{\alpha_2} \cdots \bar{e}_{i_1}^{\alpha_{i_1}} \otimes 1) = b'(1 \otimes \bar{e}_1^{\alpha_1} \cdots \bar{e}_{i_1}^{\alpha_{i_1}} \otimes 1)$$

if $\quad \|e_1\| = \cdots = \|e_{i_1}\| = 0 \quad$ *and* $\quad (\alpha_1, \dots, \alpha_{i_1}) \in \mathbb{N}^{i_1}$,

$$\Delta'(1 \otimes \bar{x}_1 \cdots \bar{x}_n \otimes 1) = (b' + b_1' + \Delta'')(1 \otimes \bar{x}_1 \cdots \bar{x}_n \otimes 1) + \phi'(\bar{x}_i)$$

for $\quad \bar{x}_i \in \{\bar{e}_1, \dots, \bar{e}_k\}$, *where*

(a) $b'(1 \otimes \bar{x}_1 \cdots \bar{x}_n \otimes 1) = \sum_{i=1}^{n} (x_i \otimes \bar{x}_1 \cdots \widehat{\bar{x}}_i \cdots \bar{x}_n \otimes 1 - 1 \otimes \bar{x}_1 \cdots \widehat{\bar{x}}_i \cdots \bar{x}_n \otimes x_i)$,

(b) $b_1'(1 \otimes \bar{x}_1 \cdots \bar{x}_n \otimes 1) = \sum_{1 \le i \le j \le n} (1 \otimes \bar{x}_1 \cdots \widehat{\bar{x}}_i \cdots \widehat{\bar{x}}_j \cdots \bar{x}_n \otimes 1) \beta([x_i, x_j])$,

(c) $\Delta''(1 \otimes \bar{x}_1 \cdots \bar{x}_n \otimes 1) = \eta' \Delta_0''(1 \otimes \bar{x}_1 \cdots \bar{x}_n \otimes 1)$, Δ_0'' *is the differential of Proposition 3.3, (3) applied to $(\wedge(e_i), D^1)$,*

(d) $\phi'(\bar{x}_i) \in F^{p-2}(M(G) \otimes \mathbb{Q}^+[\bar{e}_i] \otimes M(G))$ *if* $\quad \sum \|x_i\| = p$, *where* $\mathbb{Q}^+[e_i] = \mathbb{Q}[\bar{e}_i]/\mathbb{Q}$,

(2) *the map $\Phi' : (P' = M(G)_* \otimes \mathbb{Q}[\bar{e}_i] \otimes M(G)_*, \Delta') \to (M(G)_*, D)$ defined by $\Phi'(a \otimes a') = aa'$, $\quad \Phi'(a \otimes \xi \otimes a') = 0 \quad$ for $a, a' \in M(G)$, $\quad \xi \in \mathbb{Q}^+[\bar{e}_i]$ induces an isomorphism on homology.*

Before we prove the Theorem we observe that:

COROLLARY 3.5. *Let $(M(G)_*, D)$ be the filtered differential graded algebra associated with G. Let $P = M(G)_* \otimes \mathbb{Q}[\bar{e}_1, \dots, \bar{e}_k]$ be endowed with the product of the filtrations. Then there exists a differential Δ on P, such that*

$$HH_*(M(G)_*, D) = H_*(M(G)_* \otimes \mathbb{Q}[\bar{e}_i], \Delta),$$

$$\Delta(a \otimes \bar{x}_1 \cdots \bar{x}_n) = Da \otimes \bar{x}_1 \cdots \bar{x}_n + (-1)^{|a|} \{ \sum_{i=1}^{n} [a, x_i] \otimes \bar{x}_1 \cdots \widehat{\bar{x}}_i \cdots \bar{x}_n +$$

$$b_1(a \otimes \bar{x}_1 \cdots \bar{x}_n) + \delta''(a \otimes \bar{x}_1 \cdots \bar{x}_n) + \phi(a, \bar{x}_i) \},$$

for $\bar{x}_i \in \{\bar{e}_1, \dots, \bar{e}_k\}, a \in M(G)$, where

(a) $\phi(a, \bar{x}_i) \in F^{p-2}(M(G) \otimes \mathbb{Q}^+[\bar{e}_i])$ *with* $\quad p = l + \sum_{1 \le i \le n} \|x_i\|$ *and* $a \in F^l M(G), \quad a \notin F^{l-1} M(G)$,

(b) b_1 (*resp* δ'') *is the image of* $\quad 1 \otimes_{M(G) \otimes M(G)^{op}} b_1'$ (*resp* $1 \otimes_{M(G) \otimes M(G)^{op}} \Delta''$) *by the isomorphism*

$$\theta : M(G) \otimes_{M(G) \otimes M(G)^{op}} (M(G) \otimes \mathbb{Q}[\bar{e}_i] \otimes M(G)) \to M(G) \otimes \mathbb{Q}[\bar{e}_i],$$

$$\theta(\alpha \otimes a \otimes w \otimes a') = (-1)^{|a'|[\|a\| + |w| + |\alpha|]} a' \bullet \alpha \bullet a \otimes w,$$

for $\quad \alpha, a, a' \in M(G), \quad w \in \mathbb{Q}[\bar{e}_i].$

Proof of Theorem 3.4. $M(G) \otimes \mathbb{Q}[\bar{e}_i] \otimes M(G)$ has a natural structure of a left $M(G) \otimes M(G)^{op}$- module (see [Vi]). Thus we get on $M(G) \otimes \mathbb{Q}[\bar{e}_i] \otimes M(G)$ a structure of a differential $M(G) \otimes M(G)^{op}$- module if we define $\Delta'(1 \otimes \xi \otimes 1)$ for $\xi \in \mathbb{Q}[\bar{e}_i]$ such that $\Delta'^2 = 0$. Then we extend Δ' to P', as a derivation, by setting

$$\Delta'(a \otimes \xi \otimes a') = (-1)^{|a'|\cdot|\xi|}[D(a \otimes a'^{op})(1 \otimes \xi \otimes 1)$$
$$+ (-1)^{|a|\cdot|a'|}(a \otimes a'^{op})\Delta'(1 \otimes \xi \otimes 1)].$$

Observe, that Corollary 1.5 implies that $e_1 \cdot e_1 = 0$ and $De_1 = 0$, so that $b'^2(1 \otimes \bar{e}_1^\alpha \otimes 1) = 0$. Then we set $\Delta'(1 \otimes \bar{e}_1^\alpha \otimes 1) = b'(1 \otimes \bar{e}_1^\alpha \otimes 1)$. The proof of the theorem goes by induction on the filtering degree of the elements of $\mathbb{Q}[\bar{e}_i]$.

Let $e_1, \ldots, e_{i_1}$ be such that $\|e_j\| = 0$ for all $j \in \{1, \ldots, i_1\}$. For all $j, k \in \{1, \ldots, i_1\}$ $e_j \cdot e_j = 0, De_j = 0, e_j \cdot e_k + e_k \cdot e_j = 0$. This implies that $b'^2(1 \otimes \bar{e}_1^{\alpha_1} \cdots \bar{e}_{i_1}^{\alpha_{i_1}} \otimes 1) = 0$, for $(\alpha_1, \ldots, \alpha_{i_1}) \in \mathbb{N}^{i_1}$. We set

$$\Delta'(1 \otimes \bar{e}_1^{\alpha_1} \cdots \bar{e}_{i_1}^{\alpha_1} \otimes 1) = b'(1 \otimes \bar{e}_1^{\alpha_1} \cdots \bar{e}_{i_1}^{\alpha_{i_1}} \otimes 1).$$

For $\beta \in \mathbb{N}$, $p \in \mathbb{N}$, let $H_{p,\beta}$ stand for the following conditions (i) and (ii):

(i) for every k-tuple $(\alpha_1, \ldots, \alpha_k) \in \mathbb{N}^k$, such that $\alpha_1 + \cdots + \alpha_{i_1} \leq \beta$ and $\sum_{i=1}^k \alpha_i \|e_i\| \leq p$, we can define a differential Δ' of $M(G) \otimes M(G)^{op}$- differential module, such that

$$\Delta'(1 \otimes \bar{e}_1^{\alpha_1} \cdots \bar{e}_k^{\alpha_k} \otimes 1) = (b' + b'_1 + \Delta'')(1 \otimes \bar{e}_1^{\alpha_1} \cdots \bar{e}_k^{\alpha_k} \otimes 1) + \phi'(\bar{e}_i)$$

with

$$\phi' \in F^{(\sum \alpha_i \|e_i\|)-2}(M(G) \otimes \mathbb{Q}^+[\bar{e}_1, \ldots, \bar{e}_k] \otimes M(G)).$$

(ii) If $x \in F^h(M(G) \otimes \mathbb{Q}^{\geq n}[\bar{e}_i] \otimes M(G))$, where $h \leq p$ and $n \in \mathbb{N}$, with $\Delta'x = 0$ and $\Phi'_*[x] = 0$, then there exists $y \in F^h(M(G) \otimes \mathbb{Q}^{\geq n+1}[\bar{e}_i] \otimes M(G))$ such that $x = \Delta'y$, $\mathbb{Q}^{\geq n}[\bar{e}_i]$ is the vector space spanned by monomials in $\bar{e}_i$ of degree $\geq n$.

For $p = 0$ and any $\beta \in \mathbb{N}$, $\Delta' = b'$ satisfies (i) and (ii), according to Proposition 3.1.

Now we assume that $H_{p,\beta}$ is satisfied for fixed $p \geq 0$ and for any $\beta \geq 0$. We will show that $H_{p+1,\beta}$ is also satisfied.

First step:

We will prove $H_{p+1,0}(i)$. Let $(\alpha_{i_1+1}, \ldots, \alpha_k)$ be a sequence of positive integers such that $\sum_{i \geq i_1+1} \alpha_i \|e_i\| = p+1$. We consider $u = 1 \otimes \bar{e}_{i_1}^{\alpha_{i_1+1}} \cdots \bar{e}_k^{\alpha_k} \otimes 1$ and $z = (b' + b'_1 + \Delta'')(u)$. We have $b'(u) = \sum_i \alpha_i(e_i \otimes 1 \otimes 1 - 1 \otimes 1 \otimes e_i)(1 \otimes \bar{e}_{i_1}^{\alpha_{i_1+1}} \cdots \bar{e}_i^{\alpha_i-1} \cdots \bar{e}_k \otimes 1)$,

$$\|e_{i_1+1}^{\alpha_{i_1+1}} \cdots \bar{e}_i^{\alpha_i-1} \cdots \bar{e}_k^{\alpha_k}\| = p+1 - \|e_i\| \leq p.$$

It can be easily checked that $b'_1(u) \in F^p(M(G) \otimes \mathbb{Q}[\bar{e}_i] \otimes M(G))$. Using the fact $D^1(F^p M(G)) \subset F^{p-1}M(G)$, according to Corollary 1.5, and the exact formula for Δ''_0 ([Ha, §5]), we see that $\Delta''(u) \in F^p(M(G) \otimes \mathbb{Q}[\bar{e}_i] \otimes M(G))$. This implies that $Z = \Delta'z$ is well defined. We can write

$$Z = (\Delta' - b' - b'_1 - \Delta'')(b'(u)) + (\Delta' - b' - b'_1 - \Delta'')(b'_1 + \Delta'')(u)$$

$$+(b'^2 + b'_1 b' + b'b'_1 + b'^2_1 + \Delta''b' + b'\Delta'' + \Delta''^2)(u)$$

$$+(b'_1\Delta'' + \Delta''b'_1)(u).$$

Claim: If $\sum \alpha_i \geq 2$, then $Z \in F^{p-1}(M(G) \otimes \mathbb{Q}^+[\bar{e}_1, \dots, \bar{e}_k] \otimes M(G)) = F^{p-1}$. This claim will be proved by showing that each of the four terms in Z belongs to F^{p-1}. By the induction hypothesis $(\Delta' - b' - b'_1 - \Delta'')(b'(u)) \in F^{p-1}$, and $(\Delta' - b' - b'_1 - \Delta'')(b'_1 + \Delta'')(u) \in F^{p-2}$. Since b'_1 and Δ'' strictly decrease the filtration, it is easy to show that $(b'_1\Delta'' + \Delta''b'_1)(u) \in F^{p-1}$. A direct calculation shows that $(b'^2 + b'_1b' + b'b'_1)(u) = 0$. Since b'_1 strictly decreases the filtration, we have $b'^2_1(u) = 0$ if $\sum \alpha_i = 2$, and $b'^2_1(u) \in F^{p-1}$ if $\sum \alpha_i > 2$. It remains to study $(\Delta''b' + b'\Delta'' + \Delta''^2)(u)$. If Proposition 3.3 is applied to $(\wedge(e_1, \dots, e_k), D^1)$, then we get $(\Delta''^2_0 + b'_0\Delta''_0 + \Delta''_0 b'_0)(u) = 0$. This implies that $(b'\Delta'' + \Delta''b' + \Delta''^2)(u)$ belongs to F^{p-1}. Finally, we have proved that $Z = \Delta'z \in F^{p-1}(M(G) \otimes \mathbb{Q}^+[\bar{e}_i] \otimes M(G))$, and $\Delta'Z = 0$. Thus the induction hypothesis $H_{p-1,\beta}(\text{ii})$ implies that there exists $Y \in F^{p-1}(M(G) \otimes \mathbb{Q}^+[\bar{e}_i] \otimes M(G))$ such that $\Delta'z = \Delta'Y$. Now we define $\Delta'(u) = (b' + b'_1 + \Delta'')(u) - Y$ and $H_{p+1,0}(\text{i})$ is proved.

If $u = 1 \otimes \bar{e}_i \otimes 1$, we have $z = b'(u) + \Delta''(u)$, $Z = \Delta'z = b'^2(u) + (b'\Delta'' + \Delta''^2)(u) + \phi'(\Delta''(u))$, with $\phi'(\Delta''(u)) \in F^{p-2}(M(G) \otimes \mathbb{Q}^+[\bar{e}_i] \otimes M(G))$. Thus $Z \in F^{p-1}(M(G) \otimes \mathbb{Q}[\bar{e}_i] \otimes M(G))$, $\Delta'(Z) = 0$, $\Phi'(Z) = 0$. The conclusion follows from $H_{p-1,\beta}(\text{ii})$.

Second step:
Assume that $H_{p+1,\beta}(\text{i})$ is satisfied for all $\beta' \leq \beta$, where β is a fixed number. Let $u = 1 \otimes \bar{e}_1^{\alpha_1} \cdots \bar{e}_k^{\alpha_k} \otimes 1$ with $\sum \alpha_i = \beta + 1$, $\|u\| = p + 1$, and let $z = (b' + b'_1 + \Delta'')(u)$. Then $z = \sum_{i \leq i_1} \alpha_i(e_i \otimes 1 \otimes 1 - 1 \otimes 1 \otimes e_i)(1 \otimes \bar{e}_1^{\alpha_1} \cdots \bar{e}_i^{\alpha_i-1} \cdots \bar{e}_k^{\alpha_k} \otimes 1) + z'$, where $z' \in F^p(M(G) \otimes \mathbb{Q}[\bar{e}_i] \otimes M(G))$. Since $\Delta'(1 \otimes \bar{e}_1^{\alpha_1} \cdots \bar{e}_i^{\alpha_i-1} \cdots \bar{e}_k^{\alpha_k} \otimes 1)$ was defined in the first step, $\Delta'z$ is well defined. We proceed as in the first step to show that $H_{p+1,\beta+1}(\text{i})$ is satisfied.

Third step:
We show that $H_{p+1,\beta}(\text{ii})$ is satisfied for every β. Let $x \in F^{p+1}(M(G) \otimes \mathbb{Q}^{\geq n}[\bar{e}_i] \otimes M(G))$, $\Delta'(x) = 0$, $\Phi'_*[x] = 0$. Denote by $x_0 = \tau(x)$ the image of x in $F^{p+1}/F^p \simeq \wedge(e_i) \otimes \mathbb{Q}[\bar{e}_i] \otimes \wedge(e_i)$. Assume that $\tau(x) \neq 0$. Then $b'_0(x_0) = 0$ and $\Phi_*[x_0] = 0$. Since $(\wedge(e_i) \otimes \mathbb{Q}[\bar{e}_i] \otimes \wedge(e_i), b'_0)$ is a resolution of $\wedge(e_i)$ by $\wedge(e_i) \otimes \wedge(e_i)$-free modules, there exists $y_0 \in \wedge(e_i) \otimes \mathbb{Q}[\bar{e}_i] \otimes \wedge(e_i)$ such that $x_0 = b'_0(y_0)$. Furthermore if x_0 contains monomial in $\bar{e}_i$ of length $\geq n$, then y_0 contains monomials of length $\geq n+1$. Let $y = \eta'(y_0)$, then $y \in F^{p+1}(M(G) \otimes \mathbb{Q}^{\geq n+1}[\bar{e}_i] \otimes M(G))$, and $\tau(y) = y_0$. Now let $x' = x - \Delta'y$. Then $x' \equiv x_0 - b'_0(y_0) \equiv 0 \pmod{F^p}$, $\Delta'x' = 0$, $\Phi_*[x'] = \Phi_*[x] = 0$, so that

according to $H_{p,\beta}(ii)$, there exists $y' \in F^p(M(G) \otimes \mathbb{Q}^{\geq n+1}[\bar{e}_i] \otimes M(G))$ so that $\Delta' y' = x'$. Hence $x = \Delta'(y + y')$ and $H_{p+1,\beta}(ii)$ is proved.

This completes the proof of Theorem 3.4. $\qquad\square$

Now let $(P = M(G)_* \otimes \mathbb{Q}[\bar{e}_i], \Delta)$ be the complex (given in Corollary 3.5) which computes the Hochschild homology of $(M(G), D)$. We give on P a filtration induced by the filtrations on $M(G)_*$ and $\mathbb{Q}[\bar{e}_i]$. From the end of Part 1 it follows that $\mathrm{gr}_* M(G)_* = \wedge(e_i, \dots e_k)$ has a Poisson bracket $\{,\}$. Furthermore, Brylinski defined a differential δ' on $\Omega^*_{\wedge(e_i)} = \wedge(e_i) \otimes \mathbb{Q}[\bar{e}_i]$,

$$\delta'(a\bar{e}_{i_1}\cdots\bar{e}_{i_p}) = (-1)^{|a|}[\sum_{1\leq l\leq p}\{a, e_{i_k}\}\bar{e}_{i_1}\cdots\widehat{\bar{e}_{i_l}}\cdots\bar{e}_{i_p}+$$

$$\sum_{1\leq i_j\leq i_l\leq p} a(d\{e_{i_j}, e_{i_l}\})\bar{e}_{i_1}\cdots\widehat{\bar{e}_{i_j}}\cdots\widehat{\bar{e}_{i_l}}\cdots\bar{e}_{i_p}],$$

where $d : \wedge(e_1,\dots,e_k) \to \wedge(e_1,\dots,e_k)\otimes\mathbb{Q}[e_1,\dots,\bar{e}_k]$ is defined by

$$d(e_{i_1}\cdots e_{i_p}) = \sum_{1\leq l\leq p}(-1)^{l-1}e_{i_1}\cdots\widehat{e_{i_l}}\cdots e_{i_p}\otimes\bar{e}_{i_l} \qquad \text{for } 1\leq l\leq k.$$

Furthermore, consider the spectral sequence associated with the filtration of $(M(G), D)$. It follows that $E^0 = E^1 = \wedge(e_1,\cdots e_k)$. Denote by D^1 the differential on E^1, induced by D. Then (E^1, D^1) is a commutative differential graded algebra whose Hochschild homology is the homology of the complex $(\Omega^*_{\wedge(e_i)}, \delta)$, where $\delta_{|\wedge(e_i)} = D^1$ and $\delta d + d\delta = 0$, [B-V], [K-V].

THEOREM 3.6. *Let $(P, \Delta) = (M(G)_* \otimes \mathbb{Q}[\bar{e}_i], \Delta)$ be the model for the Hochschild homology of $(M(G)_*, D)$ given in Corollary 3.5. The filtration of this complex determines a fourth quadrant convergent spectral sequence (E^r, d^r) such that*

(1) $E^0 = \wedge(e_1,\dots,e_k)\otimes\mathbb{Q}[\bar{e}_1,\dots,\bar{e}_k] \simeq \Omega^*_{\wedge(e_i)}$, *where $\Omega^*_{\wedge(e_i)}$ is the algebra of differential forms on the commutative graded algebra $\wedge(e_1,\dots,e_k)$, and $d^0 = 0$. The differential on the forms in $\Omega^*_{\wedge(e_i)}$ is given by $de_i = \bar{e}_i$, for all i.*

(2) $E^1 \simeq \Omega^*_{\wedge(e_i)}$ *and the differential d^1 is the sum of two differentials: δ' and δ, where δ' is the Brylinski differential on the Poisson algebra $(\wedge(e_1,\dots,e_k), \{\})$ and δ is the internal differential on the algebra of differential forms of the commutative differential algebra $(E^1 M(G) = \wedge(e_1,\dots,e_k), D^1)$.*

Proof. We have $P = M(G)_* \otimes \mathbb{Q}[\bar{e}_i]$ and Δ as in Corollary 3.5. Thus it follows that $E^0 = \wedge(e_1,\dots,e_k)\otimes\mathbb{Q}[\bar{e}_1,\dots,\bar{e}_k]$, $d^0 = 0$, and that

$$d^1(a\otimes\bar{x}_1\cdots\bar{x}_n) = D^1 a\otimes\bar{x}_1\cdots\bar{x}_n$$

$$+ (-1)^{|a|}\sum_{i=1}^n\{a, x_i\}\otimes\bar{x}_1\cdots\widehat{\bar{x}_i}\cdots\bar{x}_n + \psi(a, \bar{x}_i),$$

where $\{\bar{x}_1,\ldots,\bar{x}_n\} \in \{\bar{e}_1,\ldots,\bar{e}_k\}$, and $a \in \wedge(e_1,\ldots,e_k)$ is the image in the homogeneous part of the filtering degree l of an element $\alpha \in F^l M(G)$ and $\psi(a,\bar{x}_i)$ is the image of $(b_1 + \delta'')(\alpha \otimes \bar{x}_1 \otimes \cdots \otimes \bar{x}_n)$ in $F^{p-1}P/F^{p-2}P$, with $p = l + \sum_{i=1}^{n}\|x_i\|$. It is easy to verify that b_1 is just the map that send $a\bar{x}_1\cdots\bar{x}_n$ to $\sum_{1\le j\le l\le n} ad\{x_j,x_l\}\bar{x}_1\cdots\widehat{\bar{x}}_j\cdots\widehat{\bar{x}}_l\cdots\bar{x}_n$. By the definition of Δ'', the map that sends $a\bar{x}_1\cdots\bar{x}_n$ to $D^1 a \otimes \bar{x}_1\cdots\bar{x}_n + \tau'_{p-1}\delta''(\alpha \otimes \bar{x}_1\cdots\bar{x}_n)$ is just δ; τ'_{p-1} is the quotient map $F^{p-1}(M(G) \otimes \mathbb{Q}[\bar{e}_i]) \to F^{p-1}/F^{p-2}$. $\square$

Now we come back to the filtration of the Hochschild complex of $(M(G)_*,D)$, as defined in Part 2. We have just seen that such a filtration determines a spectral sequence which, on the E^1-level, is isomorphic to the spectral sequence constructed from the filtration of the Hochschild complex of $(P(G)_*,d)$. The same argument as in the proof of Theorem 2.6 of [K-V] is used to get the following statement:

THEOREM 3.7. *There is an isomorphism between the E^1-terms of the spectral sequence constructed on the Hochschild complexes of $(M(G)_*,D)$ and $(P(G)_*,d)$, filtered by the filtering degrees of $M(G)$ and $P(G)$, and the spectral sequence on $(M(G)_* \otimes \mathbb{Q}[\bar{e}_i],\Delta)$ given in Theorem 3.6.*

4 Application

In this section we give an explicit computation of the Hochschild homology of the cochain algebra $(M(G),D)$, when G is the group of matrices

$$G = G_3 = \begin{pmatrix} 1 & a_1 & a_3/k \\ 0 & 1 & a_2 \\ 0 & 0 & 1 \end{pmatrix},$$

where k is a fixed positive integer, and $(a_1,a_2,a_3) \in \mathbb{Z}^3$. Recall, from [C-P 2], that as a $\mathbb{Q}$-module, $M(G) = H^*(S^1,\mathbb{Q})^{\otimes 3}$, and $M(G)$ is generated as a $\mathbb{Q}$-algebra, by three generators e_1,e_2,e_3 with $|e_i| = 1$, $1 \le i \le 3$, $\|e_1\| = 0, \|e_2\| = 1, \|e_3\| = 2$. The algebra structure on $M(G)$ is given by the formulas:

$$e_2 \bullet e_1 = -e_1 \bullet e_2 = e_2 \otimes e_1,$$

$$e_1 \bullet e_3 = -e_3 \otimes e_1 + ke_2 \otimes e_1, \quad e_3 \bullet e_1 = e_3 \otimes e_1,$$

$$e_2 \bullet e_3 = -e_3 \otimes e_2 + ke_2 \otimes e_1, \quad e_3 \bullet e_2 = e_3 \otimes e_2,$$

$$e_3 \bullet e_2 \bullet e_1 = e_3 \otimes e_2 \otimes e_1.$$

$M(G)$ is an associative algebra when $G = G_3$. The differential D on $M(G)$ satisfies $De_1 = De_2 = 0$, $De_3 = ke_2 \bullet e_1$. The filtration is determined by the filtering degree on $(M(G)_*,D)$. The resulting spectral sequence has $E^0 = \wedge(e_1,e_2,e_3)$, $D^0 = 0$, $E^1 = \wedge(e_1,e_2,e_3)$, $D^1 e_3 = ke_2 e_1$, $E^r = H_*(\wedge(e_1,e_2,e_3),D^1)$, for all $r \ge 2$. The Poisson bracket on $\wedge(e_1,e_2,e_3)$ induced by the commutators on $M(G)$ satisfies:

$$\{e_1,e_3\} = \{e_3,e_1\} = ke_2 e_1,$$

$$\{e_3 e_1, e_3\} = \{e_1 e_3, e_3\} = -\{e_3,e_3 e_1\} = -\{e_3, e_1 e_3\} = ke_3 e_2 e_1.$$

All other Poisson brackets are zero.

The Theorem 3.6 for $G = G_3$ has the following form.

THEOREM 4.1. *Let G be the group of matrices as described above, and let $(M(G)_*, D)$ be the almost commutative differential graded algebra over $\mathbb{Q}$ associated with G. Then there exists a filtered complex $(P, \Delta) = (M(G)_* \otimes \mathbb{Q}[\bar{e}_1, \bar{e}_2, \bar{e}_3], \Delta)$ with $|\bar{e}_i| = 0$ that gives a spectral sequence*

  (i) $(E^r, d^r) \Rightarrow HH_*(M(G)_*, D)$,
  (ii) $E^0 = \wedge(e_1, e_2, e_3) \otimes \mathbb{Q}[\bar{e}_1, \bar{e}_2, \bar{e}_3] = \Omega^*_{\wedge(e_1, e_2, e_3)}$, *where* $\Omega^*_{\wedge(e_i)}$ *is the algebra of differential forms on the commutative graded algebra $\wedge(e_1, e_2, e_3)$, in the sense of the Appendix of [K-V]; and $d^0 = 0$,*
  (iii) $E^1 = \Omega^*_{\wedge(e_1, e_2, e_3)}$ *and the differential d^1 is the sum of two differentials δ and δ', where δ is the internal differential of the algebra of differential forms on the commutative differential graded algebra $(\wedge(e_1, e_2, e_3), D^1)$, $D^1 e_1 = D^1 e_2 = 0, D^1 e_3 = k e_2 e_1$ and δ' is the Brylinski differential defined by the Poisson bracket on $\wedge(e_1, e_2, e_3)$. More precisely*

$$\delta(a \otimes \bar{e}_1^{\alpha_1} \bar{e}_2^{\alpha_2} \bar{e}_3^{\alpha_3}) = D^1 a \otimes \bar{e}_1^{\alpha_1} \bar{e}_2^{\alpha_2} \bar{e}_3^{\alpha_3} +$$

$$(-1)^{|a|} \alpha_3 a (e_2 \otimes \bar{e}_1^{\alpha_1+1} \bar{e}_2^{\alpha_2} \bar{e}_3^{\alpha_3-1} - e_1 \otimes \bar{e}_1^{\alpha_1} \bar{e}_2^{\alpha_2+1} \bar{e}_3^{\alpha_3-1}),$$

$$\delta'(a \otimes \bar{e}_1^{\alpha_1} \bar{e}_2^{\alpha_2} \bar{e}_3^{\alpha_3}) = (-1)^{|a|} [\alpha_1 \{a, e_1\} \bar{e}_1^{\alpha_1-1} \bar{e}_2^{\alpha_2} \bar{e}_3^{\alpha_3} + \alpha_2 \{a, e_2\} \bar{e}_1^{\alpha_1} \bar{e}_2^{\alpha_2-1} \bar{e}_3^{\alpha_3} +$$

$$+ \alpha_3 \{a, e_3\} \bar{e}_1^{\alpha_1} \bar{e}_2^{\alpha_2} \bar{e}_3^{\alpha_3-1} + k \alpha_1 \alpha_3 a (e_1 \otimes \bar{e}_1^{\alpha_1-1} \bar{e}_2^{\alpha_2+1} \bar{e}_3^{\alpha_3-1} - e_2 \otimes \bar{e}_1^{\alpha_1} \bar{e}_2^{\alpha_2} \bar{e}_3^{\alpha_3-1})]$$

for $a \in \wedge(e_1, e_2, e_3)$ and $(\alpha_1, \alpha_2, \alpha_3) \in \mathbb{N}^3$.

The theorem follows easily from Theorem 3.6. It will be used to compute the E^2-term of the spectral sequence and to deduce the Hochschild homology $HH_*(M(G), D)$.

The main result of this section is:

THEOREM 4.2. *Let $(M(G), D)$ be the cochain algebra over $\mathbb{Q}$ associated with G_3 and let (P, Δ) be the filtered complex described in Theorem 4.1. Then we have:*

  (1) *The resulting spectral sequence collapses on the E^2-level.*
  (2) $HH_n(M(G), D) = 0$ *if* $n \notin \{0, 1, 2, 3\}$,
     $HH_0(M(G), D) \simeq \mathbb{Q}[\bar{e}_1, \bar{e}_2]$,
     $HH_1(M(G), D) \simeq (e_3 \bar{e}_2 - e_2 \bar{e}_3) \mathbb{Q}[\bar{e}_1, \bar{e}_2] \oplus e_2 \mathbb{Q}[\bar{e}_2] \oplus e_1 \mathbb{Q}[\bar{e}_1, \bar{e}_2]$,
     $HH_2(M(G), D) \simeq e_3 e_2 \mathbb{Q}[\bar{e}_2] \oplus e_2 e_1 \bar{e}_3 \mathbb{Q}[\bar{e}_3] \oplus e_3 e_1 \mathbb{Q}[\bar{e}_1, \bar{e}_2]$,
     $HH_3(M(G), D) \simeq e_3 e_2 e_1 \mathbb{Q}[\bar{e}_3]$.

The rest of this paragraph is devoted to the proof of Theorem 4.2.

That $HH_n(M(G), D) = 0$ for $n \notin \{0, 1, 2, 3\}$ is a direct consequence of the identity $\oplus_p E^0_{p, -p+n} = (\wedge(e_i) \otimes [\bar{e}_i])_n$ and of the fact that that sum is zero if $n \notin \{-3, -2, -1, 0\}$. The computations of the E^2-term of the spectral sequence are summarized in the following

LEMMA 4.3. *The spectral sequence defined in Theorem 4.1 satisfies:*

(1) $E^2_{p,-p} = \mathbb{Q}[\bar{e}_1]\bar{e}^p_2$,

(2) $E^2_{p,-p-3} = 0$ if p is even,

$E^2_{p,-p-3} = \mathbb{Q}e_3e_2e_1\bar{e}^{\frac{p-3}{2}}_3$ if p is odd.

(3) $E^2_{p,-p-1} = (e_3\bar{e}^{p-2}_2 - e_2\bar{e}^{p-3}_2\bar{e}_3)\mathbb{Q}[\bar{e}_1] \oplus e_1\bar{e}^p_2\mathbb{Q}[\bar{e}_1] \oplus \mathbb{Q}e_2\bar{e}^{p-1}_2$,

(4) $E^2_{p,-p-2}$ is spanned, as $\mathbb{Q}$-vector space, by

(a) $e_3e_2\bar{e}^{p-3}_2, e_3e_1\bar{e}^{p-2}_2\bar{e}^n_1, n \in \mathbb{N}$, if p is even, $p \geq 4$, and by

(b) $e_2e_1\bar{e}^{\frac{p-1}{2}}_3, e_3e_2\bar{e}^{p-3}_2, e_3e_1\bar{e}^{p-2}_2\bar{e}^n_1, n \in \mathbb{N}$, if p is odd, $p \geq 3$,
$E^2_{2,-4} = e_3e_1\mathbb{Q}[\bar{e}_1], E^2_{1,-3} = 0, E^2_{0,-2} = 0$,

(5) $E^2_{p,-p-n} = 0$ for all p if $n \notin \{0, 1, 2, 3\}$.

Proof. First notice that δ preserves the sum $\alpha_1 + \alpha_2 + \alpha_3$, and that δ' decreases this sum by 1. We have

$$0 \xrightarrow{d^1} E^1_{p,-p} = \{x \in \mathbb{Q}[\bar{e}_1, \bar{e}_2, \bar{e}_3], \|x\| = p\} \xrightarrow{d^1} E^1_{p-1,-p},$$

so that $E^2_{p,-p} = \{x \in \mathbb{Q}[\bar{e}_1, \bar{e}_2, \bar{e}_3], \|x\| = p, d^1x = 0\}$. Obviously, $\mathbb{Q}[\bar{e}_1]\bar{e}^p_2 \subset$ Kerd^1. Because δ preserves the sum $\alpha_1 + \alpha_2 + \alpha_3$, and $\delta(\bar{e}^{\alpha_1}_1\bar{e}^{\alpha_2}_2\bar{e}^{\alpha_3}_3) \neq 0$ if $\alpha_3 \geq 1$, it follows that d^1 is not zero on $\bar{e}_3\mathbb{Q}[\bar{e}_1, \bar{e}_2, \bar{e}_3]$, Thus $E^2_{p,-p} = \mathbb{Q}[\bar{e}_1]\bar{e}^p_2$. On the other hand $d^1 : E^1_{p,-p-3} \rightarrow E^1_{p-1,-p-3} = 0$, so that $E^2_{p,-p-3} \simeq E^1_{p,-p-3}/d^1(E^1_{p+1,-p-3})$. We have

$$E^1_{p,-p-3} = \{x \in e_3e_2e_1\mathbb{Q}[\bar{e}_1, \bar{e}_2, \bar{e}_3], \|x\| = p\}.$$

The map $d^1 : E^1_{q,-q-2} \rightarrow E^1_{q-1,-q-2}$ should be studied for every $q \geq 0$. As a $\mathbb{Q}$-vector space, $E^1_{q,-q-2}$ is spanned by the monomials

$$x_3 = e_2e_1 \otimes \bar{e}^{\alpha_1}_1\bar{e}^{\alpha_2}_2\bar{e}^{\alpha_3}_3,$$

$$x_2 = e_3e_1 \otimes \bar{e}^{\alpha_1}_1\bar{e}^{\alpha_2}_2\bar{e}^{\alpha_3}_3, \quad x_1 = e_3e_2 \otimes \bar{e}^{\alpha_1}_1\bar{e}^{\alpha_2}_2\bar{e}^{\alpha_3}_3.$$

Furthermore

$$d^1(x_3) = \delta(x_3) + \delta'(x_3) = 0, \text{ and}$$

$$\delta(x_2) = -k\alpha_3e_3e_2e_1 \otimes \bar{e}^{\alpha_1+1}_1\bar{e}^{\alpha_2}_2\bar{e}^{\alpha_3-1}_3,$$

$$\delta'(x_2) = k\alpha_3(\alpha_1 + 1)e_3e_2e_1 \otimes \bar{e}^{\alpha_1}_1\bar{e}^{\alpha_2}_2\bar{e}^{\alpha_3-1}_3,$$

$$\delta(x_1) = -k\alpha_3e_3e_2e_1 \otimes \bar{e}^{\alpha_1}_1\bar{e}^{\alpha_2+1}_2\bar{e}^{\alpha_3-1}_3,$$

$$\delta'(x_1) = k\alpha_1\alpha_3e_3e_2e_1 \otimes \bar{e}^{\alpha_1-1}_1\bar{e}^{\alpha_2+1}_2\bar{e}^{\alpha_3-1}_3.$$

It can be checked that

$$e_3e_2e_1\bar{e}^{\alpha_1}_1\bar{e}^{\alpha_2}_2\bar{e}^{\alpha_3}_3 = \lambda e_3e_2e_1\bar{e}^{\alpha_2}_2\bar{e}^{\alpha_3}_3 + d^1u$$

for $(\alpha_1, \alpha_2, \alpha_3) \in \mathbb{N}^3$, $u \in E^2_{q,-q-2}$, and that

$$-k(\alpha_3 + 1)e_3e_2e_1\bar{e}_2^{\alpha_2+1}\bar{e}_3^{\alpha_3} = d^1(e_3e_2 \otimes \bar{e}_2^{\alpha_2}\bar{e}_3^{\alpha_3+1}).$$

Thus we have proved that any element of $\bar{e}_2\mathbb{Q}[\bar{e}_2, \bar{e}_3]$ is a boundary. Therefore $E^2_{p,-p-3}$ is isomorphic to $\{x = \lambda e_3e_2e_1\bar{e}_3^{\alpha_3}, \lambda \in \mathbb{Q}, \|x\| = p\}$. This implies that $E^2_{p,-p-3} = 0$ if p is even, and $E^2_{p,-p-3} \simeq \mathbb{Q}e_3e_2e_1\bar{e}_3^{\frac{p-3}{2}}$ if p is odd.

Now we study $\mathrm{Ker}d^1 : E^1_{p,-p-2} \to E^1_{p-1,-p-2}$. Using the preceding calculations, we see that $\mathrm{Ker}d^1$ is spanned by the monomials

$$e_3e_1\mathbb{Q}[\bar{e}_1]\bar{e}_2^{p-2}, e_3e_2\mathbb{Q}[\bar{e}_1]\bar{e}_2^{p-3}, \underset{\alpha+2\beta=p-1}{\oplus} e_2e_1\mathbb{Q}[\bar{e}_1]\bar{e}_2^{\alpha}\bar{e}_3^{\beta},$$

$$e_3e_2 \otimes \bar{e}_1^{n+1}\bar{e}_2^{\alpha}\bar{e}_3^{\beta} - e_3e_1 \otimes \bar{e}_1^{n}\bar{e}_2^{\alpha+1}\bar{e}_3^{\beta}, \quad n \in \mathbb{N}, \alpha \in \mathbb{N}, \beta \in \mathbb{N}^*, \alpha + 2\beta = p - 3.$$

The $\mathbb{Q}$-vector space $E^1_{p,-p-1}$ is spanned by elements of the form

$$x_1 = e_1 \otimes \bar{e}_1^{\alpha_1}\bar{e}_2^{\alpha_2}\bar{e}_3^{\alpha_3}, \quad x_2 = e_2 \otimes \bar{e}_1^{\alpha_1}\bar{e}_2^{\alpha_2}\bar{e}_3^{\alpha_3}, \quad x_3 = e_3 \otimes \bar{e}_1^{\alpha_1}\bar{e}_2^{\alpha_2}\bar{e}_3^{\alpha_3}.$$

We have

$$\delta(x_1) = k\alpha_3 e_2e_1 \otimes \bar{e}_1^{\alpha_1+1}\bar{e}_2^{\alpha_2}\bar{e}_3^{\alpha_3-1},$$
$$\delta'(x_1) = -k(\alpha_1 + 1)\alpha_3 e_2e_1 \otimes \bar{e}_1^{\alpha_1}\bar{e}_2^{\alpha_2}\bar{e}_3^{\alpha_3-1}$$
$$\delta(x_2) = k\alpha_3 e_2e_1 \otimes \bar{e}_1^{\alpha_1}\bar{e}_2^{\alpha_2+1}\bar{e}_3^{\alpha_3-1},$$
$$\delta'(x_2) = -k\alpha_1\alpha_3 e_2e_1 \otimes \bar{e}_1^{\alpha_1-1}\bar{e}_2^{\alpha_2+1}\bar{e}_3^{\alpha_3-1},$$
$$\delta(x_3) = ke_2e_1 \otimes \bar{e}_1^{\alpha_1}\bar{e}_2^{\alpha_2}\bar{e}_3^{\alpha_3} + k\alpha_3 e_3e_1 \otimes \bar{e}_1^{\alpha_1}\bar{e}_2^{\alpha_2+1}\bar{e}_3^{\alpha_3-1}$$
$$- k\alpha_3 e_3e_2 \otimes \bar{e}_1^{\alpha_1+1}\bar{e}_2^{\alpha_2}\bar{e}_3^{\alpha_3-1},$$
$$\delta'(x_3) = -k\alpha_1 e_2e_1 \otimes \bar{e}_1^{\alpha_1-1}\bar{e}_2^{\alpha_2}\bar{e}_3^{\alpha_3} - k\alpha_1\alpha_3 e_3e_1 \otimes \bar{e}_1^{\alpha_1-1}\bar{e}_2^{\alpha_2+1}\bar{e}_3^{\alpha_3}$$
$$+ k\alpha_1\alpha_3 e_3e_2 \otimes \bar{e}_1^{\alpha_1}\bar{e}_2^{\alpha_2}\bar{e}_3^{\alpha_3-1}.$$

This is used to prove that for $n \geq 1$, there is y and $\lambda \in \mathbb{Q}$, such that

$$e_2e_1\bar{e}_2^{n}\bar{e}_2^{\alpha}e_3^{\beta} = d^1y + \lambda e_2e_1\bar{e}_2^{\alpha}\bar{e}_3^{\beta},$$

if $\alpha \geq 1$, $e_2e_1\bar{e}_2^{\alpha}\bar{e}_3^{\beta} = d^1(\frac{1}{k(\beta+1)}e_2 \otimes \bar{e}_2^{\alpha-1}\bar{e}_3^{\beta+1})$.

We also check that

$$e_3e_2 \otimes \bar{e}_1^{n+1}\bar{e}_2^{\alpha}e_3^{\beta} - e_3e_1 \otimes \bar{e}_1^{n}\bar{e}_2^{\alpha+1}e_3^{\beta} = d^1z + z',$$

where z' is a sum of terms of type $e_2e_1\bar{e}_1^{t}\bar{e}_2^{\alpha}\bar{e}_3^{\beta+1}, 0 \leq t \leq n$. This implies that $e_3e_2\bar{e}_1^{n+1}\bar{e}_2^{\alpha} \equiv e_3e_1\bar{e}_1^{n}\bar{e}_2^{\alpha+1}$ mod $\mathrm{Im}\, d^1$. Thus, now we have a basis for $E^2_{p,-p-2} = \mathrm{Ker}d^1/\mathrm{Im}d^1$. $E^2_{p,-p-2}$ is spanned, as a $\mathbb{Q}$-vector space, by

$$e_2e_1\bar{e}_3^{\frac{p-1}{2}}, e_3e_2\bar{e}_2^{p-3}, e_3e_1\mathbb{Q}[\bar{e}_1]\bar{e}_2^{p-2} \quad \text{if } p \text{ is odd,}$$
$$e_3e_2\bar{e}_2^{p-3}, e_3e_1\mathbb{Q}[\bar{e}_1]\bar{e}_2^{p-2} \quad \text{if } p \text{ is even.}$$

It remains to study $\mathrm{Ker} d^1 : E^1_{p,-p-1} \to E^1_{p-1,-p-1}$ and $\mathrm{Im} d^1 : E^1_{p,-p} \to E^1_{p-1,-p}$.

$$\delta(\bar{e}_1^{\alpha_1}\bar{e}_2^{\alpha_2}\bar{e}_3^{\alpha_3}) = k\alpha_3(e_2 \otimes \bar{e}_1^{\alpha_1+1}\bar{e}_2^{\alpha_2}\bar{e}_3^{\alpha_3-1} - e_1 \otimes \bar{e}_1^{\alpha_1}\bar{e}_2^{\alpha_2+1}\bar{e}_3^{\alpha_3-1}),$$

$$\delta'(\bar{e}_1^{\alpha_1}\bar{e}_2^{\alpha_2}\bar{e}_3^{\alpha_3}) = k\alpha_1\alpha_3(e_1 \otimes \bar{e}_1^{\alpha_1-1}\bar{e}_2^{\alpha_2+1}\bar{e}_3^{\alpha_3-1} - e_2 \otimes \bar{e}_1^{\alpha_1}\bar{e}_2^{\alpha_2}\bar{e}_3^{\alpha_3-1}).$$

Note that $e_1\mathbb{Q}[\bar{e}_1,\bar{e}_2] \oplus e_2\mathbb{Q}[\bar{e}_1,\bar{e}_2]$ is contained in $\mathrm{Ker} d^1$. Denote $Y_{n,\alpha,\beta} = e_2 \otimes \bar{e}_1^{n+1}\bar{e}_2^{\alpha}\bar{e}_3^{\beta} - e_1 \otimes \bar{e}_1^n\bar{e}_2^{\alpha+1}\bar{e}_3^{\beta}$, $Y_{n,\alpha,\beta} \in \mathrm{Ker} d^1$, $(n,\alpha,\beta) \in \mathbb{N}^3$. We see that $\mathrm{Im} d^1$ is spanned by the family $Y_{n,\alpha,\beta} - nY_{n-1,\alpha,\beta}$. It follows that $U_{\alpha} = e_3\bar{e}_1^{\alpha+1}\bar{e}_2^{p-2} - e_1\bar{e}_1^{\alpha}\bar{e}_2^{p-2}\bar{e}_3 \in \mathrm{Ker} d^1$, $V_{\alpha} = e_3\bar{e}_1^{\alpha}\bar{e}_2^{p-2} - e_2\bar{e}_1^{\alpha}\bar{e}_2^{p-3}\bar{e}_3 \in \mathrm{Ker} d^1$, for $\alpha \in \mathbb{N}$; and that $V_{\alpha+1} = U_{\alpha} - Y_{\alpha,p-3,1}$. This implies that $E^2_{p,-p-1}$ is spanned by the family $V_{\alpha}, \alpha \in \mathbb{N}, e_1\mathbb{Q}[\bar{e}_1,\bar{e}_2], \bar{e}_2 \otimes \bar{e}_2^{\alpha}$. □

The next problem is to compute the higher differentials $d^r, r \geq 2$ of the above spectral sequence. For that we need a more precise version of Theorem 3.4 and of its Corollary 3.5. For $G = G_3$ we have:

PROPOSITION 4.4. . Let $(M(G), D)$ be the cochain algebra associated with the group G_3, and let $P = M(G)_* \otimes \mathbb{Q}[\bar{e}_1,\bar{e}_2,\bar{e}_3]$ be given together with the product of the filtrations. Then there exists a differential Δ on P, such that

$$HH_*(M(G)_*, D) = H_*(P, \Delta),$$

and for $a \in M(G), (\alpha_1, \alpha_2) \in \mathbb{N}^2, (\alpha_1, \alpha_2, \alpha_3) \in \mathbb{N}^3$

(1) $\Delta(a \otimes \bar{e}_1^{\alpha_1}\bar{e}_2^{\alpha_2}) = Da \otimes \bar{e}_1^{\alpha_1}\bar{e}_2^{\alpha_2} + (-1)^{|a|}\{\alpha_1[a,e_1] \otimes \bar{e}_1^{\alpha_1-1}\bar{e}_2^{\alpha_2}$
$+\alpha_2[a,e_2] \otimes \bar{e}_1^{\alpha_1}\bar{e}_2^{\alpha_2-1}\}$,

(2) $\Delta(a \otimes \bar{e}_1^{\alpha_1}\bar{e}_2^{\alpha_2}\bar{e}_3^{\alpha_3}) = Da \otimes \bar{e}_1^{\alpha_1}\bar{e}_2^{\alpha_2}\bar{e}_3^{\alpha_3} + (-1)^{|a|}\{\alpha_1[a,e_1] \otimes \bar{e}_1^{\alpha_1-1}\bar{e}_2^{\alpha_2}\bar{e}_3^{\alpha_3}$
$+\alpha_2[a,e_2] \otimes \bar{e}_1^{\alpha_1}\bar{e}_2^{\alpha_2-1}\bar{e}_3^{\alpha_3} + \alpha_3[a,e_3] \otimes \bar{e}_1^{\alpha_1}\bar{e}_2^{\alpha_2}\bar{e}_3^{\alpha_3-1}$
$+b_1(a \otimes \bar{e}_1^{\alpha_1}\bar{e}_2^{\alpha_2}\bar{e}_3^{\alpha_3}) + \delta''(a \otimes \bar{e}_1^{\alpha_1}\bar{e}_2^{\alpha_2}\bar{e}_3^{\alpha_3})\} + \phi(a; \alpha_1, \alpha_2, \alpha_3)$,
where

(a) b_1 and δ'' are defined as in Corollary 3.5,

(b) $\phi(a; \alpha_1, \alpha_2, \alpha_3) \in F^{p-2}(M(G) \otimes \mathbb{Q}^{\geq n-1}[\bar{e}_1,\bar{e}_2,\bar{e}_3])$ if $n = \alpha_1 + \alpha_2 + \alpha_3, p = \alpha_2 + 2\alpha_3 + l, a \in F^l M(G), a \notin F^{l-1}M(G)$,

(c) $\phi(a; \alpha_1, \alpha_2, 1) \in F^{p-2}(\wedge(e_1, e_2) \otimes \mathbb{Q}^{\geq n-1}[\bar{e}_1,\bar{e}_2])$.

Proof. This proof follows the lines of the proof of Theorem 3.4. In this case we have an exact formula for Δ'', namely

$$\Delta''(1 \otimes \bar{e}_1^{\alpha_1}\bar{e}_2^{\alpha_2}\bar{e}_3^{\alpha_3} \otimes 1) = k\alpha_3(e_2 \otimes \bar{e}_1^{\alpha_1+1}\bar{e}_2^{\alpha_2}\bar{e}_3^{\alpha_3-1} \otimes 1$$
$$- 1 \otimes \bar{e}_1^{\alpha_1}\bar{e}_2^{\alpha_2+1}\bar{e}_3^{\alpha_3-1} \otimes e_1).$$

Thus the only thing that has to be checked is the form of $\phi(a; \alpha_1, \alpha_2, \alpha_3)$. In this case b' and b'^1 decrease the sum $\alpha_1 + \alpha_2 + \alpha_3$ at most by one, and Δ'' does not decrease the sum. Using the same notation as in the proof of Theorem 3.4, we see that $Z = \Delta'z \in F^{p-1}(M(G) \otimes \mathbb{Q}^{\geq n-2}[\bar{e}_i] \otimes M(G))$

if $n = \alpha_1 + \alpha_2 + \alpha_3$. So the induction hypothesis implies that there exists $Y \in F^{p-1}(M(G) \otimes \mathbb{Q}^{\geq n-1}[\bar{e}_i] \otimes M(G))$ such that $Z = \Delta'Y$. We define

$$\Delta'(1 \otimes \bar{e}_1^{\alpha_1}\bar{e}_2^{\alpha_2}\bar{e}_3^{\alpha_3} \otimes 1) = (b' + b'_1 + \Delta'')(1 \otimes \bar{e}_1^{\alpha_1}\bar{e}_2^{\alpha_2}\bar{e}_3^{\alpha_3} \otimes 1) - Y.$$

(2,c) is proved by an inclusion into the hypothesis $H_{p,\beta}$, and by using induction. If $\alpha_1 = \alpha_2 = 0$, then

$$z = (b' + b'_1 + \Delta'')(1 \otimes \bar{e}_3 \otimes 1)$$
$$= e_3 \otimes 1 \otimes 1 - 1 \otimes 1 \otimes e_3 + k(e_2 \otimes \bar{e}_1 \otimes 1 - 1 \otimes \bar{e}_2 \otimes e_1),$$

and $\Delta'z = 0$. Thus we set $\Delta'(1 \otimes \bar{e}_3 \otimes 1) = (b' + b'_1 + \Delta'')(1 \otimes \bar{e}_3 \otimes 1)$.
Now we are ready to prove the Theorem. We observe that $\Delta(a \otimes \bar{e}_1^{\alpha_1}\bar{e}_2^{\alpha_2}\bar{e}_3^{\alpha_3}) \in M(G) \otimes \mathbb{Q}^{\geq n-1}[\bar{e}_1, \bar{e}_2, \bar{e}_3]$ if $n = \sum \alpha_i$. Therefore all the differentials d^r satisfy the same property. Since $E^r_{p,-p} = \text{Ker}(d^{r-1} : E^{r-1}_{p,-p} \to E^{r-1}_{p-r,-p+r+1}), \Delta = 0$ on $\mathbb{Q}[\bar{e}_1, \bar{e}_2]$ and $E^2_{p,-p} = \mathbb{Q}[\bar{e}_1, \bar{e}_2]$, it follows easily that $E^r_{p,-p} \subseteq E^2_{p,-p}$ and that $d^r = 0$ on $E^r_{p,-p}$ for all $r \geq 2$. Now $E^2_{p,-p-1}$ is generated by monomials $e_3\bar{e}_1^n\bar{e}_2^{-p-2} - e_2\bar{e}_1^n e_2^{p-3}\bar{e}_3, e_1\bar{e}_1^n\bar{e}_2^p, e_2\bar{e}_2^{p-1}, n \in \mathbb{N}$. From Proposition 4.4 it follows that

$$\Delta(e_3\bar{e}_1^n\bar{e}_2^{p-2} - e_2\bar{e}_1^n\bar{e}_2^{p-3}\bar{e}_3) = ke_2e_1\bar{e}_1^n\bar{e}_2^{p-2} - e_3\Delta(\bar{e}_1^n\bar{e}_2^{p-2})$$
$$+[e_2,e_3]\bar{e}_1^n\bar{e}_2^{p-3} + b_1(e_2 \otimes \bar{e}_1^n e_2^{p-3}\bar{e}_3) + \delta''(e_2\bar{e}_1^n\bar{e}_2^{p-3}\bar{e}_3) + \phi.$$

We have $\|e_2e_1\bar{e}_1^n e_2^{p-2}\| = \|e_3\Delta(\bar{e}_1^n\bar{e}_2^{p-2})\| = \|\delta''(e_2\bar{e}_1^n\bar{e}_2^{p-3}\bar{e}_3)\| = p - 1$, $[e_2,e_3]\bar{e}_1^n\bar{e}_2^{p-3} +b^1(e_2 \otimes \bar{e}_1^n\bar{e}_2^{p-3}\bar{e}_3) \in e_2e_1\mathbb{Q}^{\geq n+p-3}[\bar{e}_1,\bar{e}_2], \phi \in \wedge(e_2,e_1) \otimes \mathbb{Q}^{\geq n+p-3}[\bar{e}_1,\bar{e}_2]$. This shows that $d^2(e_3\bar{e}_1^n\bar{e}_2^{p-2} - e_2\bar{e}_1^n\bar{e}_2^{p-3}\bar{e}_3)$ is a linear combination of elements belonging to $e_2e_1\mathbb{Q}^{\geq n+p-3}[\bar{e}_1,\bar{e}_2]$. These elements are zero in $E^2_{p-2,-p}$. This implies that $d^2 = 0$ on $E^2_{p-2,-p}$. The same argument shows that $d^r = 0$, for all $r \geq 2$. Now, $E^2_{p,-p-2}$ is generated by

$$z_1 = e_3e_2\bar{e}_2^{p-3}, \qquad \Delta z_1 = 0,$$
$$z_2 = e_3e_1\bar{e}_1^n\bar{e}_2^{p-2}, \qquad \Delta z_2 = 0,$$
$$z_3 = e_2e_1\bar{e}_3^{\frac{p-1}{2}} \quad \text{if} \quad p \text{ is odd and } \geq 3,$$
$$\Delta z_3 = \frac{p-1}{2}[e_2e_1,e_3]\bar{e}_3^{\frac{p-3}{2}} + b_1(e_2e_1 \otimes \bar{e}_3^{\frac{p-1}{2}}) + \delta''(e_2e_1 \otimes \bar{e}_3^{\frac{p-1}{2}}) + \phi,$$
$$\Delta z_3 = \phi(e_2e_1\bar{e}_3^{\frac{p-1}{2}}).$$

By Proposition 4.4, $\Delta z_3 \in M(G) \otimes \mathbb{Q}^{\geq \frac{p-1}{2}-1}[\bar{e}_1, \bar{e}_2\bar{e}_3]$, but $d^2 z_3 \in E^2_{p-2,-p-1} = \mathbb{Q}e_3e_2e_1\bar{e}_3^{\frac{p-3}{2}-1}$. This implies that $d^2 : E^2_{p,-p-2} \to E^2_{p-2,-p-1}$ is zero. The same argument shows that $d^r = 0$, for all $r \geq 2$. This completes the proof of the Theorem . $\qquad\square$

5 The negative cyclic homology of $(M(G), D)$.

In this section is computed the negative cyclic homology of $(M(G), D)$, when $M(G)$ is an associative algebra. First we study the case when $G = H$ is an abelian group on k generators.

Recall that M_*H is an exterior algebra on k generators e_i of degree -1. denote by d the derivation on the algebra $\wedge(e_i) \otimes \mathbb{Q}[\bar{e}_i]$ defined by setting $d(e_i) = \bar{e}_i, d(\bar{e}_i) = 0$. Then we get

PROPOSITION 5.1. [C-V]. Let $M(H) = \wedge(e_1,\dots,e_k)$ be the cochain algebra on k generators of degree 1. Then we have

(i) $HC_n^-(M(H)) = 0$ if n is odd, $n \notin [0, k-1]$,

(ii) $HC_n^-(M(H)) = \mathbb{Q}$ if n is even, $n \notin [0, k-1]$,

(iii) $HC_0^-(M(H)) \simeq \mathbb{Q}[\bar{e}_1,\dots,\bar{e}_k]$,

(iv) $HC_n^-(M(H)) \simeq \wedge^n(e_1,\dots,e_k) \otimes \mathbb{Q}[\bar{e}_1,\dots,\bar{e}_k]\cap\mathrm{Kerd}$, if n is odd, $0 < n < k$,

(v) $HC_n^-(M(H)) = \mathbb{Q} \oplus (\wedge^n(e_1,\dots,e_k) \otimes \mathbb{Q}[\bar{e}_1,\dots,\bar{e}_k]\cap\mathrm{Kerd})$, if n is even, $0 < n < k$.

($\wedge^n(e_i)$ is spanned by words of length n in $(e_1,\dots,e_k)$.)

When (A, d) is a filtered algebra, we know how to filter the Hochschild complex $(C_*(A), d+b)$ in Section 2. Then we can filter the Connes complex. We get

THEOREM 5.2. Let G be a finitely generated torsion free nilpotent group, such that $K(G, 1)$ is a k-dimensional nilmanifold. Assume that $(M(G), D)$ is associative, then the filtration of $M(G)$ induces a filtration on the Connes complex of $(M(G), D)$. The resulting spectral sequence $\tilde{E}^r$ satisfies:

(1) $\tilde{E}^r \Rightarrow HC_*^-(M(G), D)$,

(2) $\tilde{E}^1 = HC_*^-(M(H))$, where $H = \mathbb{Z}^k$ and $\tilde{E}^1$ is given by the Proposition 5.1.

(3) $\displaystyle\bigoplus_{p+q=n} \tilde{E}^r_{p,q} = 0$ if n is odd and $n \notin \{0,\dots,k-1\}, r \geq 1$.

$\displaystyle\bigoplus_{p+q=n} \tilde{E}^r_{p,q} = \mathbb{Q}$ if n is even and $n \notin \{0,\dots,k-1\}, r \geq 1$.

(4) $HC_0^-(M(G), D) = HH_0(M(G), D)$

$HC_{k-1}^-(M(G), D) = \mathbb{Q} \oplus HH_k(M(G), D)$ if k is odd, and

$HC_{k-1}^-(M(G), D) = HH_k(M(G), D)$ if k is even.

There is a bounded exact sequence:

$$0 \to HC_1^-(M(G), D) \to HH_1(M(G), D) \to HC_0^-(M(G), D) \xrightarrow{S^-}$$

$$HC_2^-(M(G), D) \to \cdots \to HH_{k-1}(M(G), D) \to \overline{HC}_{k-2}^-(M(G), D) \to 0$$

where $\overline{HC}_{k-2}^- = HC_{k-2}^-$ if k is odd, and

where $\overline{HC}_{k-2}^- = HC_{k-2}^-/\mathbb{Q}$ if k is even.

Now assume that $G = G_3$ is the Heisenberg group from Part 4. Let

$$\epsilon : (M(G), D) \to \wedge(e_1)$$

be the $\mathbb{Z}$-module map given by $\epsilon(e_1) = e_1$, $\epsilon(e_3^{u_3} \otimes e_2^{u_2} \otimes e_1^{u_1}) = 0$ if $u_3 + u_2 \geq 1$, $\epsilon(1) = 1$. It follows that ϵ is an algebra map, and $\epsilon_0 D = 0$. So we have an injection $HH_*(\wedge(e_1)) \to HH_*(M(G), D)$ and $HC_*^-(\wedge(e_1)) \to HC_*^-(M(G), D)$. When applied to the group G_3 we get

THEOREM 5.3. *Let G_3 be the Heisenberg group defined above, and let $(M(G_3), D)$ be the finite cochain algebra associated with G_3. Then*

(1) $HC_n^-(M(G_3), D) = 0$ *if n is odd, $n \geq 3$ or $n < 0$.*
   $HC_n^-(M(G_3), D) = \mathbb{Q}$ *if n is even, $n \geq 3$ or $n < 0$.*
(2) $HC_0^-(M(G_3), D) = HH_0(M(G_3), D) \simeq \mathbb{Q}[\bar{e}_1, \bar{e}_2]$.
(3) $HC_2^-(M(G_3), D) = \mathbb{Q} \oplus HH_3(M(G_3), D)$.
(4) *There is an exact sequence:*

$$0 \to HC_1^-(M(G_3), D) \to HH_1(M(G_3))/e_1 \cdot \mathbb{Q}[\bar{e}_1] \to HC_0^-(M(G_3))/\mathbb{Q}[\bar{e}_1]$$

$$\xrightarrow{S^-} HC_2^-(M(G_3))/\mathbb{Q} \to HH_2(M(G_3)) \to HC_1^-(M(G_3)) \to 0.$$

References

[A-H]   L. Avramov and S. Halperin, *Through the looking glass*, Lecture Notes in Math., vol. 1183, Springer-Verlag, Berlin and New York, 1986, pp. 1–27.

[Br]   J.L. Brylinski, *A differential complex for Poisson manifold*, Journ. Diff. Geom. **28** (1988), 93–114.

[B-V]   D. Burghelea and M. Vigué-Poirrier, *Cyclic homology of commutative algebras*, Lecture Notes in Math., vol. 1318, Springer-Verlag, Berlin and New York, 1988, pp. 51–72.

[Ce]   B. Cenkl, *Hochschild homology for a polynomial cochain algebra of a nilmanifold*, Topology Hawaii (Honolulu 12-18 August 1990) (K.H. Dowermann, eds.), World Scientific, P O Box 128, Farrer Road, Singapore 9128, 1992, pp. 41–52.

[C-P 1]   B. Cenkl and R. Porter, *Polynomial cochains on nilmanifolds* (preprint 1989).

[C-P 2]   B. Cenkl and R. Porter, *Algorithm for the computation of the cohomology of T-groups*, Lecture Notes in Math., (Algebraic Topology, Homotopy and Group Cohomology, Barcelona 1990), 1509 Springer-Verlag, Berlin and New York, 1992, pp. 79–94.

[C-P 3]   B. Cenkl and R. Porter, *The spectral sequence for polynomial cochains*, (preprint 1990).

[C-P 4]   B. Cenkl and R. Porter, *Lazard completion of a group and free differential graded algebra models over subrings of the rationals*, Topology **23** (1984), 445–464.

[C-V]   B. Cenkl and M. Vigué-Poirrier, *The cyclic homology of $P(G)$*, Suppl. Rend. Circ. Mat. di Palermo, Serie II (Proc. Geometry and Topology – Srní, 1992) **1993**, no. 32, 195–199.

[Fg-T]   P. Feng and B. Tsygan, *Hochschild and cyclic homology of quantum groups*, Commun. Math. Phys. **140** (1991), 481–521.

[Go]   T. Goodwillie, *Cyclic homology, derivations and the free loop space*, Topology **24** (1985), 187–215.

[G-H-V]   K. Grove, S. Halperin and M. Vigué-Poirrier, *The rational homotopy theory of certain path spaces*, Acta Math. **140** (1978), 277–303.

[Ha]   S. Halperin, Lectures on minimal models, Mémoire S.M.F. 9-10, 1983.

[H-K]   J. Huebschmann and T. Kadeishvili, *Minimal models for chain algebras over a local ring*, (to appear), Math. Zeit.

[Jo]   J.D.S. Jones, *Cyclic homology and equivariant homology*, Inv. Math. **87** (1987), 403–423.

[Ka]   C. Kassel, *Homologie cyclique des algèbres enveloppantes*, Inv. Math. **91** (1988), 221–251.

[K-V]   C. Kassel and M.Vigué-Poirrier, *Homologie des quotients de $U(sl_2)$*, Math. Annalen **294** (1992), 483–502.

[L-S]   L. Lambe and J. Stasheff, *Applications of perturbation theory to iterated fibrations*, Manuscripta Math. **58** (1987), 363–376.

[L]   L. Lambe, *Homological perturbation theory, Hochschild homology and fundamental groups*, Contemporary Mathematics, (Proceedings of the Conference on Deformation Theory and Quantization with Applications to Physics, Amherst, June 1990), 134 AMS, Providence, RI, 1992, pp. 183–218.

[Ma]   A. Malcev, *On a class of homogeneous spaces*, Math. USSR - Izv. **39** (1949).

[St]   R. Sternberg, Lectures on Chevalley groups, Yale University, New Haven, Connecticut, 1967.

[Ta]   L.A. Takhtadjian, *Non commutative homology of quantum tori*, Funct. Anal. Appl. **23** (1988), 147–149.

[Vi]   M. Vigué-Poirrier, *Homologie de Hochschild et homologie cyclique des algèbres différentielles graduées*, Astérisque **191** (1990), 255–267.

Department of Mathematics, Northeastern University, 360 Huntington Avenue, Boston, Massachusetts 02115, U.S.A.
cenkl@neu.edu

Département de Mathématiques, Université Paris-Nord, Avenue J.-B. Clément, 93430 Villetaneuse, France
vigue@math.univ-paris13.fr

Progress in Mathematics, Vol. 136
© 1996 Birkhäuser Verlag Basel/Switzerland

Closed Classes

Wojciech Chachólski

1. Introduction

A non empty class C of connected spaces is said to be a **closed class** if it is closed under weak equivalences and **pointed** homotopy colimits. A closed class can be characterized as a non empty class of connected spaces which is closed under weak equivalences and is closed under certain simple operations: arbitrary wedges, homotopy push-outs and homotopy sequential colimits. The notion of a closed class was introduced by E. Dror Farjoun [6].

Two important constructions give rise to examples of closed classes. The first one is the Bousfield-Dror periodization functor P_A [2]. The class of those spaces X, such that $P_A X$ is weakly contractible, forms a closed class. By looking just at the properties of this class we can prove, for example, that $P_A \Omega X$ is weakly equivalent to $\Omega P_{\Sigma A} X$ (see [2], [4]). The second construction is E. Dror Farjoun's colocalization functor CW_A. The class of those spaces X, for which there exists a space Y, such that X is weakly equivalent to $CW_A Y$, forms a closed class. This class is denoted by $C(A)$ and is called the class of A-cellular spaces. By looking just at the properties of the class $C(A)$ we can prove, for example, that $CW_A \Omega X$ is weakly equivalent to $\Omega CW_{\Sigma A} X$ (see [4], [6]).

We say that a closed class C is closed under extensions by fibrations, if for every fibration sequence $(Z \to E \to B)$, such that Z and B belong to C, E belongs to C. A closed class C is closed under extensions by fibrations if and only if for every diagram $F : I \to C$, such that the classifying space BI belongs to C, the unpointed homotopy colimit $hocolim_I F$ belongs to C.

The purpose of this paper is to understand to what extent a closed class is closed under extensions by fibrations and under taking unpointed homotopy colimits. We start with proving a theorem that, in particular, implies:

- Let $F : I \to Spaces_*$ be a pointed diagram, such that the classifying space BI belongs to C. If for every $i \in I$, $F(i)$ belongs to C, then so does the unpointed homotopy colimit $hocolim_I F$.
- Let $(Z \to E \to B)$ be a fibration sequence with a section. If Z and B belong to C, then so does E.
- Let $F : I \to C$ and $G : I \to C$ be diagrams and $\Psi : F \to G$ be a natural transformation. If $hocolim_I F$ belongs to C, then so does $hocolim_I G$.

Surprisingly these and many other results are the consequences of just one statement, see theorem 5.1.

We continue with investigating the properties of a base space B (respectively of the classifying space BI), which will guarantee that a closed class C is closed under extensions by fibrations with base B (respectively C is closed under taking the unpointed homotopy colimit of diagrams $F : I \to C$). We study the following class:

$$D(C) = \{BI \mid \text{if } F : I \to C \text{ is a diagram, then } hocolim_I F \in C\}$$

The main result of this paper is:

Theorem. *The class $D(C)$ is a closed class and it is closed under extensions by fibrations.*

Using this theorem, we can characterize the class $D(C)$ as follows:

$$D(C) = \{B \mid \text{if } Z \to E \to B \text{ is a fibration sequence and } Z \in C, \text{ then } E \in C\}$$

Since $D(C)$ is a closed class, it is closed under weak equivalences. This is a very non trivial fact itself. It is obvious that $\star$ belongs to $D(C)$. What is not clear at all is that for any diagram $F : I \to C$ over a contractible category I, the homotopy colimit $hocolim_I F$ belongs to C. As a result we get a new characterization of a closed class:

A non empty class C of connected spaces is a closed class if and only if it is closed under weak equivalences and for any unpointed diagram $F : I \to C$ over a contractible category I, the homotopy colimit $hocolim_I F$ belongs to C.

The techniques that are used to prove the main theorem involve studying the homotopy fiber of a map $f : X \to Y$ through inverse images of simplices in Y (a simplicial analogues of point inverse images). We prove, roughly, that if point inverse images belong to a closed class, then so does the homotopy fiber, see corollary 7.9. One consequence of this is that if the point inverse images of f are acyclic with respect to some homology theory, then so is the homotopy fiber of f.

Techniques, we have introduced, and properties of $D(C)$ are applied to prove a generalization of a theorem of E. Dror Farjoun (see 9.1 and [7, theorem I]):

Theorem. *Let $\Psi : E \to B$ be a natural transformation between **unpointed** diagrams $E : I \to Spaces$ and $B : I \to Spaces$. If for every $i \in I$, the homotopy fiber $Fib(\Psi_i : E(i) \to B(i))$ belongs to a closed class C, then so does $Fib(\Psi : hocolim_I E \to hocolim_I B)$.*

2. Notation

The symbol Δ denotes the simplicial category [9, §2], in which the objects are the ordered sets $[n] = \{0, 1, \dots, n\}$, and morphisms are weakly monotone maps of sets. The morphisms of Δ are generated by codegeneracy maps $s_i : [n] \to [n+1]$ $(i = 0, 1, \dots, n)$ and coface maps $d_i : [n-1] \to [n]$, subject to well-known cosimplicial identities (see [3]). A simplicial set is a functor $K : \Delta^{op} \to Sets$, where $Sets$ denotes the category of sets [9, §2]. The set $K([n])$ is usually denoted by K_n. A map between two simplicial sets is by definition a natural transformation of functors. A simplicial set K can be interpreted as a collection of sets K_n together with face maps $d_i : K_n \to K_{n-1}$ and degeneracy maps $s_i : K_{n+1} \to K_n$ $(i = 1, 2, \ldots, n)$ which satisfy the duals of the cosimplicial identities (see [3]). For description of haw to do homotopy theory with simplicial sets see [3] and [9].

If $\sigma \in K_n$, then σ is called an n-dimensional simplex of K. The dimension of σ will be denoted by $dim(\sigma)$. The object $\Delta[n]$ is the standard n-simplex, given by $\Delta[n]_k = mor_\Delta([k], [n])$ (see [3]). There is a distinguish n-dimensional simplex $\tau \in \Delta[n]_n$, the one that comes from the identity map $[n] \to [n]$. It is easy to check that for any simplicial set K, the assignment $f \to f(\tau)$ gives a bijection between the set of maps $\Delta[n] \to K$ and K_n. If $\sigma \in K_n$, we will denote the corresponding map by $\sigma : \Delta[n] \to K$.

A pointed simplicial set is a pair (K, k), where k is a chosen simplex of dimension zero in K. We will refer to this 0-dimensional simplex as the basepoint of (K, k). A map between pointed simplicial sets is a map of simplicial sets which preserves the basepoints. We will use the following notation for some categories which frequently occur:

- *Spaces* denotes the category of simplicial sets.
- *cSpaces* denotes the category of connected simplicial sets.
- *Spaces*$_\star$ denotes the category of pointed simplicial sets.
- *cSpaces*$_\star$ denotes the category of pointed and connected simplicial sets.

If I is a small category, the nerve of I, denoted by $N(I)$, is the simplicial set whose n-simplices are n-tuples $(i_0 \to i_1 \to \cdots i_n)$ of composable morphisms in I (see [3]).

If C is a category and K is an object in C, then by C/K we denote the category of objects of C over K [8, 1§6]. This is the category whose objects are morphisms $f : X \to K$ in C and maps from $f : X \to K$ to $g : Y \to K$ are those morphisms $h : X \to Y$ in C such that $g \circ h = f$.

3. The homotopy colimit

In this section we describe the notion of a diagram indexed by a simplicial set and define the homotopy colimit of such a diagram. The particular form of the homotopy colimit which is going to be used was introduced by E. Dror Farjoun Various properties of the homotopy colimit are also listed in the appendix. The reference for the proofs is [5].

The motivation for these constructions comes from the fact that any map $f : X \to K$ of simplicial sets, can be reconstructed (up to a weak equivalence) from the homotopy colimit of a diagram indexed essentially by the range of K. The constituents of this diagram are the analogues (in the simplicial category) of the point inverse images of f. The ability to build a map in a homotopically meaningful way from its range and its point inverse images is going to be explored in the following sections.

3.1. Definition. Let $K : \Delta^{op} \to Sets$ be a simplicial set. The category associated to K, sometimes called the transport category of K or Grothendieck construction on K, is the category whose objects are pairs $([n], \sigma)$ where $[n]$

is an object of Δ^{op} and $\sigma \in K_n$. A morphism $([n], \sigma) \rightarrow ([m], \tau)$ is a map $\varphi : [m] \rightarrow [n]$ in Δ (or $\varphi : [n] \rightarrow [m]$ in Δ^{op}) such that $K(\varphi)(\sigma) = \tau$.

To avoid introducing to many notation we will denote the category associated to a simplicial set K by the same symbol K, and speak of "functors with domain K". One can think about the category K as having as objects the simplices of K and morphisms generated by the arrows $d_i : \sigma \rightarrow d_i\sigma$ $s_i : \sigma \rightarrow s_i\sigma$, subject to some relations that come from the simplicial structure of K. The morphisms $s_i : \sigma \rightarrow s_i\sigma$ is called the degeneracy morphisms and $d_i : \sigma \rightarrow d_i\sigma$ the boundary morphisms.

The notion of the category associated to a simplicial set can be used to define the subdivision of K (see [11]):

3.2. Definition. *If K is a simplicial set, the subdivision of K is the nerve (see section 2) of the category associated to K. The subdivision of K will be denoted by sdK.*

3.3. Example. The category associated to $\Delta[0] = \star$ is isomorphic to Δ^{op}. Diagrams over $\star$ are simplicial spaces.

If K is a simplicial set, a functor $F : K \rightarrow Spaces$ over the category associated to K will be called a diagram with shape K. Diagrams are our main object of interest. As example 3.3 suggests, the category associated to a simplicial set can be quite complicated. Since we care about diagrams, in order to simplify the situation, we will distinguish the class of bounded diagram which are technically more manageable. It turns out that all the examples of diagrams, we are going to consider, are bounded.

3.4. Definition. *$F : K \rightarrow Spaces$ is a bounded diagram if for any degeneracy morphism $s_i : \sigma \rightarrow s_i\sigma$, $F(\sigma) = F(s_i\sigma)$ and $F(s_i) = id_{F(\sigma)}$*

3.5. Example. A bounded diagram with shape $\star$ is determined by the value on the only zero dimensional simplex of $\star$. The category of bounded diagrams with shape $\star$ is equivalent to the category of simplicial sets.

3.6. Example. The category of bounded diagrams over $\Delta[1]$ is equivalent to the category of diagrams of the form $A \leftarrow B \rightarrow C$, so called push-out diagrams. Out of a bounded diagram $F : \Delta[1] \rightarrow Spaces$, a push-out diagram can be extracted in the following way:

$$F(0) \leftarrow F(0,1) \rightarrow F(1)$$

3.7. Example. Let $f : X \rightarrow K$ be a map. For every simplex $\sigma \in K$, let $F_f(\sigma)$ be the space that fits into the following pull-back square:

$$
\begin{array}{ccc}
F_f(\sigma) & \longrightarrow & X \\
\downarrow & & \downarrow f \\
\Delta[dim(\sigma)] & \stackrel{\sigma}{\longrightarrow} & K
\end{array}
$$

Roughly speaking $F_f(\sigma)$ is the inverse image in X of the simplex σ of K. This construction clearly defines a functor $F_f : K^{op} \rightarrow Spaces$. Out of F_f we can built a bounded diagram $df : sdK \to Spaces$ such that for an n-dimensional simplex $v = \sigma_0 \overset{\varphi_0}{\to} \sigma_1 \overset{\varphi_1}{\to} ... \overset{\varphi_{n-1}}{\to} \sigma_n$ in sdK:

$$df(v) = F_f(\sigma_n)$$

$$df(d_i : v \to d_i v) = \begin{cases} id : F_f(\sigma_n) \to F_f(\sigma_n) & \text{if } i < n \\ F_f(\varphi_{n-1}) : F_f(\sigma_n) \to F_f(\sigma_{n-1}) & \text{if } i = n \end{cases}$$

The construction d is natural. If $f : X \to K$, $g : Y \to K$ and $h : X \to Y$ are maps such that $g \circ h = f$, then there is a natural transformation $df \to dg$. In particular there is a natural transformation $\Psi : df \to d(id)$ induced by $f : X \to K$ itself. It turns out that:

$$\left(colim_{sdK} \Psi : colim_{sdK} df \to colim_{sdK} d(id)\right) = (f : X \to K)$$

We will see in example 3.12 that the diagram df has nice homotopic properties with respect to the map f.

If $f : X \to K$ is a fibration, then the functor $F_f : K^{op} \to Spaces$ has the property that for every morphism φ, $F_f(\varphi)$ is a weak equivalence. In this case $F_f(\sigma)$ is weakly equivalent to the homotopy fiber of f. It is clear from the definition that the diagram $df : sdK \to Spaces$ inherits the same properties. This motivates the following definition:

3.8. Definition. $F : K \to Spaces$ *is a* good diagram *if it is a bounded diagram and for every morphism* $\varphi \in K$, $F(\varphi)$ *is a weak equivalence.*

The construction d defines a functor:

$$d : Spaces/K \to \{\text{bounded diagrams over } sdK\}$$

$$(f : X \to K) \mapsto (df : sdK \to Spaces)$$

such that fibrations are carried out to good diagrams.

We will now introduce a construction which will allow us to recover, up to homotopy, a map f from the diagram df. One can think about this construction as sort of an "inverse" of d:

3.9. Definition. *The* homotopy colimit *is the following functor:*

$$\oint_K : \{\text{diagrams over } K\} \to Spaces$$

$$\oint_K F = \left(\bigsqcup_{\sigma \in K} \Delta[dim(\sigma)] \times F(\sigma)\right)\Big/ \sim$$

where $\sim$ *is an equivalence relation generated by:*

$$\text{let } \varphi \in mor_\Delta([n],[m]) \ , \ \tau \in K_m \ , \ x \in F(\tau) \ , \ t \in \Delta[n]$$

$$(\Delta[\varphi](t), x) \sim (t, F(\varphi)(x))$$

3.10. Definition. *The pointed homotopy colimit is the following functor:*

$$\int_K : \{pointed\ diagrams\ over\ K\} \to Spaces_\star$$

$$\int_K F = \left(\bigvee_{\sigma \in K} \left(\Delta[dim(\sigma)] \times F(\sigma) \right) / \left(\Delta[dim(\sigma)] \times \{\star\} \right) \right) \Big/ \sim$$

where $\sim$ *is an equivalence relation generated by:*

$$\text{let } \varphi \in mor_\Delta([n],[m]) \ , \ \tau \in K_m \ , \ x \in F(\tau) \ , \ t \in \Delta[n]$$
$$(\Delta[\varphi](t), x) \sim (t, F(\varphi)(x))$$

3.11. Example. If $F : K \to Spaces_\star$ is a constant diagram $F(\sigma) = X$ and $F(\varphi) = id_X$, then:

$$\oint_K F = K \times X \ , \ \int_K F = (K \times X)/(K \times \{\star\}) \ = K \ltimes X$$

In case $X = \star$, we get that $\oint_K \star = K$, where $\star : K \to Spaces$ denotes the constant diagram whose value is $\star$.

Let $F : K \to Spaces$ be a diagram. There is a natural transformation $F \to \star$ between F and the constant diagram $\star$. This natural transformation induces the following map:

$$\oint_K F \to \oint_K \star = K$$

Let $F : K \to Spaces_\star$ be a pointed diagram. This means that there is a natural transformation $\star \to F$ between the constant diagram $\star$ and F. This transformation induces the map $K = \oint_K \star \to \oint_K F$, which is a section of the map $\oint_K F \to K$.

As a consequence we get that the homotopy colimit can be seen as a functor with values in $Spaces/K$:

$$\oint_K : \{diagrams\ over\ K\} \to Spaces/K$$

$$(F : K \to Spaces) \mapsto \left(\oint_K F \to K \right)$$

which carries out pointed diagrams into maps with a section.

If $F : K \to Spaces_\star$ is a pointed diagram, then the following is a cofibration sequence:

$$K \to \oint_K F \to \int_K F$$

As a consequence we get that if $F : K \to Spaces_\star$ is a pointed diagram over weakly contractible simplicial set, then the unpointed and the pointed homotopy colimits of F are weakly equivalent.

3.12. Example. Let $f : X \to K$ be a map. Out of f we have constructed a diagram $df : sdK \to Spaces$ (see example 3.7). The main property of df is that in the following commutative diagram, all the horizontal arrows are weak equivalences:

$$
\begin{array}{ccccccc}
\displaystyle\oint_{sdK} df & \stackrel{id}{\longleftarrow} & \displaystyle\oint_{sdK} df & \longrightarrow & colim_{sdK}\, df & = & X \\
\downarrow & & \downarrow & & \downarrow & & \downarrow f \\
sdK & \longleftarrow & \displaystyle\oint_{sdK} d(id) & \longrightarrow & colim_{sdK}\, d(id) & = & K
\end{array}
$$

It implies that every map $f : X \to K$ is weakly equivalent to a map of the form $\oint_L F \to L$ for some bounded diagram $F : L \to Spaces$. Since every map is weakly equivalent to a fibration, we can assume that $F : L \to Spaces$ is a good diagram whose values are weakly equivalent to the homotopy fiber of f (see example 3.7).

3.13. Example. Let $\dot\Delta[n] \to L$ be a map, where $\dot\Delta[n]$ is the boundary of $\Delta[n]$. We are going to consider diagrams over $L \cup_{\dot\Delta[n]} \Delta[n]$.

- Let $F : L \cup_{\dot\Delta[n]} \Delta[n] \to Spaces$ be a diagram. One can show that:

$$
\oint_{L\cup_{\dot\Delta[n]}\Delta[n]} F = colim\Big(\oint_L F \leftarrow \oint_{\dot\Delta[n]} F \hookrightarrow \oint_{\Delta[n]} F\Big)
$$

- Let $F : L \cup_{\dot\Delta[n]} \Delta[n] \to Spaces$ be a bounded diagram. If $\tau \in (\Delta[n])_n$ is the only non degenerate simplex and $F(\tau) = X$, one can conclude that in this case:

$$
\oint_{L\cup_{\dot\Delta[n]}\Delta[n]} F = colim\Big(\oint_L F \leftarrow \dot\Delta[n] \times X \hookrightarrow \Delta[n] \times X\Big)
$$

3.14. Example. A functor $F : I \to Spaces$ over a small category I defines a bounded diagram $F_{sd} : N(I) \to Spaces$ over the nerve of I. Let:

$$
\sigma = (a_0 \stackrel{\varphi_0}{\to} a_1 \stackrel{\varphi_1}{\to} \ldots \stackrel{\varphi_{n-1}}{\to} a_n) \in N(I)_n
$$

F_{sd} is defined as follows:

$$
F_{sd}(\sigma) = F(a_0), \qquad F_{sd}(d_i : \sigma \to d_i\sigma) = \begin{cases} F(\varphi_0) : F(a_0) \to F(a_1) & \text{if } i = 0 \\ id : F(a_0) \to F(a_0) & \text{if } i > 0 \end{cases}
$$

It can be shown that in this case:

$$
\oint_{N(I)} F_{sd} = hocolim_I F
$$

where $hocolim_I F$ denotes the homotopy colimit of F in the sense of Bousfield and Kan. In case of a pointed functor $F : I \to Spaces_*$ there is a similar equality:

$$
\int_{N(I)} F_{sd} = phocolim_I F
$$

where $phocolim_I F$ denotes the **pointed** homotopy colimit of F in the sense of Bousfield and Kan.

4. Closed Classes

In this section we state the definition and give some examples and basic properties of closed classes. The notion of a closed class was introduced by E. Dror Farjoun [6], [7]. The definition presented in this papers is slightly different from the one given by E. Dror Farjoun [6, definition 2.1]. We think about a closed class as a class of **unpointed** and connected simplicial sets. A good example, to keep in mind, is the class of acyclic spaces with respect to some homology theory.

4.1. Definition. (E. Dror Farjoun [6]). *A non empty class C of connected simplicial sets is said to be a closed class if it is closed under weak equivalences and taking **pointed** homotopy colimits. If $F : K \to Spaces_\star$ is a pointed diagram such that for every simplex $\sigma \in K$, $F(\sigma) \in C$, then $\int_K F \in C$.*

Observe that a closed class is assumed to be non-empty. Notice also that since the empty space is not connected, it does not belong to any closed class.

4.2. Notation.

- Throughout this article C always denotes a closed class.
- By $F : K \to C$ we denote a diagram such that for every simplex $\sigma \in K$, $F(\sigma)$ belongs to C.
- Let $f : X \to Y$ be a map. We say that the homotopy fiber $Fib(X \xrightarrow{f} Y)$ belongs to a closed class C, if the homotopy fibers of f over every component of Y belong to C. In particular, the homotopy fibers of f, over various components, are connected and f induces an isomorphism on π_0. If $Fib(X \xrightarrow{f} Y)$ belongs to C, then we will write $Fib(X \xrightarrow{f} Y) \in C$.

4.3. Remark. The definition of the pointed homotopy colimit 3.10 implies that a class C is closed if and only if:

- C is non-empty.
- Let X and Y be weakly equivalent simplicial sets. If $X \in C$, then $Y \in C$.
- Let $(X_i)_{i \in I}$ be a family of simplicial sets. If $X_i \in C$, then for any choice of basepoints in X_i, $\bigvee_{i \in I} X_i \in C$.
- Let $X_\star$ be a simplicial space. If for every $n \geq 0$, $X_n \in C$, then the realization $|X_\star| \in C$.

It follows that a closed class can be characterized as a class of connected simplicial sets, such that:

- C is non-empty.
- C is closed under weak equivalences.
- C is closed under taking arbitrary wedges.
- Let $X_1 \leftarrow X_2 \to X_3$ be a diagram. If $X_i \in C$, then the following simplicial set belongs to C:

$$hocolim(X_1 \leftarrow X_2 \to X_3)$$

- Let $(X_0 \to X_1 \to X_2 \to \cdots)$ be a diagram. If $X_i \in C$, then the following simplicial set belongs to C:

$$hocolim(X_0 \to X_1 \to X_2 \to \cdots)$$

4.4. Examples. Here is a list of some examples of closed classes:

- Let A be a connected simplicial set. The smallest closed class $C(A)$ such that $A \in C(A)$. This class is called the class of A-cellular spaces. This class was introduced by E. Dror Farjoun, see [4],[6] and [7].
- The class of acyclic spaces with respect to some homology theory.
- The class $C(\star)$ of weakly contractible spaces.
- The class $C(S^{n+1})$ of n-connected spaces.
- $\{X \in cSpaces \mid \tilde{H}^i(X, G)$ is trivial for $i \leq n\}$.
- Let A be a pointed and connected Kan simplicial set.
  $\{X \in cSpaces \mid$ for any choice of basepoints in X, $map_\star(X, A) \simeq \star\}$.
- Let A be a connected simplicial set.
  $\{X \in cSpaces \mid$ if Y is Kan and the basepoint evaluation map $map(A, Y) \to Y$ is a weak equivalence, then $map(X, Y) \to Y$ is also a weak equivalence$\}$.
  This class is called the class A-acyclic spaces, see [4, definition 16.1].

The following two propositions give some examples of elements of a closed class.

4.5. Proposition. (E. Dror Farjoun [6, section 2.3]). *If C is a closed class, then $\star \in C$.*

Proof. Let $X \in C$ and $X \xrightarrow{\star} X$ be the constant map. Notice that the following space is contractible and belongs to C:

$$hocolim(X \xrightarrow{\star} X \xrightarrow{\star} X \xrightarrow{\star} X \cdots)$$

It implies that $\star \in C$.   □

4.6. Proposition. (E. Dror Farjoun [6, theorem 2.8]). *Let K and X be simplicial sets. If $X \in C$, then for any choice of basepoint in X, $K \ltimes X \in C$ (see example 3.11).*

Proof. Lets consider the constant diagram $X : K \to C$, $X(\sigma) = X$ and $X(\varphi) = id_X$. Since C is a closed class, $K \ltimes X = \int_K X \in C$.   □

5. Closed classes and unpointed homotopy colimits

Closed classes are not usually closed under unpointed homotopy colimits (see [6, section 2.3]). This section contains the first approach to this question, to what extent a closed class is closed under unpointed homotopy colimits.

The motivation for the following theorem can be found in the next section. This theorem is going to be applied to prove various properties of closed classes with respect to fibrations. Surprisingly all those properties are just particular cases of this one statement.

5.1. Theorem. *Let $F : K \to C$, $G : K \to C$ be diagrams and $\Psi : F \to G$ be a natural transformation. If $h : \oint_K F \to Y$ is a map such that $Y \in C$, then:*

$$hocolim\left(\oint_K G \leftarrow \oint_K F \xrightarrow{h} Y\right) \in C$$

5.2. Lemma. *For every diagram $F : \Delta[n] \to C$, $\oint_{\Delta[n]} F$ belongs to C.*

Proof. Let $\tau \in (\Delta[n])_n$ be the non-degenerate simplex. Observe that by choosing a vertex in $F(\tau)$, we can think about $F : \Delta[n] \to C$ as a pointed diagram (this chosen vertex determines a basepoint in $F(\sigma)$, for all $\sigma \in \Delta[n]$). Since $\Delta[n]$ is contractible, pointed and unpointed homotopy colimits are weakly equivalent. It proves the lemma.   □

Proof of the theorem. the proof will be by the induction on the dimension of K. It is obvious that the theorem is true for K such that $dim(K) = 0$. Lets assume that the theorem is true for K such that $dim(K) < n$. Let $dim(L) < n$ and $\dot{\Delta}[n] \to L$ be a map. We prove that the theorem holds for $K = L \cup_{\dot{\Delta}[n]} \Delta[n]$.

Lets consider the following commutative diagram:

$$
\begin{array}{ccccc}
Y & \xleftarrow{id} & Y & \xrightarrow{id} & Y \\
\uparrow h & & \uparrow h & & \uparrow h \\
\oint_L F & \longleftarrow & \oint_{\dot{\Delta}[n]} F & \longrightarrow & \oint_{\Delta[n]} F \\
\downarrow & & \downarrow & & \downarrow \\
\oint_L G & \longleftarrow & \oint_{\dot{\Delta}[n]} G & \longrightarrow & \oint_{\Delta[n]} G
\end{array}
$$

By the inductive assumption:

$$hocolim\left(\oint_L G \leftarrow \oint_L F \xrightarrow{h} Y\right) \in C$$

$$hocolim\left(\oint_{\dot{\Delta}[n]} G \leftarrow \oint_{\dot{\Delta}[n]} F \xrightarrow{h} Y\right) \in C$$

According to lemma 5.2, $\oint_{\Delta[n]} F$ and $\oint_{\Delta[n]} G$ belong to C. Because C is closed under homotopy push-outs we get:

$$hocolim\left(\oint_K G \leftarrow \oint_K F \xrightarrow{h} Y\right) \in C \qquad \qquad □$$

5.3. Corollary. *Let $F : K \to C$, $G : K \to C$ be diagrams and $\Psi : F \to G$ be a natural transformation. If $\oint_K F \in C$, then $\oint_K G \in C$.*

Proof. Apply theorem 5.1 to the case when $Y = \oint_K F$ and $h = id$.   □

5.4. Corollary. *Let K be a simplicial set. If $F : K \to C$ is a diagram such that $\oint_K F \in C$, then $K \in C$.*

Proof. Since there is a natural transformation $F \to \star$ between F and the constant diagram $\star : K \to Spaces$, whose value is $\star$, corollary 5.3 implies that $K = \oint_K \star$ belongs to C.   □

6. Closed classes and fibrations

The behavior of a closed class with respect to fibrations has been studied by E. Dror Farjoun [6],[7]. This section contains generalizations of his results.

6.1. Definition. We say that a closed class C is closed under extensions by fibrations if for every fibration sequence $Z \to E \to B$ such that Z and B belong to C, E belongs to C. 2mm
Closed classes are not usually closed under extensions by fibrations

6.2. Examples. Here is a list of some examples of closed classes that are closed under extensions by fibrations:

- The class of acyclic spaces with respect to some homology theory.
- The class $C(\star)$ of weakly contractible spaces.
- The class $C(S^{n+1})$ of n-connected spaces.
- Let A be a connected simplicial set.
  $\{X \in cSpaces \mid$ if Y is Kan and the basepoint evaluation map $map(A, Y) \to Y$ is a weak equivalence, then $map(X, Y) \to Y$ is also a weak equivalence$\}$.

The next theorem is a geometric interpretation of theorem 5.1.

6.3. Theorem. *Let $p_1 : E_1 \to B$, $p_2 : E_2 \to B$ and $s : E_1 \to E_2$ be maps such that $p_1 = p_2 \circ s$ and the homotopy fibers $Fib(E_1 \overset{p_1}{\to} B)$, $Fib(E_2 \overset{p_2}{\to} B)$ belong to C. For any map $h : E_1 \to Y$, where $Y \in C$:*

$$hocolim(E_2 \overset{s}{\leftarrow} E_1 \overset{h}{\to} Y) \in C$$

Proof. Example 3.12 implies that p_1 and p_2 are weakly equivalent, respectively to maps of the form:

$$\oint_{sdB} F \to sdB \ , \ \oint_{sdB} G \to sdB$$

where F has values weakly equivalent to the homotopy fibers of p_1 and G has values weakly equivalent to the homotopy fibers of p_2. Since the definitions were natural, $s : E_1 \to E_2$ induces a natural transformation $\Psi : F \to G$. Theorem 5.1 implies then:

$$hocolim(E_2 \leftarrow E_1 \to Y) \simeq hocolim\left(\oint_{sdB} G \leftarrow \oint_{sdB} F \to Y \right) \in C \qquad □$$

The following corollaries are particular cases of theorem 6.3.

6.4. Corollary. Let $p_1 : E_1 \to B$, $p_2 : E_2 \to B$ and $s : E_1 \to E_2$ be maps such that $p_1 = p_2 \circ s$. If the homotopy fibers $Fib(E_1 \xrightarrow{p_1} B)$, $Fib(E_2 \xrightarrow{p_2} B)$ and E_1 belong to C, then so does E_2.

Proof. Apply theorem 6.3 to the case when $Y = E_1$ and $h = id$. ☐

6.5. Corollary. (E. Dror Farjoun [7]). Let $p : E \to B$ be a map such that the homotopy fiber $Fib(E \xrightarrow{p} B)$ belongs to C. If $E \in C$, then $B \in C$.

Proof. Apply corollary 6.4 to the case when $p_1 = p$, $p_2 = id_B$ and $s = p$. ☐

6.6. Corollary. (E. Dror Farjoun [7]). Let $F \to E \xrightarrow{p} B$ be a fibration sequence. If ΩB and F belong to C, then so does E.

Proof. Since $\Omega B \to F \to E$ is a fibration sequence such that ΩB and F belong to C, corollary 6.5 implies that E belongs to C. ☐

6.7. Corollary. Let $F \to E \xrightarrow{p} B$ be a fibration sequence and $B \xrightarrow{s} E$ be a section of p. If $h : B \to Y$ is a map such that $Y \in C$, then:

$$colim(E \xleftarrow{s} B \xrightarrow{h} Y) \in C$$

Proof. Apply theorem 6.3 to the case when $p_1 = id_B$, $p_2 = p$. ☐

6.8. Corollary. Closed classes are closed under split extensions. Let $F \to E \xrightarrow{p} B$ be a fibration sequence such that p has a section. If $F \in C$ and $B \in C$, then $E \in C$.

Proof. Apply corollary 6.7 to the case when $Y = B$ and $h = id_B$ and s is a section of p. ☐

6.9. Corollary. (W. Dwyer [6]). Closed classes are closed under products. If $X \in C$ and $Y \in C$, then $X \times Y \in C$.

Proof. Notice that $X \to X \times Y \to Y$ is a fibration sequence with a section. According to corollary 6.8, $X \times Y$ belongs to C. ☐

7. Homotopy properties of shapes of diagrams

In this section the behavior of a closed class under unpointed homotopy colimits is going to be investigated further (see section 5).

A diagram $F : K \to Spaces$ consist of bunch of spaces which are related to each other by various maps. Those relations are coming from the geometry of K. We will try to understand how the geometry of K effects the homotopy colimit functor of diagrams over K.

7.1. Definition. $D(C) = \{K \mid \text{if } F : K \to C \text{ is a diagram, then } \int_K F \in C\}$

Class $D(C)$ consists of those simplicial sets that carry enough information so by gluing elements of class C, according to K, we get back a space in C.

7.2. Proposition. $D(C) \subseteq C$

Proof. Let $K \in D(C)$. Since $\star \in C$, according to the definition, $K = \oint_K \star$ belongs to C.

Notice that corollary 5.4 is stronger than this proposition. It says that if there exist a diagram $F : K \to C$ such that $\oint_K F \in C$, then automatically $K \in C$. $\square$

Observe that lemma 5.2 implies:

7.3. Proposition. *For every n, $\Delta[n] \in D(C)$.*

It turns out that $D(C)$ has nice homotopic properties. The next theorem suggests that to some extant, not geometry but the homotopy type of a simplicial set K plays a crucial role toward the homotopic properties of the homotopy colimit functor of diagrams over K.

7.4. Theorem. *Class $D(C)$ is closed under weak equivalences.*

7.5. Lemma.
- Let $K \leftarrow L \hookrightarrow M$ be a push-out diagram such that $L \hookrightarrow M$ is a cofibration. If K, L and M belong to $D(C)$, then so does:
$$K \cup_L M = colim(K \leftarrow L \hookrightarrow M)$$
- Let Θ be the category associated with an ordinal number Θ (see [8, page 11]). Let $G : \Theta \to Spaces$ be a functor such that for every morphism $\varphi \in \Theta$, $G(\varphi)$ is a cofibration. If for every $\theta \in \Theta$, $G(\theta)$ belongs to $D(C)$, then so does $colim_\Theta G$.

Proof. We will prove only the first part of the lemma. The second part can be proven in the same way.

Let $F : K \cup_L M \to C$ be a diagram. According to example 3.13:

$$\oint_{K \cup_L M} F = colim\left(\oint_K F \leftarrow \oint_L F \hookrightarrow \oint_M F \right)$$

By the assumption $\oint_K F$, $\oint_L F$ and $\oint_M F$ belong to C. Since any closed class is closed under taking homotopy push-outs:

$$hocolim\left(\oint_K F \leftarrow \oint_L F \hookrightarrow \oint_M F \right) \in C$$

Notice that the cofibration assumption implies that the following map is a weak equivalence:

$$hocolim\left(\oint_K F \leftarrow \oint_L F \hookrightarrow \oint_M F \right) \to colim\left(\oint_K F \leftarrow \oint_L F \hookrightarrow \oint_M F \right)$$

It implies that $\oint_{K \cup_L M} F$ belongs to C. $\square$

7.6. Lemma. *Let* $0 \leq k \leq n$. $\Delta[n, k]$ *belongs to* $D(C)$, *where if* $\tau \in \Delta[n]_n$ *is the non degenerate simplex,* $\Delta[n, k]$ *is the simplicial subset of* $\Delta[n]$, *generated by simplices* $\{d_i\tau\}_{i \neq k}$.

Proof. We are going to present $\Delta[n, k]$ as a sequence of push-outs of standard simplices. In order to do this we have to introduce some notation:

- Let $i \in \{0, 1, \cdots, n\}$. Δ_i denotes the simplicial subset of $\Delta[n]$ generated by the simplex $d_i\tau$.
- $\{i\}$ denotes the simplicial subset of $\Delta[n]$ generated by the vertex $\{i\}$.
- Let $\{i, j\} \in \{0, 1, \cdots, n\}$. $\Delta_{i,j}$ denotes the simplicial subset of $\Delta[n]$ generated by the simplex $d_{i-1}d_j\tau$ if $i > j$, or by $d_{j-1}d_i\tau$ if $i < j$.

There are obvious inclusions $\Delta_{i,j} \to \Delta_i \to \Delta[n]$. Let X be the colimit of the following diagram:

$$
\begin{array}{ccccccccc}
\Delta_{k+2} & \xrightarrow{id} & \Delta_{k+2} & \xrightarrow{id} & \Delta_n & & \Delta_1 & \xrightarrow{id} & \Delta_{k-1} \\
\uparrow & & \uparrow & & \uparrow & & \uparrow & & \uparrow \\
\Delta_{k+1,k+2} & & \Delta_{k+2,k+3} & \cdots & \Delta_{n,0} & & \Delta_{0,1} & \cdots & \Delta_{k-2,k-1} \\
\downarrow & & \downarrow & & \downarrow & & \downarrow & & \downarrow \\
\Delta_{k+1} & & \Delta_{k+3} & \xrightarrow{id} & \Delta_0 & \xrightarrow{id} & \Delta_0 & \xrightarrow{id} & \Delta_{k-2}
\end{array}
$$

Out of the construction of X, we have two natural inclusions $\Delta_{k-1} \to X$, $\Delta_{k+1} \to X$. Notice that there is a cofibration map $\Delta_{k-1,k+1} \vee_{\{k\}} \Delta_{k-1,k+1} \to X$ which is the wedge of the following maps:

$$\Delta_{k-1,k+1} \to \Delta_{k-1} \to X$$

$$\{k\} \to \Delta_{k-1} \to X$$

$$\Delta_{k-1,k+1} \to \Delta_{k+1} \to X$$

By laborious but straightforward calculation one can show that:

$$\Delta[n, k] = colim\left(\Delta_{k-1,k+1} \xleftarrow{id \vee id} \Delta_{k-1,k+1} \vee_{\{k\}} \Delta_{k-1,k+1} \to X\right)$$

Since $\Delta[n, k]$ is built from standard simplices by push-out process, where the maps involved are cofibrations, according to the lemma 7.5, $\Delta[n, k]$ belongs to $D(C)$. $\qquad\square$

Proof of the Theorem. The proof will be divided into several steps.

STEP 1. *Let* $E \xrightarrow{\simeq} B$ *be a fibration and a weak equivalence. If* $E \in D(C)$, *then* $B \in D(C)$.

Proof. Let $F : B \to C$ be a diagram. We have to show that $\oint_B F \in C$. According to section A.1, the following is a pull-back square:

$$
\begin{array}{ccc}
\displaystyle\oint_E F \circ p & \longrightarrow & \displaystyle\oint_B F \\
\downarrow & & \downarrow \\
E & \xrightarrow{\quad p \quad} & B
\end{array}
$$

Since p is a fibration and a weak equivalence, $\oint_E F{\circ}p \to \oint_B F$ is a weak equiva-

lence as well. Because $E \in D(C)$, $\oint_E F{\circ}p$ belongs to C. It implies that $\oint_B F \in C$.

STEP 2. *Let $f : X \to Y$ be a weak equivalence. If $X \in D(C)$, then $Y \in D(C)$.*

Proof. We are going to construct by the induction a sequence of spaces and inclusions:

$$(X^0 \to X^1 \to X^2 \to \cdots)$$

together with a sequence of maps:

$$\{i_l : X \to X^l\}_{l \geq 0} , \ \{p_l : X^l \to Y\}_{l \geq 0}$$

such that $X^l \in D(C)$, i_l is a cofibration and a weak equivalence, $i_l{\circ}p_l = f$ and the map $colimit(p_l) : colimit(X^l) \to Y$ is a fibration. We denote $colimit(p_l)$ by $p : \bar{X} \to Y$, in particular $\bar{X} = colimit(X^l)$.

Let $X_0 = X$, $i_0 = id_X$ and $p_0 = f$. Lets assume that the construction has been carried out for $i < l$. Let J be the set of all commutative diagrams of the form:

$$
\begin{array}{ccc}
\Delta[n,k] & \longrightarrow & X^{l-1} \\
\downarrow & & \downarrow {\scriptstyle p_{l-1}} \\
\Delta[n] & \longrightarrow & Y
\end{array}
$$

where $\Delta[n,k] \to \Delta[n]$ is the canonical inclusion. X_l is defined to be the simplicial set that fits into the following push-out square:

$$
\begin{array}{ccc}
\bigsqcup_J \Delta[n,k] & \longrightarrow & X^{l-1} \\
\downarrow & & \downarrow \\
\bigsqcup_J \Delta[n] & \longrightarrow & X^l
\end{array}
$$

i_l is defined to be the following composition:

$$X \overset{i_{l-1}}{\to} X^{l-1} \to X^l$$

p_l is defined to be the push-out of the following maps:

$$X^{l-1} \overset{p_{l-1}}{\to} Y , \ \bigsqcup_J \Delta[n,k] \to Y , \ \bigsqcup_J \Delta[n] \to Y$$

By the inductive assumption $X^{l-1} \in D(C)$. Since X^l is built by gluing lots of $\Delta[n]$ along $\Delta[n,k]$ to X^{l-1}, according to lemma 7.5, $X^l \in D(C)$. Observe that the natural map $i : X = X^0 \to \bar{X}$ is a weak equivalence. Notice also that $p{\circ}i = f$. By Quillen's small object argument (see [10]), p is a fibration. Since f and i are weak equivalences, so is p.

Lemma 7.5 implies that $\bar{X} \in D(C)$. Since $p : \bar{X} \to Y$ is a fibration and a weak equivalence, according to step 1, $Y \in D(C)$.

STEP 3. *If X is contractible, then $X \in D(C)$.*

Proof. Since $\star \in D(C)$ and $\star \to X$ is a weak equivalence, step 2 implies that $X \in D(C)$.

STEP 4. *Let $F : K \to C$ be a diagram. The homotopy fiber $Fib(\oint_K F \to K)$ belongs to C.*

Proof. Lets choose a connected component of K and a fibration $PK \to K$ such that PK is contractible and the image of PK is in the chosen component. According to corollary A.2, the homotopy fiber of $\oint_K F \to K$ over the chosen component is weakly equivalent to $\oint_{PK} F$. Since PK is contractible, according to step 3, $\oint_{PK} F$ belongs to C.

STEP 5. *Let $Z \to E \to B$ be a fibration sequence. If $B \in D(C)$ and $Z \in C$, then $E \in C$.*

Proof. We can assume that p is a fibration. Example 3.12 implies that $E \to B$ is weakly equivalent to $\oint_{sdB} dp \to sdB$, where $dp : sdB \to Spaces$ is a good diagram whose values are weakly equivalent to Z. Proposition A.5 gives the following weak equivalence:

$$\oint_B \oint_{N(B/\sigma)} (dp)\circ l_\sigma \to \oint_{sdB} dp$$

Since $N(B/\sigma)$ is contractible $\oint_{N(B/\sigma)} (dp)\circ l_\sigma \in C$. The assumption $B \in D(C)$ implies:

$$E \simeq \oint_B \oint_{N(B/\sigma)} (dp)\circ l_\sigma \in C$$

STEP 6. *Let $f : X \to Y$ be a weak equivalence. If $Y \in D(C)$, then $X \in D(C)$.*

Proof. Let $F : X \to C$ be a diagram. Notice that the homotopy fiber of the composition $\oint_X F \to X \xrightarrow{f} Y$ is weakly equivalent to the homotopy fiber $Fib(\oint_X F \to X)$. According to step 4, it belongs to C. Since $Y \in D(C)$, Step 5 implies that $\oint_X F \in C$. This proves that $X \in D(C)$. $\square$

Theorem 7.4 implies an interesting characterization of a closed class (see also[1]):

7.7. Corollary. *Non empty class C of connected simplicial sets is closed if it is closed under weak equivalences and for every, not necessarily pointed diagram,*

$F : K \to C$ *over a contractible simplicial set* K, $\displaystyle\oint_K F \in C$.

The definition of a closed class says that it is closed under pointed homotopy colimits. It means that for any pointed diagram $F : K \to Spaces_*$ the homotopy cofiber:

$$Cof\left(K \to \oint_K F\right) = \int_K F$$

belongs to C. The next corollary implies that the dual statement is also true (see [6] and [7] for discussion of similar statements).

7.8. Corollary. *Let $F : K \to C$ be a diagram.* $Fib\left(\displaystyle\oint_K F \to K\right) \in C$.

The following corollary says that if the the pre-images of simplices have certain properties (belong to a closed class), then so does the homotopy fiber (see also [7]).

7.9. Corollary. *Let $f : X \to K$ be a map. If for every simplex $\sigma \in K$:*

$$pullback\left(X \xrightarrow{f} K \leftarrow \Delta[dim(\sigma)]\right) \in C$$

then $Fib(X \xrightarrow{f} Y) \in C$.

Proof. According to example 3.12, $f : X \to K$ is weakly equivalent to a map of the form $\displaystyle\oint_{sdK} df \to sdK$, where for $v = (\sigma_0 \to \cdots \to \sigma_n) \in sdK$, df is a diagram such that:

$$df(v) = pullback\left(X \xrightarrow{f} K \leftarrow \Delta[dim(\sigma_n)]\right)$$

Corollary 7.8 implies $Fib\left(\displaystyle\oint_{sdK} df \to sdK\right) \in C$. $\qquad\qquad\square$

8. Class $D(C)$

In this section we present other characterizations of the class $D(C)$. We will restrict the class of diagrams on which a simplicial set should be tested in order to find out if it belongs to $D(C)$. We will show also that the class $D(C)$ is a closed class and is closed under extensions fibrations.

8.1. Proposition.

$$D(C) = \{K \mid \text{if } F : K \to C \text{ is a bounded diagram, then } \oint_K F \in C\}$$

Proof. Let:

$$D' = \{K \mid \text{if } F : K \to C \text{ is a bounded diagram, then } \oint_K F \in C\}$$

Inclusion $D(C) \subset D$ is obvious.

By the same arguments as in theorem 7.4, we can show that the class D' is closed under weak equivalences. Let $K \in D'$ and $F : K \to C$ be a diagram. According to remark A.6, $\oint_K F$ is weakly equivalent to $\oint_{sdK} F_{sd}$. Since sdK is weakly equivalent to K, it belongs to D'. Notice that F_{sd} is a bounded diagram, therefore $\oint_{sdK} F_{sd} \in C$. It implies that $\oint_K F \in C$ and $K \in D(C)$.   □

8.2. Proposition.

$D(C) = \{B \mid$ if $Z \to E \to B$ is a fibration sequence and $Z \in C$, then $E \in C\}$

Proof. Let:

$D' = \{B \mid$ if $Z \to E \to B$ is a fibration sequence and $Z \in C$, then $E \in C\}$

Let $B \in D$ and $F : B \to C$ be a diagram. Since $\oint_{PB} F \to \oint_B F \to B$ is a fibration sequence (see A.2) and $\oint_{PB} F \in C$, we get $\oint_B F \in C$. It implies the inclusion $D' \subset D(C)$.

Let $K \in D(C)$ and $Z \to E \xrightarrow{p} K$ be a fibration sequence. According to example 3.12, $E \to K$ is weakly equivalent to a map of the form $\oint_L F \to L$, where the values of F are weakly equivalent to Z. Since L is weakly equivalent to K, it belongs to $D(C)$. As a consequence we get $E \simeq \oint_L F \in C$. It proves that $K \in D'$ and $D \subset D'$.   □

8.3. Corollary. *A closed class C is closed under extensions by fibrations if and only if $C = D(C)$.*

The next corollary says that class $D(C)$ is usually quite big.

8.4. Corollary. *If B is such that $\Omega B \in C$, then $B \in D(C)$.*

Proof. See corollary 6.6.   □

8.5. Proposition.

$D(C) = \{K \mid$ if $F : K \to C$ is a good diagram, then $\oint_K F \in C\}$

Proof. Let:

$D' = \{K \mid$ if $F : K \to C$ is a good diagram, then $\oint_K F \in C\}$

Inclusion $D(C) \subset D'$ is obvious.

By the same arguments as in the theorem 7.4, we can show that the class D' is closed under weak equivalences. Let $B \in D'$ and $p : E \to B$ be a fibration such that the fiber of p belongs to C. According to example 3.12, $p : E \to B$ is weakly equivalent to a map of the form $\oint_{sdB} dp \to sdB$, where dp is a good diagram whose values are weakly equivalent to the fiber of p. It implies that $E \simeq \oint_{sdB} dp \in C$. This proves the proposition. $\qquad\square$

8.6. Theorem. *Let $G : K \to D(C)$ be a diagram. If K belongs to $D(C)$, then so does $\oint_K G$.*

Proof. According to remark A.6, $\oint_K G$ is weakly equivalent to $\oint_{sdK} G_{sd}$. Theorem 7.4 implies that $\oint_K G$ belongs to $D(C)$ if and only if $\oint_{sdK} G_{sd}$ does. Since G_{sd} is a bounded diagram, without loss of generality, it is enough to prove the theorem for a bounded diagram $G : K \to D(C)$.

Let $G : D(C) \to$ be a bounded diagram and $F : \oint_K G \to C$ be a diagram. Theorem A.9 implies that $\oint_{\oint_K G} F$ is weakly equivalent to $\oint_{sdK} \oint_{\Delta G(v)} F$. Since G has values in $D(C)$, then so does ΔG, therefore $\oint_{\Delta G(v)} F$ belongs to C. Because $K \in D(C)$, $sdK \in D(C)$ and it follows that $\oint_{sdK} \oint_{\Delta G(v)} F$ belongs to C. This proves $\oint_{\oint_K G} F \in C$. $\qquad\square$

8.7. Corollary. *$D(C)$ is a closed class and $D(D(C)) = D(C)$, therefore $D(C)$ is closed under extensions by fibrations.*

9. Theorem of E. Dror Farjoun

9.1. Theorem. *Let $\Psi : E \to B$ be a natural transformation between diagrams $E : K \to Spaces$ and $B : K \to Spaces$. If for every simplex $\sigma \in K$ the homotopy fiber $Fib(E(\sigma) \xrightarrow{\Psi_\sigma} B(\sigma))$ belongs to C, then:*

$$Fib\left(\oint_K E \xrightarrow{\Psi} \oint_K B\right) \in C$$

9.2. Lemma. *Lets consider the following commutative diagram:*

$$
\begin{array}{ccccc}
E_1 & \longleftarrow & E_2 & \longrightarrow & E_3 \\
\downarrow{\scriptstyle p_1} & & \downarrow{\scriptstyle p_2} & & \downarrow{\scriptstyle p_3} \\
B_1 & \xleftarrow{\ f\ } & B_2 & \xrightarrow{\ g\ } & B_3
\end{array}
$$

where the maps $E_2 \to E_3$, $B_2 \xrightarrow{g} B_3$ are cofibrations. If $Fib(p_1)$, $Fib(p_2)$ and $Fib(p_3)$ belong to C, then:

$$Fib(E_1 \cup_{E_2} E_3 \to B_1 \cup_{B_2} B_3) \in C$$

Proof. Without loss of generality we can assume that p_1, p_2 and p_3 are fibrations. Let $p = colim(p_1 \leftarrow p_2 \rightarrow p_3)$. According to corollary 7.9, it is enough to prove that for every simplex, $\sigma \in B_1 \cup_{B_2} B_3$:

$$F(\sigma) = pullback(E_1 \cup_{E_2} E_3 \xrightarrow{p} B_1 \cup_{B_2} B_3 \leftarrow \Delta[dim(\sigma)]) \in C$$

Let $\sigma \in B_1 \cup_{B_2} B_3$. Either σ lies in the image of B_1 or B_2. Lets assume that it belongs to the image of B_1. Let $K = pullback(\Delta[dim(\sigma)] \rightarrow B_1 \xleftarrow{f} B_2)$. There is a natural map $K \rightarrow B_2$. Let X_1, X_2 and X_3 be simplicial sets that fit into the following pull-back squares:

$$
\begin{array}{ccc}
X_1 \longrightarrow E_1 & \quad X_2 \longrightarrow E_2 & \quad X_3 \longrightarrow E_3 \\
\downarrow \qquad \downarrow p_1 & \quad \downarrow \qquad \downarrow p_2 & \quad \downarrow \qquad \downarrow p_3 \\
\Delta[dim(\sigma)] \longrightarrow B_1 & \quad K \longrightarrow B_2 & \quad K \longrightarrow B_3
\end{array}
$$

Observe that the definition gives natural maps $X_2 \rightarrow X_1$ and $X_2 \rightarrow X_3$. By straightforward combinatorial calculation one can show:

$$F(\sigma) = colim(X_3 \leftarrow X_2 \rightarrow X_1)$$

Notice that the maps $X_3 \rightarrow K$, $X_2 \rightarrow K$ and $X_2 \rightarrow X_1$ satisfy the assumptions of theorem 6.3, therefore $hocolim(X_3 \leftarrow X_2 \rightarrow X_1) \in C$. Cofibration assumption of the lemma implies:

$$hocolim(X_3 \leftarrow X_2 \rightarrow X_1) \simeq colim(X_3 \leftarrow X_2 \rightarrow X_1)$$

It proves the lemma. □

Proof of the theorem. Instead of $E : K \rightarrow Spaces$, $B : K \rightarrow Spaces$ we can consider bounded diagrams $E_{sd} : sdK \rightarrow Spaces$, $B_{sd} : sdK \rightarrow Spaces$. Since the homotopy fibers $Fib(\oint_K E \rightarrow \oint_K B)$ and $Fib(\oint_{sdK} E_{sd} \rightarrow \oint_{sdK} B_{sd})$ are weakly equivalent, it is enough to prove the theorem for bounded diagrams.

The proof will be by the induction on the dimension of K. If $dim(K) = 0$, the theorem is obvious. Lets assume that the theorem is true for K such that $dim(K) < n$. Let L be a simplicial of dimension less than n and $\dot\Delta[n] \rightarrow L$ be a map. We prove that the theorem holds for $K = L \cup_{\dot\Delta[n]} \Delta[n]$.

Let $\tau \in (\Delta[n])_n$ be the only non degenerate simplex. Lets consider the following diagram:

$$
\begin{array}{ccccccc}
\oint_K E & = & colim & (& \oint_L E & \leftarrow & \dot\Delta[n] \times E(\tau) & \hookrightarrow & \Delta[n] \times E(\tau) &) \\
\downarrow \Psi & & & & \downarrow \Psi & & \downarrow id \times \Psi_\tau & & \downarrow id \times \Psi_\tau & \\
\oint_K B & = & colim & (& \oint_L B & \leftarrow & \dot\Delta[n] \times B(\tau) & \hookrightarrow & \Delta[n] \times B(\tau) &)
\end{array}
$$

By the inductive assumption $Fib\left(\oint_L E \xrightarrow{\Psi} \oint_L B\right)$ belongs to C. Since the homotopy fiber $Fib(E(\tau) \xrightarrow{\Psi_\tau} B(\tau))$ also belongs to C, according to lemma 9.2:

$$Fib\left(\oint_K E \xrightarrow{\Psi} \oint_K B\right) \in C \qquad \square$$

As a corollary we get the theorem of E. Dror Farjoun

9.3. Corollary. (E. Dror Farjoun [7, theorem I]). *Let $E : K \to Spaces_\star$ and $B : K \to Spaces_\star$ be pointed diagrams and $\Psi : E \to B$ be a natural transformation. If for every simplex $\sigma \in K$, $Fib(E(\sigma) \xrightarrow{\Psi_\sigma} B(\sigma)) \in C$, then:*

$$Fib\left(\int_K E \xrightarrow{\Psi} \int_K B\right) \in C$$

Proof. Lets consider the following diagram:

$$
\begin{array}{ccccccc}
\int_K E & \simeq & hocolim & (& \star & \leftarrow & K & \to & \oint_K E &) \\
\downarrow{\scriptstyle\Psi} & & & & \downarrow & & \downarrow{\scriptstyle id} & & \downarrow{\scriptstyle\Psi} & \\
\int_K B & \simeq & hocolim & (& \star & \leftarrow & K & \to & \oint_K B &)
\end{array}
$$

According to theorem 9.1, $Fib\left(\oint_K E \xrightarrow{\Psi} \oint_K B\right)$ belongs to C. Since the homotopy fiber $Fib(K \xrightarrow{id} K)$ belongs to C, applying once again theorem 9.1 we get:

$$Fib\left(\int_K E \xrightarrow{\Psi} \int_K B\right) \in C \qquad \square$$

9.4. Theorem. *Let the following be a homotopy push-out square:*

$$
\begin{array}{ccc}
A & \xrightarrow{\;f\;} & B \\
\downarrow{\scriptstyle i} & & \downarrow \\
X & \xrightarrow{\;g\;} & Y
\end{array}
$$

If the homotopy fiber $Fib(A \xrightarrow{f} B)$ belongs to C, then so does the homotopy fiber $Fib(X \xrightarrow{g} Y)$.

Proof. Lets consider the following diagram:

$$
\begin{array}{ccccccc}
X & \simeq & hocolim & (& X & \xleftarrow{i} & A & \xrightarrow{id} & A &) \\
\downarrow{\scriptstyle g} & & & & \downarrow{\scriptstyle id} & & \downarrow{\scriptstyle id} & & \downarrow{\scriptstyle f} & \\
Y & \simeq & hocolim & (& X & \xleftarrow{i} & A & \xrightarrow{f} & B &)
\end{array}
$$

Since $Fib(A \xrightarrow{f} B)$ belongs to C, theorem 9.1 implies that $Fib(X \xrightarrow{g} Y) \in C$.
$\square$

9.5. Corollary. *Let $f : X \to Y$ be a map. If X belongs to C, then so does the homotopy fiber $Fib\big(Y \to Cof(X \xrightarrow{f} Y)\big)$.*

Proof. Apply theorem 9.4 to the following homotopy push-out square:

$$
\begin{array}{ccc}
X & \longrightarrow & \star \\
\downarrow & & \downarrow \\
Y & \longrightarrow & Cof(X \xrightarrow{f} Y)
\end{array}
\qquad \square
$$

Appendix A. The homotopy colimit

The reference for the proofs of the statements, listed in the appendix, is [5].

A.1. Pulling-back of diagrams. Let $F : K \to Spaces$ be a diagram over K and $f : L \to K$ be a map. We can pull-back F into a diagram $F{\circ}f$ over L ($F{\circ}f$ will be often denoted simply by F). Let τ and σ be simplices in L and $\varphi : \tau \to \sigma$ be a morphism in the category associated with L. $F{\circ}f : L \to Spaces$ is defined as follows:

$$
F{\circ}f(\sigma) = F(f(\sigma))
$$
$$
F{\circ}f(\varphi) = F(\varphi)
$$

The basic property of the pull-back diagram $F{\circ}f$ is that the following is a pull-back square:

$$
\begin{array}{ccc}
\displaystyle\oint_L F{\circ}f & \longrightarrow & \displaystyle\oint_K F \\
\downarrow & & \downarrow \\
L & \xrightarrow{\;\;f\;\;} & K
\end{array}
$$

As corollary of the this property we get:

A.2. Corollary. *Let $F : K \to Spaces$ be a diagram and $PK \to K$ be a fibration such that PK is contractible. The following is a fibration sequence:*

$$
\oint_{PK} F \to \oint_K F \to K
$$

A.3. Diagrams over $colim_I G$. Let $G : I \to Space$, $F : colim_I G \to Spaces$ be diagrams. There is a family of maps $\{l_i : G(i) \to colim_I G\}_{i \in I}$ which satisfies the universal property of the colimit of the diagram $G : I \to Space$ (see [8, 3§3]). Out of this data we can construct a functor:

$$
I \longrightarrow Spaces
$$
$$
i \longmapsto \oint_{G(i)} F{\circ}l_i
$$
$$
(a \xrightarrow{\varphi} b) \longmapsto \Big(\oint_{G(a)} F{\circ}l_a \xrightarrow{\;\oint_{G(\varphi)} F{\circ}l_b\;} \oint_{G(b)} F{\circ}l_b \Big)
$$

This functor has the following property:

$$\oint_{colim_I G} F = colim_G \oint_{G(i)} F \circ l_i$$

A.4. Diagrams over sdK. Let σ be a simplex in K. By K/σ we denote the over category of K (see section 2). There is a functor:

$$K/\sigma \to K$$
$$(\tau \to \sigma) \mapsto \tau \; , \; (\tau_0 \to \tau_1 \to \sigma) \mapsto (\tau_0 \to \tau_1)$$

This functor induces a map between simplicial sets:

$$l_\sigma : N(K/\sigma) \to N(K) = sdK$$
$$(\tau_0 \to \cdots \to \tau_n \to \sigma) \overset{l_\sigma}{\mapsto} (\tau_0 \to \cdots \to \tau_n)$$

One can verify that the family of maps $\{l_\sigma : N(K/\sigma) \to sdK\}_{\sigma \in K}$ satisfies the universal property of the colimit of the functor:

$$K \to Spaces \; , \; \sigma \mapsto N(K/\sigma)$$

It implies:

$$sdK = colim_K N(K/\sigma)$$

Let $F : sdK \to Spaces$ be a diagram. According to subsection A.3:

$$\oint_{sdK} F = \oint_{colim_K N(K/\sigma)} F = colim_K \oint_{N(K/\sigma)} F \circ l_\sigma$$

A.5. Proposition. *The natural map:*

$$\oint_K \oint_{N(K/\sigma)} F \circ l_\sigma \to colim_K \oint_{N(K/\sigma)} F \circ l_\sigma = \oint_{sdK} F$$

is a weak equivalence.

A.6. Remark. Let $F : K \to Spaces$ be a diagram. This means that F is a functor over the category associated to K. According to example 3.14, it defines a bounded diagram $F_{sd} : sdK \to Spaces$. It turns out that $\oint_K F$ is weakly equivalent to $\oint_{sdK} F_{sd}$ and $\oint_{sdK} F_{sd} = hocolim_K F$, where $hocolim_K F$ is the homotopy colimit of $F : K \to Spaces$ in the sense of Bousfield-Kan.

A.7. Diagrams over $\oint_K G$. Let $G : K \to Spaces$ be a diagram. Out of G we can construct a new diagram $\Delta G : sdK \to Spaces$. Let:

$$u = (\tau_0 \overset{\psi_0}{\to} \tau_1 \overset{\psi_1}{\to} \ldots \overset{\psi_{m-1}}{\to} \tau_m) \in (sdK)_m$$
$$v = (\sigma_0 \overset{\varphi_0}{\to} \sigma_1 \overset{\varphi_1}{\to} \ldots \overset{\varphi_{n-1}}{\to} \sigma_n) \in (sdK)_n$$
$$\eta : u \to v \text{ be a morphism in } sdK$$

By definition 3.1, η is a morphism in Δ such that $\eta : [n] \to [m]$ and $sdK(\eta)(u) = v$. $\Delta G : sdK \to Spaces$ is a diagram defined as follows:

$$\Delta G(v) = \Delta[dim(\sigma_n)] \times G(\sigma_0)$$

$$\Delta G(\eta) = \begin{cases} id \times id & \text{if } \eta(0) = 0 \ , \ \eta(n) = m \\ id \times G(\psi_{\eta(0)-1} \circ \cdots \circ \psi_0) & \text{if } \eta(0) > 0 \ , \ \eta(n) = m \\ \Delta[\psi_{m-1} \circ \cdots \circ \psi_{\eta(n)}] \times id & \text{if } \eta(0) = 0 \ , \ \eta(n) < m \\ \Delta[\psi_{m-1} \circ \cdots \circ \psi_{\eta(n)}] \times G(\psi_{\eta(0)-1} \circ \cdots \circ \psi_0) & \text{if } \eta(0) > 0 \ , \ \eta(n) < m \end{cases}$$

Observe that ΔG has the values weakly equivalent to the values of G.

A.8. Proposition.

$$\oint_K G = colim_{sdK} \Delta G$$

A.9. Theorem. *Let* $G : K \to Spaces$ *and* $F : \oint_K G \to Spaces$ *be diagrams. If* G *is a bounded diagram, then the following natural map is a weak equivalence:*

$$\oint_{sdK} \oint_{\Delta G(v)} F \to colim_{sdK} \left(\oint_{\Delta G(v)} F \right) = \oint_{\oint_K G} F$$

References

[1] A. Amit, *Direct limits over diagrams with contractible nerve*, Master thesis, Hebrew Univ. (1993).

[2] A.K. Bousfield, *Localization and periodicity in unstable homotopy theory*, preprint.

[3] A.K. Bousfield and D.M. Kan, *Homotopy Limits, Completions and Localizations*, Lect. Notes in Math. 304, Springer (1972)

[4] W. Chachólski, *Functors* CW_A *and* P_A, Ph.D. thesis, Univ. of Notre Dame (1995).

[5] W. Chachólski, *Homotopy properties of shapes of diagrams*, report No. 6, 1993/94, Institut Mittag-Leffler.

[6] E. Dror Farjoun, *Cellular spaces*, preprint.

[7] E. Dror Farjoun, *Cellular inequalities*, Proc. to the conf. in Alg. Top. Northeastern Univ. June 1993, Springer Verlag.

[8] S. MacLane, *Categories for working mathematician*, Grad. Texts in Math. 5, Springer (1971).

[9] J. Peter May, *Simplicial objects in algebraic topology*, Van Nostrand Math. Studies 11, (1987).

[10] D. Quillen, *Homotopical algebra*, Lect. Notes in Math. 43, Springer (1967)

[11] G. Segal, *Classifying spaces and spectral sequences*, Inst. haut. Etul. sci., Publ. math. 34, 105–112 (1968).

Wojciech Chachólski, Department of Mathematics, University of Notre Dame
Mail Distribution Center, Notre Dame, Indiana 46556-5683
wchacho1@kenna.math.nd.edu

Progress in Mathematics, Vol. 136
© 1996 Birkhäuser Verlag Basel/Switzerland

On mapping class groups from a homotopy theoretic point of view

F.R. COHEN[*]

The mapping class group for a fixed oriented surface is the group of path-components for the group of orientation preserving diffeomorphisms of the surface. These groups are traditionally encountered in the realms of low dimensional topology, classical algebraic geometry of curves, complex analysis and elsewhere. In a somewhat different direction, there are intimate connections to both classical non-stable and stable homotopy theory from several points of view. These connections in turn inform on the mapping class groups. It is the purpose of this article (1) to explain some of these connections and (2) to obtain some new results.

Some of the topics considered are
(1) connections to vector bundles and to the J-homomorphism of stable homotopy theory,
(2) connections to cyclic homology and the homology of function spaces, and
(3) bundles obtained from the classical Hopf bundle and flag varieties.

These constructions are applied to obtain some new homology classes for the mapping class groups. In addition, these bundles in turn give fibrations which provide a relation between the classical EHP sequence in non-stable homotopy theory and the homology of the mapping class group. A few applications are given.

The following is a table of contents.

We would like to thank the organizers of the BCAT '94. We would also like to thank Hans-Werner Henn, Henry Glover, Guido Mislin and Yining Xia.

---

[*] Partially supported by the NSF

1 Definitions and preliminaries

Let S_g denote a closed and orientable Riemann surface of genus g.

Let $Diff^+(S_g)$ denote the group of orientation preserving diffeomorphisms. If k is a fixed non-negative integer, then $Diff^+(S_g; k)$ is the subgroup of $Diff^+(S_g)$ which leaves a given set of k points $\{x_1, \ldots, x_k\}$ in S_g invariant.

The mapping class group Γ_g is defined to be $\pi_0 Diff^+(S_g)$ the group of components of $Diff^+(S_g)$. Similarly, Γ_g^k is $\pi_0 Diff^+(S_g; k)$.

These groups are tied closely to configuration spaces as considered in work of Fadell and Neuwirth [FN]. Namely $F(M, k)$ is the subspace $\{(m_1, \ldots, m_k) \in M^k \mid m_i \neq m_j \text{ if } i \neq j\}$. Evidently $Diff^+(S_g)$ acts on $F(S_g, k)$ by the diagonal action. The symmetric group on k-letters Σ_k also acts on $F(S_g, k)$ and commutes with the $Diff^+(S_g)$-action; let $B(S_g, k)$ denote $F(S_g, k)/\Sigma_k$.

The spaces $B(S_g, k)$ arise in studying the "difference" between Γ_g and Γ_g^k, $k \geq 1$, and is described below:

(1) If $q \geq 3$, then

$$ESO(3) \times_{SO(3)} B(S^2, q) = K(\Gamma_0^q, 1) \quad [\text{Co, BCP}].$$

(2) If $q \geq 2$, then

$$EDiff^+(S^1 \times S^1) \times_{Diff^+(S^1 \times S^1)} B(S^1 \times S^1, q) = K(\Gamma_1^q, 1) \quad [\text{CH}].$$

(3) If $g \geq 2$, then

$$EDiff^+(S_g) \times_{Diff^+(S_g)} B(S_g, k) = K(\Gamma_g^k, 1) \quad [\text{Co, CH}].$$

These results which are almost certainly "folk theorems" follow at once from the natural transitive action of $Diff^+(S_g)$ on $B(S_g; k)$. In this case, there is a principle $Diff^+(S_g, k)$ fibration given by $Diff^+(S_g; k) \to Diff(S_g) \to B(S_g, k)$. The proofs of (1)–(3) are gotten by applying the long exact homotopy sequence of a fibration together with results in [St]. The groups Γ_0^k and Γ_g^0 will be denoted by Γ^k and Γ_g respectively.

There are stabilization maps for related groups $\Gamma_{g,1}$, given by $\pi_0 Diff^+(S_g; D^2)$ the path components of diffeomorphisms which fix a chosen disk in S_g [M]. Namely, there are maps

$$\Gamma_{g,1} \times \Gamma_{h,1} \to \Gamma_{g+h,1}$$

obtained by taking connected sums along their fixed disks. Let

$$\Gamma = \varinjlim_g \Gamma_{g,1}.$$

Recall that Γ_g acts on $H_1(S_g; R)$ for any ring R. Thus there are homomorphisms

$$\rho : \Gamma_g \to Sp(2g, R) \qquad \text{and} \qquad \rho : \Gamma_{g,1} \to Sp(2g, R)$$

as the Γ_g-action preserves the cup-product form. These representations are called the homology representation. Passage to limits gives a homomorphism

$$\rho : \Gamma \to Sp(R).$$

The mapping class groups Γ_g have a large number of accessible subgroups. One useful such subgroup is the centralizer of the class of the hyperelliptic involution Δ_g [Co]. If $g = 2$, then $\Delta_2 = \Gamma_2$. However, Δ_g is a proper subgroup of Γ_g if $g > 2$. Let Ω denote the class of the hyperelliptic involution in Γ_g. A diffeomorphism which represents Ω is gotten by regarding S_g as a sausage which is given a half-twist [Bir].

2 On "image of J spaces" for mapping class groups

This section describes relations between classifying spaces for stable vector bundles and the classifying space for the mapping class group. In particular certain elements in the cohomology of the mapping class group arise from Chern classes of representations [CL, GM, GM_2]. On the other hand, these classes sometimes arise from a more geometric setting.

First of all, it has been known for some time that the stable homotopy groups of spheres split into a sum of groups given by (1) the image of the J-homomorphism, and (2) an, as yet mysterious, second summand given by the cokernel. One attractive way of seeing this decomposition is through J. Tørnehave's thesis [Tør]. In particular, he exhibits a homotopy equivalence

$$SG \to J \times \operatorname{coker} J$$

where (1) $SG = \Omega_{(1)}^{\infty} S^{\infty}$ with $\pi_i SG$ isomorphic to the i-th stable homotopy group of the sphere, (2) J is a connected H-space with torsion homology groups and is thus homotopy equivalent to a product of its' localizations at each prime p, $J_{(p)}$, (3) if $p = 2$, $J_{(2)}$ is homotopy equivalent to the localization at 2 of the fibre of $\Psi^3 - 1 : BSO \to BSO$, and (4) if $p > 2$, $J_{(p)}$ is homotopy equivalent to the localization at p of the fibre of $\Psi^q - 1 : BU \to BU$ where q is an odd prime power with $q^i - 1 \not\equiv 0(p)$ if $1 \le i \le p - 2$, and $\nu_p(q^{p-1} - 1) = 1$.

By inspecting the p-Sylow subgroups of $GL(2g, \mathbb{F}_q)$ and using a theorem of Quillen [Q], the following result was shown in [CC].

THEOREM 2.1. *If p is an odd prime, then $J_{(p)}$ is a stable retract of $B\Gamma$. Thus the natural map*

$$B\Gamma \to BGL(\mathbb{F}_q)$$

induced by the homology representation of Γ with coefficients in $\mathbb{F}_q$ induces a split
epimorphism after localization at p on any homology theory.

Consider the Adams operation Ψ^q in quaternionic K-theory $\tilde{K}(Sp(\cdot))$ to obtain a map

$$\Psi^q - 1 : BSp \to BSp.$$

As in [FP], let $JSp(q)$ denote the homotopy theoretic fibre of $\Psi^q - 1$.

THEOREM 2.2 [Q, FP].

    If q is as above with $(p, q) = 1$, there is a map

$$BSp(\mathbb{F}_q) \to JSp(q)$$

which induces a mod-p homology isomorphism. Furthermore $JSp(q)_p^\wedge$ is homotopy equivalent to $J_{(p)}$.

    Define the space $J(\Gamma)$ by

$$J(\Gamma) = (\prod_{p>2} J_{(p)})$$

A restatement of Theorem 2.1 is the following

THEOREM 2.3.

    The homology representation

$$B\rho : B\Gamma \to J(\Gamma)$$

induces a split epimorphism for any homology theory. Furthermore, there is a stable map σ from $J(\Gamma)$ to $B\Gamma$ such that $(B_\rho) \cdot \sigma$ is a stable homotopy equivalence. Thus $J(\Gamma)$ is a stable retract of $B\Gamma$.

    Whether there is a 2-primary analogue of this last theorem is an open question. One form of this question is to consider

$$B\rho : B\Gamma \to BSp(\mathbb{F}_q)$$

for q an odd prime power and ask whether (1) this map gives a surjection on homology with $\mathbb{Z}_{(2)}$-coefficients or (2) this map has a 2-local stable cross-section.

    An application of the fact that $K_3(\mathbb{Z}) = \pi_3 BGL(\mathbb{Z})^+ = \mathbb{Z}/16\mathbb{Z} \oplus \mathbb{Z}/3\mathbb{Z}$ [LS] together with the Riemann-Hurwitz equations gives some partial answers below.

PROPOSITION 2.4. If q is an odd prime power and the map $B\rho : B\Gamma \to BSp(\mathbb{F}_q)$ has a 2-local stable section, then $H_3(BSp(\mathbb{F}_q); \mathbb{Z}_{(2)}) = \mathbb{Z}/16\mathbb{Z}$ and $\nu_2(q^2 - 1) = 4$.

    Thus by this last proposition it suffices to restrict attention to the cases for which $\nu_2(q^2 - 1) = 4$. In these cases, the 2-Sylow subgroup of $Sp(2, \mathbb{F}_q)$ is Q_3 the quaternion group of order 16; indeed the 2-Sylow subgroup of $Sp(2^k, \mathbb{F}_q)$ is the wreath product $\mathbb{Z}/2 \int_{k-1} Q_3$ [FP]. Thus to carry out the methods for stably splitting $B\Gamma$ in [CC], one must find an action of Q_3 on some surface S_g with the property that the natural map

$$BQ_3 \to B\Gamma_g \to BSp(2g, \mathbb{F}_q)$$

induces an isomorphism on $H_3(; \mathbb{Z}_{(2)}) = \mathbb{Z}/16\mathbb{Z}$.

Namely, it is known that there are isomorphisms

$$\mathbb{Z}/16\mathbb{Z} = H_3(BQ_3; \mathbb{Z}_{(2)}) \to H_3(BSp(2g, \mathbb{F}_q); \mathbb{Z}_{(2)}).$$

Thus if there are 2-local stable sections for $B\rho$, it is natural to ask whether such a map arises from an action of Q_3 on a surface. This was the situation which worked well at odd primes. Unfortunately, the next result gives that such actions do not exist. Thus the methods for giving a stable cross-section fail; this failure translates directly to the value of $K_3(\mathbb{Z})$.

PROPOSITION 2.5. *There does not exist an action of Q_3 on $S_g, g \geq 0$, such that the composite*

$$\mathbb{Z}/16\mathbb{Z} = H_3(BQ_3; \mathbb{Z}_{(2)}) \to H_3(B\Gamma_g; \mathbb{Z}_{(2)}) \to H_3(BSp(\mathbb{F}_q); \mathbb{Z}_{(2)})$$

is an isomorphism.

In short, the methods of [CC] fail at $p = 2$. A little can be salvaged by using naturality; however this result was obtained earlier in work of Glover and Mislin [GM2].

PROPOSITION 2.6. *There is an isomorphism*

$$H_3(B\Gamma; \mathbb{Z}) \to \mathbb{Z}/2^r\mathbb{Z} \oplus \mathbb{Z}/3\mathbb{Z} \oplus A$$

where $2 \leq r$ and A is a finite (possibly zero) abelian group.

It seems likely that F. Waldhausen's work on $A(X)$ would give an immediate improvement of 2.6.

The remainder of this section is speculative and concerns the homology of $B\Gamma$. Consider the homology representation $\rho : \Gamma \to Sp(\mathbb{Z})$ with the inclusion $Sp(\mathbb{Z}) \to Sp(\mathbb{R}) = \varinjlim_n Sp(2n, \mathbb{R})$. Thus $Sp(\mathbb{R})$ is homotopy equivalent to U and by real Bott periodicity, there is a multiplicative fibration

$$BO \to BU \to B(U/0),$$

where $B(U/0) = Sp/U$ (again by real Bott periodicity) [B, Car]. Thus there is a map
$$B\Gamma \to B(U/O).$$

Next, consider $Sp(n)$ the group of quaternionic linear isomorphisms of $\mathbb{R}^{4n}$ (which is not $Sp(2q, \mathbb{R})$ for any q). Let Sp denote the natural colimit of the $Sp(n)$. Thus the cohomology of BSp is given by a polynomial algebra on classes in all positive degrees which are congruent to zero mod 4.

A question which arises in studying $J(\Gamma)$ above is whether there is a map

$$\theta : B\Gamma \to J(\Gamma) \times B(U/O) \times BSp$$

which gives (1) an epimorphism on integral homology or (2) an isomorphism in homology with $\mathbb{Z}_{(\frac{1}{2})}$-coefficients. The projection to BSp cannot arise from the homology representation. The existence of a map $B\Gamma \to BSp$ will be addressed elsewhere. In addition, one might guess that $H_3(\Gamma; \mathbb{Z})$ is isomorphic to $\mathbb{Z}/2^s \oplus \mathbb{Z}/3$ for $s \leq 4$.

If $\mathbb{C}P^\infty$ were a retract of $B\Gamma^+$, then $B\Gamma^+$ is homotopy equivalent to the product $\mathbb{C}P^\infty \times (B\Gamma^+)\langle 2 \rangle$. Thus $\pi_3 B\Gamma^+$ would be isomorphic to $H_3(B\Gamma^+; \mathbb{Z})$. In this case an argument analogous to that used in the proof of Proposition 2.4 gives that $r = 2$ in the statement of Proposition 2.6. If $B\Gamma^+$ were an infinite loop space, then the proof of Theorem 2.3 gives a homotopy equivalence

$$B\Gamma^+ \to J(\Gamma) \times \mathbb{X}$$

for some space $\mathbb{X}$. By work in [M, Mo, Mu], the rational cohomology of $B(U/0) \times BSp$ injects in the rational cohomology of $B\Gamma$.

3 On Γ_g-homology, an analogue of cyclic homology

Consider the space of all continuous functions from a surface of genus g to a space X, $map(S_g, X)$. The diffeomorphism group of S_g, $\text{Diff}^+(S_g)$, acts on S_g and thus on $map(S_g, X)$. Of course there is a version with $\text{Diff}^+(S_g)$ replaced by $\text{Diff}^+(S_g; *)$ and $map(S_g, X)$ replaced by the pointed mapping space $map_*(S_g, X)$.

For a fixed g, consider

$$(i) \quad M_g(X) = E \times_{Diff^+(S_g)} map(S_g, X) \text{ and}$$
$$(ii) \quad N_g(X) = E \times_{Diff^+(S_g;*)} map_*(S_g, X)$$

where E is either a free $\text{Diff}^+(S_g)$ or a free $\text{Diff}^+(S_g; *)$ space which is contractible.

These spaces in (i) and (ii) give direct analogues of

$$E \times_{SO(2)} map(S^1, X)$$

where E is now a free contractible $SO(2)$-space. This last construction has homology (with any field coefficients $\mathbb{F}$) which is isomorphic to the cyclic homology of $H_*(\Omega X; \mathbb{F})$ [G]. In the special case for which $g = 0$ and S_0 is the 2-sphere S^2, choose an equatorial embedding of S^1 in S^2 which is compatible with the inclusion of $SO(2)$ in $SO(3)$. Given a map $f : S^1 \to X$, there is a map $\Sigma(f) : S^2 \to \Sigma X$ where Σ is the (unreduced) suspension. Thus there is a map of bundles

$$\lambda : E \times_{SO(2)} map(S^1, X) \to E \times_{SO(3)} map(S^2, \Sigma X).$$

The homology of the left-hand side (with field coefficients $\mathbb{F}$) gives the cyclic homology of $H_*(\Omega X; \mathbb{F})$ while the homology of the right-hand side has useful features some of which are given below.

In the case that X is itself a suspension, then the space $E \times_{SO(2)} map(S^1, X)$ is stably homotopy equivalent to a certain bouquet and thus the cyclic homology of the tensor algebra $H_*(\Omega X; \mathbb{F})$ is naturally bigraded [B, Cn]. There are analogous results (some of which date back to the middle '80's) in the case of $M_g(X)$ and $N_g(X)$ which have naturally bigraded homology when X is a triple suspension. These observations follow at once from theorems which are known. It is the purpose of this section to collect these observations.

To explain this natural bigrading, one needs more information about the mapping class groups Γ_g^k [Co4, BCP]. In particular, there are surjections

$$\Gamma_g^k \to \Sigma_k \to 1$$

to the symmetric groups obtained by the action of Γ_g^k on punctures. If V is a graded vector space over a field $\mathbb{F}$, then

$$V^{\otimes k} = V \otimes_{\mathbb{F}} \cdots \otimes_{\mathbb{F}} V, \quad k - \text{times}$$

is a module over the group ring $\mathbb{F}[\Sigma_k]$. Thus $V^{\otimes k}$ is a module over $\mathbb{F}[\Gamma_g^k]$ and one might consider

$$H_*(\Gamma_g^k; V^{\otimes k}) = Tor_*^{\mathbb{F}[\Gamma_g^k]}(\mathbb{F}, V^{\otimes k}).$$

For example if V is the reduced homology of the n-sphere, $H_n(S^n; \mathbb{F})$, then

(i) for n even, $V^{\otimes k}$ is a copy of the trivial representation concentrated in degree nk, and

(ii) for n odd, $V^{\otimes k}$ is a copy of the sign representation concentrated in degree nk.

The next lemma then follows at once.

LEMMA 3.1.

(i) If $V = \bar{H}_{2n}(S^{2n}; \mathbb{F})$, then there are isomorphisms

$$H_{q+2nk}(\Gamma_g^k; V^{\otimes k}) \to H_q(\Gamma_g^k; \mathbb{F})$$

where $\mathbb{F}$ is the trivial Γ_g^k-module.

(ii) If $V = H_{2n+1}(S^{2n+1}; \mathbb{F})$, then there are isomorphisms

$$H_{t+(2n+1)k}(\Gamma_g^k; V^{\otimes k}) \to H_t(\Gamma_g^k; \mathbb{F}(-1))$$

where $\mathbb{F}(-1)$ is a copy of $\mathbb{F}$ and where Γ_g^k acts on $\mathbb{F}$ by the sign of a permutation.

In general, the module $V^{\otimes k}$ is given by a direct sum of Coxeter representations with a shift of degrees. Indeed, this remark is a special case of the results in [Co3] where $B\Gamma_g^k$ is replaced by configuration spaces.

THEOREM 3.2 [BCP].
   If $g = 0, X = \Sigma^3 Y$, and $V = \bar{H}_*(\Sigma Y, \mathbb{F})$, then the homology of $M_0(X)$
and $N_0(X)$ are naturally bigraded by $H_{t,k}$. There are isomorphisms

$$H_{t,k}(M_0(X)) \longrightarrow H_t(\Gamma_0^k; V^{\otimes k})$$

for all $k \geq 3$.

   The anomalous cases for which $k < 3$ in Theorem 3.2 above arise from the
fact that $Diff^+(S^2)$ is not contractible as it is homotopy equivalent to $SO(3)$
[S]. An analogous situation arises in case $g = 1$, but "stabilizes" for $g \geq 2$;
again the reason for this anomaly is that the components of $Diff^+(S^1 \times S^1)$
are not contractible while the components of $Diff^+(S_g)$ are contractible for
$g \geq 2$ [EE].

THEOREM 3.3 [CH].
   (i) If $g = 1$, $X = \Sigma^3 Y$, and $V = \bar{H}_*(\Sigma Y; \mathbb{F})$, then the homology of $M_1(X)$
        and $N_1(X)$ is naturally bigraded by $H_{t,k}$. There are isomorphisms

$$H_{t,k} M_1(X) \longrightarrow H_t(\Gamma_1^k; V^{\otimes k})$$

        for all $k \geq 2$.

   (ii) If $g \geq 2$, $X = \Sigma^3 Y$, and $V = \bar{H}_*(\Sigma Y; \mathbb{F})$, then the homology of $M_g(X)$
        and $N_g(X)$ is naturally bigraded by $H_{t,k}$. There are isomorphisms

$$H_{t,k} M_g(X) \longrightarrow H_t(\Gamma_g^k; V^{\otimes k}).$$

   One application of these theorems has been to work out the homology
of the group Γ_0^k with either trivial coefficients or coefficients in the sign representation [Co, Co4, BCP]; the answers when $\mathbb{F} = \mathbb{F}_2$ are given in terms of the
homology of (i) $BSO(3)$ and (ii) Artin's braid groups. Results for the mod-2
homology of Γ_1^k with coefficients in $\mathbb{F}_2$ or $\mathbb{Q}$ are given in [CH]; the answers are
in terms of automorphic forms by applying work in [FTY]. In any case, the
point of these theorems is that the homology of the groups Γ_g^k can be given all
at once by the homology of certain function spaces. Furthermore, the homology
of the function spaces can be worked out in favorable situations.
   In view of the above results, it is natural to make the following definition

DEFINITION 3.4. Given an algebra $A = H_*(\Omega^2 X; \mathbb{F})$ define

$$H\Gamma(g)_*(A) = H_*(E \times_{Diff^+(S_g)} map(S_g, X); \mathbb{F}).$$

   It appears likely that the definition of $H\Gamma(g)_*(A)$ can be given using the
methods of [Ge] via an algebraic process which does not necessarily depend on
a choice of topological space X. However, this remark will not be addressed
here.

In the cases for which $X = \Sigma Y$ (for "reasonable" spaces Y), there are isomorphisms

$$A' = H_*(\Omega X; \mathbb{F}) \approx T[V] \text{ the tensor algebra}$$

with $V = \bar{H}_*(Y; \mathbb{F})$. If in addition $Y = \Sigma(B)$ and $B = \Sigma(C)$ (and thus $X = \Sigma^3(C)$), there are isomorphisms

$$A = H_*(\Omega^2 \Sigma^2(B); \mathbb{F}) \approx W_1(U)$$

with $U = \bar{H}_*(B; \mathbb{F})$ [Co$_3$, p. 226]. With these assumptions, the following equalities hold where $HC_*(A'; \mathbb{F})$ denotes the cyclic homology of the algebra A':

(1) $HC_*(A'; \mathbb{F}) = HC_*(\mathbb{F}) \oplus (\oplus_{m \geq 1} H_*(\mathbb{Z}/m\mathbb{Z}; V^{\otimes m}))$ [G]
(2) $H\Gamma(0)_*(A; \mathbb{F}) = D_0(U) \oplus_{m \geq 3} H_*(\Gamma_0^m; U^{\otimes m})$ [Co$_4$, BCP]
(3) $H\Gamma(1)_*(A; \mathbb{F}) = D_1(U) \oplus_{m \geq 2} H_*(\Gamma_1^m; U^{\otimes m})$ [CH]
(4) $H\Gamma(g)_*(A; \mathbb{F}) = \mathbb{F} \oplus_{m \geq 0} H_*(\Gamma_g^m; U^{\otimes m})$ [CH]

The modules $D_0(U)$ and $D_1(U)$ appear because of the anomalous behavior for Γ_g^k when (1) $g = 0$ and $k < 3$ or (2) $g = 1$ and $k < 2$. In particular, $D_0(U)$ and $D_1(U)$ are the homology groups for 1 and 2 adic constructions if $g = 0$ and the 1 adic construction if $g = 1$.

Let $A = H_*(\Omega^2 \Sigma^2 X; \mathbb{F}_p)$ for a path-connected CW complex X. The natural map

$$A \rightarrow H\Gamma(g)_*(A; \mathbb{F}_p)$$

has a kernel if $\text{rank}_{\mathbb{F}_p}(\bar{H}_*(X; \mathbb{F}_p)) > 1$. One wonders whether there is a description of the kernel of this map in terms of invariants of graded Lie algebras.

The homology groups $H\Gamma(0)_*(A)$ for $A = H_*(\Omega^2 X; \mathbb{F})$ admit another relation to cyclic homology. Again choose an embedding of S^1 in S^2 which is compatible with the inclusion of $SO(2)$ in $SO(3)$. Let $\Lambda^n X$ denote the space of (unpointed) continuous maps from S^n to X.

There is a fibre bundle

$$S^2 \rightarrow ESO(2) \times_{SO(2)} \Lambda^2 X \rightarrow ESO(3) \times_{SO(3)} \Lambda^2 X.$$

Recall that if G is a topological group which acts on a space Y, then the G-equivariant homology of Y, $H_*^G(Y; \mathbb{F})$, is the homology of the homotopy orbit space $EG \times_G Y$. Thus by the Gysin sequence, there is a long exact sequence

(1) $\quad \ldots \rightarrow H_q^{SO(2)}(\Lambda^2 X; \mathbb{F}) \rightarrow H\Gamma(0)_q(A; \mathbb{F}) \rightarrow$

$$\rightarrow H\Gamma(0)_{q-3}(A; \mathbb{F}) \rightarrow H_{q-3}^{SO(2)}(\Lambda^2 X; \mathbb{F}) \rightarrow \ldots.$$

Furthermore $H_*^{SO(2)}(\Lambda^2 X; \mathbb{F})$ can be given by the Serre spectral sequence for the fibration

$$(\Omega^2 X)^2 \rightarrow ESO(2) \times_{SO(2)} \Lambda^2 X \rightarrow ESO(2) \times_{SO(2)} \Lambda X.$$

Thus setting B equal to the algebra $H_*(\Omega X; \mathbb{F})$ and letting $(A \otimes_{\mathbb{F}} A)_t$ denote the summand of $A \otimes_{\mathbb{F}} A$ in total degree t, one has the Serre spectral sequence with

(2) $\qquad E_{s,t}^2 = HC_s(B; (A \otimes_{\mathbb{F}} A)_t)$ which abuts to $H_{s+t}^{SO(2)}(\Lambda^2 X; \mathbb{F})$.

The above observations are collected in the next result.

PROPOSITION 3.5.

(i) *With algebras A and B above, there is a spectral sequence (the Serre spectral sequence) with*

$$E^2_{**} = HC_*(B; A \otimes A)$$

which abuts to $H^{SO(2)}_*(\Lambda^2 X; \mathbb{F})$.

(ii) *There is a long exact sequence*

$$\cdots \to H^{SO(2)}_q(\Lambda^2 X; \mathbb{F}) \to H\Gamma(0)_q(A; \mathbb{F}) \to$$

$$\to H\Gamma(0)_{q-3}(A; \mathbb{F}) \to H^{SO(2)}_{q-3}(\Lambda^2 X; \mathbb{F}) \to \cdots .$$

To finish this section $H\Gamma(0)_*(\Omega^2 G; \mathbb{Q})$ for a 2-connected topological group of the homotopy type of a finite complex is listed as an example. In this case G is rationally equivalent to a product of odd spheres $\prod_{k \in I} S^{2k+1}$. Thus consider the projection $\prod_{k \in I} S^{2k+1} \to S^{2k+1}$ together with the morphisms of fibrations

$$E \times_{SO(3)} map(S_o, G) \longrightarrow E \times_{SO(3)} map(S_o, S^{2k+1})$$

$$\downarrow \qquad\qquad\qquad\qquad\qquad \downarrow$$

$$BSO(3) \quad\xrightarrow{\ \ 1\ \ }\quad BSO(3).$$

As these fibrations admit sections, the Serre spectral sequence for the right-hand side (in rational cohomology) collapses by a comparison of degrees. On the other hand $map(S_0, G)$ is rationally equivalent to a product $\prod_{k \in I}(S^{2k+1} \times S^{2k-1})$. Thus the fibre in the left-hand fibration is totally non-homologous to zero. The next proposition follows.

PROPOSITION 3.6. *Let G be a 2-connected topological group which has the homotopy type of a finite complex with $A = H_*(\Omega^2 G; \mathbb{Q})$. Then there is an isomorphism*

$$H_*(BSO(3); \mathbb{Q}) \otimes H_*(G; \mathbb{Q}) \otimes H_*(\Omega^2 G; \mathbb{Q}) \to H\Gamma(0)_*(A; \mathbb{Q}).$$

The torsion is of course more complicated. The cases for which $g > 0$ are also more complicated. The case of $H\Gamma(g)_*(B^3 O; \mathbb{F}_2)$ provides "an explanation" of the image of the mod-2 homology of the braid group in the mapping class group (as seen in section 6 here).

In the cases above the map

$$\lambda_* : HC_*(A'; \mathbb{F}) \to H\Gamma(0)_*(A; \mathbb{F})$$

induced by the map λ in paragraph 3 above arises from naive geometric maps with an intermediate factorization given by Artin's braid group modulo its' center. Namely, the standard embeddings $S^1 \subset \mathbb{R}^2 \subset S^2$ give maps on the level of configuration spaces: $F(S^1, k) \subset F(R^2, k) \subset F(S^2, k)$. Passage to orbit spaces gives

$$E \times_{SO(2)} F(S^1, k) \to E \times_{SO(2)} F(R^2, k) \to E \times_{SO(3)} F(S^2, k).$$

These maps are precisely the ones which induce λ_*. The intermediate space $Ex_{SO(2)}(F(R^2, k)/\Sigma_k)$ is a $K(\pi, 1)$ where π is the k-th Artin braid group modulo its' center [Co].

Furthermore, the work in [Ge] gives that $H\Gamma(g)_*(A; \mathbb{F})$ provides invariants which appear in two-dimensional topological field theories. There is some hold on these invariants if $g \leq 2$ as seen by the remarks in sections 4 and 6.

4 On sub-bundles of the Hopf bundle

Consider the Hopf circle bundle

$$\pi_n : S^{2n+1} \to \mathbb{C}P^n$$

with fibre S^1. The purpose of this section is to describe a sub-bundle of π_n, X_n, together with a $U(2)$-action on X_n such that the homotopy orbit space $EU(2) \times_{U(2)} X_n$ is a $K(G_n, 1)$. Furthermore, the groups G_n which appear here are frequently isomorphic to certain natural subgroups of the mapping class group with G_6 isomorphic to Γ_2.

Regard $\mathbb{C}P^n$ as the n-fold symmetric product $Sp^n(S^2)$ and let $B(S^2, n)$ denote the subspace of $Sp^n(S^2)$ given by n-tuples $(z_1, \ldots, z_n)$ in $Sp^n(S^2)$ with $z_i \neq z_j$ if $i \neq j$. Let X_n denote the pull-back. Thus there is a morphism of bundles

$$
\begin{array}{ccc}
X_n & \longrightarrow & S^{2n+1} \\
\pi_n \downarrow & & \downarrow \\
B(S^2, n) & \longrightarrow & \mathbb{C}P^n
\end{array}
$$

which is an isomorphism on fibres.

Also, notice that $S^3 \times SO(2)$ acts on X_n by $(\alpha, \beta)((z_i), \lambda) = ((p\alpha)(z_i), \beta^2 \cdot \lambda)$ where (1) $p : S^3 \to SO(3)$ is the standard double cover, (2) $p(\alpha)(z_i)$ is given by the usual action of $SO(3)$ on S^2, and (3) $\beta^2 \cdot z$ is given by complex multiplication. Notice that this action of $S^3 \times SO(2)$ on X_n descends to an action of the central product $S^3 \times_{\mathbb{Z}/2\mathbb{Z}} SO(2)$ where $\mathbb{Z}/2\mathbb{Z}$ is identified with the center of S^3 and as $\{\pm 1\}$ in $SO(2)$. The central product $S^3 \times_{\mathbb{Z}/2} SO(2)$ is also written as $\mathrm{Spin}^c(3)$ in [ABS] and is isomorphic to $U(2)$.

THEOREM 4.1 [Co].
If $n \geq 3$, the homotopy orbit space

$$EU(2) \times_{U(2)} X_n = Y_n$$

is an Eilenberg-Mac Lane space of type $K(G_n, 1)$. Furthermore if $n = 2g + 2$ with $g \equiv 0(2)$, then G_n is isomorphic to the centralizer of a hyperelliptic involution in Γ_g; if $n = 6$, then G_6 is isomorphic to Γ_2.

As $U(2)$ is the maximal compact subgroup of $Sp(4, \mathbb{R})$, there is a map

$$EU(2) \times_{U(2)} X_n \xrightarrow{\alpha_n} BU(2) \xrightarrow{B(i)} BSp(4, \mathbb{R})$$

where α_n is given by first coordinate projection and $i : U(2) \to Sp(4, \mathbb{R})$ is the natural inclusion.

THEOREM 4.2 [Co].
    The map α_n is not induced by a representation of G_n in $U(2)$.

Guido Mislin pointed out that it is possible that the composite $B(i) \circ \alpha_n$ is in fact induced by a representation. For example if $n = 6$, then $EU(2) \times_{U(2)} X_n$ is a $K(\Gamma_2, 1)$. There is of course the natural homology representation of Γ_2 in $Sp(4, \mathbb{R})$ given by the usual composites $\Gamma_2 \to Sp(4, \mathbb{Z}) \to Sp(4, \mathbb{R})$. At this writing, it is not known whether α_n is homotopic to the map $B\Gamma_2 \to BSp(4, \mathbb{R})$ which is induced by the previous representation.

The spaces X_n and Y_n have non-trivial fundamental groups which surject to the symmetric groups on n letters. The coverings of Y_n obtained in this way are particularly nice.

THEOREM 4.4 [Co$_2$].
    If $n \geq 3$, the homotopy theoretic fibre of $Y_n \to B\Sigma_n$ is

$$\mathbb{R}P^\infty \times \mathbb{F}(\mathbb{C} - \{0, 1\}, n - 2).$$

These bundles have been used to work out the integral homology of Γ_2 [Co] and corroborate the mod-p calculations in [BC]. In addition, these bundles were used to work out the homology of Δ_g with coefficients in $\mathbb{Z}_{(2)}$ for all $g \equiv 0 (mod 2)$. A sample result is that there are isomorphisms of $H^*(BU(2); \mathbb{F}_2)$-modules given by

$$H^*(BU(2); \mathbb{F}_2) \otimes H^*(X_6; \mathbb{F}_2) \to H^*(B\Gamma_2; \mathbb{F}_2), \text{ and}$$

$$H^*(BU(2); \mathbb{F}_2) \otimes H^*(X_{2g+2}; \mathbb{F}_2) \to H^*(B\Delta_g; \mathbb{F}_2), g \equiv 0 (mod 2).$$

In addition Theorem 4.4 gives a classifying space for the kernel of the composite when $n = 2g + 2$:

$$\Delta_g \to \Gamma_g \to Sp(2g, \mathbb{F}_2).$$

Thus Theorem 4.4 provides a little information about the Torelli group, the kernel of $\rho : \Gamma_g \to Sp(2g, \mathbb{Z})$.

5 On $K(\pi, 1)$'s obtained from flag varieties

The constructions of $K(G_n, 1)$ and $K(\Delta_g, 1)$ given in section 4 are analogues of an "incidence" bundle obtained from flags of linear subspaces of $\mathbb{C}^{n+1}$. As these bundles fit naturally with mapping class groups, and will be used elsewhere, they are described here.

Consider the space of flags

$$\phi = (L_n, L_{n-1}, \dots, L_2)$$

where

(1) L_i is a linear subspace of $\mathbb{C}^{n+1}$ of dimension i,
(2) $L_i \subset L_{i+1}$ for each i.
    Next, fix an integer $q \geq 1$ and define

$$Z(n,q) = \{(\phi, (x_1, \dots, x_q))\}$$

    where
(3) ϕ is a flag described above and
(4) $(x_1, \dots, x_q)$ is an ordered q-tuple of vectors in the orthogonal complement $L_n^{\perp}$ such that x_i and x_j are distinct if $i \neq j$.

Notice that $U(n+1)$ acts on $\mathbb{C}^{n+1}$ and on the space of flags. Furthermore this action extends to one on $Z(n,q)$. Let $E = EU(n+1)$ and consider the homotopy orbit space

$$E \times_{U(n+1)} Z(n,q).$$

Let T^{n+1} denote the maximal torus in $U(n+1)$. The next lemma follows at once from [FN].

PROPOSITION 5.1. *There is a fibre bundle*

$$E \times_{U(n+1)} Z(n,q) \xrightarrow{1 \times p} E \times_{U(n+1)} Z(n,1)$$

where p is first coordinate projection. The fibre is homeomorphic to $F(\mathbb{R}^2 - \{0\}, q-1)$ and is thus a $K(\pi, 1)$.

Proof. Consider $p_k : Z(n,q) \to Z(n,k)$ given by projection onto the first k-coordinates. By the proof in [FN], p_k is the projection in a fibre bundle with fibre homeomorphic to $F(\mathbb{R}^2 - Q_k, q-k)$ where Q_k is a subset of $\mathbb{R}^2$ having cardinality k. If $k = 1$, the fibre of p is thus homeomorphic to $F(\mathbb{R}^2 - \{0\}, q-1)$. Since p_k is $U(n+1)$-equivariant, the result follows.   □

Notice that $Z(n,1) = U(n+1)/T^{n+1}$ and so $E \times_{U(n+1)} Z(n,1)$ is BT^{n+1}. In addition, $F(\mathbb{R}^2, q-1)$ is a $K(\pi, 1)$. Thus if A is any discrete subgroup of T^{n+1}, the next result follows at once.

PROPOSITION 5.2. *The pull-back $\xi(n,q)$ in the cartesian square*

$$
\begin{array}{ccc}
\xi(n,q) & \longrightarrow & E \times_{U(n+1)} Z(n,q) \\
\downarrow & & \downarrow \\
BA & \longrightarrow & BT^{n+1}
\end{array}
$$

is an Eilenberg-Mac Lane space $K(\hat{\pi}, 1)$.

6 On the EHP sequence and $H_*\Gamma_g^k$

The homology groups of $\Gamma_g^k, k \geq 0$, with coefficients in $\mathbb{F}$ (the trivial representation) or $\mathbb{F}(-1)$ (the sign representation) are given in terms of the homology
of $M_g(S^n) = E \times_{Diff^+(S_g)} map(S_g, S^n)$, $n \geq 2$, by the remarks of section 3.
The pointed versions $N_g(S^n) = E \times_{Diff^+(S_g,*)} map_*^0(S_g, S^n)$ provide information about $H_*\Gamma_g^k$ with $k \geq 1$ (where $map_*^0(\ )$ denotes the component of the
base-point when $n = 2$). In these cases, $map_*(S_g, S^n)$ and $N_g(S^n)$ fit with constructions in non-stable homotopy theory which themselves give the classical
EHP sequence. The purpose of this section is to describe this relation.

First recall that there are maps

$$h_k : \Omega S^{n+1} \to \Omega S^{nk+1}$$

called the k-th James-Hopf invariant. Let $J_t(S^n)$ denote the (nt)-skeleton of
ΩS^{n+1}. One of the most important features of the maps h_k in homotopy theory
lies in the following well-known theorem.

THEOREM 6.1 [J, T, SE].
   (i) If all spaces are localized at the prime 2, then the homotopy theoretic
fibre of $h_{2^t} : \Omega S^{n+1} \to \Omega S^{2^t n+1}$ is $J_{2^t-1}(S^n)$.
   (ii) If all spaces are localized at any prime p, then the homotopy theoretic
fibre of $h_{p^k} : \Omega S^{2n+1} \to \Omega S^{2p^k n+1}$ is $J_{p^k-1}(S^{2n})$.

Thus consider the function space of pointed degree zero maps of S_g to S^2,
$map_*^0(S_g, S^2)$. Since there is a cofibration

$$S^1 \xrightarrow{\alpha} \vee_{2g} S^1 \to S_g,$$

there is a fibration

$$\alpha^* : (\Omega S^2)^{2g} \to \Omega S^3$$

with homotopy theoretic fibre $map_*^0(S_g, S^2)$.
   Furthermore, the space $map_*^0(S_g, S^2)$ is homotopy equivalent to

$$(\Omega S^3)^{2g} \times Y_g$$

where Y_g is the homotopy theoretic fibre of a map

$$\alpha^{**} : (S^1)^{2g} \to \Omega S^3$$

as described in [BCM]. In addition, the composite β given by $(S^1)^{2g} \xrightarrow{\alpha^{**}}$
$\Omega S^3 \xrightarrow{i} \mathbb{C}P^\infty$ for i a map of degree one is specified by the equation

$$\beta^*(i_2) = \sum_{i=1}^{g} x_{2i-1} x_{2i}$$

with $H^*(S^1)^{2g} = E[x_1, \dots, x_{2g}]$.
   Analogous splitting results apply to $map_*(S_g, S^{2n})$ if $n \geq 2$ with the
primes 2 and 3 are inverted.

The map α^{**} can be described more precisely. There are projections

$$\pi_i : (S^1)^{2g} \to S^1 \times S^1, 1 \le i \le g$$

which send $(x_1, \dots, x_{2g})$ to (x_{2i-1}, x_{2i}). The natural collapse map $q : S^1 \times S^1 \to S^2$ composed with π_i gives maps $\bar{q}_i : (S^1)^{2g} \to S^2$. Composing $\bar{q}_i$ with the suspension $E : S^2 \to \Omega S^3$ provides a map $e_i : (S^1)^{2g} \to \Omega S^3$. The map α^{**} is then given as the composite

$$(S^1)^{2g} \xrightarrow{\Delta^g} (S^1)^{2g} \times \cdots \times (S^1)^{2g} \xrightarrow{e_1 \times e_2 x \dots x e_g} (\Omega S^3)^g \xrightarrow{\mu} \Omega S^3$$
$$\overset{\leftarrow g \rightarrow}{}$$

where μ is the product (in a fixed order).

If $n = 1$ in Theorem 6.1, there is a fibration

$$h_{p^t} : \Omega S^3 \to \Omega S^{2p^t+1}$$

with p-local homotopy theoretic fibre $J_{p^t-1}(S^2)$. If $p^t > g$, then the composite $h_{p^k} \circ \alpha^{**}$ is null-homotopic. Thus there is a morphism of p-local fibrations

$$
\begin{array}{ccccc}
\Xi(g, p^k) & \longrightarrow & (S^1)^{2g} & \xrightarrow{\alpha^{**}} & J_{p^t-1}S^2 \\
\downarrow & & 1 \downarrow & & \downarrow \\
Y_g & \longrightarrow & (S^1)^{2g} & \xrightarrow{\alpha^{**}} & \Omega S^3 \\
\downarrow & & \downarrow & & \downarrow \\
\Omega^2 S^{2p^k+1} & \longrightarrow & * & \longrightarrow & \Omega S^{2p^t+1}
\end{array}
$$

where $\Xi(g, p^k)$ is the homotopy theoretic fibre of $\alpha^{**} : (S^1)^{2g} \to J_{p^t-1}(S^2)$. Thus the map $\Omega h_{p^t} : \Omega^2 S^3 \to \Omega^2 S^{2p^t+1}$ factors through Y_g up to homotopy. Furthermore, the induced map $(\Omega h_{p^t})_* : H_*(\Omega^2 S^3; \mathbb{F}_p) \to H_*(\Omega^2 S^{2p^t+1}; \mathbb{F}_p)$ is a split surjection of Hopf algebras; this splitting is unique up to a choice of non-zero scalars. Let $D(p^t)$ denote $H_*(\Omega^2 S^{2p^t+1}; \mathbb{F}_p)$. This construction contributes classes to the homology of Γ_g^k if $p \ge 2$.

PROPOSITION 6.2. If $p = 2$, then $D(2^0)$ injects in $H_*(N_g(S^2); \mathbb{F}_2)$.

SKETCH. This is direct as there is a map $B_k \to \Gamma_g^k$ where B_k denotes Artin's braid group. This last map is obtained by embedding a disk in S_g and then applying the configuration space constructions given in section 3. The composite $B_k \to \Gamma_g^k \to \Sigma_k$ is the natural epimorphism which gives an injection in mod-2 homology.

It seems likely that if $p > 2$ and $p^t > g$, then $D(p^t)$ injects in $H_*(N_g(S^2); \mathbb{F}_p)$. In addition, notice that $F(S_g, k)$ is a finite dimensional $K(\pi, 1)$ if $g \ge 1$ [FN]. Thus $\pi_1(F(S_g, k)/\Sigma_k)$ has no elements of finite order in these cases. However, the homology of these spaces have lots of torsion, namely "strange torsion" as described in work of Glover and Mislin [GM, GM$_2$]. It follows directly from these remarks that the homology of Γ_g^k contains torsion which might be regarded as "strange torsion".

7 Actions of quaternion groups on S_g

This section as well as sections 8–9 are used in the proofs of the statements in section 2. Recall that the quaternion group of order $2^{t+1}, Q_t$, has presentation given by

(1) generators x and y with
(2) relations $x^{2^{t-1}} = y^2, yxy^{-1} = x^{-1}$, and $x^{2^t} = 1$ [ ].

Thus (3) $[x, y] = x^2$, (4) there is exactly one element of order 2 given by y^2 for $t \geq 2$, and (5) the center of Q_t is cyclic of order 2 if $t \geq 2$. The cohomology of Q_t is given in [FP, p. 330].

If G is a finite group, then actions of G on S_g are described in [Br] as follows: Write $S_\lambda = S_g/G$.

(i) Branch data $\vec{b} = (\lambda : m_1, \dots , m_r)$ where λ is an integer with $\lambda \geq 0$, and m_i are positive (non-decreasing) integers with m_i given by orders of certain elements in G,
(ii) A $(2\lambda + r)$-tuple of elements in G $\vec{v} = (a_1, \dots, a_\lambda, b_1, \dots, b_\lambda, c_1, \dots, c_r)$

$$\text{such that } (\prod_1^\lambda [a_i, b_i])(\prod_1^r c_i) = 1,$$

order $(c_i) = m_i$, and the elements of $\vec{v}$ generate G.
(iii) In addition, the Riemann-Hurwitz equation

$$\frac{2g - 2}{|G|} = 2\lambda - 2 + \sum_1^r (1 - \frac{1}{m_i}) \text{ must be satisfied.}$$

THEOREM 7.1. *If $G = Q_3$ acts on a surface of genus g and q is an odd prime power, then the representation of the central $\mathbb{Z}/2\mathbb{Z}$ in Q_3 acting on $(\mathbb{F}_q)^{2g} = H_1(S_g; \mathbb{F}_q)$ is conjugate to a multiple of 8 times the sign representation plus trivial representations.*

Since $H_4(BSp(2g; \mathbb{F}_q); \mathbb{F}_2)$ is isomorphic to $\mathbb{F}_2$ with generator obtained from an odd multiple of the natural representtion of $Q_3[FP]$, the next corollary follows at once.

COROLLARY 7.2. *Let G and S_g be as in 7.1. Then the map $H_4(BQ_3; \mathbb{F}_2) \to H_4(B\Gamma_g; \mathbb{F}_2) \to H_4(BSp(2g, \mathbb{F}_q); \mathbb{F}_2)$ is zero.*

Proof. Assume that $G = Q_3$ acts on S_g with $S_g/G = S_\lambda$ and with branch data $\vec{b} = (\lambda : m_1, \dots , m_r)$. By the Riemann- Hurwitz equation

$$g = 16\lambda - 15 + 16(\sum_i (1 - \frac{1}{m_i})).$$

Furthermore if z is any non-identity element in Q_3, then $z^i = y^2$ for some i. Thus the number of fixed points of the element $\chi = y^2$ in Q_3, $|S_g^\chi|$, is given by

$$|S_g^\chi| = 16(\sum_i \frac{1}{m_i})$$

as (a) the normalizer of χ in Q_3 is all of Q_3, and (b) $\delta_i(\chi) = 1$ as used in [Br]. $\square$

By the Lefschetz fixed point theorem [See [EE] for example.],

$$|S_g^\chi| = 2 - Tr$$

where Tr is the trace of the matrix for

$$\chi_* : H_1(S_g; \mathbb{Q}) \to H_1(S_g; \mathbb{Q}).$$

Diagonalizing the action of χ_* on $H_1(S_g; \mathbb{Z}_{(\frac{1}{2})})$ as $\mathbb{Z}_{(\frac{1}{2})}[\mathbb{Z}/2\mathbb{Z}]$ is semi-simple and $\chi^2 = 1$, let β be the number of (-1)'s appearing (where all diagonal entries are ± 1). Thus

$$\beta = -Tr + \frac{2g + Tr}{2} = g - \frac{Tr}{2}.$$

Then

$$\beta = (16g - 15 + 16(\sum_i (1 - \frac{1}{m_i}) - \frac{1}{2}[16(\sum_i \frac{1}{m_i} - 2)] \text{ and}$$

$$\beta = 16g - 16 + 8(\sum_i 1).$$

Thus $\beta \equiv 0 (\mod 8)$. Theorem 7.1 and its' corollary follow.

A similar result applies in case t is arbitrary:

(1) The branch data for a $G = Q_t$ action is $(\rho : m_1, \ldots, m_r)$ where G acts on S_σ with $S\sigma/G = S\rho$.

(2) $\dfrac{2\sigma - 2}{|G|} = 2\rho - 2 + \sum_i (1 - \frac{1}{m_i})$ by the Riemann-Hurwitz equation.

(3) $\sigma = 2^{t+1}\rho - 2^{t+1} + 1 + 2^t \sum_i (1 - \frac{1}{m_i})$

(4) If $\chi = y^2$, then

$$|S_\sigma^\chi| = 2^{t+1}(\sum_i \frac{1}{m_i}) = 2 - Tr.$$

(5) The matrix of $\chi_* : H_1(S_\sigma; \mathbb{Z}_{(\frac{1}{2})}) \to H_1(S_\sigma; \mathbb{Z}_{(\frac{1}{2})})$ may be assumed to be diagonal with the number of (-1)'s on the diagonal given by

$$\beta = -Tr + \frac{2\sigma + Tr}{2} = \sigma - \frac{Tr}{2}.$$

(6) $\beta = 2^{t+1}\rho - 2^{t+1} + 1 + 2^t(\Sigma_i(1 - \frac{1}{m_i})) + \frac{1}{2}(2^{t+1}(\Sigma_i \frac{1}{m_i}) - 2)$ and thus

$$\beta = 2^{t+1}\rho - 2^{t+1} + 2^t r.$$

Notice that β is zero mod 2^t. Also, if r is odd then there are maps $BQ_t \to B\Gamma_\sigma$ which are non-trivial on $H_{\ell 2^{t+1}-1}(\quad ; \mathbb{Z})$.

8 On quaternion and dihedral groups in Γ_g

Let Q_2 be the quaternion group of order 8 as given in section 7.

Let D_t be the dihedral group of 2^t. The groups Q_2 and D_3 are embedded in the centralizer of the hyperelliptic involution, Δ_g, for various values of g. Some properties of these are listed below for future use. In what follows, Ω denotes the class of the hyperelliptic involution.

The next lemma is well-known and is given without proof.

LEMMA 8.1. *There is a choice of symplectic basis for $H_1(S_g; \mathbb{Z})$ such that the image of Ω in $Sp(2g, \mathbb{Z})$ is $(-1)I_{2g}$ where I_{2g} denotes the identity matrix.*

PROPOSITION 8.2.

(i) *If $g \equiv 2(\mathrm{mod}\ 4)$, then Q_2 is isomorphic to a subgroup of Δ_g. Furthermore, the center of Q_2, $\mathbb{Z}/2\mathbb{Z}$, is generated by Ω.*

(ii) *If $g \equiv 1(\mathrm{mod}\ 2)$ with $g \geq 1$, then D_3 is isomorphic to a subgroup of Δ_g. Furthermore, the center of D_3, $\mathbb{Z}/2\mathbb{Z}$, is generated by Ω.*

Let T_2 denote the $\mathbb{Z}/2$-torus in $SO(3)$ with generators

$$\alpha = \begin{pmatrix} -1 & 0 & 0 \\ 0 & -1 & 0 \\ 0 & 0 & 1 \end{pmatrix} \text{ and } \beta = \begin{pmatrix} 1 & 0 & 0 \\ 0 & -1 & 0 \\ 0 & 0 & -1 \end{pmatrix}.$$

Fix a natural number ℓ together with the subset V_ℓ of S^2 given by

$$V_\ell = \{x_{k,\ell} | 0 \leq k < 4\ell\} \cup \{y_{k,\ell} | 0 \leq k < 4\ell\}$$

with

$$x_{k,\ell} = (\cos(\frac{k\pi}{2\ell}), \sin(\frac{k\pi}{2\ell}), 0) \text{ and}$$

$$y_{k,\ell} = (0, \sin(\frac{k\pi}{2\ell}), \cos(\frac{k\pi}{2\ell})).$$

Notice that V_ℓ has cardinality $8\ell - 2$ as $x_{\ell,\ell} = y_{\ell,\ell}$ and $x_{3\ell,\ell} = y_{3\ell,\ell}$.

The elements α and β leave V_ℓ invariant and thus there is a homomorphism $e : T_2 \to Diff^+(S^2; 8\ell - 2)$. Composition with the "component map" to $\pi_0 Diff^+(S^2; 8\ell - 2)$ gives a homomorphism $e : T_2 \to \Gamma^{8\ell-2}$.

Next recall that

$$K(\Gamma^q, 1) = ESO(3) \times_{SO(3)} B(S^2, q)$$

if $q \geq 3$. However, the projection

$$\pi : K(\Gamma^q, 1) \to BSO(3) = ESO(3)/SO(3)$$

for $q \geq 6$ does not arise from a homomorphism and is not homotopic to $B(r)$ for any representation r [BC].

LEMMA 8.3. *If $\ell \geq 1$, the composite*

$$BT_2 \xrightarrow{B(e)} B\Gamma^{8\ell-2} \xrightarrow{\pi} BSO(3)$$

is given by $B(i)$ where i is the natural inclusion of T_2 in $SO(3)$.

Proof. Let $E = ESO(3)$ (as in [M]). Define

$$g : E \to E \times B(S^2, 8\ell - 2)$$

by

$$g(e) = (e, V_\ell)$$

where the set V_ℓ of cardinality $8\ell - 2$ is regarded as a point in $B(S^2, 8\ell - 2)$. Since T_2 leaves V_ℓ invariant, the map g is T_2-equivariant. Thus the composite

$$E/T_2 \xrightarrow{g} E \times_{T^2} B(S^2, 8\ell - 2) \to E \times_{SO(3)} B(S^2, 8\ell - 2) \to BSO(3)$$

is $B(i)$ and the lemma follows.   □

Next consider the composite

$$T_2 \xrightarrow{e} \Gamma^{8\ell-2} \xrightarrow{j} \Sigma_{8\ell-2}$$

with $j : \Gamma^{8\ell-2} \to \Sigma_{8\ell-2}$ the natural surjection. Notice that $je(\alpha)$ is a product of $(2\ell) + (2\ell - 1)$ disjoint transpositions as α leaves both sets $\{x_{k\ell}\}$ and $\{y_{k\ell}\}$ invariant while fixing $\{(0, 0, \pm 1)\}$. Similarly $je(\beta)$ is a product of $4\ell - 2$ disjoint transpositions.

LEMMA 8.4. *The sign of $je(\alpha)$ or $je(\beta)$ is $+1$.*

Next consider the central extension

$$1 \to \mathbb{Z}/2 \to \Delta_g \to \Gamma^{2g+2} \to 1.$$

Thus there is a group π_ℓ obtained as a pull-back in the diagram

$$
\begin{array}{ccc}
\pi_\ell & \longrightarrow & T_2 \\
\downarrow & & \downarrow{\scriptstyle e} \\
\Delta_g & \longrightarrow & \Gamma^{8\ell-2}
\end{array}
$$

with $2g + 2 = 8\ell - 2$.

LEMMA 8.5. *If $g = 4\ell - 2$ for $\ell \geq 1$, then π_ℓ is isomorphic to Q_2.*

Proof. Recall

(i) $H^*(BT_2; \mathbb{F}_2) = \mathbb{F}_2[a, b]$ where a and b are of degree 1. The characteristic class of the $\mathbb{Z}/2$-central extension giving Q_2 is

$$B(i)^*(w_2) = a^2 + b^2 + ab \quad [MP]$$

where $i : T_2 \to SO(3)$ is the previous inclusion.

(ii) The cohomology of $B\Gamma^{8\ell-2}$ is known and satisfies the following properties [Co]:

(1): If $n \geq 2$, then $H^1(\Gamma^{2n}; \mathbb{F}_2) = \mathbb{F}_2$.

(2): If $n \geq 2, j^* : H^1(\Sigma_{2n}; \mathbb{F}_2) \to H^1(\Gamma^{2n}; \mathbb{F}_2)$ is an isomorphism where y_1 denotes the generator of $H^1(\Sigma_{2n}; \mathbb{F}_2)$.

(3): If $n \geq 2, H^2(\Gamma^{2n}; \mathbb{F}_2) = \oplus_3 \mathbb{F}_2$. Furthermore, a choice of generators is given by

$$\pi^*(w_2), j^*(y_1^2), \text{ and } j^*(y_2)$$

where y_2 is dual to the unique primitive homology class in $H_2(\Sigma_{2n}; \mathbb{F}_2)$.

(4): The characteristic class, χ_g, for the extension

$$1 \to \mathbb{Z}/2 \to \Delta_g \to \Gamma^{2g+2} \to 1$$

is given by

$$x_g = \begin{cases} \pi^* w_2 + j^*(y_1^2) \text{ if } g \equiv 2 (\text{mod } 4) \\ j^*(y_1^2) \text{ if } g \equiv 0 (\text{mod } 4). \end{cases}$$

Consider $(Be)^*(\chi_g)$ with $g \equiv 2 (\text{mod } 4)$:

(a) $(Be)^*(\pi^* w_2) = a^2 + b^2 + ab$ by Lemma 8.3.

(b) $(Be)^*(j^* y_1) = 0$ by Lemma 8.4.

(c) $(Be)^*(j^* y_1^2) = 0$ by (b).

(d) $(Be)^*(\chi_g) = a^2 + b^2 + ab$ by (a)–(c).

Thus π_ℓ is isomorphic to Q_8.   □

The proof of 8.2(ii) is analogous and is sketched below. Consider the points

$$w_{\ell,k} = (e^{\ell(\frac{2\pi i}{2k})}, 0), \quad 0 \leq \ell < k.$$

Thus T_2 preserves the set $\{w_{\ell,k} | 0 \leq \ell < k\}$ and there is an induced homomorphism from T_2 to Γ^{2k}. The sign of the permutation associated to α and β is $(-1)^{k-1}$. Thus if k is even, y_1 restricts to $a + b$ in $H^1(T_2; \mathbb{F}_2)$. If $g \equiv 1 (mod 2)$ and $g > 1$, then $\chi_g = (j^* y_1)^2$. Thus the pull-back BH in the Cartesian square

$$\begin{array}{ccc} BH & \longrightarrow & BT_2 \\ \downarrow & & \downarrow \\ B\Delta_g & \longrightarrow & B\Gamma^{2g+2} \end{array}$$

has characteristic class $a^2 + b^2$ when $k = g + 1$ is even. Thus H is isomorphic to D_3 [MP] and 8.2(ii) follows.

9 Proofs for section 2

Recall that $K_3(\mathbb{Z}) = \pi_3 BGL(\mathbb{Z})^+ = \mathbb{Z}/16 \oplus \mathbb{Z}/3$ by [LS]. This information together with naturality is used to give the next proof.

Proof of Proposition 2.4. The statement to be proven is that if the homology representation

$$B\rho : B\Gamma \to BSp(\mathbb{F}_q)$$

admits a stable 2-local section for q odd then $H_3(BSp(\mathbb{F}_q); \mathbb{Z}_{(2)}) = \mathbb{Z}/16$. Thus assume that $\sigma : BSp(\mathbb{F}_q) \to QB\Gamma$ is the adjoint of a stable section for $B\rho$. There is a homotopy commutative diagram

$$
\begin{array}{ccccccc}
BSp(\mathbb{F}_q) & \xrightarrow{\ \sigma\ } & QB\Gamma & \xrightarrow{\ Q(B\gamma)\ } & QBGL(\mathbb{Z}) & \longrightarrow & BGL(\mathbb{Z})^+ \\
{\scriptstyle 1}\downarrow & & \downarrow{\scriptstyle QB\rho} & & \downarrow{\scriptstyle QB\rho'} & & \downarrow{\scriptstyle B\rho'} \\
BSp(\mathbb{F}_q) & \xrightarrow[\ E\ ]{} & QBSp(\mathbb{F}_q) & \xrightarrow[\ Q(Bi)\ ]{} & QBGL(\mathbb{F}_q) & \longrightarrow & BGL(\mathbb{F}_q)^+
\end{array}
$$

where (1) $\gamma : \Gamma \to GL(\mathbb{Z})$ is the composite $\Gamma \xrightarrow{\rho} Sp(\mathbb{Z}) \subset GL(\mathbb{Z})$, (2) $\rho' : GL(\mathbb{Z}) \to GL(\mathbb{F}_q)$ is given by reduction, (3) $i : Sp(\mathbb{F}_q) \to GL(\mathbb{F}_q)$ is the natural inclusion, and (4) $E : X \to QX$ is the stabilization map. The bottom composite

$$BSp(\mathbb{F}_q) \to BGL(\mathbb{F}_q)^+$$

induces an isomorphism

$$\pi_3 BSp(\mathbb{F}_q)_2^\wedge \to \pi_3 BGL(\mathbb{F}_q)_2^\wedge \quad [\text{FP, p. 10}]$$

which factors through $\pi_3 BGL(\mathbb{Z})^+ = K_3(\mathbb{Z})$. Thus

$$\pi_3 BSp(\mathbb{F}_q)^+ \cong H_3(BSp(\mathbb{F}_q)^+; \mathbb{Z})$$

by the Hurewicz theorem and $H_3(BSp(\mathbb{F}_q); \mathbb{Z}_{(2)})$ is a retract of $\mathbb{Z}/16 \oplus \mathbb{Z}/3$. Thus the 2-primary component of $H_3(BSp(\mathbb{F}_q); \mathbb{Z})$ is either $\mathbb{Z}/16$ or $\{0\}$. Since q is an odd prime, $H_3 BSp(\mathbb{F}_q)_2^\wedge = \mathbb{Z}/16$ and Proposition 2.4 follows. $\qquad\square$

Proof of Proposition 2.5. Let $G = Q_3$ act on S_g. The inclusion of the center $\mathbb{Z}/2$ in G induces an isomorphism on $H_4(\ ; \mathbb{F}_2) = \mathbb{F}_2$. By Proposition 2.4, $v_2(q^2 - 1) = 4$. Thus the representation of the central $\mathbb{Z}/2$ in $Sp(2g, \mathbb{F}_q)$ is a multiple of 8 times the sign representation by Theorem 7.1, and the composite $H_4(BG :; \mathbb{F}_2) \to H_4(B\Gamma_{2g}; \mathbb{F}_2) \to H_4(BSp(2g, \mathbb{F}_q); \mathbb{F}_2)$ is the zero map. Thus the map on $H^4(\ ; \mathbb{Z}_{(2)})$

$$\mathbb{Z}/16 = H^4(BSp(2g, \mathbb{F}_q); \mathbb{Z}_{(2)}) \to H^4(BG; \mathbb{Z}_{(2)}) = \mathbb{Z}/16$$

must have a kernel by the universal coefficient theorem. The proposition follows.

$\qquad\qquad\qquad\qquad\qquad\qquad\qquad\qquad\qquad\qquad\qquad\qquad\qquad\qquad\square$

Proof of Proposition 2.6. If $g \equiv 1(\mathrm{mod}\ 2)$, there is a morphism of group extensions

$$
\begin{array}{ccccccccc}
1 & \longrightarrow & \mathbb{Z}/2\mathbb{Z} & \longrightarrow & D_3 & \longrightarrow & \mathbb{Z}/2\mathbb{Z} \oplus \mathbb{Z}/2\mathbb{Z} & \longrightarrow & 1 \\
& & \Big\downarrow{\scriptstyle 1} & & \Big\downarrow & & \Big\downarrow & & \Big\downarrow \\
1 & \longrightarrow & \mathbb{Z}/2\mathbb{Z} & \longrightarrow & \Delta_g & \longrightarrow & \Gamma^{2g+2} & \longrightarrow & 1
\end{array}
$$

by Proposition 7.1. Furthermore the generator of $\mathbb{Z}/2\mathbb{Z}, \Omega$, has image $(-1)I_{2g}$ in $Sp(2g, \mathbb{F}_q)$, for q odd. An argument analogous to that given in [FP, Lemma 5.6, p. 279] implies that the composite $\mathbb{Z}/2\mathbb{Z} \to Q_2 \to Sp(2g, \mathbb{F}_q)$, sends the generator x_{4i} in degree $4i$ to the element σ_i of [FP, p. 277].

The image of x_{4i} in $H_{4i}(D_3; \mathbb{F}_2)$ supports a second Bockstein and thus $H_{4i-1}(D_3; \mathbb{Z})$ contains a $\mathbb{Z}/4\mathbb{Z}$. The image of x_{4i} supports a higher order Bockstein in $H_*(Sp(2g; \mathbb{F}_q); \mathbb{F}_2)$. Thus by the Universal Coefficient Theorem, this copy of $\mathbb{Z}/4\mathbb{Z}$ in $H_{4i-1}(BD_3; \mathbb{Z})$ injects in $H_{4i-1}(BSp(2g; \mathbb{F}_q); \mathbb{Z}_{(2)})$, q odd.

Thus $H_{4i-1}(B\Gamma_g; \mathbb{Z})$ contains an element of order 4 for all $i \geq 1$ with $g \equiv 1(mod\,2), g > 1$. In particular, $H_3(B\Gamma; \mathbb{Z})$ contains a cyclic 2-group of order at least 4.

Since $H_3(B\Gamma; \mathbb{Q}) = \{0\}$ by [H], and $H_3(B\Gamma; \mathbb{Z})$ contains $\mathbb{Z}/3\mathbb{Z}$, the proposition follows. $\square$

References

[ABS] M. Atiyah, R. Bott, and A. Shapiro, *Clifford modules*, Topology **3 (supplement 1)**, 3–38.

[BC]   D. Benson and F.R. Cohen, *Mapping class groups of low genus and their cohomology*, Memoirs of the A.M.S. **(443)90** (1991).

[Bir] J. Birman, *Braids, Links, and Mapping Class Groups*, Ann. of Math. Studies **75** (1982), Princeton Univ. Press, Princeton.

[B]   C.-F. Bödigheimer, *Stable splittings of mapping spaces*, S.L.M. **1286** (1985), 174–187.

[BCM] C.-F. Bödigheimer, F.R. Cohen, and R.J. Milgram, *Truncated symmetric products and configuration spaces*, Math. Zeit. **214** (1993), 179–216.

[BCP] C.-F. Bödigheimer, F.R. Cohen and M. Peim, preprint (1986).

[Bt]   R. Bott, *The stable homotopy of the classical groups*, Proc. Nat. Acad. Sc. U.S.A. **43** (1957), 933–935.

[BR]   S.A. Broughton, *Classifying finite group actions on surfaces of low genus*, JPAA **69** (1991), 233–270.

[BL]   A. Brownstein and R. Lee, *Cohomology of the symplectic group $Sp_4(\mathbb{Z})$ I The odd torsion case*, T.A.M.S. **334** (1992 pages 575–596).

[Car] H. Cartan, *Démonstration homologiques des Théorèms de périodicité de Bott II*, Seminaire H. Cartan **no. 17** (1959).

[CC]   R.M. Charney and F.R. Cohen, *A stable splitting for the mapping class group*, Mich. J. Math. **35** (1988), 269–284.

[CL]   R.M. Charney and R. Lee, *An application of homotopy theory to mapping class groups*, J. Pure App. Algebra **44** (1987), 127–135.

[Co]   F.R. Cohen, *On the hyperelliptic mapping class groups, $SO(3)$ and $Spin^c(3)$*, Amer. J. Math. **115** (1993), 389–434.

[Co2]   F.R. Cohen, *A course in some aspects of classical homotopy theory*, SLM **1286** (1986), 1–93.

[Co3]   F.R. Cohen, *The homology of C_n-spaces*, SLM **533** (1976), 208–351.

[Co4]   F.R. Cohen, *Applications of loop spaces to some problems in topology*, Lond. Math. Soc. Lecture Notes **139**, 11–20.

[CH]   F.R. Cohen and H.W. Henn, *in preparation*.

[Cn]   R.L. Cohen, *A model for the free loop space of a suspension*, SLM **1286** (1985), 193–207.

[EE]   C.J. Earle and C. Eells, *The diffeomorphism group of a compact Riemann surface*, B.A.M.S. **73** (1967), 557–559.

[EdE]   A.L. Edmonds and J.H. Ewing, *Surface symmetry and homology*, Math. Proc. Camb. Phil. Soc. **99** (1986), 73–77.

[FN]   E. Fadell and L.P. Neuwirth, *Configuration spaces*, Math. Scand. **10** (1962), 111–118.

[FP]   Z. Fiedorowicz and S.B. Priddy, *Homology of classical groups over finite fields and their associated infinite loop spaces*, SLM **674** (1978), Springer-Verlag, Berlin.

[FTY]   M. Furusawa, M. Tezuka and N. Yagita, *On the cohomology of classifying spaces of torus bundles and automorphic forms*, J. Lond. Math. Soc. **(2)37** (1988), 520–534.

[Ge]   E. Getzler, *Batalin-Vilkovisky algebras and two-dimensional topological field theories*, Comm. Math. Phy. **159** (1994), 265–285.

[GM]   H. Glover and G. Mislin, *On the stable cohomology of the mapping class group*, SLM **1172** (1985), 80–84.

[GM2]   H. Glover and G. Mislin, *Torsion in the mapping class group and its cohomology*, JPPA (1987), 177–189.

[G]   T. Goodwillie, *Cyclic homology, derivations and the free loop space*, Topology **24** (1985), 187–215.

[H]   J. Harer, *The third homology group of the moduli space of curves*, Duke Math. J. **(63)1** (1991), 25–55.

[J]   I.M. James, *Reduced product spaces*, Ann. of Math. **62** (1953), 170–197.

[LS]   R. Lee and R. Szczarba, *The group $K_3(\mathbb{Z})$ is cyclic of order 48*, Ann. of Math. **164** (1976), 31–60.

[L]   J.-L. Loday, *Cyclic homology*, Ergebnisse series (1994), Springer-Verlag, Berlin.

[M]   E.Y. Miller, *The homology of the mapping class group*, J. Diff. Geom. **24** (1986), 1–145.

[Mi]   G. Mislin, *Mapping class groups, characteristic numbers and Bernoulli numbers*, preprint.

[Mo]   S. Morita, *Characteristic classes of surface bundles*, Invent. Math. **90** (1987), 551–577.

[Mu]   D. Mumford, *Towards an enumerative geometry of the moduli space of curves*, Shafarevich volume, Birkhäuser Press, 271–328.

[Q]   D. Quillen, *The Adams conjecture*, Topology **19** (1971), 67–80.

[Q2]   D. Quillen, *On the cohomology and K-theory of the general linear groups over finite fields*, Ann. of Math. **96** (1972), 552–586.

[S]   P.S. Selick, *A spectral sequence concerning the double suspension*, Inv. Math. **64** (1981), 15–24.

[S]   S. Smale, *Diffeomorphisms of the 2-sphere*, P.A.M.S. **10** (1959), 621–626.

[St]   N.E. Steenrod, The Topology of Fibre Bundles, Princeton Univ. Press, Princton, 1957.

[T]   H. Toda, Composition Methods in Homotopy Groups of Spheres, Annals of Math. Studies, Princeton Univ. Press, Princeton.

[Tor]   J. Tørnhave, *On the splitting of spherical fibations at odd primes*, Aarhus University, preprint.

[X]   Y. Xia, *The p-periodicity of the mapping class group and the estimate of its p-period*, P.A.M.S. **116** (1992), 1161– 1169.

Department of Mathematics, University of Rochester, Rochester, N.Y. 14627, U.S.A.

Progress in Mathematics, Vol. 136

Representations of the Homology of BV and the Steenrod Algebra II

M. C. Crabb and J. R. Hubbuck

1. Introduction

Let V be an elementary Abelian 2-group of rank n, which it will usually be more useful to consider as an n-dimensional vector space over $\mathbb{F}_2$, the field with 2 elements. We write $H^*(V)$ and $H_*(V)$ for the cohomology and homology of BV with $\mathbb{F}_2$-coefficients. Of course, BV is homotopy equivalent to the cartesian product of n copies of $\mathbb{R}P^\infty$. The importance of $H^*(V)$ as a module or algebra over the mod 2 Steenrod algebra, $\mathcal{A}$, in unstable homotopy is now well established. It is therefore to be expected that $H^*(V)$ itself has much inner subtlety.

We consider $M^*(V)$, the space of generators of $H^*(V)$ over $\mathcal{A}$, which is $\mathbb{F}_2 \otimes_{\mathcal{A}} H^*(V)$, and more explicitly its dual $M_*(V)$, the subring of elements of the Pontrjagin ring $H_*(V)$ which are mapped to zero by all Steenrod squares of positive degrees, frequently denoted by $\mathrm{Ann}_{\overline{\mathcal{A}}} H_*(V)$.

It was Frank Peterson who first raised the problem of determining $M^*(V)$. Initially much attention concentrated on what became known as the Peterson conjecture:

1.1. $M^d(V) \neq 0$ if and only if $d = (2^{\alpha_1} - 1) + (2^{\alpha_2} - 1) + \cdots + (2^{\alpha_n} - 1)$ for some integers $\alpha_i \geq 0$.

This was established by Reg Wood in [11].

The general linear group $GL(V)$ acts on V and therefore on $H^*(V)$ and $H_*(V)$. This action commutes with that of the Steenrod algebra and so acts on $M^*(V)$ and $M_*(V)$. The polynomial algebra $H^*(V)$ has been used to study the modular representation theory of $GL(n, \mathbb{F}_2)$ as every composition factor occurs in some degree, see for example [6]. The key observation that the latter property is shared by $M^*(V)$ is due to Wood. But whereas the vector space dimension of $H^d(V)$ is unbounded as d increases, Carlisle and Wood [4] have proved that there is a constant $\kappa(V)$ such that

$$\dim M^d(V) \leq \kappa(V). \tag{1.2}$$

There is a third aspect to the study of $M^*(V)$ and $M_*(V)$ as $GL(V)$-modules, originating in work of Bill Singer [10]; this is the relationship between the invariants of $M^*(V)$ or the co-invariants of $M_*(V)$ and Tor or Ext terms in the Adams Spectral Sequence. For example, it has been shown by Mike Boardman [3] that when $\dim V = 3$ there is an isomorphism

$$\mathbb{F}_2 \otimes_{GL(3,\mathbb{F}_2)} M_*(V) \to \operatorname{Ext}_{\mathcal{A}}^{3,*+3}(\mathbb{F}_2, \mathbb{F}_2).$$

This has been investigated further by Minami [9].

But there is a major problem with the programme outlined above; it is very difficult to determine $M^*(V)$ or $M_*(V)$. Complete results are known only for $n \leq 3$ and for $n = 3$ the computations are lengthy [7,1,3]. The most extensive computations are due to M. Kameko in [8], but none of his work appears to be published. On the basis of his results, he has conjectured that in (1.2) one can take $\kappa(V) = \prod_{1 \leq i \leq n}(2^i - 1)$. Further, in the notation of (1.1), we believe that he has proved that when

$$\alpha_1 - \alpha_2 > n - 1, \ \alpha_2 - \alpha_3 > n - 2, \ldots, \ \alpha_{n-1} - \alpha_n > 1, \ \alpha_n \geq 0,$$

$$\dim M^d(V) = \prod_{1 \leq i \leq n} (2^i - 1). \tag{1.3}$$

In this note, following the approach though not the notations of [1], we consider an approximation $L_*(V)$ to $M_*(V)$ which we call the "ring of lines". This is defined without reference to the Steenrod algebra. The calculation of $L_*(V)$ as a $GL(V)$-module is not without problems but is easier than the calculation of $M_*(V)$ and, assuming (1.3), "generically" $L_d(V) = M_d(V)$. (See Example 3.12.) The treatment we give of $L_*(V)$ is incomplete. In Section 3 we describe two types of linear relations satisfied by a particular linear spanning set for $L_*(V)$. The first type are easily handled but the second type are discussed only briefly because we do not know how to calculate them in general other than by tedious elementary computations. In Section 4 we say something about the relationship between $L_*(V)$ and $M_*(V)$, concentrating on recent work of D. Anick and F. Peterson [2].

The reader is referred to [5] for related work at odd primes.

The second author thanks the CRM Barcelona for its hospitality during January–July 1994 when he worked on this paper.

2. The ring of lines $L_*(V)$

In standard notation, $H_*(V) = \Gamma(V)$, the divided polynomial algebra on V over $\mathbb{F}_2$. We identify $H_1(V)$ with V. It is useful to have two slightly different descriptions of $\Gamma(V)$. In the first, we identify $H_k(V)$, for all k, with the subspace of $V^{\otimes k}$ fixed by the permutation action of the symmetric group S_k. The product $\cdot : H_k(V) \times H_l(V) \to H_{k+l}(V)$ is then given by:

$$a \cdot b = \sum_{\sigma(S_k \times S_l) \in S_{k+l}/(S_k \times S_l)} \sigma(a \otimes b).$$

(To be more explicit, we can sum over those permutations $\sigma \in S_{k+l}$, with $\sigma(1) < \cdots < \sigma(k)$ and $\sigma(k+1) < \cdots < \sigma(k+l)$.) The definition of the "ring of lines" depends upon the abundance of zero divisors in $\Gamma(V)$.

Let v be an element of V, $k \geq 0$. Then we write $v^{(k)}$ for the element $v \otimes \cdots \otimes v \in H_k(V)$, with the convention that $v^{(0)} = 1$. The product is given by

$$v^{(k)} \cdot v^{(l)} = \binom{k+l}{k} v^{(k+l)}.$$

If $u, v \in V$, then

$$(u+v)^{(k)} = \sum_{i+j=k} u^{(i)} \cdot v^{(j)}. \tag{2.1}$$

When $V = \ell$ has dimension one, with non-zero element v, $\Gamma(\ell)$ is the free, associative, commutative graded algebra over $\mathbb{F}_2$ on generators $v^{(2^r)}$ of degree 2^r, $r \geq 0$, modulo the ideal generated by $\left(v^{(2^r)}\right)^2$, $r \geq 0$. (The mod 2 exterior algebra on generators $v^{(2^r)}$.) When ℓ is a 1-dimensional subspace or line in V, the inclusion of ℓ in V enables us to identify $\Gamma(\ell)$ as a sub-(Hopf-)algebra of $\Gamma(V)$. In particular, let $w_1, w_2, \ldots, w_n$ be a basis for V and let ℓ_i be the line spanned by w_i. Then

$$\bigotimes_{1 \leq i \leq n} \Gamma(\ell_i) = \Gamma(V).$$

When considering $\Gamma(V)$ in this way, it is convenient to have an alternative notation for elements. We set $v^r = v^{(2^r)}$, of degree 2^r, for $r \geq 0$ —the power notation causes no problems as the square of any element of positive degree in $\Gamma(V)$ is zero. If $k = 2^{r_1} + 2^{r_2} + \cdots + 2^{r_m}$, with $0 \leq r_1 < r_2 < \cdots < r_m$, then

$$v^{(k)} = v^{r_1} v^{r_2} \ldots v^{r_m}.$$

The **ring of lines** $L_*(\ell)$ is defined to be the subring of $H_*(\ell)$ generated by 1 and all classes x such that $xy = 0$ whenever $0 < |y| < |x|$ for $y \in H_*(\ell)$.

It is clear that a basis for $L_*(\ell)$ consists of 1 and the classes

$$a_s(\ell) = a_s\langle v\rangle = v^{(2^s-1)} = v^0 v^1 \ldots v^{s-1}, \quad s \geq 1, \tag{2.2}$$

of degree $2^s - 1$. An elementary and well known computation confirms that

$$L_*(\ell) = M_*(\ell). \tag{2.3}$$

The **ring of lines** $L_*(V) \subseteq H_*(V)$ is defined, for general V, to be the subring generated by 1 and all classes $a_s(\ell)$, $s \geq 1$, ℓ a line in V.

Two properties of $L_*(V)$ are easily established.

$$L_*(V) \subseteq M_*(V). \tag{2.4}$$

This follows from (2.3) and the Cartan formula for Steenrod squares in homology.

2.5. $L_d(V) \neq 0$ *if and only if* $d = (2^{\alpha_1} - 1) + (2^{\alpha_2} - 1) + \cdots + (2^{\alpha_n} - 1)$ *for some integers* $\alpha_i \geq 0$.

The observation here is that $|a_s(\ell)| = 2^s - 1$, but if one takes a product of more than n such classes, then expressing each in the form (2.2) one has a product of more than n classes $v \in V$, which is necessarily zero.

The essential motivation for the definition of $L_*(V)$ comes from (1.1), (2.4) and (2.5).

Next we introduce an element $a_1(V) \in L_n(V)$; we shall extend the definition in Section 3. Let $a_1(V) = w_1 \cdot w_2 \cdots w_n$ or, in other words, $a_1(\ell_1) \cdots a_1(\ell_n)$, where ℓ_i is the line spanned by w_i. The element $a_1(V)$ is clearly independent of the choice of basis and so fixed by $GL(V)$. Indeed, it is the image of the generator of $\Lambda^n V$ under the (skew) symmetrization map

$$\Lambda^n V \to H_n(V) = \Gamma_n(V) = (V \otimes \cdots \otimes V)^{S_n}.$$

We shall need later

Lemma 2.6. *The subspace of* $\Gamma_k(V)$ *fixed by* $GL(V)$ *is equal to*

$$(\Gamma_k(V))^{GL(V)} = \begin{cases} \mathbb{F}_2 & \text{if } k = 0, \\ 0 & \text{if } 0 < k < n, \\ \mathbb{F}_2 a_1(V) & \text{if } k = n. \end{cases}$$

Proof. The proof is an easy induction on the dimension of V. Let $v \in V$ be non-zero. Write $V = \mathbb{F}_2 v \oplus W$. We have

$$\Gamma_k(V) = \Gamma_k(W) \cdot 1 \oplus \Gamma_{k-1}(W) \cdot v^{(1)} \oplus \cdots \oplus \mathbb{F}_2 v^{(k)}.$$

We are assuming the result for W. Since $GL(W) \leq GL(V)$, we see that the subspace of $\Gamma_k(V)$ fixed by $GL(V)$ is contained in $\mathbb{F}_2 v^{(k)}$ for $0 < k < n - 1$, $\mathbb{F}_2 a_1(W) \oplus \mathbb{F}_2 v^{(n-1)}$ for $k = n - 1$, and $\Gamma_n(W) \cdot 1 \oplus \mathbb{F}_2 a_1(V) \oplus \mathbb{F}_2 v^{(n)}$ for $k = n$. This holds for any v, and if $w_1, \ldots, w_n$ is a basis, then the element $w_1^{(k)} + \cdots + w_n^{(k)}$ is not fixed by $GL(V)$, $k \neq 0$. The result follows. $\square$

We end this section by discussing periodicity homomorphisms. Let $\mathcal{F} : H^*(V) \to H^*(V)$ be defined by $\mathcal{F}(a) = a^2$ for all a, and let $c : H_*(V) \to H_*(V)$ be the degree-halving dual homomorphism. Then c is a surjective ring homomorphism. In particular, in the notation of (2.2), $c(v^r) = v^{r-1}$ if $r > 0$ and $c(v^0) = 0$. The kernel of c is the ideal generated by $V = H_1(V)$, so that

$$y \in \ker c \iff a_1(V) \cdot y = 0. \tag{2.7}$$

Also $c \cdot Sq_{2i} = Sq_i \cdot c$ and $c \cdot Sq_{2i+1} = 0$ (where Sq_i is the dual of Sq^i). So $c(x) \in M_*(V)$ if and only if $a_1(V) \cdot x \in M_*(V)$.

The periodicity homomorphism

$$f_V : M_*(V) \rightarrow M_*(V) \tag{2.8}$$

is characterized by the property that $f_V(c(x)) = a_1(V) \cdot x$ whenever $c(x) \in M_*(V)$. It is a well defined monomorphism of $GL(V)$-modules, $f_V : M_d(V) \rightarrow M_{2d+n}(V)$. (This homomorphism f_V was also used in the calculations of [8], where it is called Sq^0; see [3,9].)

By (1.2), all monomorphisms are isomorphisms beyond a certain point in the sequence

$$M_d(V) \xrightarrow{f_V} M_{2d+n}(V) \xrightarrow{f_V} \cdots \xrightarrow{f_V} M_{2^t d + (2^t-1)n}(V) \xrightarrow{f_V} M_{2^{t+1} d + (2^{t+1}-1)n}(V) \xrightarrow{f_V} .$$

In addition $f_V(L_d(V)) \subseteq L_{2d+n}(V)$; an explicit formula establishing this is given in Section 3.

3. Linear relations in $L_*(V)$

As a graded vector space $L_*(V)$ is spanned by classes $a_{s_1}(\ell_1) a_{s_2}(\ell_2) \cdots a_{s_r}(\ell_r)$ where $0 \le r \le n$. There are many linear relations among them, of two types.

First we show that distinct members of this set are indexed by flags in V. The key ingredient of the proof is an elementary lemma.

Lemma 3.1. Let $v, w \in V - \{0\}$ and $v \neq w$, and let $s \ge t$. Then

$$a_s \langle v \rangle a_t \langle v + w \rangle = a_s \langle v \rangle a_t \langle w \rangle.$$

Proof. We compute

$$v^{(2^s-1)} \cdot (v+w)^{(2^t-1)} = \sum_{0 \le i \le 2^t - 1} v^{(2^s-1)} \cdot v^{(i)} \cdot w^{(2^t-1-i)}$$

$$= v^{(2^s-1)} \cdot w^{(2^t-1)} + \sum_{0 < i < 2^t} \binom{i + 2^s - 1}{2^s - 1} v^{(i+2^s-1)} \cdot w^{(2^t-1+i)}.$$

But the binomial coefficent $\binom{2^s + (i-1)}{2^s - 1}$ is zero (mod 2) if $i < 2^s$. □

The lemma implies that the element $a_s \langle v \rangle a_t \langle w \rangle$ is determined by the plane $\langle v, w \rangle$ and the line $\langle v \rangle$ and, when $s = t$, by the plane alone. But by elementary linear algebra, what is true in two dimensions generalizes immediately to n dimensions as follows.

Let r integers be given, $s_1 > s_2 > \cdots > s_r > 0$ with $r \leq n$, and r non-zero subspaces of V

$$W_1 \subset W_2 \subset \cdots \subset W_r \tag{3.2}$$

where $\dim W_i = k_i$ and $0 < k_1 < k_2 < \cdots < k_r \leq n$. We again choose a basis $w_1, \ldots, w_n$ for V such that the w_j for $1 \leq j \leq k_i$ span W_i and write $\alpha_j = s_i$ where i is the smallest integer in the range $1 \leq i \leq r$ such that $j \leq k_i$ if $j \leq k_r$, and $\alpha_j = 0$ if $j > k_r$. Let

$$a_{s_1,s_2,\ldots,s_r}(W_1, W_2, \ldots, W_r) = \prod_{1 \leq j \leq k_r} a_{\alpha_j}\langle w_j\rangle. \tag{3.3}$$

By Lemma 3.1, this element is independent of the choice of basis. It has degree d given by

$$d = \sum_{1 \leq i \leq r} (2^{s_i} - 1)(k_i - k_{i-1}) = \sum_{1 \leq j \leq n} (2^{\alpha_j} - 1), \tag{3.4}$$

with the convention that $k_0 = 0$.

The families of elements so constructed can be indexed either by the data $(r; s_1, \ldots, s_r; k_1, \ldots, k_r)$, with $r \geq 0$, $s_1 > \cdots > s_r > 0$ and $0 < k_1 < \cdots < k_r \leq n$ or by n-tuples $(\alpha_1, \ldots, \alpha_n)$, with $\alpha_1 \geq \cdots \geq \alpha_n \geq 0$. (If $r = 0$ the element is taken to be 1.)

Example 3.5. In the notation above let $n = 5$, $(s_1, s_2, s_3) = (6, 4, 2)$ and $(k_1, k_2, k_3) = (2, 3, 5)$, so that $(\alpha_1, \ldots, \alpha_5) = (6, 6, 4, 2, 2)$, and write the basis as a, b, c, d, e. Then

$$a_{6,4,2}(W_1, W_2, W_3) = a^0 a^1 a^2 a^3 a^4 a^5 b^0 b^1 b^2 b^3 b^4 b^5 c^0 c^1 c^2 c^3 d^0 d^1 e^0 e^1$$

of dimension $(2^6 - 1)2 + (2^4 - 1)1 + (2^2 - 1)2$.

The set of all elements of type (3.3) span $L_*(V)$. The action of the periodicity operator f_V on these classes is given by:

$$f_V(a_{s_1,s_2,\ldots,s_r}(W_1, W_2, \ldots, W_r))$$
$$= \begin{cases} a_{s_1+1,s_2+1,\ldots,s_r+1}(W_1, W_2, \ldots, W_r) & \text{if } W_r = V, \\ a_{s_1+1,s_2+1,\ldots,s_r+1,1}(W_1, W_2, \ldots, W_r, V) & \text{if } W_r \neq V. \end{cases}$$

We see, in particular, that $f_V(L_d(V)) \subseteq L_{2d+n}(V)$.

The elements constructed in this way are distinct.

Lemma 3.6. *We have*

$$a_{s_1,s_2,\ldots,s_r}(W_1, W_2, \ldots, W_r) = a_{s_1',s_2',\ldots,s_{r'}'}(W_1', W_2', \ldots, W_{r'}')$$

if and only if $r = r'$, $W_i = W_i'$ and $s_i = s_i'$ for $1 \leq i \leq r$.

Proof. Suppose that $x = a_{s_1, s_2, \ldots, s_r}(W_1, W_2, \ldots, W_r)$, where $r > 0$. Then W_r is characterized as the subspace

$$W = \{w \in V \mid x \cdot w = 0\}.$$

Now, by the formulae above, $x \in L_*(W) \subseteq L_*(V)$ is in the image of f_W. Since the periodicity operator f_W is injective, the result follows by induction on the degree of x. □

It is clear that the product of two elements of type (3.3) is non-zero if and only if the intersection of the two largest subspaces W_r and $W'_{r'}$ is trivial.

Let $G_{k_1, k_2, \ldots, k_r}$ be the variety of all flags of type (3.2); the group $GL(V)$ acts on it in the standard manner. Let $\mathbb{F}_2\langle S \rangle$ denote the free vector space on the set S. So $\mathbb{F}_2\langle G_{k_1, \ldots, k_r} \rangle$ is a permutation representation of $GL(V)$. (It can also be described as the representation induced from the trivial 1-dimensional representation of the parabolic subgroup of $GL(V)$ which is the stabilizer of a given flag. This is the description which generalizes from 2 to an odd prime.) We have thus constructed for each $(r; s_1, \ldots, s_r; k_1, \ldots, k_r)$ as above a $GL(V)$-equivariant map

$$\mathbb{F}_2\langle G_{k_1, \ldots, k_r} \rangle \to L_d(V), \tag{3.7}$$

where d is given by (3.4).

For any fixed degree d only certain flag varieties $G_{k_1, \ldots, k_r}$ will arise: those for which there exist exponents $s_1, \ldots, s_r$ satisfying (3.4). We have therefore described $L_d(V)$ as a quotient of a permutation representation of $GL(V)$ as follows.

Theorem 3.8. *There exists a surjective $GL(V)$-equivariant homomorphism*

$$\bigoplus \mathbb{F}_2\langle G_{k_1, \ldots, k_r} \rangle \to L_d(V),$$

where the direct sum runs over all $(r; s_1, \ldots, s_r; k_1, \ldots, k_r)$ satisfying (3.4).

The kernel of this homomorphism, I_d say, gives the second type of linear relations satisfied by the classes of (3.3) and implicitly contains much of the representation theory of $GL(n, \mathbb{F}_2)$.

Example 3.9. We quote from the final section of [1]. Let $n = 3$ and $s > 3$. Then as $GL(V)$-modules

$$L_{2^s-2}(V) = M_{2^s-2}(V) \cong \mathbb{F}_2\langle G_2 \rangle \cong \mathbb{F}_2\langle GL(3, \mathbb{F}_2)/H_2 \rangle$$

and

$$L_{2^{s+1}-1}(V) = M_{2^{s+1}-1}(V) \cong \mathbb{F}_2\langle G_1 \rangle \oplus \mathbb{F}_2\langle G_2 \rangle$$
$$\cong \mathbb{F}_2\langle GL(3, \mathbb{F}_2)/H_1 \sqcup GL(3, \mathbb{F}_2)/H_2 \rangle,$$

where H_1, H_2 are the subgroups of matrices of the form

$$\begin{bmatrix} 1 & * & * \\ 0 & * & * \\ 0 & * & * \end{bmatrix}, \qquad \begin{bmatrix} * & * & * \\ * & * & * \\ 0 & 0 & 1 \end{bmatrix}$$

respectively.

In the sequence

$$M_{2^s-2}(V) \xrightarrow{f_V} M_{2^{s+1}-1}(V) \xrightarrow{f_V} \cdots \xrightarrow{f_V} M_{2(2^s+u-1)+(2^{1+u}-1)}(V) \xrightarrow{f_V} \cdots,$$

the first homomorphism is the inclusion and thereafter f_V is an isomorphism.

Generically $I_d = 0$.

Proposition 3.10. *Suppose that $k_i = i$, $1 \le i \le r$, and that $2^{s_1-s_2} > n$, $2^{s_2-s_3} > n-1,\ldots,$ $2^{s_{r-1}-s_r} > n-r+2$, $2^{s_r} > n-r+1$. Then the map*

$$\mathbb{F}_2\langle G_{1,2,\ldots,r}\rangle \to L_d(V)$$

is injective.

Proof. We begin with the case $r = 1$, writing $s = s_1$, where $2^s > n$. The classes $a_s\langle v\rangle$, $v \in V$, $v \ne 0$, are asserted to be linearly independent. Thinking of $a_s\langle v\rangle$ as $v^{(2^s-1)} \in V\otimes\cdots\otimes V$, we fix a non-zero vector w in V. Then it is easy to find n linear forms λ_i on V, $1 \le i \le n$, such that $\lambda_i(w) = 1$ for each i, but every other vector $v \ne w$ lies in the kernel of some λ_i. Now extending the indexing to $i \ge 1$ with $\lambda_{i+n} = \lambda_i$ we write

$$\phi = \lambda_1 \otimes \lambda_2 \otimes \cdots \otimes \lambda_{2^s-1}.$$

Then $\phi \in V^* \otimes \cdots \otimes V^* = (V \otimes \cdots \otimes V)^*$ takes the value 1 on $a_s\langle w\rangle$ and is zero on every other $a_s\langle v\rangle$. This verifies linear independence.

The general case is similar. It is convenient to write $y = y_1 \otimes \cdots \otimes y_d$ for a decomposable tensor of degree d, where $y_j \in V$, and to set $y' = y_1 \otimes\cdots\otimes y_{2^{s_1}-1}$, $y'' = y_{2^{s_1}} \otimes \cdots \otimes y_d$ (so that $y = y' \otimes y''$).

Consider a flag $W_1 \subset \cdots \subset W_r$ and choose a basis $w_1,\ldots,w_r$ of W_r such that $w_1,\ldots,w_i$ span W_i, $1 \le i \le r$. Then the class $a_{s_1,\ldots,s_r}(W_1,\ldots,W_r) \in \Gamma_d(V) \subseteq V \otimes \cdots \otimes V$ is the sum of all decomposable tensors y as above in which exactly $2^{s_i} - 1$ of the factors y_j are equal to w_i.

We construct, by induction on r, a multilinear form μ of degree d on V such that, for any y, (i) $\mu(y) = 0$ if 2^{s_1} of the factors y_j are equal, and (ii) if $u_1,\ldots,u_r$ are linearly independent and exactly $2^{s_i} - 1$ of the factors y_j are equal to u_i, then $\mu(y) = 1$ if and only if

$$y_j - w_i \in W_{i-1} \tag{3.11}$$

when $(2^{s_1} - 1) + \cdots + (2^{s_{i-1}} - 1) < j \le (2^{s_1} - 1) + \cdots + (2^{s_i} - 1)$,

for $1 \le i \le r$, $1 \le j \le d$.

We apply the inductive hypothesis to the vector space V/W_1 to produce a form ν of degree $d - 2^{s_1} + 1$ such that (i') $\nu(y'') = 0$ whenever 2^{s_2} of the factors in y'' are equal or one of them is equal to w_1 and (ii') if exactly $2^{s_i} - 1$ of the factors, for $1 < i \leq r$, are equal to u_i, where $u_2, \ldots, u_r$ are linearly independent, then $\nu(y'') = 0$ if and only if the condition (3.11) above holds for $2 \leq i \leq r$, $2^{s_1} \leq j \leq d$.

Let ϕ be the form of degree $2^{s_1} - 1$ constructed in the first part of the proof with $w = w_1$. Because of the condition $2^{s_1-s_2} > n$, if $2^{s_1} - 2^{s_2}$ of the factors of y' are equal and $\phi(y') = 1$, then all the factors of y' are equal to w_1.

We claim that $\mu = \phi \otimes \nu$ has the required properties (i) and (ii) above. Suppose that $\mu(y) = \phi(y')\nu(y'') = 1$ and that $2^{s_1} - 1$ of the factors of y are equal. Then, by property (i') for y'', at most $2^{s_2} - 1$ of these factors occur in y'' and so at least $2^{s_1} - 2^{s_2}$ in y'. It follows that the first $2^{s_1} - 1$ factors of y are equal to w_1 and no others. This establishes the property (i), and (ii) follows easily from the induction hypothesis. □

Example 3.12. Let $d = (2^{\alpha_1} - 1) + (2^{\alpha_2} - 1) + \cdots + (2^{\alpha_n} - 1)$ where $2^{\alpha_1-\alpha_2} > n$, $2^{\alpha_2-\alpha_3} > n-1, \ldots, 2^{\alpha_{n-1}-\alpha_n} > 2$. Then by Theorem 3.8 and Proposition 3.10, $\mathbb{F}_2\langle G_{1,2,\ldots,n}\rangle \cong L_d(V)$ if $\alpha_n > 0$ or $\mathbb{F}_2\langle G_{1,2,\ldots,n-1}\rangle \cong L_d(V)$ if $\alpha_n = 0$. In either case we have

$$L_d(V) \cong \mathbb{F}_2\langle GL(n, \mathbb{F}_2)/U\rangle,$$

where U is the Sylow 2-subgroup of upper triangular matrices. In particular, $L_d(V)$ has dimension $\prod_{1 \leq i \leq n}(2^i - 1)$. Therefore, referring to the paragraph above (1.3), if the conjecture that one can take $\kappa(V) = \prod_{1 \leq i \leq n}(2^i - 1)$ is correct, by (2.4), we have $L_d(V) = M_d(V)$ as $GL(V)$-module. Assuming (1.3) we have $L_d(V) = M_d(V)$ as a $GL(V)$-module provided that $\alpha_1 - \alpha_2 > n - 1$, $\alpha_2 - \alpha_3 > n - 2, \ldots, \alpha_{n-1} - \alpha_n > 1$.

Finally in this section we comment on I_d, that is relations among classes (3.3) of the second type. It is clear that when $W \subseteq V$,

$$\sum_{\ell \subseteq W} a_1(\ell) = \begin{cases} 0 & \text{if dim } W > 1, \\ a_1(W) & \text{if dim } W = 1. \end{cases}$$

But one can add lines in any degree of the form $2^s - 1$.

Proposition 3.13. *Let $W \subseteq V$ and $\dim W = m$. Then*

$$\sum_{\ell \subseteq W} a_s(\ell) = \begin{cases} a_1(W) & \text{if } m = 2^s - 1, \\ 0 & \text{if } m \geq 2^s, \end{cases}$$

where the sum is over all lines ℓ in W.

Proof. The sum is fixed by $GL(W)$. By Lemma 2.6 the sum is zero if $2^s - 1 < m$. If $m = 2^s - 1$, from Proposition 3.10, the classes $a_s(\ell)$ are linearly independent and so, by Lemma 2.6, their sum is $a_1(W)$. □

There are similar relations for sums of planes, 3-dimensional subspaces, etc. Further, these basic relations give rise to others between elements indexed by different flag varieties. For example, using the notations above, let $3 = k_1 < k_2 < k_3$ and $W_1 \subset W_2 \subset W_3$. Then the identity $f_{W_3}^{t_1} f_{W_2}^{t_2} (\sum_{\ell \subseteq W_1} a_2(\ell) + a_1(W_1)) = 0$ is

$$\sum_{\ell \subseteq W_1} a_{t_1+t_2+2,t_1+t_2,t_1}(\ell, W_2, W_3) = a_{t_1+t_2+1,t_1+t_2,t_1}(W_1, W_2, W_3),$$

a linear relation between elements indexed by G_{1,k_2,k_3} and G_{3,k_2,k_3}.

4. The relationship between $L_*(V)$ and $M_*(V)$

We know that $L_d(V) = 0$ if and only if $M_d(V) = 0$, (1.1) and (2.5), $L_*(V) \subseteq M_*(V)$, (2.4), and generically, $L_d(V) = M_d(V)$ by (3.12). If $n = 1$ or 2, then $L_*(V) = M_*(V)$. When $n = 3$, $L_d(V) = M_d(V)$ unless $d = (2^{s+3} - 1) + (2^{s+1} - 1) + (2^s - 1)$, $s \geq 0$; in these latter dimensions $f_V^s : M_8(V) \rightarrow M_d(V)$ and the restricted homomorphism $f_V^s : L_8(V) \rightarrow L_d(V)$ are isomorphisms and $M_8(V)/L_8(V)$ has dimension 1 by [1]. The discrepancy between $L_*(V)$ and $M_*(V)$ certainly increases as the dimension of V increases; the periodicity homomorphisms imply that the discrepancy when $\dim V = 3$ permeates higher dimensions. But there are also genuinely new elements which arise in larger dimensions in $M_*(V)$ but are not in $L_*(V)$; these appear to be highly asymmetric elements. We give examples below. It is not yet clear if the entire discrepancy between $L_*(V)$ and $M_*(V)$ arises from a small number of such elements in small degrees.

Rather than trying to relate $L_*(V)$ and $M_*(V)$ directly, one can consider their intersections with certain subalgebras of $H_*(V)$.

Let $N_*(V, r)$ be the subalgebra of $H_*(V)$ generated by $\bigoplus_{0 \leq i \leq 2^r} H_i(V)$.

Theorem 4.1. $L_*(V) \cap N_*(V, 1) = M_*(V) \cap N_*(V, 1)$.

This result is essentially the main theorem of [2] and is non-trivial. We will give what is superficially a different proof, but the idea of the proof was extracted from that of Anick and Peterson.

Let $w_1, w_2, \ldots, w_n$ be a basis for V. To avoid irrelevant subscripts, it will soon be convenient to use the letters x, y, z for general elements of the basis and to write a for w_1. So $N_*(V, 1)$ has a basis of distinct monomials $\{(w_1^0)^{e_1}(w_1^1)^{f_1}(w_2^0)^{e_2}(w_2^1)^{f_2} \ldots (w_n^0)^{e_n}(w_n^1)^{f_n}\}$ where e_i and f_i are 0 or 1. We need to prove that $M_*(V) \cap N_*(V, 1) \subseteq L_*(V) \cap N_*(V, 1)$. We proceed by induction on n. The inductive hypothesis and the Cartan formula imply that it is sufficient to restrict attention to elements which, when expressed in terms of the basis, are sums of monomials in which $e_i + f_i = 1$ for each i. Let $p \in H_d(V)$ be such an element and set $e(p) = \sum_{1 \leq i \leq n}(e_i - f_i)$, which is easily checked to be equal to $3n - 2d$.

We prove the following statement inductively on $\dim V = n$.

4.2.   *Let a non-zero p satisfy $Sq_i(p) = 0$ for $i > \max\{0, -e(p)\}$. Then p is a sum of elements which are products of commutators of the form $(y^0 z^1 + z^0 y^1)$ and either $e(p)$ classes of type x^0 when $e(p) \geq 0$ or $-e(p)$ classes of type x^1 when $e(p) < 0$. If, in addition, $e(p) = -r < 0$, then $Sq_r(p) \neq 0$.*

This statement implies Theorem 4.1 as

$$y^0 z^1 + z^0 y^1 = y^0 y^1 + z^0 z^1 + (y^0 + z^0)(y^1 + z^1 + y^0 z^0).$$

When $n = 1$, (4.2) is trivial, so we assume that the result is established for dimension $n - 1$. Let p be as above and suppose that $e(p) = -r < 0$. We write $p = a^0 p' + a^1 p''$. So $e(p') = -(r + 1)$, $e(p'') = -(r - 1)$ and $Sq_i(p) = a^0 Sq_i p' + a^1 Sq_i p'' + a^0 Sq_{i-1} p''$. In particular, as $Sq_i p = 0$ for $i > r$, $Sq_i p'' = 0$ for $i > r$ and $Sq_i p' = 0$ for $i > r + 1$. By the induction hypothesis, p' satisfies the conclusion of (4.2) for $n - 1$ and we can write $p' = \sum x^1 p_x$ where $e(p_x) = -r$ and $Sq_i p_x = 0$ for $i > r$. Therefore (4.2) applies to each p_x. So $p = \sum (a^0 x^1 + a^1 x^0) p_x + a^1 p'''$ where $e(p''') = -(r - 1)$ and $Sq_i(a^1 p''') = 0$ for $i > r$. Hence $Sq_i p''' = 0$ for $i > r - 1$ and again (4.2) applies. Therefore p is a product of $-e(p)$ classes of type x^1 and commutators.

In addition $Sq_r p \neq 0$; the effect of applying Sq_r to p is to interchange the 0 and 1 superscripts everywhere.

When $e(p) = r \geq 0$, the proof of (4.2) is similar. We write $p = a^1 p' + a^0 p''$. So $e(p') = r + 1$, $e(p'') = r - 1$ and $Sq_i p' = 0$ for $i > 0$. Using the inductive hypothesis, we can write $p = \sum (a^1 x^0 + a^0 x^1) p_x + a^0 p'''$, where $p_x \in M_*$ and $e(p''') = r - 1$. So $p''' \in M_*$ and, if $r \geq 1$, p''' has the required form. If $r = 0$, $p''' = 0$, for otherwise $Sq_1 p''' \neq 0$ by the induction hypothesis. Thus p has the required form.   □

It is not true that $N_*(V, 2) \cap L_*(V) = N_*(V, 2) \cap M_*(V)$ (for $n \geq 3$); see Proposition 3.7 (c) of [1]. But from our viewpoint the main interest in Theorem 4.1 is that, using the periodicity homomorphisms and the homomorphism c, one can use inductive arguments as was done in [1]. There remain major combinatorial problems to overcome before one can hope to understand the relationship between L_* and M_*.

The proof of Theorem 4.1 gives a hint on how to construct families of elements in $M_*(V)$ which are not in $L_*(V)$. Let $\{w_1, w_2, \ldots, w_{2^{k+1}-1}, y\}$ be linearly independent elements of V, $k > 0$, and set

$$u_k = \sum \sigma(w_1^0 w_2^0 \ldots w_{2^k}^0 w_{2^k+1}^1 \ldots w_{2^{k+1}-1}^1),$$

where σ runs over $S_{2^{k+1}-1}$ and acts on the indices. Then $Sq_i u_k \neq 0$ if and only if $i = 0$ or 2^k; $Sq_{2^k} u_k = w_1^0 w_2^0 \ldots w_{2^{k+1}-1}^0$. We now consider

$$v_k = y^0 y^1 \ldots y^k y^{k+1} u_k + y^0 y^1 \ldots y^{k-1} y^{k+1} w_1^0 w_2^0 \ldots w_{2^{k+1}-1}^0.$$

Then $v_k \in M_*$, but $v_k \notin L_*$. The first such element occurs with dim $V = 4$ and has degree 8.

References

[1] M. A. ALGHAMDI, M. C. CRABB and J. R. HUBBUCK, *Representations of the Homology of BV and the Steenrod Algebra I,* Adams Memorial Symposium on Algebraic Topology 2, ed. N. Ray and G. Walker, London Math. Soc. Lect. Note Series **176** (1992), 217–234.

[2] D. ANICK and F. P. PETERSON, A_2-*Annihilated elements in* $H_*(\Omega\Sigma\mathbb{R}P^2)$, Proc. Amer. Math. Soc. **117** (1993), 243–250.

[3] J. M. BOARDMAN, *Modular representations on the homology of powers of real projective space,* Algebraic Topology: Oaxtepec 1991, ed. M. C. Tangora, Contemp. Math. **146** (1993), 49–70.

[4] D. P. CARLISLE and R. M. W. WOOD, *The boundedness conjecture for the action of the Steenrod algebra on polynomials,* Adams Memorial Symposium on Algebraic Topology 2, ed. N. Ray and G. Walker, London Math. Soc. Lect. Note Series **176** (1992), 203–216.

[5] M. D. CROSSLEY, Results concerning the Steenrod algebra, Thesis, University of Aberdeen, 1994.

[6] D. J. GLOVER, *A study of certain modular representations,* J. Algebra **51** (1978), 425–475.

[7] M. KAMEKO, Products of projective spaces as Steenrod modules, manuscript, June 1990.

[8] M. KAMEKO, Products of projective spaces as Steenrod modules, Thesis, Johns Hopkins University.

[9] N. MINAMI, The Adams spectral sequence and the triple transfer, manuscript.

[10] W. M. SINGER, *The transfer in homological algebra,* Math. Z. **202** (1989), 493–524.

[11] R. M. W. WOOD, *Steenrod squares of polynomials and the Peterson conjecture,* Math. Proc. Camb. Phil. Soc. **105** (1989), 307–309.

M. C. Crabb and J. R. Hubbuck, Department of Mathematical Sciences
University of Aberdeen, Aberdeen AB9 2TY, Scotland

Progress in Mathematics, Vol. 136
© 1996 Birkhäuser Verlag Basel/Switzerland

Polynomial structures for iterated central extensions of abelian-by-nilpotent groups

KAREL DEKIMPE*AND PAUL IGODT

1. Polynomial representations and iterated extensions

A famous question of John Milnor [10] dealt with the problem of whether or not any torsion-free, polycyclic-by-finite group Γ occurs as the fundamental group of a compact, complete, affinely flat manifold. This is equivalent to saying that Γ admits a faithful affine representation, making it acting properly discontinuously on $\mathbb{R}^K$ ($K =$ Hirsch length of Γ), with compact quotient. This question was answered negatively by Y. Benoist [1] and D. Burde & F. Grunewald [2], even in the nilpotent case. However, it is known that any torsion-free, polycyclic-by-finite group Γ admits a smooth action on $\mathbb{R}^K$ with compact quotient (see [3] and [7]). We might look at an affine mapping as being a polynomial of degree 1, while a smooth map, having in mind a power series expansion, can be regarded as "polynomial" of infinite degree. This paper investigates if it is possible to find anything in between those two. So our question becomes:

> *Does any (torsion-free) polycyclic-by-finite group Γ of rank K admit a properly discontinuous action on $\mathbb{R}^K$, expressed by polynomial functions, such that the quotient space is compact?*

We remark that the kernel of any properly discontinuous action of a polycyclic-by-finite group Γ on $\mathbb{R}^K$ ($K =$ Hirsch length of Γ) is exactly the maximal finite normal subgroup of Γ ([6], see also [5]) Everything we will say, makes perfectly sense in case Γ is finite ($K = 0$, $\mathbb{R}^K$ is a point), but this is of course not interesting, except perhaps as the starting point for an iteration.

If Γ is an infinite polycyclic-by-finite group, then it is well known that Γ has a non-trivial free abelian, normal (even characteristic, but this will not be needed here) subgroup, say $\mathbb{Z}^{k_1}$, for some $k_1 > 0$. Therefore, Γ can be seen as the result of a series of extensions:

---

*Research Assistant National Fund For Scientific Research (Belgium)

$$1 \longrightarrow \mathbb{Z}^{k_1} \longrightarrow \Gamma = \Gamma_1 \longrightarrow \Gamma_2 \longrightarrow 1 \qquad \text{(top)}$$

$$1 \longrightarrow \mathbb{Z}^{k_2} \longrightarrow \Gamma_2 \longrightarrow \Gamma_3 \longrightarrow 1$$

$$\vdots$$

$$1 \longrightarrow \mathbb{Z}^{k_{i-1}} \longrightarrow \Gamma_{i-1} \longrightarrow \Gamma_i \longrightarrow 1$$

$$1 \longrightarrow \mathbb{Z}^{k_i} \longrightarrow \Gamma_i \longrightarrow \Gamma_{i+1} \longrightarrow 1 \qquad (i\text{-th layer}) \qquad (1)$$

$$1 \longrightarrow \mathbb{Z}^{k_{i+1}} \longrightarrow \Gamma_{i+1} \longrightarrow \Gamma_{i+2} \longrightarrow 1$$

$$\vdots$$

$$1 \longrightarrow \mathbb{Z}^{k_c} \longrightarrow \Gamma_c \longrightarrow \Gamma_{c+1} = F \longrightarrow 1 \qquad \text{(bottom)}$$

where F is a finite group. So the Hirsch length of Γ, $h(\Gamma) = k_1 + k_2 + \cdots + k_c = K$. We will use also $K_i = k_i + k_{i+1} + \cdots + k_c$. Let us call a collection of extensions as in (1), a **tower** (of extensions). The top (bottom) of the tower will be the upper (lower) extension, and the i-th layer is meant to be the extension $1 \to \mathbb{Z}^{k_i} \to \Gamma_i \to \Gamma_{i+1} \to 1$. We will call c the height of the tower.

It is our aim to build up a polynomial representation of Γ by starting at the bottom of a tower like (1) and going up stepwise. The procedure for building up this representation is known as an iterated Seifert Fiber Space construction, with typical fiber a torus ([9]). Let us digress for a moment and recall the algebraic setting of this construction in our specific situation.

Write $P(\mathbb{R}^K, \mathbb{R}^k)$ (resp. $\mathrm{Aff}(\mathbb{R}^K, \mathbb{R}^k)$) for the real vector space of polynomial (resp. affine) mappings from $\mathbb{R}^K$ to $\mathbb{R}^k$. We will use $P(\mathbb{R}^K)$ (resp. $\mathrm{Aff}(\mathbb{R}^K)$) to denote the group of polynomial (resp. affine) diffeomorphisms of $\mathbb{R}^K$; here, the group-law is composition of mappings. The reader should remark that, while $\mathrm{Aff}(\mathbb{R}^K)$ is a finite dimensional Lie-group, $P(\mathbb{R}^K)$ is highly non trivial and far from "finite dimensional". Elements of $P(\mathbb{R}^K)$ are polynomial self-diffeomorphisms λ of $\mathbb{R}^K$ for which the inverse mapping $\mu = \lambda^{-1}$ is again polynomial. E.g. $\lambda, \mu : \mathbb{R}^2 \to \mathbb{R}^2$ s.t.

$$\lambda(x, y) = (x + x^2 + 4xy + 4y^2, x + 2y) \text{ and } \mu(x, y) = (x - y^2, \frac{1}{2}(y - x + y^2))$$

are each others inverse in $P(\mathbb{R}^2)$.

$P(\mathbb{R}^K, \mathbb{R}^k)$ is made into a $Gl(k, \mathbb{R}) \times P(\mathbb{R}^K)$-module, via

$$\forall g \in Gl(k, \mathbb{R}), \ \forall h \in P(\mathbb{R}^K), \ \forall \lambda \in P(\mathbb{R}^K, \mathbb{R}^k) : \ ^{(g,h)}\lambda = g \circ \lambda \circ h^{-1}.$$

The resulting semi-direct product $P(\mathbb{R}^K, \mathbb{R}^k) \rtimes (Gl(k, \mathbb{R}) \times P(\mathbb{R}^K))$ embeds into $P(\mathbb{R}^{k+K})$ as follows: $\forall \lambda \in P(\mathbb{R}^K, \mathbb{R}^k), \ \forall g \in Gl(k, \mathbb{R}), \ \forall h \in P(\mathbb{R}^K)$:

$$\forall x \in \mathbb{R}^k, \ \forall y \in \mathbb{R}^K : \ (\lambda, g, h)(x, y) = (g(x) + \lambda(h(y)), h(y)).$$

Now, assume $1 \to \mathbb{Z}^k \to \Gamma \to \bar{\Gamma} \to 1$ is an extension of a polycyclic-by-finite group $\bar{\Gamma}$ and $\bar{\rho} : \bar{\Gamma} \to P(\mathbb{R}^K)$ is a polynomial representation of $\bar{\Gamma}$. Suppose also that the action of $\bar{\Gamma}$ on $\mathbb{Z}^k$, induced by conjugation in Γ, is denoted by φ. Then, we investigate when $\bar{\rho}$ can be lifted to a polynomial representation $\rho : \Gamma \to P(\mathbb{R}^K, \mathbb{R}^k) \rtimes (Gl(k, \mathbb{R}) \times P(\mathbb{R}^K))$, making the following diagram commutative:

$$
\begin{array}{ccccccccc}
1 & \longrightarrow & \mathbb{Z}^k & \longrightarrow & \Gamma & \longrightarrow & \bar{\Gamma} & \longrightarrow & 1 \\
& & \downarrow{\scriptstyle i} & & \downarrow{\scriptstyle \rho} & & \downarrow{\scriptstyle \varphi \times \bar{\rho}} & & \\
1 & \longrightarrow & P(\mathbb{R}^K, \mathbb{R}^k) & \longrightarrow & P(\mathbb{R}^K, \mathbb{R}^k) \rtimes (Gl(k, \mathbb{R}) \times P(\mathbb{R}^K)) & \longrightarrow & Gl(k, \mathbb{R}) \times P(\mathbb{R}^K) & \longrightarrow & 1
\end{array}
$$

Here, $i : \mathbb{Z}^k \hookrightarrow P(\mathbb{R}^K, \mathbb{R}^k)$ is defined by considering each element $z \in \mathbb{Z}^k$ as a constant mapping in $P(\mathbb{R}^K, \mathbb{R}^k)$. The problem of the existence (and uniqueness) of ρ has been translated in terms of the vanishing of certain cohomology classes. To explain this, remark that $P(\mathbb{R}^K, \mathbb{R}^k)$ is a $\bar{\Gamma}$-module via $\varphi \times \bar{\rho}$. The short exact sequence of $\bar{\Gamma}$-modules

$$ 0 \to \mathbb{Z}^k \to P(\mathbb{R}^K, \mathbb{R}^k) \to P(\mathbb{R}^K, \mathbb{R}^k)/\mathbb{Z}^k \to 0 $$

induces a long exact sequence of cohomology groups (where we omit the notations for the module structures):

$$ \to H^1(\bar{\Gamma}, P(\mathbb{R}^K, \mathbb{R}^k)) \to H^1(\bar{\Gamma}, P(\mathbb{R}^K, \mathbb{R}^k)/\mathbb{Z}^k) \xrightarrow{\delta} H^2(\bar{\Gamma}, \mathbb{Z}^k) \to H^2(\bar{\Gamma}, P(\mathbb{R}^K, \mathbb{R}^k)) \to \tag{2} $$

Let $\langle f \rangle \in H^2_\varphi(\bar{\Gamma}, \mathbb{Z}^k)$ correspond to the extension $1 \to \mathbb{Z}^k \to \Gamma \to \bar{\Gamma} \to 1$. Then Γ can be seen as the group with underlying set $\mathbb{Z}^k \times \bar{\Gamma}$ and with group structure given by:

$$ \forall (a, \alpha), (b, \beta) \in \mathbb{Z}^k \times \bar{\Gamma} : \quad (a, \alpha) \cdot (b, \beta) = (a + \varphi(\alpha)b + f(\alpha, \beta), \alpha\beta). $$

There exists a ρ making the above diagram commutative iff $\langle f \rangle$ belongs to the image of the connecting homomorphism δ. More explicitly, if $f = \delta\bar{\lambda}$ for a 1-cocycle $\bar{\lambda} : \bar{\Gamma} \to P(\mathbb{R}^K, \mathbb{R}^k)/\mathbb{Z}^k$, then $\rho : \Gamma = \mathbb{Z}^k \times \bar{\Gamma}$ can be chosen as follows:

$$ \forall (a, \alpha) \in \mathbb{Z}^k \times \bar{\Gamma} : \quad \rho(a, \alpha) = (a + \lambda(\alpha), \varphi(\alpha), \bar{\rho}(\alpha)), $$

where $\lambda : \bar{\Gamma} \to P(\mathbb{R}^K, \mathbb{R}^k)$ denotes any normalized lift of the 1-cocycle $\bar{\lambda}$.

Remark that, if ρ exists, we say that the representation $\bar{\rho} : \bar{\Gamma} \to P(\mathbb{R}^K)$ has been *lifted* to Γ; moreover, this lifting happens in such a way that $\mathbb{Z}^k$ acts via pure translations on the first k components of $\mathbb{R}^{k+K}$ and such that $\forall \gamma \in \Gamma, \forall x \in \mathbb{R}^k, \forall y \in \mathbb{R}^K :$

$$ {}^\gamma(x, y) = \rho(\gamma)(x, y) = (\varphi_i(\bar{\gamma})(x) + \lambda(\bar{\gamma})(y), \bar{\rho}(\bar{\gamma})(y)). $$

Analogous statements can be made for affine representations. One simply replaces every occurrence of $P(\mathbb{R}^K)$ by $\text{Aff}(\mathbb{R}^K)$ while $P(\mathbb{R}^K, \mathbb{R}^l)$ is replaced by $\text{Aff}(\mathbb{R}^K, \mathbb{R}^l)$ (see also [4] and [8]).

Coming back to our tower of extensions (1), we can now formulate more precisely our aim: assuming Γ_c admits a polynomial representation $\bar{\rho} : \Gamma_c \to$

$P(\mathbb{R}^{K_c})$, we will try to lift this representation iteratively according to the above described picture until we arrive at a representation $\rho : \Gamma \to P(\mathbb{R}^K)$.

To be able to state a precise definition of a canonical type polynomial representation, we need the notion of a torsion-free filtration as introduced in [6].

DEFINITION 1.1 *A torsion-free filtration* Γ_* *of a polycyclic-by-finite group* Γ *is an ascending sequence of normal subgroups* Γ'_i $(0 \le i \le c+1)$ *of* Γ

$$\Gamma_* : \quad \Gamma'_0 = 1 \subset \Gamma'_1 = \mathbb{Z}^{k_1} \subset \Gamma'_2 \subset \cdots \subset \Gamma'_{c-1} \subset \Gamma'_c \subset \Gamma'_{c+1} = \Gamma$$

for which

$$\Gamma'_i / \Gamma'_{i-1} \cong \mathbb{Z}^{k_i} \text{ for } 1 \le i \le c \text{ and some } k_i \in \mathbb{N}_0 \text{ and}$$

$$\Gamma / \Gamma'_c \text{ is finite.}$$

Observe that, associated to the given tower, there is a torsion-free filtration Γ_* of Γ, where Γ'_i $(1 \le i \le c)$ is defined by the condition $\Gamma / \Gamma'_i = \Gamma_{i+1}$. Indeed, one verifies that $\Gamma'_i / \Gamma'_{i-1} \cong \mathbb{Z}^{k_i}$ $(1 \le i \le c)$, while Γ / Γ'_c is finite.
Conversely, given a torsion-free filtration Γ_*, there is a unique tower which is associated to this torsion-free filtration. So, towers and torsion-free filtrations are just two different ways of looking at the same objects.

We can now recall the definition of a canonical type polynomial representation, as given in [6].

DEFINITION 1.2 *Assume* Γ *is polycyclic-by-finite with a torsion-free filtration* Γ_*. *A polynomial representation* $\rho = \rho_0 : \Gamma \to P(\mathbb{R}^K)$ *will be called of canonical type with respect to* Γ_* *iff it induces a sequence of representations:*

$$\rho_i : \Gamma / \Gamma'_i \to P(\mathbb{R}^{K_{i+1}}), \quad (1 \le i \le c)$$

such that for all i *the following diagram commutes:*

$$
\begin{array}{ccccccccc}
1 & \to & \mathbb{Z}^{k_i} \cong \Gamma'_i / \Gamma'_{i-1} & \to & \Gamma / \Gamma'_{i-1} & \to & \Gamma / \Gamma'_i & \to & 1 \\
 & & \downarrow j & & \downarrow \rho_{i-1} & & \downarrow \varphi_i \times \rho_i & & (3) \\
1 & \to & P(\mathbb{R}^{K_{i+1}}, \mathbb{R}^{k_i}) & \to & P \rtimes (G \times H) & \to & \mathrm{GL}(k_i, \mathbb{Z}) \times P(\mathbb{R}^{K_{i+1}}) & \to & 1
\end{array}
$$

where

- $P \rtimes (G \times \mathcal{H})$ *stands for* $P(\mathbb{R}^{K_{i+1}}, \mathbb{R}^{k_i}) \rtimes (\mathrm{GL}(k_i, \mathbb{Z}) \times P(\mathbb{R}^{K_{i+1}}))$,

- $j(z) : \mathbb{R}^{K_{i+1}} \to \mathbb{R}^{k_i} : x \mapsto z, \forall z \in \mathbb{Z}^{k_i}$ *and*

- $\varphi_i : \Gamma / \Gamma'_i \to \mathrm{GL}(k_i, \mathbb{Z})$ *denotes the action of* Γ / Γ'_i *induced on* $\mathbb{Z}^{k_i}$ *by conjugation in* Γ / Γ'_{i-1}.

As it is not hard to see that any polynomial representation of Γ which is of canonical type with respect to a given torsion-free filtration Γ_* can be obtained by the stepwise lifting procedure in the associated tower of extensions, we will often speak of representations which are "of canonical type with respect to a given tower".

As already said before, it is known that smooth and continuous representations can always be found by this method ([3], [9]). On the other hand, the counter-examples of Benoist and Burde & Grunewald show that there exist torsion-free, finitely generated nilpotent groups of rank 11, admitting no properly discontinuous action on $\mathbb{R}^{11}$ via affine motions. A fortiori, this implies that no canonical type affine representation exists for such a nilpotent group. Therefore, the best we can hope for in the general case of polycyclic-by-finite groups, is a canonical type polynomial representation, if possible such that its image in $P(\mathbb{R}^K)$ consists entirely of polynomials of degree lower than or equal to some upper-bound (probably dependent from the Hirsch length of the group). We remark that it is inherent to the notion of a canonical type representation, that the resulting action is properly discontinuous and that the quotient space is compact.

In [6], the problem of finding a canonical type polynomial representation was solved in case every extension involved in the tower is almost-central.

DEFINITION 1.3 *An extension* $1 \to A \to G \to H \to 1$ *of a group H by an abelian group A is called* **almost-central** *if there exists a subgroup H' of finite index in H, such that the action of H' on A, induced by conjugation in G, is trivial. In this case, we call the action of H on A almost-trivial.*

If $H' = H$, we recover the usual notion of a central extension. If every layer of a tower is almost-central, then the group obtained is virtually nilpotent (use induction on the height of the tower).

The main results of this paper improve the result of [6]:

- by showing that a canonical type polynomial representation *can always be lifted at a layer which is almost central* (independent of the (almost) centralness of lower layers) (Theorem 2.3).

- by treating (certain) towers with exactly one non almost-central layer (Theorem 3.1). This will prove the existence of polynomial structures for a fairly larger class than the virtually nilpotent groups.

In the closing section we translate Milnor's original question in terms of the *affine defect* number of a polycyclic-by-finite group. A list of unsolved problems is included.

2. Iterating almost-central extensions

First of all, we will need the following lemma.

LEMMA 2.1 *Let Γ be a polycyclic-by-finite group equiped with a canonical type polynomial representation $\rho : \Gamma \to P(\mathbb{R}^K)$. If $1 \to \mathbb{Z}^k \to \Gamma \to \bar{\Gamma} \to 1$ is the top of the tower, with respect to which ρ is canonical, then we have the following isomorphism of $\bar{\Gamma}$-modules ($\forall l \in \mathbb{N}_0$):*

$$P(\mathbb{R}^K, \mathbb{R}^l)^{\mathbb{Z}^k} \cong P(\mathbb{R}^{K-k}, \mathbb{R}^l),$$

where the Γ-module structure of $P(\mathbb{R}^K, \mathbb{R}^l)$ is given by the map $1 \times \rho : \Gamma \to Gl(l, \mathbb{R}) \times P(\mathbb{R}^K)$.

Proof: Note that $P(\mathbb{R}^K, \mathbb{R}^l)^{\mathbb{Z}^k}$ inherits a $\Gamma/\mathbb{Z}^k = \bar{\Gamma}$-module structure, as being a fixed point set. The $\bar{\Gamma}$-module structure of $P(\mathbb{R}^{K-k}, \mathbb{R}^l)$ is given by the map $1 \times \rho_1 : \bar{\Gamma} \to Gl(l, \mathbb{R}) \times P(\mathbb{R}^K)$. (The reader might wish to recall the meaning of ρ_1, which can be found in definition 1.2).

Let $p(x) \in P(\mathbb{R}^K, \mathbb{R}^l)$ and recall the action of $\gamma \in \Gamma$ on $p(x)$: $(^\gamma p)(x) = (p \circ \rho^{-1}(\gamma))(x) = p(^{\gamma^{-1}}x)$. One verifies that $^z p = p$ for all $z \in \mathbb{Z}^k$ iff $p(x)$ is a polynomial which is independent of the first k components of $\mathbb{R}^K$. Consequently, we can identify p with an element of $P(\mathbb{R}^{K-k}, \mathbb{R}^l)$, and, this identification is compatible with the $\bar{\Gamma}$-module structure. $\qquad\square$

We will now show that it is always possible to lift a canonical type polynomial structure in the case of almost-central extensions. The main reason for this lies in a cohomology-vanishing theorem, which is a generalisation of theorem 3.8 in [6]. To prove this theorem we need the following proposition, for the proof of which we refer to the first part in the proof of theorem 3.8 in [6].

PROPOSITION 2.2 *Let the free abelian group $\mathbb{Z}^k$ act faithfully as translations on the first k components on $\mathbb{R}^K$, with $K \geq k$. I.e. for an appropriate choice of coordinates we have $\forall z = (z_1, z_2, \ldots, z_k) \in \mathbb{Z}^k$, $\forall x = (x_1, x_2, \ldots, x_K) \in \mathbb{R}^K$:*

$$^z x = (x_1 + z_1, \ldots, x_k + z_k, x_{k+1}, \ldots, x_K).$$

Then

$$H^i(\mathbb{Z}^k, P(\mathbb{R}^K, \mathbb{R})) = 0, \quad \forall i > 0,$$

where $\mathbb{Z}^k$ acts on $P(\mathbb{R}^K, \mathbb{R})$, via

$$\forall z \in \mathbb{Z}^k, \ \forall x \in \mathbb{R}^K, \ \forall p \in P(\mathbb{R}^K, \mathbb{R}) : \ (^z p)(x) = p(^{-z}x). \qquad\square$$

We are now ready for

THEOREM 2.3 *Let Γ be a polycyclic-by-finite group and suppose that $\rho : \Gamma \to P(\mathbb{R}^K)$ is a canonical type polynomial representation of Γ. If $\varphi : \Gamma \to \text{Aut}(\mathbb{Z}^k)$ denotes an almost-trivial action of Γ on $\mathbb{R}^k$, then*

$$H^i_{\varphi \times \rho}(\Gamma, P(\mathbb{R}^K, \mathbb{R}^k)) = 0, \quad \forall i > 0.$$

Proof: We argue by induction on the Hirsch length $h(\Gamma) = K$ of Γ. If $h(\Gamma) = 0$, then Γ is finite and so $H^i_{\varphi \times \rho}(\Gamma, P(\mathbb{R}^0, \mathbb{R}^k)) = H^i_{\varphi}(\Gamma, \mathbb{R}^k) = 0$ for all $i > 0$.

Now let $K > 0$ and assume $1 \to \mathbb{Z}^{k_1} \to \Gamma \to \Gamma_2 \to 1$ is the top extension of the tower with respect to which ρ is of canonical type. Write Γ' for $\mathrm{Ker}(\varphi)$; so $[\Gamma : \Gamma'] < \infty$.

We will first prove the theorem for this Γ'. The restriction of ρ to Γ' (which we denote also by ρ) is clearly also of canonical type with respect to the obvious tower of extensions, having top

$$1 \to A = \mathbb{Z}^{k_1} \cap \Gamma' \to \Gamma' \to \Gamma' \cdot \mathbb{Z}^{k_1}/\mathbb{Z}^{k_1} \cong \Gamma'/(\mathbb{Z}^{k_1} \cap \Gamma') = \Gamma'_2 \to 1.$$

By the induction hypothesis we know that

$$H^i_{1 \times \bar{\rho}}(\Gamma'_2, P(\mathbb{R}^{K_2}, \mathbb{R}^k)) = 0, \quad \forall i > 0,$$

where $\bar{\rho} \, (= \rho_1) : \Gamma'_2 \to P(\mathbb{R}^{K_2})$ denotes the induced canonical type polynomial representation of Γ'_2.
By proposition 2.2, $H^i_{1 \times \rho_{|A}}(A, P(\mathbb{R}^K, \mathbb{R}^k)) = 0$, for all $i > 0$. Consequently, the inflation-restriction 5-term exact sequences

$$0 \to H^i_{1 \times \rho}(\Gamma'_2, P(\mathbb{R}^K, \mathbb{R}^k)^A) \to H^i_{1 \times \rho}(\Gamma', P(\mathbb{R}^K, \mathbb{R}^k)) \to \underbrace{H^i_{1 \times \rho_{|A}}(A, P(\mathbb{R}^K, \mathbb{R}^k))^{\Gamma'}}_{=0} \to$$

and lemma 2.1, induce isomorphisms

$$0 = H^i_{1 \times \rho}(\Gamma'_2, P(\mathbb{R}^K, \mathbb{R}^k)^A) \xrightarrow{\cong} H^i_{1 \times \rho}(\Gamma', P(\mathbb{R}^K, \mathbb{R}^k)).$$

This proves the theorem for Γ'.
The proof finishes now by realizing that

$$\mathrm{res} : H^i_{\varphi \times \rho}(\Gamma, P(\mathbb{R}^K, \mathbb{R}^k)) \to H^i_{1 \times \rho}(\Gamma', P(\mathbb{R}^K, \mathbb{R}^k))$$

is injective, since Γ' is of finite index in Γ and $P(\mathbb{R}^K, \mathbb{R}^k)$ is divisible.   □

COROLLARY 2.4 *Let Γ be any polycyclic-by-finite group admitting a canonical type polynomial representation $\rho : \Gamma \to P(\mathbb{R}^K)$. If $1 \to \mathbb{Z}^k \to \tilde{\Gamma} \to \Gamma \to 1$ is an almost-central extension of Γ, then ρ lifts to a polynomial representation of $\tilde{\Gamma}$, which is of canonical type with respect to the tower obtained by putting the extension $1 \to \mathbb{Z}^k \to \tilde{\Gamma} \to \Gamma \to 1$ on top of the tower of Γ.*

Proof: By theorem 2.3, the connecting homomorphism δ in the long exact cohomology sequence (2), is an isomorphism.   □

We remark that also the injectivity of the connecting homomorphism has its own meaning. In fact, it implies that the lift of ρ is unique up to a polynomial conjugation (see [6]). However, as we will not be able to prove the injectivity of δ for non-central extensions (next section), we do not exploit this uniqueness property here.

We obtain also

COROLLARY 2.5 *Any finitely generated, virtually nilpotent group Γ admits a canonical type, polynomial representation $\rho : \Gamma \rightarrow P(\mathbb{R}^K)$, where K denotes the Hirsch length or rank of Γ.*

3. Polynomial structures for abelian-by-nilpotent groups

Here we explain how to obtain a canonical type polynomial representation for any abelian-by-nilpotent group (within the category of polycyclic-by-finite groups). Let A be any finitely generated free abelian group and N a finitely generated nilpotent group. We may assume that $A = \mathbb{Z}^{l_1}$ for some $l_1 \in \mathbb{N}$. Let E denote any extension of N by A, so there is a short exact sequence

$$1 \rightarrow \mathbb{Z}^{l_1} \rightarrow E \rightarrow N \rightarrow 1. \tag{4}$$

By the previous section, we know that we can find a canonical type polynomial representation of N, with respect to some tower of N. This tower of N will be used as lower layers for the tower we want to build for E. We will not simply use extension (4) as the top of the tower for E, but we will cut this extension into pieces, namely a bunch of central ones, which are easy to handle by the previous section and one (very) non-central extension.
Suppose $(\mathbb{Z}^{l_1})^N \neq 0$. Then $\mathbb{Z}^{l_1}/(\mathbb{Z}^{l_1})^N$ is also free abelian, say of rank $l_2 < l_1$. Now consider the short exact sequence

$$1 \rightarrow \mathbb{Z}^{l_2} = \mathbb{Z}^{l_1}/(\mathbb{Z}^{l_1})^N \rightarrow E_2 = E/(\mathbb{Z}^{l_1})^N \rightarrow N \rightarrow 1$$

and repeat the same considerations about this extension. This implies that we inductively build up extensions ($E = E_1$)

$$1 \rightarrow \mathbb{Z}^{l_i} \rightarrow E_i \rightarrow N \rightarrow 1.$$

Since $l_1 > l_2 > l_3 > \cdots$ this procedure has to stop at some point, say n. So, the last extension obtained, is of the form

$$1 \rightarrow \mathbb{Z}^{l_n} \rightarrow E_n \rightarrow N \rightarrow 1 \tag{5}$$

with $(\mathbb{Z}^{l_n})^N = 0$. We remark that l_n might be zero, but this will play no role in what follows. Given the extension (5), we can return to the group $E = E_1$, by means of iterated central extensions:

$$\begin{aligned}
1 \rightarrow (\mathbb{Z}^{l_1})^N &\rightarrow E_1 \rightarrow E_2 \rightarrow 1 \\
1 \rightarrow (\mathbb{Z}^{l_2})^N &\rightarrow E_2 \rightarrow E_3 \rightarrow 1 \\
&\vdots \\
1 \rightarrow (\mathbb{Z}^{l_{n-1}})^N &\rightarrow E_{n-1} \rightarrow E_n \rightarrow 1
\end{aligned} \tag{6}$$

The tower we want to use for E is the following: the top $n-1$ layers are given by (6); (5) is the n-th layer and finally we add a tower of N, used to construct a canonical type polynomial representation of N at the bottom.

Postcard

Stamp

Sender:

Birkhäuser Verlag AG
Marketing
P.O. Box 133

CH–4010 Basel/Switzerland

By the results of the previous section, E will admit a canonical type polynomial representation with respect to this tower if and only if we succeed in the lifting procedure at the n-th layer. The following theorem shows that this lifting is always possible, even for affine representations.

THEOREM 3.1 *Let N be any finitely generated, nilpotent group equiped with a canonical type polynomial (resp. affine) representation $\bar{\rho} : N \to P(\mathbb{R}^K)$ (resp. $\mathrm{Aff}(\mathbb{R}^K)$). Suppose, that $1 \to \mathbb{Z}^k \to E \to N \to 1$ is an extension of groups, such that $(\mathbb{Z}^k)^N = 0$, then $\bar{\rho}$ can be lifted to a canonical type polynomial (resp. affine) representation $\rho : E \to P(\mathbb{R}^{K+k})$ (resp. $\mathrm{Aff}(\mathbb{R}^{K+k})$).*

Proof: Let $M(\mathbb{R}^K, \mathbb{R}^k)$ denote either $P(\mathbb{R}^K, \mathbb{R}^k)$ or $\mathrm{Aff}(\mathbb{R}^K, \mathbb{R}^k)$, depending on whether N allows just a polynomial or even stronger, an affine representation. Let $\langle f \rangle \in H^2(N, \mathbb{Z}^k)$ denote the cohomology class of the extension under consideration. We have to show that $\langle f \rangle$ belongs to the image of the connecting homomorphism

$$\delta : H^1(N, M(\mathbb{R}^K, \mathbb{R}^k)/\mathbb{Z}^k) \to H^2(N, \mathbb{Z}^k).$$

or, equivalently that $\langle f \rangle$ lies in the kernel of $H^2(N, \mathbb{Z}^k) \to H^2(N, M(\mathbb{R}^K, \mathbb{R}^k))$. Since $H^2(N, M(\mathbb{R}^K, \mathbb{R}^k))$ is torsion-free, it is enough to prove that $\langle f \rangle$ is a torsion element of $H^2(N, \mathbb{Z}^k)$. By a result of D. Robinson (see [11, Theorem 1, page 44]) there exists a subgroup N' of finite index in N, such that the restriction map res:$H^2(N, \mathbb{Z}^k) \to H^2(N', \mathbb{Z}^k)$ maps $\langle f \rangle$ onto 0. Using the fact that cor $\circ$ res is the same as multiplication by the index $[N : N']$, we see that $[N : N']\langle f \rangle = \mathrm{cor}(0) = 0$, which finishes the proof.   □

COROLLARY 3.2 *Let Γ be a polycyclic-by-finite group which can be obtained via iterated almost-central extensions of a (free abelian)-by-nilpotent group, then Γ admits a canonical type polynomial representation.*

COROLLARY 3.3 *Any polycyclic-by-finite, abelian-by-nilpotent group Γ acts properly discontinuous and with compact quotient on $\mathbb{R}^{h(\Gamma)}$ via polynomial diffeomorphisms.*

Proof: Γ fits in a short exact sequence $1 \to A \to \Gamma \to B \to 1$, where A and B are finitely generated, A abelian and B nilpotent. The set of torsion elements of A, $\tau(A)$, forms a finite characteristic subgroup of A and so $\tau(A)$ is normal in Γ. By the previous corollary, there is a canonical type polynomial representation $\rho : \Gamma/\tau(A) \to P(\mathbb{R}^{h(\Gamma)})$. Of course, the action of Γ on $\mathbb{R}^{h(\Gamma)}$ is then given by composing the natural projection $\Gamma \to \Gamma/\tau(A)$ with ρ.   □

4. The affine defect and questions

Let us call a quotient manifold $_E\backslash\mathbb{R}^K$, where E acts freely, properly discontinuously, via polynomial diffeomorphisms on $\mathbb{R}^K$ a *polynomial manifold*. Moreover, if all diffeomorphisms involved are of degree $\leq s$, we say that the (polynomial) manifold is of degree $\leq s$. We can now extend the definition of affine defect as introduced in [6] to all polycyclic-by-finite groups, in an obvious way:

DEFINITION 4.1 *Let E be a polycyclic-by-finite group with Hirsch rank $h(E)$. The **affine defect** of E, denoted by $d(E)$ is defined as*

$$Min\left\{s \in \mathbb{N} \middle\| \begin{array}{l} E \text{ acts properly discontinuous and via polynomial} \\ \text{diffeomorphisms of degree} \leq s+1 \text{ on } \mathbb{R}^{h(E)} \end{array} \right\},$$

if there exists such an action for E; $d(E) = \infty$ if E does not allow any properly discontinuous action via polynomial diffeomorphisms of bounded degree on $\mathbb{R}^{h(E)}$.

As is easily seen, a torsion-free, polycyclic-by-finite group E has affine defect zero iff E occurs as the fundamental group of a compact, complete affinely flat manifold. Therefore, this affine defect number somehow measures the obstruction for E to be realized as the fundamental group of a compact, complete affinely flat manifold.

In [6], we showed that the affine defect of a finitely generated, virtually nilpotent group is always finite. It is clear at once, that, given this concept, given the results above and in [6] and given the counter-examples of Benoist and Burde & Grunewald, many questions arise on affine defects and their meaning (behaviour).

We end this paper with a number of questions related to this approach.

Q-1 As a variation of Milnor's original question we might ask: *Is the affine defect of a polycyclic-by-finite group always finite?*

Note that we did not show this for the groups occurring in this paper. It might be that there is no upper-bound for the degrees of the polynomials $\rho(\gamma)$ if γ varies over the whole group Γ. A less stronger version of the above question is *Can any torsion-free, polycyclic-by-finite group be realized as the fundamental group of a compact polynomial manifold?*

Q-2 Another number measuring the obstruction for a torsion-free polycyclic-by-finite group Γ to be realized as the fundamental group of a compact, complete affinely flat manifold is the following:

$$\tilde{d}(\Gamma) = m(\Gamma) - h(\Gamma),$$

where $m(\Gamma)$ denotes the minimal dimension m in which Γ occurs as the fundamental group of a (not necessarily compact) complete affinely flat manifold. This number satisfies $0 \leq \tilde{d}(\Gamma) < \infty$ by the work of Milnor [10]. In view of the existence of affine structures, both numbers, $d(\Gamma)$ and $\tilde{d}(\Gamma)$, have analogous interpretations. *Is there a relation between $d(\Gamma)$ and $\tilde{d}(\Gamma)$?*

Q-3 *Given $c \in \mathbb{N}$, is there a number $d_c \geq 0$, such that each group Γ containing a finitely generated nilpotent group of class $\leq c$ as a subgroup of finite index, has affine defect $\leq d_c$?* Remark that it is known that $d_1 = d_2 = d_3 = 0$ (see [6]). *If the answer to this question is yes, is it possible to give a description of how d_c varies with c? In a linear, polynomial, exponential way?*

Q-4 *Given $h \in \mathbb{N}$, is there a number d_h, such that all polycyclic-by-finite groups of Hirsch length $\leq h$, have affine defect $\leq d_h$? If so, how does d_h vary with h?*

Q-5 *If Γ is a polycyclic-by-finite group of finite affine defect, does the same hold for every finite extension of Γ? Or even stronger: If Γ contains Γ' as a subgroup of finite index, is it possible that $d(\Gamma) \neq d(\Gamma')$?*

Q-6 *Are there finitely generated, (virtually) 4-step nilpotent groups of affine defect $\neq 0$?*

Q-7 *What is the smallest possible nilpotency class of finitely generated nilpotent groups with affine defect $\neq 0$?*

Q-8 As it is known that the affine defect of any finitely generated torsion-free nilpotent group of class c is less than or equal to $\mathrm{Max}\{0, c - 2\}$, on the one hand, and there are no examples known of such groups having affine defect ≥ 2, it is relevant to ask: *Do there exist finitely generated, torsion-free nilpotent groups of class c with affine defect strictly greater than one?*

Q-9 *Is there an algorithm to compute the affine defect for (a class of) polycyclic-by-finite groups?*

Q-10 *Does any polycyclic-by-finite group with affine defect $= 0$ admit a* **canonical type** *affine representation?*

References

[1] Benoist, Y. *Une nilvariété non affine.* C. R. Acad. Sci. Paris Sér. I Math., 1992, 315 pp. 983–986.

[2] Burde, D. and Grunewald, F. *Modules for certain Lie algebras of maximal class.* Preprint, 1993. to appear in J. Pure and Applied Algebra.

[3] Conner, P. E. and Raymond, F. *Deforming Homotopy Equivalences to Homeomorphisms in Aspherical Manifolds.* Bull. A.M.S., 1977, 83 (1), pp. 36–85.

[4] Dekimpe, K. and Igodt, P. *Computational aspects of affine representations for torsion free nilpotent groups via the Seifert construction.* J. Pure Applied Algebra, 1993, 84, pp. 165–190.

[5] Dekimpe, K. and Igodt, P. *The structure and topological meaning of almost-torsion free groups.* Comm. Algebra, 1994, 22 (7), pp. 2547–2558.

[6] Dekimpe, K., Igodt, P., and Lee, K. B. *Polynomial structures for nilpotent groups.* 1994. Preprint.

[7] Kamishima, Y., Lee, K. B., and Raymond, F. *The Seifert construction and its applications to infra-nilmanifolds.* Quarterly J. of Math. (Oxford), 1983, 34, pp. 433–452.

[8] Lee, K. B. *Aspherical manifolds with virtually 3-step nilpotent fundamental group.* Amer. J. Math., 1983, 105 pp. 1435–1453.

[9] Lee, K. B. and Raymond, F. *Geometric realization of group extensions by the Seifert construction.* Contemporary Math. A. M. S., 1984, 33 pp. 353–411.

[10] Milnor, J. *On fundamental groups of complete affinely flat manifolds.* Adv. Math., 1977, 25 pp. 178–187.

[11] Segal, D. *Polycyclic Groups.* Cambridge University Press, 1983.

Katholieke Universiteit Leuven, Campus Kortrijk
Universitaire Campus, B-8500 Kortrijk, Belgium

E-mail: Paul.Igodt@kulak.ac.be
Karel.Dekimpe@kulak.ac.be

Progress in Mathematics, Vol. 136
© 1996 Birkhäuser Verlag Basel/Switzerland

The centralizer decomposition of BG

W.G. DWYER*

1 Introduction

Let G be a compact Lie group and p a fixed prime number. Recall that an *elementary abelian p-group* is an abelian group isomorphic to $(\mathbf{Z}/p)^r$ for some r. Jackowski and McClure showed in [10] how to decompose the classifying space BG at the prime p as a homotopy colimit of spaces of the form $B\mathcal{C}_G(V)$, where V is a nontrivial elementary abelian p-subgroup of G and $\mathcal{C}_G(V)$ is the centralizer of V in G (see §2). If the center of G is trivial then each of the centralizers $\mathcal{C}_G(V)$ is a *proper* subgroup of G, and so in this case the decomposition theorem gives an explicit way of gluing together BG, at least at p, from the classifying spaces of smaller groups. In this paper we will use this decomposition to give parallel inductive proofs of three theorems about BG; the first two theorems are already known but the third is probably new.

The prime p will be fixed in everything that follows. If X is a space, let $L_{\mathbf{Z}/p}X$ denote the *$\mathbf{HZ}/p$-localization* of X constructed by Bousfield [2]. A space is said to be *$\mathbf{HZ}/p$-local* if the natural map $X \to L_{\mathbf{Z}/p}X$ is an equivalence, or alternatively if any map $f : A \to B$ which induces an isomorphism on mod p homology also induces an equivalence

$$f^{\#} : \mathrm{Map}(B, X) \xrightarrow{\simeq} \mathrm{Map}(A, X) \ .$$

If W and X are spaces, say that X is *W-null* if every map from W to X is canonically homotopic to a constant, in the sense that the map

$$\kappa : X \to \mathrm{Map}(W, X) \tag{1.1}$$

given by inclusion of constant maps is an equivalence (see [3] and [7]).

1.2 THEOREM. (*cf.* [9], [16]) *Let G be a compact Lie group and X a space which is $\mathbf{HZ}/p$-local and $B\mathbf{Z}/p$-null. Then X is BG-null.*

1.3 Miller's theorem. Suppose that Y is a finite complex. Miller [13] shows that Y is BG-null for any finite group G (this is the "Sullivan Conjecture"). Theorem 1.2 implies (see below) that the Bousfield-Kan p-completion $Y_{\hat{p}}$ is BG-null for any compact Lie group G, and in this sense gives a generalization of Miller's theorem to compact Lie groups. Some kind of completion is definitely

---

*)   The author was supported in part by the National Science Foundation.

necessary here; one way to see this is to use 1.2 and the arithmetic square [4] to compute that, if G is a connected compact Lie group, the 3-sphere S^3 is usually not BG-null.

To derive the fact that $Y_p^\wedge$ is BG-null from 1.2, observe that $Y_p^\wedge$ is $H\mathbf{Z}/p$-local by [2] and $B\mathbf{Z}/p$-null by Miller's arguments. Note that $Y_p^\wedge$ is equivalent to $L_{\mathbf{Z}/p}Y$ if Y is simply connected or more generally "$\mathbf{Z}/p$-good" [1]. If Y is a finite complex which is *not* $\mathbf{Z}/p$-good, it seems to be unknown whether or not $L_{\mathbf{Z}/p}Y$ is $B\mathbf{Z}/p$-null.

If G is a (topological) group, Z is a space, and $f : BG \to Z$ is a map, say that f *is null on finite p-groups* if $f \cdot (B\rho)$ is null homotopic for every finite p-group P and homomorphism $\rho : P \to G$.

1.4 THEOREM. *(cf. [9, 3.3], [11, 3.11]) Let G be a compact Lie group and Z a pointed connected space such that Z is $H\mathbf{Z}/p$-local and ΩZ is $B\mathbf{Z}/p$-null. Then a map $f : BG \to Z$ is null homotopic if and only if it is null on finite p-groups.*

Remark. If H be a compact Lie group such that $\pi_0 H$ is a p-group, the hypotheses of 1.4 apply to the space $Z = L_{\mathbf{Z}/p}(BH)$. To see this use 1.3 and note that BH is $\mathbf{Z}/p$-good [1, VII, §5], so that by the fibre lemma [1, Ch. II] there is an equivalence

$$\Omega L_{\mathbf{Z}/p}(BH) = \Omega((BH)_p^\wedge) \simeq H_p^\wedge .$$

Theorem 1.4 thus gives a criterion for maps between classifying spaces to be null homotopic at p.

1.5 The functor P_W. The statement of the final theorem requires some more terminology from Bousfield [3] and Farjoun [7]. Suppose that W is some fixed space. A map $f : A \to B$ is said to be a P_W-*equivalence* if f induces an equivalence

$$\mathrm{Map}(B, X) \xrightarrow{f^\#} \mathrm{Map}(A, X)$$

for every W-null space X. Bousfield and Farjoun show that for any space X there is an associated W-null space $P_W(X)$ together with a natural P_W-equivalence $X \to P_W(X)$. It is easy to check from the definitions that if X' is any other W-null space with a P_W-equivalence $X \to X'$, then up to homotopy there is a unique equivalence $X' \to P_W(X)$ which makes the appropriate diagram involving X commute. This implies that a map f is a P_W-equivalence if and only if $P_W(f)$ is an equivalence.

1.6 A natural map. Suppose now that $W = B\mathbf{Z}/p$ and that q is a prime different from p. Let $L_{\mathbf{Z}[1/p]}X$ denote Bousfield's $H\mathbf{Z}[1/p]$-localization of the space X [2]. Direct checking with the definition shows that the Eilenberg-Mac Lane spaces $K(\mathbf{Z}/q, n)$ and $K(\mathbf{Q}, n)$ $(n \geq 0)$ are W-null. It follows that if $f : X \to Y$ is a P_W-equivalence then $\mathrm{H}^*(f, \mathbf{Z}/q)$ and $\mathrm{H}^*(f, \mathbf{Q})$ are isomorphisms, hence that $\mathrm{H}_*(f, \mathbf{Z}[1/p])$ is an isomorphism, and hence that $L_{\mathbf{Z}[1/p]}f : L_{\mathbf{Z}[1/p]}X \to L_{\mathbf{Z}[1/p]}Y$ is an equivalence. In particular, $L_{\mathbf{Z}[1/p]}X \to L_{\mathbf{Z}[1/p]}P_W(X)$ is an equivalence, and so the natural map $P_W(X) \to L_{\mathbf{Z}[1/p]}P_W(X)$ gives up to homotopy a natural map $P_W(X) \to L_{\mathbf{Z}[1/p]}X$.

1.7 THEOREM. *Let G be a compact Lie group such that $\pi_0 G$ is a p-group, and let $W = \mathbf{B}\mathbf{Z}/p$. Then the natural map $P_W(BG) \to L_{\mathbf{Z}[1/p]}(BG)$ is an equivalence.*

Remark. Miller's theorem (1.3) implies that in the above situation the space BG is ΣW-null and hence that the map $BG \to P_{\Sigma W}(BG)$ is an equivalence. This shows that there is a large difference between the spaces $P_W(BG)$ and $P_{\Sigma W}(BG)$. The functors P_W and $P_{\Sigma W}$ are related somewhat more closely if W itself is a suspension [8].

Remark. The three theorems above have corresponding forms that apply to p-compact groups. For a version of 1.2 see [5, §9] and for a version of 1.4 see [14, §5]. The analogue of 1.7 is proved by an argument very similar to the one in §6.

Remark. It is an interesting exercise to derive 1.2 from 1.7 (and 6.3), at least in the case in which the group G involved has $\pi_0 G$ a finite p-group.

Notation and terminology. We assume that all spaces have been replaced if necessary by weakly equivalent CW-complexes (for instance, by the geometric realizations of their singular complexes). The word *equivalence* means "homotopy equivalence". A space is said to be $\mathbf{Z}/p$-*acyclic* if it has the mod p homology of a point; a map is a $\mathbf{Z}/p$-*equivalence* if it induces an isomorphism on mod p homology.

The author would like to thank E. Farjoun and the referee for their suggestions.

2 The Jackowski-McClure theorem

In this section we will briefly describe the main theorem of [10] and indicate the general way in which it can be used in inductive arguments.

Suppose that G is a compact Lie group. Let $\mathbf{A}_G$ be the category in which the objects are the nontrivial elementary abelian p-subgroups V of G; a morphism $V \to V'$ in $\mathbf{A}_G$ is a group homomorphism $f : V \to V'$ with the property that there exists $g \in G$ such that $f(x) = gxg^{-1}$ for all $x \in V$. Note that specifying a morphism in $\mathbf{A}_G$ does *not* involve choosing a particular such g. There is a functor α'_G from the opposite category $\mathbf{A}_G^{\mathrm{op}}$ to G-spaces which assigns to V the coset space $G/\mathcal{C}_G(V)$; if $f : V \to V'$ is realized by conjugation with $g \in G$, then $\alpha'_G(f)$ assigns to the coset $x\mathcal{C}_G(V')$ the coset $xg\mathcal{C}_G(V)$. Let EG be the total space of a universal principal G-bundle and $\alpha_G : \mathbf{A}_G \to \mathbf{Top}$ the functor $(EG \times \alpha'_G)/G$. The following two properties of this functor are easy to check.

(1) For each object V of $\mathbf{A}_G$, $\alpha_G(V)$ is homeomorphic to $EG/\mathcal{C}_G(V)$ and thus equivalent to $B\mathcal{C}_G(V)$.
(2) The unique G-maps $\alpha'_G(V) \to *$ pass to compatible maps $\alpha_G(V) \to BG$. These induce a map $a_G : \operatorname{hocolim} \alpha_G \to BG$.

See [1, Ch. XII] for a discussion of homotopy colimits. Recall from [1, XII, 3.2] that if $\mathbf{C}$ is a category, the *classifying space* $\mathbf{BC}$ is defined to be $\mathrm{hocolim}(*_\mathbf{C})$, where $*_\mathbf{C} : \mathbf{C} \to \mathbf{Top}$ assigns to each object of $\mathbf{C}$ the one-point space. The classifying spaces $\mathbf{BC}$ and $\mathbf{BC}^{\mathrm{op}}$ are homotopy equivalent [15, p. 86].

2.1 THEOREM. [10] *Suppose that G is a compact Lie group. Then*

(1) *the map $a_G : \mathrm{hocolim}\,\alpha_G \to BG$ is a $\mathbf{Z}/p$-equivalence, and*
(2) *the classifying space $\mathbf{BA}_G$ is $\mathbf{Z}/p$-acyclic.*

Remark. Jackowski and McClure actually show that a certain cohomology spectral sequence for $\mathrm{H}^*(\mathrm{hocolim}\,\alpha_G, \mathbf{Z}/p)$ collapses; this is much sharper than 2.1(1). In the course of this they show that the higher limits $\lim^i \mathrm{H}^j(\alpha_G)$ vanish for $i > 0$. For $j = 0$ these are the higher limits of the constant functor on $\mathbf{A}_G$ with value $\mathbf{Z}/p$, which by [1, XI, §5] are just the mod p cohomology groups of $\mathbf{BA}_G$. This gives 2.1(2).

Theorem 2.1 more or less states that at the prime p the homotopy type of BG can be constructed from the homotopy types of classifying spaces of smaller compact Lie groups, in such a way that the shape of the gluing diagram, in other words the classifying space of $\mathbf{A}_G$, is trivial at p. This suggests using the theorem to prove statements about BG by induction on the size (e.g., the dimension) of G. There is one minor problem with this: the values of the functor α_G, which up to homotopy are the spaces $BC_G(V)$ for elementary abelian p-subgroups of G, are *not* necessarily the classifying spaces of Lie groups smaller than G. In fact, if G contains a central subgroup of order p then the space BG appears among the values of the functor α_G and Theorem 2.1 amounts to a complicated but essentially circular construction of BG in terms of itself. This indicates that any inductive argument using 2.1 must treat centers in some special way. In practice this also involves treating disconnected groups in a special way, since in general it is only if G is connected that dividing out by the center of G gives a quotient group with trivial center. One inductive scheme that fits this situation is described in the following proposition.

2.2 PROPOSITION. *Suppose that $\mathcal{C}$ is class of (topological) groups which has the following three closure properties with respect to compact Lie groups G, H:*

(1) *If G is connected, the center of G is trivial, and $H \in \mathcal{C}$ for all H of smaller dimension than G, then $G \in \mathcal{C}$.*
(2) *If G is connected with center C and $G/C \in \mathcal{C}$, then $G \in \mathcal{C}$.*
(3) *If G_0 is the identity component of G and $G_0 \in \mathcal{C}$ then $G \in \mathcal{C}$.*

Then $\mathcal{C}$ contains every compact Lie group.

Proof. We prove by induction on the dimension of the compact Lie group G that $G \in \mathcal{C}$. Condition (1) implies that the trivial group is in $\mathcal{C}$. Suppose that that G is of dimension d and assume inductively that $H \in \mathcal{C}$ for each compact Lie group H of dimension less than d (this is certainly true if $d=0$). Let G_0 be the identity component of G and C the center of G_0. Then $G_0/C \in \mathcal{C}$ by property (1) and induction, $G_0 \in \mathcal{C}$ by (2), and hence $G \in \mathcal{C}$ by (3). $\square$

2.3 FINITE GROUPS. In the examples that come up in this paper, C will typically be the class of all topological groups G such that BG has some appropriate property. Statement 2.2(1) is then proved using 2.1. Statements 2.2(2)-(3) are proved by more direct arguments involving facts about finite groups; for instance, obtaining 2.2(3) usually involves knowing something about the finite group G/G_0. We will prove the necessary statements about finite groups by a subsidiary initial induction that depends on the following elementary proposition.

2.4 PROPOSITION. *Suppose that C is class of (topological) groups which has the following two closure properties with respect to finite groups G, H:*

(1) *If G has no central subgroup of order p, and $H \in C$ for all H of smaller order than G, then $G \in C$.*

(2) *If G has a central subgroup C of order p and and $G/C \in C$, then $G \in C$.*

Then C contains every finite group.

3 The fibration principle

In this section we will discuss some basic homotopy theoretic observations that turn out to be useful in proving Theorems 1.2, 1.4 and 1.7. We first discuss mapping spaces for which the domain is the total space of a fibration (3.1), and then mapping spaces for which the domain is a homotopy colimit (3.6). We end by explaining how to tie these two discussions together (3.11).

3.1 MAPS FROM THE TOTAL SPACE OF A FIBRATION.

3.2 PROPOSITION. *(Fibration Principle) Suppose that $f : E \to B$ is a fibration over a connected base B with fibre F, and that X is some space. Then there is another naturally associated fibration $f_X : E_X \to B$ with fibre $\mathrm{Map}(F, X)$ such that the space of sections of f_X is equivalent to $\mathrm{Map}(E, X)$.*

The way to understand this proposition is to picture E as a fibre bundle over B and notice that giving a map $E \to X$ amounts to giving, for each point $b \in B$, a map from a copy of F to X. We will sketch a direct proof (see also 3.11). For a topological treatment of fibrewise function spaces, see [12, Ch. 9].

Proof of 3.2 (Sketch). Let $G = \mathrm{Aut}(F)$ denote the monoid of self homotopy equivalences of F. The monoid G acts from the left on F, and associated to this action is a universal fibration $u : E(F) \to BG$ with fibre F. Since u is universal there is a map $c : B \to BG$, unique up to homotopy, such that the pullback of u over c is equivalent to the fibration f. We denote such a pullback $B \times_{BG} E(F)$; the map c is understood in this notation.

The monoid G also acts from the right (by composition) on the mapping space $M = \mathrm{Map}(F, X)$. Associated to this action is a fibration $v : E(M) \to BG$ with fibre M. The evaluation map $F \times M \to X$ induces a map

$$e : E(F) \times_{BG} E(M) \to X . \tag{3.3}$$

Consider now the category **C** of spaces over BG. An object in this category is a space Y together with a map $g : Y \to BG$; a morphism $h : Y \to Y'$ is a map of spaces such that $g'h = g$. For any object Y of $\mathbf{C}$ let $\Psi(Y)$ denote the mapping space

$$\mathrm{Map}(Y \times_{BG} E(F), X)$$

and $\Phi(Y)$ the space of sections of the fibration $Y \times_{BG} E(M) \to Y$. The map e above (3.3) induces a map

$$t(Y) : \Phi(Y) \to \Psi(Y)$$

which gives a natural transformation between the two indicated functors $\mathbf{C}^{\mathrm{op}} \to \mathbf{Top}$. To prove the proposition we will show that $t(B)$ is an equivalence.

The map $t(Y)$ is an equivalence if Y is a point or more generally if Y is a contractible space; in this case the domain and range of $t(Y)$ are each equivalent to $M = \mathrm{Map}(F, X)$. Suppose that

$$
\begin{array}{ccc}
Y_1 & \longrightarrow & Y_2 \\
\downarrow & & \downarrow \\
Y_3 & \longrightarrow & Y_4
\end{array}
$$

is a homotopy pushout diagram of spaces over BG. It is not hard to see that the induced diagram

$$
\begin{array}{ccc}
Y_1 \times_{BG} E(F) & \longrightarrow & Y_2 \times_{BG} E(F) \\
\downarrow & & \downarrow \\
Y_3 \times_{BG} E(F) & \longrightarrow & Y_4 \times_{BG} E(F)
\end{array}
$$

is also a homotopy pushout diagram. The natural transformation t then gives a map of squares

$$
\begin{array}{ccc}
\Phi(Y_4) \longrightarrow \Phi(Y_2) \\
\downarrow \qquad\qquad \downarrow \\
\Phi(Y_3) \longrightarrow \Phi(Y_1)
\end{array}
\quad \xrightarrow{\;t\;} \quad
\begin{array}{ccc}
\Psi(Y_4) \longrightarrow \Psi(Y_2) \\
\downarrow \qquad\qquad \downarrow \\
\Psi(Y_3) \longrightarrow \Psi(Y_1)
\end{array}
$$

in which each square is a homotopy pullback square (because mapping constructions like Φ and Ψ convert homotopy pushouts to homotopy pullbacks [1, XII 4.1]). It follows that if $t(Y_i)$ is an equivalence for $i \leq 3$ then $t(Y_4)$ is also an equivalence. Both Φ and Ψ convert disjoint unions to products, so if $\{Y_\alpha\}$ is a collection of spaces over BG with disjoint union $Y = \coprod_\alpha Y_\alpha$, then $t(Y)$ is an equivalence if each $t(Y_\alpha)$ is. Let $\mathcal{C}$ be the smallest homotopy invariant class of spaces over BG which contains all contractible spaces, is closed under homotopy pushouts, and is closed under disjoint unions. By the discussion above, $t(Y)$ is an equivalence for each space Y in $\mathcal{C}$. It is clear by induction on dimension that $\mathcal{C}$ contains every space Y over BG such that the underlying space of Y is a finite dimensional CW-complex. Suppose that Y is an infinite complex over

BG, and let Y_n be the n-skeleton of Y. The fact that $Y \in \mathcal{C}$ then follows from the fact that there is a homotopy pushout diagram

$$
\begin{array}{ccc}
(\coprod_n Y_n) \amalg (\coprod_n Y_n) & \xrightarrow{\mathrm{id}+\mathrm{id}} & \coprod_n Y_n \\
{\scriptstyle \mathrm{id}+s} \downarrow & & \downarrow \\
\coprod_n Y_n & \longrightarrow & Y
\end{array}
$$

in which the map s is a shift map derived from the inclusions $Y_n \to Y_{n+1}$. $\square$

One case is particularly interesting. The following proposition is closely related to work of Zabrodsky as reformulated by Miller [13, 9.5].

3.4 PROPOSITION. *Suppose that $f : E \to B$ is a fibration over a connected base B with fibre F, and that X is a space. If the map $\kappa : X \to \mathrm{Map}(F, X)$ is an equivalence (1.1), then the restriction map $f^\# : \mathrm{Map}(B, X) \to \mathrm{Map}(E, X)$ is an equivalence.*

Proof. Let $g : B \to B$ be the identity fibration. Under the stated hypotheses the map f itself induces an equivalence between the fibration f_X of 3.2 and the fibration g_X (which is the projection $X \times B \to B$). The induced map between spaces of sections, which is essentially $f^\#$, is then also an equivalence. $\square$

Restricting attention to individual mapping space components gives a more specialized variant of 3.4. If F and X are spaces, let $\mathrm{Map}(F, X)_{[F]}$ denote the space of maps $F \to X$ which are homotopic to constant maps. More generally, if $f : E \to B$ is a fibration over a connected base B with fibre F, let $\mathrm{Map}(E, X)_{[F]}$ denote the space of those maps $E \to X$ which are homotopic to constant maps when restricted to F.

3.5 PROPOSITION. *Suppose that $f : E \to B$ is a fibration over a connected base B with fibre F, and that X is a space. If the map $\kappa : X \to \mathrm{Map}(F, X)_{[F]}$ is an equivalence, then the restriction map $f^\# : \mathrm{Map}(B, X) \to \mathrm{Map}(E, X)_{[F]}$ is an equivalence.*

3.6 MAPS FROM A HOMOTOPY COLIMIT. There are mapping space results roughly parallel to the above ones with the notion "total space of a fibration" replaced by the notion "homotopy colimit". The analogue of 3.2 is the following proposition of Bousfield and Kan.

3.7 PROPOSITION. [1, XII, §4] *Suppose that $\mathbf{C}$ is a small category, X a space, and $\gamma : \mathbf{C} \to \mathbf{Top}$ a functor. Let $\mathrm{Map}(\gamma, X) : \mathbf{C}^{\mathrm{op}} \to \mathbf{Top}$ be the functor which assigns to each object c the space $\mathrm{Map}(\gamma(c), X)$. Then there is an equivalence*

$$
\mathrm{Map}(\mathrm{hocolim}\, \gamma, X) \xrightarrow{\sim} \mathrm{holim}\, \mathrm{Map}(\gamma, X) .
$$

In a situation like that of 3.4 this gives the following.

3.8 PROPOSITION. *Let* **C** *be a small category,* X *a space, and* $\gamma : \mathbf{C} \to \mathbf{Top}$ *a functor. Assume that for each object* c *of* **C**, *the map* $\kappa : X \to \mathrm{Map}(\gamma(c), X)$ *is an equivalence (1.1). Then the unique natural transformation* $\gamma \to *_{\mathbf{C}}$ *induces an equivalence*

$$\mathrm{Map}(B\mathbf{C}, X) = \mathrm{Map}(\mathrm{hocolim}(*_{\mathbf{C}}), X) \xrightarrow{\sim} \mathrm{Map}(\mathrm{hocolim}\,\gamma, X) \ .$$

Proof. By assumption the functor $\mathrm{Map}(\gamma, X)$ (see 3.7) is equivalent to the constant functor $\mathrm{Map}(*_{\mathbf{C}}, X)$. The homotopy limit of this constant functor is $\mathrm{Map}(\mathrm{hocolim}(*_{\mathbf{C}}), X)$.   □

3.9. As above, restricting attention to individual mapping space components gives a specialized variant. If **C** is a small category and $\gamma : \mathbf{C} \to \mathbf{Top}$ is a functor, then by the definition of homotopy colimit [1, XII, §2] there is a natural map $\gamma(c) \to \mathrm{hocolim}\,\gamma$ for each object c of **C**. If X is a space, let $\mathrm{Map}(\mathrm{hocolim}\,\gamma, X)_{[\gamma]}$ denote the space of maps $f : \mathrm{hocolim}\,\gamma \to X$ such that for each object c of **C** the restriction of f to $\gamma(c)$ is homotopic to a constant map.

3.10 PROPOSITION. *Let* **C** *be a small category,* X *a space, and* $\gamma : \mathbf{C} \to \mathbf{Top}$ *a functor such that for each object* c *of* **C**, *the map* $\kappa : X \to \mathrm{Map}(\gamma(c), X)_{[\gamma(c)]}$ *is an equivalence. Then the unique natural transformation* $\gamma \to *_{\mathbf{C}}$ *induces an equivalence*

$$\mathrm{Map}(B\mathbf{C}, X) = \mathrm{Map}(\mathrm{hocolim}(*_{\mathbf{C}}), X) \xrightarrow{\sim} \mathrm{Map}(\mathrm{hocolim}\,\gamma, X)_{[\gamma]} \ .$$

3.11 TOTAL SPACES VS. HOMOTOPY COLIMITS. Propositions 3.2 and 3.7 are tied together by the fact that the total space of a fibration over B with fibre F is equivalent to the homotopy colimit of a functor whose values are all equivalent to F and whose domain category has classifying space equivalent to B. Suppose for simplicity that B is the geometric realization of a simplicial complex K. Let $\mathbf{C}_B$ be the category in which an object is a (closed) simplex σ of B and there is a single morphism $\sigma \to \sigma'$ if σ is contained in σ' (there are no other morphisms). The classifying space $B\mathbf{C}_B$ is the geometric realization of the barycentric subdivision of K and so is homeomorphic to B. If $f : E \to B$ is a fibration with fibre F, then there is a functor $\gamma_E : \mathbf{C}_B \to \mathbf{Top}$ which sends σ to $f^{-1}(\sigma)$, and it is possible to check that $\mathrm{hocolim}\,\gamma_E$ is equivalent to E. This is obvious if B itself is a simplex, and in general one can make an induction, based on homotopy pushouts, over the skeletal filtration of B (cf. proof of 3.2). By 3.7 there is an equivalence

$$\mathrm{Map}(E, X) \to \mathrm{holim}\,\mathrm{Map}(\gamma_E, X) \ .$$

Now consider the following proposition.

3.12 PROPOSITION. *Suppose that* **C** *is a small category and that* $\gamma : \mathbf{C} \to \mathbf{Top}$ *is a functor which sends each object of* **C** *to a space equivalent to* Z *and each morphism of* **C** *to an equivalence. Then*

(1) *the natural map* $\operatorname{hocolim} \gamma \to \mathbf{BC}$ *is up to homotopy a fibration with fibre* Z, *and*

(2) *the space of sections of this fibration is equivalent to* $\operatorname{holim} \gamma$.

Applying this proposition to the functor $\operatorname{Map}(\gamma_E, X)$ shows that, up to homotopy, the space $\operatorname{hocolim} \gamma_E$ is the total space of a fibration over B with fibre $\operatorname{Map}(F, X)$ and space of sections equivalent to $\operatorname{Map}(E, X)$. (Note [15, p. 91] that the classifying space $\mathbf{BC}_B^{\mathrm{op}}$ is equivalent to $\mathbf{BC}_B$, and hence to B). This gives a proof of 3.2 which uses 3.7.

Remark. The first statement of 3.12 is a form of Quillen's Theorem B [15, p. 97]. Statement 3.12(2) can be proved by using the interpretation of homotopy limit in [6, 2.12]. This identifies $\operatorname{holim} \gamma$ up to homotopy as the mapping space $\operatorname{Map}(\tilde{*}_{\mathbf{C}}, \gamma)$, where $\tilde{*}_{\mathbf{C}}$ is a "free resolution" of the functor $*_{\mathbf{C}}$ (i.e. a CW-functor [6, 1.16] weakly equivalent to $*_{\mathbf{C}}$) and the maps are computed in the category of functors $\mathbf{C} \to \mathbf{Top}$. The space $\operatorname{hocolim}(\tilde{*}_{\mathbf{C}})$ is equivalent to $\mathbf{BC}$. One proves by skeletal induction ([6, 1.16], cf. proof of 3.2) that if $A : \mathbf{C} \to \mathbf{Top}$ is any CW-functor then the space $\operatorname{Map}(A, \gamma)$ is equivalent in a natural way to the space of sections of the fibration $E_A \to \operatorname{hocolim} A$, where E_A is determined by the (homotopy) pullback diagram

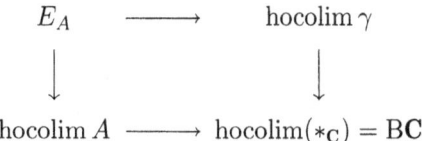

4 Maps into spaces which are $\mathbf{BZ}/p$-null

In this section we will prove 1.2. Suppose that X is a space which is $\mathbf{HZ}/p$-local and $\mathbf{BZ}/p$-null. Let $\mathcal{C}$ be the class of all topological groups G with the property that $\kappa : X \to \operatorname{Map}(BG, X)$ is an equivalence; it is necessary to prove that $\mathcal{C}$ contains all compact Lie groups. Recall that, by the definition of what it means to be "$\mathbf{HZ}/p$-local", any $\mathbf{Z}/p$-equivalence $A \to B$ induces an equivalence $\operatorname{Map}(B, X) \xrightarrow{\sim} \operatorname{Map}(A, X)$.

4.1 LEMMA. *If* G *is a discrete locally finite group (i.e. a union of finite groups) then* $G \in \mathcal{C}$.

Remark. The following argument is a prototype of the argument below for compact Lie groups.

Proof of 4.1. We prove using 2.4 and induction on the order of G that any finite group G belongs to $\mathcal{C}$. The group $\mathbf{Z}/p$ belongs to $\mathcal{C}$ by definition. If G has a central subgroup C of order p then applying the 3.4 to the fibration

$$BC \to BG \to B(G/C)$$

and using the induction hypothesis shows that $G \in \mathcal{C}$. If G has no central subgroup C of order p, then for each object V of $\mathbf{A}_G$ the space $\alpha_G(V)$ has the homotopy type of BH, where H is of smaller order than G. By 2.1(1), the map hocolim $\alpha_G \to BG$ induces an equivalence

$$\mathrm{Map}(BG, X) \xrightarrow{\sim} \mathrm{Map}(\mathrm{hocolim}\,\alpha_G, X).$$

By induction and 3.8 the natural transformation $\alpha_G \to *_{\mathbf{A}_G}$ induces an equivalence

$$\mathrm{Map}(B\mathbf{A}_G, X) \xrightarrow{\sim} \mathrm{Map}(\mathrm{hocolim}\,\alpha_G, X) .$$

However 2.1(2) guarantees that $B\mathbf{A}_G$ is $\mathbf{Z}/p$-acyclic, so that the map

$$\kappa : X \to \mathrm{Map}(B\mathbf{A}_G, X)$$

is an equivalence. Tracing through the various identifications shows that

$$\kappa : X \to \mathrm{Map}(BG, X)$$

is also an equivalence.

The passage to general locally finite groups is by a standard homotopy colimit argument [13, proof of 9.8]. □

4.2 LEMMA. *If G is an abelian compact Lie group, then $G \in \mathcal{C}$.*

Proof. The group G is isomorphic to the product of a finite abelian group with a torus. Let $D \subset G$ be the group of elements of order a power of p, considered as a discrete group. It is not hard to see by explicit calculation that the map $BD \to BG$ induces an isomorphism on mod p homology and therefore an equivalence $\mathrm{Map}(BG, X) \xrightarrow{\sim} \mathrm{Map}(BD, X)$. The desired result follows from 4.1. □

There are now three steps to carry out, which correspond to the three hypotheses of 2.2.

Step I. Suppose that G is a connected compact Lie group of dimension d with a trivial center, and that $H \in \mathcal{C}$ for all compact Lie groups H of dimension less than d. It is necessary to prove that $G \in \mathcal{C}$. By 2.1(1), the map hocolim $\alpha_G \to BG$ induces an equivalence

$$\mathrm{Map}(BG, X) \xrightarrow{\sim} \mathrm{Map}(\mathrm{hocolim}\,\alpha_G, X).$$

Since G is connected and has trivial center, for each object V of $\mathbf{A}_G$ the space $\alpha_G(V)$ has the homotopy type of BH, where H is of dimension less than d. By induction and 3.8 the natural transformation $\alpha_G \to *_{\mathbf{A}_G}$ induces an equivalence

$$\mathrm{Map}(\mathrm{B}\mathbf{A}_G, X) \xrightarrow{\sim} \mathrm{Map}(\mathrm{hocolim}\,\alpha_G, X) \ .$$

However 2.1(2) guarantees that B$\mathbf{A}_G$ is $\mathbf{Z}/p$-acyclic, so that the map

$$\kappa : X \to \mathrm{Map}(\mathrm{B}\mathbf{A}_G, X)$$

is an equivalence. Tracing through the various identifications shows that

$$\kappa : X \to \mathrm{Map}(\mathrm{B}G, X)$$

is also an equivalence.   □

Step II. Suppose that G is connected with center C, and that $G/C \in \mathcal{C}$. It is necessary to show that $G \in \mathcal{C}$. There is a fibration sequence

$$\mathrm{B}C \to \mathrm{B}G \to \mathrm{B}(G/C)$$

in which by 4.2 the fibre BC belongs to $\mathcal{C}$. By 3.4, then, the restriction map

$$\mathrm{Map}(\mathrm{B}(G/C), X) \to \mathrm{Map}(\mathrm{B}G, X)$$

is an equivalence. The result now follows from the fact that $G/C \in \mathcal{C}$.   □

Step III. Suppose that the identity component G_0 of G belongs to $\mathcal{C}$; it is necessary to show that $G \in \mathcal{C}$. There is a fibration sequence

$$\mathrm{B}G_0 \to \mathrm{B}G \to \mathrm{B}\pi_0 G \ .$$

By 3.4 and the assumption on G_0 the restriction map

$$\mathrm{Map}(\mathrm{B}\pi_0 G, X) \to \mathrm{Map}(\mathrm{B}G, X)$$

is an equivalence. The result now follows from the fact (4.1) that $\pi_0 G \in \mathcal{C}$.
                                                                            □

5 Maps null on finite p-groups

In this section we will prove 1.4, or more accurately a slight generalization of it. If G is a topological group and X is a space, let $\mathrm{Map}(\mathrm{B}G, X)_{[p]}$ denote the space of all maps $f : \mathrm{B}G \to X$ which are null on finite p-groups. Let Z be a pointed connected space such that Z is $\mathbf{HZ}/p$-local and ΩZ is $\mathrm{B}\mathbf{Z}/p$-null. Define $\mathcal{C}$ to be the class of all topological groups G with the property that the map

$$\kappa : Z \to \mathrm{Map}(\mathrm{B}G, Z)_{[p]}$$

is an equivalence (1.1). What we will prove is the following.

5.1 THEOREM. *The class* $\mathcal{C}$ *contains all compact Lie groups* G.

In particular, if G is a compact Lie group then the space $\mathrm{Map}(\mathrm{B}G, Z)_{[p]}$ is connected. This implies that every map $\mathrm{B}G \to Z$ which is null on finite p-groups is homotopic to a constant map, which is 1.4.

5.2 LEMMA. *If* G *is a locally finite group (i.e., a union of finite p-groups) then* $G \in \mathcal{C}$.

Remark. As in §4, this is a prototype of the proof below for compact Lie groups.

Proof. We prove using 2.4 and induction on the order of G that any finite group G belongs to $\mathcal{C}$. If G has a central subgroup C of order p then applying the 3.5 to the fibration

$$\mathrm{B}C \to \mathrm{B}G \to \mathrm{B}(G/C)$$

and using the induction hypothesis shows that $G \in \mathcal{C}$. If G has no central subgroup of order p, for each object V of $\mathbf{A}_G$ the space $\alpha_G(V)$ has the homotopy type of $\mathrm{B}K$ for a group K of order smaller than G. By 2.1(1), the map $\mathrm{hocolim}\,\alpha_G \to \mathrm{B}G$ induces an equivalence $a_G^{\#} : \mathrm{Map}(\mathrm{B}G, Z) \to \mathrm{Map}(\mathrm{hocolim}\,\alpha_G, Z)$. Moreover, a check with the definitions shows that the composite maps (cf. 3.9)

$$\alpha_G(V) \to \mathrm{hocolim}\,\alpha_G \xrightarrow{\ a_G\ } \mathrm{B}G$$

are obtained up to homotopy by applying the classifying space construction to homomorphisms $K \to G$. It follows in the notation of 3.9 (with $\gamma = \alpha_G$) that $a_G^{\#}$ induces an equivalence from $\mathrm{Map}(\mathrm{B}G, Z)_{[p]}$ to a union of components of the space $\mathrm{Map}(\mathrm{hocolim}\,\alpha_G, Z)_{[\gamma]}$. By induction, 3.10, and 2.1(2), the map

$$\kappa : Z \to \mathrm{Map}(\mathrm{hocolim}\,\alpha_G, Z)_{[\gamma]}$$

is an equivalence. Tracing through the various identifications gives the desired equivalence $Z \to \mathrm{Map}(\mathrm{B}G, Z)_{[p]}$.

The result follows for arbitrary locally finite groups by a homotopy colimit calculation [13, proof of 9.8]. □

5.3 LEMMA. *If* G *is an abelian compact Lie group, then* $G \in \mathcal{C}$.

Proof. This follows from 5.2: as in 4.2, there is a locally finite p-group D and a homomorphism $D \twoheadrightarrow G$ which induces an equivalence

$$\mathrm{Map}(\mathrm{B}G, Z) \xrightarrow{\ \sim\ } \mathrm{Map}(\mathrm{B}D, Z) \, . □$$

The proof of 5.1 now has three steps, which correspond to the three steps of 2.2.

Step I. Suppose that G is a connected compact Lie group of dimension d with a trivial center, and that $K \in \mathcal{C}$ for all compact Lie groups K of dimension less than d. It is necessary to prove that $G \in \mathcal{C}$. By 2.1(1), the map $\operatorname{hocolim} \alpha_G \to BG$ induces an equivalence $a_G^{\#} : \operatorname{Map}(BG, Z) \to \operatorname{Map}(\operatorname{hocolim} \alpha_G, Z)$. Since G is connected and has trivial center, for each object V of $\mathbf{A}_G$ the space $\alpha_G(V)$ has the homotopy type of BK, where $K \in \mathcal{C}$. Moreover, a check with the definitions shows that the composite maps (cf. 3.9)

$$\alpha_G(V) \to \operatorname{hocolim} \alpha_G \xrightarrow{a_G} BG$$

are obtained up to homotopy by applying the classifying space construction to homomorphisms $K \to G$. It follows in the notation of 3.9 (with $\gamma = \alpha_G$) that $a_G^{\#}$ induces an equivalence from $\operatorname{Map}(BG, Z)_{[p]}$ to a union of components of the space $\operatorname{Map}(\operatorname{hocolim} \alpha_G, Z)_{[\gamma]}$. By induction, 3.10, and 2.1(2), the map

$$\kappa : Z \to \operatorname{Map}(\operatorname{hocolim} \alpha_G, Z)_{[\gamma]}$$

is an equivalence. Tracing through the various identifications gives the desired equivalence $Z \to \operatorname{Map}(BG, Z)_{[p]}$. $\qquad\square$

Step II. Suppose that G is connected with center C, and that $G/C \in \mathcal{C}$. It is necessary to show that $G \in \mathcal{C}$. This is the same as the second step in the proof of 1.2, but uses 5.3 and 3.5 instead of 4.2 and 3.4. $\qquad\square$

Step III. Suppose that the identity component G_0 of G belongs to $\mathcal{C}$; it is necessary to show that $G \in \mathcal{C}$. This is the same as the third step in the proof of 1.2, but uses 5.2 and 3.5 instead of 4.1 and 3.4. $\qquad\square$

6 Calculating $P_W(BG)$

In this section W will denote the fixed space $B\mathbf{Z}/p$. The goal is to compute $P_W(BG)$ when G is a compact Lie group such that $\pi_0 G$ is a p-group.

6.1 Remark. As described in [3], the space $P_W(BG)$ is built by constructing a nested collection of spaces X_λ, one for each countable ordinal λ. The space X_0 is BG; if λ is a limit ordinal then $X_\lambda = \cup_{\lambda' < \lambda} X_{\lambda'}$; if $\lambda = \lambda' + 1$ is a successor ordinal then X_λ is obtained from $X_{\lambda'}$ by adjoining cones on all maps of W and its suspensions into $X_{\lambda'}$. The space $P_W(BG)$ is then the union or colimit $\cup_\lambda X_\lambda$ (note that the number of spaces in this union is uncountable).

6.2 LEMMA. For a 1-connected space X the following three conditions are equivalent:

(1) X is $\mathbf{Z}/p$-acyclic.
(2) For each $i \geq 2$, $\pi_i X$ is a module over $\mathbf{Z}[1/p]$.
(3) The natural map $X \to L_{\mathbf{Z}[1/p]} X$ is an equivalence.

Moreover, these conditions imply

(4) X is W-null.

Proof. The equivalence of (1) and (2) follows from Serre "mod-$\mathcal{C}$" theory, since X is $\mathbf{Z}/p$-acyclic if and only if the reduced integral homology groups of X are uniquely p-divisible, which if $\pi_1 X$ is trivial is the case if and only if the homotopy groups of X are uniquely p-divisible. The equivalence between (2) and (3) is from [1, V, §3] (note that since X is simply connected the Bousfield localization $L_{\mathbf{Z}[1/p]}X$ is equivalent to the Bousfield-Kan space $\mathbf{Z}[1/p]_\infty(X)$ [2]). The fact that (2) implies (4) results for instance from a direct calculation with obstruction theory. Note that (4) does not imply the others, e.g., the n-sphere $(n \geq 2)$ is 1-connected and satisfies (4) (by 1.3), but does not satisfy (1). $\square$

6.3 Remark. If X is a connected space with the property that $\mathrm{H}_1(X, \mathbf{Z}[1/p]) = 0$, e.g., $X = BG$ with $\pi_0 G$ a p-group, then $L_{\mathbf{Z}[1/p]}X$ is 1-connected [1, VII, 3.2] [2] and satisfies the conditions given in 6.2 [1, V, §3]. In particular it follows from Theorem 1.7 that if G is a compact Lie group with $\pi_0 G$ a p-group then all of the mod p homology of BG can be killed by the iterated cone adjunction process of 6.1.

6.4 Remark. Lemma 6.2 leads to the following recognition principle. Let X be a connected space such that $\mathrm{H}_1(X, \mathbf{Z}[1/p])$ is trivial, and suppose that we can construct a P_W-equivalence $f : X \to Y$ such that Y is 1-connected and $\mathbf{Z}/p$-acyclic. Then the natural map $P_W(X) \to L_{\mathbf{Z}[1/p]}X$ is an equivalence. To see this, note first that $\mathrm{H}_*(f, \mathbf{Z}[1/p])$ is an isomorphism (1.6). The desired result now follows from the commutative diagrams

$$
\begin{array}{ccc}
X & \xrightarrow{\ f\ } & Y \\
\downarrow & & \sim\downarrow \\
L_{\mathbf{Z}[1/p]}X & \xrightarrow{\ \sim\ } & L_{\mathbf{Z}[1/p]}Y
\end{array}
\qquad
\begin{array}{ccc}
X & \xrightarrow{\ f\ } & Y \\
\downarrow & & \sim\downarrow \\
P_W(X) & \xrightarrow{\ \sim\ } & P_W(Y)
\end{array}
$$

in which by 6.2 the indicated vertical arrows are equivalences (for instance, 6.2(4) implies that $Y \sim P_W(Y)$).

6.5 LEMMA. [3, 2.5] *The class of P_W-equivalences is closed under homotopy colimits, in the sense that if $\mathbf{C}$ is a small category, $\gamma, \gamma' : \mathbf{C} \to \mathbf{Top}$ are functors, and $\tau : \gamma \to \gamma'$ is a natural transformation which gives a P_W-equivalence $\tau_c : \gamma(c) \to \gamma'(c)$ for each object c of $\mathbf{C}$, then $\operatorname{hocolim} \tau : \operatorname{hocolim} \gamma \to \operatorname{hocolim} \gamma'$ is a P_W-equivalence.*

6.6 Remark. Note that 6.5 is proved with 3.7. This lemma implies in particular that P_W-equivalences are stable under "cobase change", i.e., if $X \to X'$ is a P_W-equivalence and $X \to Y$ is a map, then the natural inclusion of Y in the homotopy pushout of the diagram $X' \leftarrow X \to Y$ is also a P_W-equivalence.

6.7 Remark. The class of $\mathbf{Z}/p$-equivalences is also closed under arbitrary homotopy colimits [1, XII, 5.7].

6.8 LEMMA. *Let* **C** *be a small category such that* **BC** *is connected,* $\gamma, \gamma' : \mathbf{C} \to$ **Top** *functors, and* $\tau : \gamma \to \gamma'$ *a natural transformation. Suppose that for each object* c *of* **C** *the map* $\tau_c : \gamma(c) \to \gamma'(c)$ *is a map between connected spaces which induces a surjection of fundamental groups. Then the map* $\mathrm{hocolim}\,\tau :$ $\mathrm{hocolim}\,\gamma \to \mathrm{hocolim}\,\gamma'$ *is also a map between connected spaces which induces a surjection of fundamental groups.*

Proof. This follows from the van Kampen theorem and the explicit construction of the homotopy colimit in [1, XII, §5]. □

6.9 LEMMA. [3, 2.9] *Suppose that* X *is a connected space. Then* $P_W(X)$ *is connected, and the map* $X \to P_W(X)$ *induces a surjection of fundamental groups.*

Remark. This is clear from the description of $P_W(X)$ in 6.1.

Let $\mathcal{C}$ be the class of (topological) groups with the property that the map $P_W(BG) \to L_{\mathbf{Z}[1/p]}(BG)$ is an equivalence. As usual, there are three steps (cf. 2.2) involved in proving that $\mathcal{C}$ contains every compact Lie group G such that $\pi_0 G$ is a p-group. We have to modify Step I slightly to handle a technical issue connected with the center (see Step II). The following theorem guarantees that the main inductive step does not leave the class of compact Lie groups G such that $\pi_0 G$ is a p-group.

6.10 THEOREM. [11, A.4] *Suppose that* G *is a compact Lie group such that* $\pi_0 G$ *is a* p-group, and that K *is a finite* p-subgroup of G (for example, K might be an elementary abelian p-subgroup of G). Then $\pi_0 C_G(K)$ *is also a* p-group.*

Step I. Suppose that G is a connected compact Lie group of dimension d with no element of order p in its center. Assume that $H \in \mathcal{C}$ for all H of dimension less than d such that $\pi_0 H$ is a p-group. We have to show that $G \in \mathcal{C}$. Let Y be the space which fits into the homotopy pushout diagram

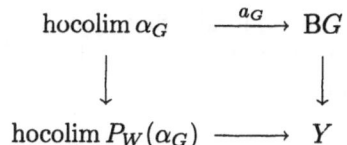

in which the left hand vertical arrow is induced by the natural map $\alpha_G \to P_W(\alpha_G)$. By 6.9, 6.8, and the van Kampen theorem, Y is 1-connected. By 6.5 the left hand vertical map is a P_W-equivalence, and so (6.6) the map $BG \to Y$ also is. Part (1) of 2.1 gives that the upper horizontal arrow in the above diagram is a $\mathbf{Z}/p$-equivalence. Since G has no central element of order p, each space $\alpha_G(V)$ (V an object of $\mathbf{A}_G$) is of form BH for some H which by induction (6.10) belongs to $\mathcal{C}$, and so (6.3) each space $P_W(\alpha_G(V))$ is $\mathbf{Z}/p$-acyclic. It follows from 6.7, applied to the unique natural transformation $P_W(\alpha_G) \to *_{\mathbf{A}_G}$, that the induced map $\mathrm{hocolim}\,P_W(\alpha_G) \to B\mathbf{A}_G$ is a $\mathbf{Z}/p$-equivalence, and thus by 2.1(2) that $\mathrm{hocolim}\,P_W(\alpha_G)$ is $\mathbf{Z}/p$-acyclic. A Meyer-Vietoris sequence calculation now gives that Y is $\mathbf{Z}/p$-acyclic, and so the desired result is a consequence of 6.4. □

6.11 P_W and fibrations. Recall from [3, §4] that given a fibration sequence $F \to$ $E \to B$ over a connected base B, it is possible to apply P_W fibrewise to obtain another fibration sequence $P_W(F) \to \bar{E} \to B$. There is a P_W-equivalence $E \to \bar{E}$ of spaces over B which on fibres gives the natural map $F \to P_W(F)$. If $E \to B$ is a principal fibration, then so is $\bar{E} \to B$ (cf. [3, §3]); in particular, taking $E = *$ and $F = G$ shows that if G is a topological group or more generally a loop space then $P_W(G)$ is also a loop space.

6.12 LEMMA. *Let $G \to E \to B$ be a principal fibration sequence over a connected base B. Assume that the classifying space $\mathrm{B}P_W(G)$ is W-null. Then the sequence $P_W(G) \to P_W(E) \to P_W(B)$ is also up to homotopy a fibration sequence.*

Proof. This is a restatement of [3, 4.3].   □

6.13 LEMMA. *If G is a locally finite p-group then $P_W(\mathrm{B}G)$ is contractible. In particular, $G \in \mathcal{C}$.*

Proof. Note that $\tilde{\mathrm{H}}_*(\mathrm{B}G, \mathbf{Z}[1/p])$ vanishes (because G is a union of finite p-groups) and so $L_{\mathbf{Z}[1/p]}(\mathrm{B}G)$ is contractible. This implies that $G \in \mathcal{C}$ if and only if $P_W(\mathrm{B}G)$ is contractible. We first prove by induction on the order of G that if G is a finite p-group then $P_W(\mathrm{B}G)$ is contractible. This is clear if G is trivial or if $G = \mathbf{Z}/p$. Otherwise, there exists a cyclic group σ of order p in the center of G and a corresponding fibration sequence

$$\mathrm{B}\sigma \to \mathrm{B}G \to \mathrm{B}(G/\sigma) = \mathrm{B}K$$

with $P_W(\mathrm{B}K)$ contractible by induction. Applying P_W fibrewise (6.11) thus gives a P_W-equivalence $\mathrm{B}G \to \mathrm{B}K$, which shows that $P_W(\mathrm{B}G)$ is contractible too.

   The statement for a general locally finite group G follows 6.5 and the fact that $\mathrm{B}G$ can be expressed as the filtered homotopy colimit of the classifying spaces of the finite subgroups of G (cf. [1, XII, 3.5] or [13, 9.8]).   □

6.14 LEMMA. *If G is an abelian compact Lie group such that $\pi_0 G$ is a p-group, then $G \in \mathcal{C}$.*

Proof. As in the proof of 4.2, let $D \subset G$ be the group of elements of order a power of p, considered as a discrete group. Let $f : \mathrm{B}D \to \mathrm{B}G$ be the map induced by the inclusion $D \to G$. Then F is a $\mathbf{Z}/p$-equivalence and (by the assumption on $\pi_0 G$) $\pi_1(f)$ is a surjection. Let Y be the space with fits into the homotopy pushout diagram

$$\begin{array}{ccc} \mathrm{B}D & \longrightarrow & \mathrm{B}G \\ \downarrow & & \downarrow \\ P_W(\mathrm{B}D) \sim * & \longrightarrow & Y \end{array}$$

in which $P_W(BD)$ is contractible by 6.13. By the van Kampen theorem Y is 1-connected. Clearly Y is $\mathbf{Z}/p$-acyclic, and by 6.6 the map $BG \to Y$ is a P_W-equivalence. The desired result follows from 6.4. $\qquad\square$

Step II. Suppose that G is a connected compact Lie group. Let C' be the center of G, and $C \subset C'$ the inverse image in C' of the p-torsion subgroup of $\pi_0 C'$. Assume that $G/C \in \mathcal{C}$. (Observe that the center of G/C, which is isomorphic to C'/C, has no nontrivial element of order p, so that in the modified inductive scheme we are following the group G/C will be handled in Step I.) It is necessary to prove that $G \in \mathcal{C}$. Since C is abelian, the fibration sequence

$$BC \to BG \to B(G/C)$$

is principal. By 6.14 and 6.3, $P_W(BC)$ is a 1-connected $\mathbf{Z}/p$-acyclic space. By a Serre spectral sequence argument, the classifying space $BP_W(BC)$ (see 6.11) is also $\mathbf{Z}/p$-acyclic, and so (6.2) is W-null. Lemma 6.12 thus shows that there is a fibration sequence

$$P_W(BC) \to P_W(BG) \to P_W(B(G/C))$$

From 6.3 it is clear that the base and fibre here are $\mathbf{Z}/p$-acyclic. This implies the space $P_W(BG)$ is $\mathbf{Z}/p$-acyclic. The space $P_W(BG)$ is 1-connected (6.9) and so the desired result follows from 6.4. $\qquad\square$

Step III. Suppose that G is a compact Lie group such that $\pi_0 G$ is a p-group, and assume that the identity component $G_0 \in \mathcal{C}$. It is necessary to prove that $G \in \mathcal{C}$. Consider the fibration

$$P_W(BG_0) \to \bar{X} \to B\pi_0 G \qquad\qquad (6.15)$$

obtained by applying P_W fibrewise (6.11) to the fibration $BG_0 \to BG \to B\pi_0 G$. There is a P_W-equivalence $BG \to \bar{X}$. By induction and 6.3, the space $P_W(BG_0)$ satisfies the conditions given in 6.2, and in particular the higher homotopy groups of this space are uniquely p-divisible. This implies that the (twisted) cohomology groups $H^i(B\pi_0 G, \pi_j P_W(BG_0))$ vanish for $i > 0$ and $j \geq 2$ and hence by obstruction theory that the fibration 6.15 has a section $s : B\pi_0 G \to \bar{X}$. Since $P_W(BG_0)$ is $\mathbf{Z}/p$-acyclic (6.2), $H_*(s, \mathbf{Z}/p)$ is an isomorphism; since $P_W(BG_0)$ is 1-connected, $\pi_1(s)$ is an isomorphism. Let Y be the space which fits into the homotopy pushout diagram

$$
\begin{array}{ccc}
B\pi_0 G & \xrightarrow{\;s\;} & \bar{X} \\
\downarrow & & \downarrow \\
P_W(B\pi_0 G) \sim * & \longrightarrow & Y
\end{array}
$$

in which $P_W(B\pi_0 G)$ is contractible by 6.13. Clearly Y is simply connected and $\mathbf{Z}/p$-acyclic; by 6.6 the map $\bar{X} \to Y$ is a P_W-equivalence. The composite $BG \to \bar{X} \to Y$ is then also a P_W-equivalence, and the desired result follows from 6.4. $\qquad\square$

References

[1] A.K. Bousfield and D.M. Kan, Homotopy Limits, Completions and Localizations, Lecture Notes in Math. 304, Springer, Berlin, 1972.

[2] A. K. Bousfield, *The localization of spaces with respect to homology*, Topology **14** (1975), 133–150.

[3] A. K. Bousfield, *Localization and periodicity in unstable homotopy theory*, preprint (University of Illinois, Chicago) 1992.

[4] W. G. Dwyer, E. Dror, and D. M. Kan, *An arithmetic square for virtually nilpotent spaces*, Illinois J. Math. **21** (1977), 242–254.

[5] W. G. Dwyer and C. W. Wilkerson, *The center of a p-compact group*, preprint (Notre Dame) 1993.

[6] E. Dror Farjoun, *Homotopy and homology of diagrams of spaces*, Algebraic Topology, Proceedings, Seattle (1985), H. R. Miller and D. C. Ravenel, eds., Lect. Notes in Math. 1286, Springer, Berlin, 1987, pp. 93–134.

[7] E. Dror Farjoun, *Localization with respect to a map and v_1-periodicity*, Proceedings 1990 Barcelona Conference on Algebraic Topology, Lect. Notes in Math. 1509, Springer, Berlin, 1991.

[8] E. Dror Farjoun and J. H. Smith, *Homotopy localization nearly preserves fibrations*, Topology (to appear).

[9] E. Friedlander and G. Mislin, *Locally finite approximations of Lie groups, II*, Math. Proc. Camb. Phil. Soc. **106** (1986), 505–517.

[10] S. Jackowski and J.E. McClure, *Homotopy decomposition of classifying spaces via elementary abelian p-groups*, Topology **31** (1992), 113–132.

[11] S. Jackowski, J.E. McClure and R. Oliver, *Homotopy classification of self-maps of BG via G-actions*, Annals of Math. **135** (1992), 227–270.

[12] I. M. James, Fibrewise Topology, Cambridge University Press, Cambridge, 1989.

[13] H. R. Miller, *The Sullivan conjecture on maps from classifying spaces*, Ann. of Math. **120** (1984), 39–87.

[14] J. M. Møller, *Rational isomorphisms of p-compact groups*, preprint (Mathematisk Institut, Kobenhavn) 1994.

[15] D. G. Quillen, *Higher algebraic K-theory: I*, Algebraic K-theory I, Lect. Notes in Math. 341, Springer, Berlin, 1973, pp. 85–147.

[16] A. Zabrodsky, *Maps between classifying spaces*, Algebraic topology and algebraic K-theory, Ann. Math. Studies 113, Princeton Univ. Press, Princeton, 1987, pp. 228–246.

University of Notre Dame, Notre Dame, Indiana 46556
dwyer.1@nd.edu

Progress in Mathematics, Vol. 136
© 1996 Birkhäuser Verlag Basel/Switzerland

The Yagita invariant of general linear groups

H. H. Glover, I. J. Leary and C. B. Thomas

We give a definition of the Yagita invariant $p°(G)$ of an arbitrary group G, and compute $p°(GL_n(\mathcal{O}))$ for each prime p, where $\mathcal{O}$ is any integrally closed subring of the complex numbers $\mathbb{C}$ (i.e., $\mathcal{O}$ is integrally closed in its field of fractions F, which is also a subring of $\mathbb{C}$). We also show that $p°(SL_n(\mathcal{O}))$ is equal to $p°(GL_n(\mathcal{O}))$ for $n \geq 2$ except that possibly $p°(SL_n(\mathcal{O}))$ is $1/2 p°(GL_n(\mathcal{O}))$ for some small n and 'small' rings $\mathcal{O}$. Our definition of $p°(G)$ extends both Yagita's original definition for G finite [9] and the definition given by one of us for G of finite virtual cohomological dimension (or vcd) [8]. For G of finite vcd such that the Tate-Farrell cohomology $\hat{H}^*(G)$ of G is p-periodic, $p°(G)$ is equal to the p-period. Hence our results may be viewed as a generalisation of those of Bürgisser and Eckmann [1,2], who compute the p-periods of $GL_n(\mathcal{O})$ and $SL_n(\mathcal{O})$ for various $\mathcal{O}$ such that these groups have finite vcd and for all n such that these groups are p-periodic. The methods we use are similar to those used in [1].

Notation. Throughout this paper p shall be a fixed prime number, $\mathcal{O}$ an integrally closed subring of $\mathbb{C}$, and F the field of fractions of $\mathcal{O}$. ζ_p shall be a primitive pth root of unity in $\mathbb{C}$, and l shall be the degree of $F[\zeta_p]$ as an extension of F. In [1,2,4], the notation $\phi_F(p)$ is used for l.

Definition. Let G be a group. If C is an order p subgroup of G, then

$$H^*(BC; \mathbb{Z}) = \mathbb{Z}[x]/(px)$$

for x a generator of H^2. Define $n(C)$ (either a positive integer or infinity) to be the supremum of the integers n such that the image of $H^*(BG)$ in $H^*(BC)$ is contained in the subalgebra generated by x^n. Now define the Yagita invariant $p°(G)$ to be

$$p°(G) = \text{l.c.m.}\{2n(C) : C \leq G, \quad |C| = p\},$$

where the least common multiple of the empty set is equal to 1, and the least common multiple of an unbounded set of integers, or of a set containing infinity, is infinity. The invariant $p°(G)$ depends on the prime p as well as on the group G, but we have resisted the temptation to write, for example, $2°(G)$ for $p°(G)$ in the case when $p = 2$.

The following result is immediate from the definition.

Proposition 1. *Let $f: H \to G$ be a group homomorphism whose kernel contains no element of order p. (For example, f might be injective, or the kernel of f might be torsion-free.) Then $p°(H)$ divides $p°(G)$.* □

It is easy to see that $p^\circ(G)$ is finite if G is finite, and Proposition 1 applied in the case when G is the quotient of H by a torsion-free normal subgroup of finite index implies that $p^\circ(H)$ is finite if H is a group of finite vcd (see [5] for a more detailed proof of this fact). Using Chern classes, we shall show in Corollary 7 that if G admits a faithful finite-dimensional representation over $\mathbb{C}$, then $p^\circ(G)$ is finite.

Some remarks concerning the origin of the definition of $p^\circ(G)$ and its geometric motivation are in order. N. Yagita made the above definition for finite groups [9], and proved that if a finite group G acts freely on a (finite) product $S^{m-1} \times \ldots \times S^{m-1}$ of spheres of the same dimension $m-1$, with trivial action on their homology, then $p^\circ(G)$ divides m. This result may be viewed as a generalisation of the theorem bounding the dimension of a sphere with a free G-action in the case when the cohomology of G is periodic. One of us suggested extending the definition to groups of finite vcd in [8]. The connection with actions on products of spheres extends to this case too. If G is a group of finite vcd acting freely, properly discontinuously and with trivial action on homology, on a finite product $S^{m-1} \times \ldots \times S^{m-1} \times \mathbb{R}^n$, then $p^\circ(G)$ divides m, by the same argument as used in [9] for the case when G is finite. For arbitrary groups, the connection with the geometry of group actions is less clear. One reason for extending the definition to arbitrary groups is that this makes our results easier to state and no harder to prove.

To motivate our work we start by stating the main theorems and some corollaries. The proofs of the main theorems will occupy the remainder of the paper.

THEOREM 2. *As explained under the title 'Notation', let $\mathcal{O}$ be an integrally closed subring of $\mathbb{C}$ with field of fractions F, and let $l = |F[\zeta_p] : F|$. Define $\psi(t)$, for t a positive real, to be the greatest integer power of p less than or equal to t. The Yagita invariant $p^\circ(GL_n(\mathcal{O}))$ is given by the following table.*

$$p^\circ(GL_n(\mathcal{O})) = \begin{cases} 1 & \text{for } n < l \\ 2\text{l.c.m.}\{m : l|m, \quad m|(p-1), \quad m \le n\} & \text{for } l \le n \le p-1 \\ 2(p-1)\psi(n/l) & \text{for } n \ge p \end{cases}$$

REMARK. The above theorem holds for all primes p, but the statement can be simplified slightly for $p = 2$. In this case $l = 1$, and $p^\circ(GL_n(\mathcal{O})) = 2\psi(n)$ for all $n \ge 1$. Note that $p^\circ(GL_n(\mathcal{O}))$ depends only on F, so that if R is any subring of $\mathbb{C}$ such that $\mathcal{O} \le R \le F$, then $p^\circ(GL_n(R)) = p^\circ(GL_n(\mathcal{O}))$.

Replacing $GL_n(\mathcal{O})$ by $SL_n(\mathcal{O})$ does not change the Yagita invariant, except for small values of n and 'small' rings $\mathcal{O}$, when it may be necessary to introduce a factor of $1/2$. Thus one obtains the following.

THEOREM 3. *With notation as above, for $n \ge 2$, $p^\circ(SL_n(\mathcal{O}))$ is equal to $p^\circ(GL_n(\mathcal{O}))$ except that $p^\circ(SL_n(\mathcal{O}))$ may be equal to $\frac{1}{2}p^\circ(GL_n(\mathcal{O}))$ in the following cases.*

a)$p = 2$, $n = 2$, and $\mathcal{O}$ contains no square root of -1.

b)p is odd, n has the form $n = 2^r l$ for 2^r dividing $(p-1)/l$, and $\mathcal{O}$ contains no nth root of -1.

COROLLARY 4. *Define the function ψ as in the statement of Theorem 2. Then for $n \geq 2$ and any prime p, the Yagita invariant of $GL_n(\mathbb{Z})$ is as follows.*

$$p^\circ(GL_n(\mathbb{Z})) = \begin{cases} 1 & \text{for } n < p - 1, \\ 2(p-1)\psi(n/(p-1)) & \text{for } n \geq p - 1. \end{cases}$$

For all $n \geq 2$ and all p, $p^\circ(SL_n(\mathbb{Z})) = p^\circ(GL_n(\mathbb{Z}))$, except that for p odd, $p^\circ(SL_{p-1}(\mathbb{Z})) = p - 1$, and for $p = 2$, $p^\circ(SL_2(\mathbb{Z})) = 2$.

Proof. The statement for the general linear groups is a special case of Theorem 2. The only cases when $p^\circ(SL_n(\mathbb{Z}))$ is not completely determined by Theorem 3 are those when $n = p - 1$ and when $n = p = 2$, but in these cases $SL_n(\mathbb{Z})$ is p-periodic and the p-period is determined in [1]. □

The following statement may be proved in the same way as Theorems 2 and 3, and could be considered as a corollary but for the weaker conditions imposed on the subring R of $\mathbb{C}$.

THEOREM 5. *Let R be any subring of $\mathbb{C}$ containing ζ_p, a primitive pth root of 1, and let ψ be as in the statement of Theorem 2. Then the Yagita invariant of $GL_n(R)$ is given by the following table.*

$$p^\circ(GL_n(R)) = \begin{cases} 2\text{l.c.m.}\{m : m|(p-1), \quad m \leq n\} & \text{for } n \leq p - 1 \\ 2(p-1)\psi(n) & \text{for } n \geq p \end{cases}$$

For R as above, $p^\circ(SL_n(R)) = p^\circ(GL_n(R))$ if any of the following conditions is satisfied:

a) $n \geq \max\{p, 3\}$,
b) $n \geq 2$, p is odd, and R contains a $(p-1)$st root of -1,
c) $n = 2$, $p = 2$, and R contains a square root of -1.

Our upper bound for $p^\circ(GL_n(\mathcal{O}))$ uses the Chern classes of the natural representation in $GL_n(\mathbb{C})$, and relies on the following proposition.

PROPOSITION 6. *Let $f(X)$ be a polynomial over the field $\mathbb{F}_p$, all of whose roots lie in $\mathbb{F}_p^\times$. If there is a polynomial g and integer n such that $f(X) = g(X^n)$, then n has the form mp^q for some m dividing $p - 1$ and some positive integer q.*

Proof. Let $n = mp^q$, where p does not divide m. It remains to prove that m divides $p - 1$. Now $g(X^n) = g(X^m)^{p^q}$, so without loss of generality we may assume that $q = 0$. Now if $g(Y) = 0$ has roots $y_1, \ldots, y_k$, then the roots of $g(X^m) = 0$ are the roots of $y_i - X^m = 0$ for each i, and because p does not divide m these polynomials have no repeated roots. If y_i is not an element of $\mathbb{F}_p^\times$ then $y_i - X^m$ can have no roots in $\mathbb{F}_p$. It has exactly m roots in $\mathbb{F}_p$ if and only if the inverse image of y_i under the map $x \mapsto x^m$ from $\mathbb{F}_p^\times$ to itself has order m, which can only happen if m divides $p - 1$. □

We will use the following Corollary in the proof of Theorem 2.

COROLLARY 7. *With notation as in the statement of Theorem 2, let G be a subgroup of $GL_n(F)$. Then the Yagita invariant $p^\circ(G)$ divides the number given for $p^\circ(GL_n(\mathcal{O}))$ in Theorem 2.*

Proof. For each order p subgroup C of G, we bound the number $n(C)$ occuring in the definition of $p^\circ(G)$ by considering the image of the cohomology $H^*(BGL_n(\mathbb{C}))$ of the Lie group $GL_n(\mathbb{C})$ in $H^*(BC)$ under the inclusion of C in $GL_n(\mathbb{C})$. Recall (or see for example [7]) that $H^*(BGL_n(\mathbb{C}))$ is a free polynomial algebra on generators $c_1, \ldots, c_n$, the *universal Chern classes*, where c_i has degree $2i$. For a representation ρ of a group H in $GL_n(\mathbb{C})$, the element $\rho^*(c_i)$ is usually written $c_i(\rho)$, and called the ith Chern class of the representation ρ. The sum $c.(\rho)$ of all the $c_i(\rho)$ is known as the total Chern class of ρ, and has the property that $c.(\rho \oplus \rho') = c.(\rho)c.(\rho')$.

If $H^*(BC) = \mathbb{Z}[x]/(px)$, then the total Chern classes of the p distinct 1-dimensional complex representations of C are $1 + ix$ for $i \in \mathbb{F}_p$, where the case $i = 0$ corresponds to the trivial representation. If we fix a generator of C such that the 1-dimensional representation of C sending this generator to ζ_p has total Chern class $1 + x$, then the representation sending this same generator to $\zeta_p{}^i$ has total Chern class $1 + ix$. For F as in the statement, there are $1 + (p-1)/l$ irreducible F-representations of C, the trivial representation of dimension one and $(p-1)/l$ others of dimension l. Over $\mathbb{C}$ the faithful representations split as a direct sum of one copy each of the representations sending a fixed generator of C to $\zeta_p{}^{ri^j}$, for some r, where i generates a subgroup of $\mathbb{F}_p^\times$ of order l and $0 \le j < l$. It follows that the total Chern classes of these representations have the form $1 - ix^l$, where i ranges over the $(p-1)/l$ distinct lth roots of unity in $\mathbb{F}_p$.

Now the ring $H^*(BC) \otimes \mathbb{F}_p$ is isomorphic to the free polynomial ring $\mathbb{F}_p[x]$, and the total Chern class of the inclusion of C in $GL_n(\mathbb{C})$ is a polynomial $f(x)$ of degree less than or equal to n satisfying the hypotheses of Proposition 6. Moreover, since this inclusion factors through $GL_n(F)$, the total Chern class can be viewed as a polynomial $\tilde{f}(y)$ in $y = x^l$ of degree less than or equal to n/l satisfying the hypotheses of the Proposition. This polynomial is not the trivial polynomial because the inclusion of C is not the trivial representation. The integer $n(C)$ (resp. $n(C)/l$) is the greatest integer m such that $f(x)$ (resp. $\tilde{f}(y)$) can be expressed as a polynomial in x^m (resp. y^m), and so by Proposition 2 each of $n(C)$ and $n(C)/l$ is a divisor of $p - 1$ multiplied by a power of p. Note also that $n(C)$ is at most n. This implies that for each C, the number $n(C)$ divides the bound given in Theorem 2, and hence so does the l.c.m. of all such $n(C)$.   □

REMARK.   The discussion of the total Chern classes of F-representations of the cyclic group of order p could be avoided by quoting the general bounds on the orders of Chern classes of F-representations given in [4]. Corollary 7 also holds for groups G having a representation in $GL_n(F)$ which is faithful on every order p subgroup, by the same proof. Corollary 7 may be compared with Lemma 1.2 of [5].

Our lower bounds for $p^\circ(GL_n(\mathcal{O}))$ and $p^\circ(SL_n(\mathcal{O}))$ use various finite subgroups, which we define below.

DEFINITION.   For p an odd prime, and m a divisor of $p-1$, let $G_1(p, m)$ be the split metacyclic group $C_p : C_m$, where C_m acts faithfully on C_p, and let $G_2(p, m)$

be the split metacyclic group $C_p : C_{2m}$, where C_{2m} acts via the faithful action of its quotient C_m. For p an odd prime let $E(p, 1)$ be the non-abelian group of order p^3 and exponent p, and let $E(2, 1)$ be the dihedral group of order eight. For any p, and any posititve integer m, let $E(p, m)$ be the central product of m copies of $E(p, 1)$, so that $E(p, m)$ is one of the two extraspecial groups of order p^{2m+1}. In ATLAS notation [3], the group $E(p, m)$ is called p_+^{1+2m}. Since the symbol p is already overworked in this paper we shall not use the ATLAS notation here.

LEMMA 8. *For p an odd prime the Yagita invariants of the groups defined above are as follows.*

$$p^\circ(G_1(p, m)) = p^\circ(G_2(p, m)) = 2m, \qquad p^\circ(E(p, m)) = 2p^m$$

For $p = 2$, let Q_8 be the quaternion group of order 8. Then

$$p^\circ(E(2, m)) = 2^{m+1} \quad and \quad p^\circ(Q_8) = 4.$$

Proof. The groups $G_1(p, m)$ and $G_2(p, m)$ are p-periodic, and their p-periods are shown to be as claimed in Proposition 3.1 and Section 3.3 of [2]. The Yagita invariant for the extraspecial groups was computed in Example B of Section 2 of [9], which in turn used results of [6]. □

LEMMA 9. *For p an odd prime, and m a divisor of $p - 1$, let $n = ml/(m, l)$. The group $G_1(p, m)$ embeds in $GL_n(\mathcal{O})$, and in $SL_{n+1}(\mathcal{O})$. If either $l/(m, l)$ is even or m is odd (and greater than 1) then $G_1(p, m)$ embeds in $SL_n(\mathcal{O})$. If m is even and $\mathcal{O}$ contains an mth root of -1, then $G_2(p, m)$ embeds in $SL_n(\mathcal{O})$. The group $E(p, m)$ embeds in $SL_{lp^m}(\mathcal{O})$.*

For $p = 2$ and $m > 1$ the group $E(2, m)$ embeds in $SL_{2^m}(\mathbb{Z})$. The group $E(2, 1)$ (the dihedral group of order 8) embeds in $GL_2(\mathbb{Z})$ and in $SL_3(\mathbb{Z})$. The quaternion group of order eight embeds in $SL_2(\mathbb{Z}[i])$.

Proof. First consider the case when p is odd. For $G_1(p, m)$ and $G_2(p, m)$ this is essentially contained in [2] Section 3.2. If m divides m', then $G_1(p, m)$ is a subgroup of $G_1(p, m')$, so without loss of generality we may assume that m is divisible by l and show that in this case $G_1(p, m)$ embeds in $GL_m(\mathcal{O})$. As an $\mathcal{O}$-module, $\mathcal{O}[\zeta_p]$ is free of rank l, which is closed under multiplication by ζ_p and under the action of $\mathrm{Gal}(F[\zeta_p]/F)$, which together generate a group of $\mathcal{O}$-linear automorphisms of $\mathcal{O}[\zeta_p]$ isomorphic to $G_1(p, l)$. If m is a proper multiple of l, view $G(p, l)$ as a subgroup of $G(p, m)$, and the induced representation V coming from the above representation is an m-dimensional faithful representation of $G(p, m)$ over $\mathcal{O}$.

It is easy to check that the determinant of the action of an element of order p of $G_1(p, m)$ on V is equal to 1, and that $V \otimes F$ restricts to a cyclic subgroup of $G_1(p, m)$ of order m as a sum of $l/(m, l)$ copies of the regular representation (use the normal basis theorem). Thus if either m is odd or $l/(m, l)$ is even, the determinant of an element of order m acting on V is 1, and so $G_1(p, m)$ is contained in $SL(V) \cong SL_n(\mathcal{O})$. In any case, the determinant of the action of $G_1(p, m)$ on V has image contained in $\{\pm 1\} \subseteq \mathcal{O}^\times$, so the action of $G_1(p, m)$ on $V \oplus \Lambda^n(V)$ gives an embedding from $G_1(p, m)$ into $SL_{n+1}(\mathcal{O})$.

If m is even and $l/(m, l)$ is odd, then $\mathcal{O}$ contains an mth root of -1 if and only if $\mathcal{O}$ contains an nth root of -1. In this case, if μ is an nth root of -1 in $\mathcal{O}$, and A of order p and B of order m with $\det(B) = -1$ generate a subgroup of $GL_n(\mathcal{O})$ isomorphic to $G_1(p, m)$, then A and μB generate a subgroup of $SL_n(\mathcal{O})$ isomorphic to $G_2(p, m)$.

Now consider the extraspecial group $E(p, m)$ for p odd. Note that the centre Z of $E(p, m)$ is cyclic of order p, and that $E(p, m)$ has a subgroup of index p^m containing Z as a direct factor. This subgroup has an l-dimensional $\mathcal{O}$-representation which is faithful on Z, and the corresponding induced $E(p, m)$-module is an lp^m-dimensional representation which must be faithful (since any nontrivial normal subgroup of a p-subgroup meets the centre nontrivially). Over $\mathbb{C}$, $E(p, m)$ has $(p-1)$ faithful irreducible representations, each of dimension p^m (arising as induced modules in the above way for different choices of 1-dimensional modules for Z). Using characters it is easy to see that these representations restrict to any non-central subgroup of order p as a sum of copies of the regular representation and to the centre as a sum of copies of a single irreducible representation, and hence that they have image in $SL_{p^m}(\mathbb{C})$ (note that $E(p, m)$ has exponent p). The representations over $\mathcal{O}$ constructed above split over $\mathbb{C}$ into a sum of some of the faithful irreducibles, so have determinant 1.

For $p = 2$, there is a unique faithful irreducible complex representation of $E(2, m)$, which has dimension 2^m, and an argument similar to the above shows that it is realisable over $\mathbb{Z}$. Using characters one can show that the restriction of this representation to any cyclic subgroup of order four lies in $SL_{2^m}(\mathbb{Z})$, and that the restriction to any non-central subgroup of order two is isomorphic to a sum of 2^{m-1} copies of the regular representation, so lies in $SL_{2^m}(\mathbb{Z})$ provided that $m \geq 2$.

The left action of the quaternion group Q_8 on $\mathbb{Z}[i, j, k]$ commutes with the right action of $\mathbb{Z}[i]$, giving a faithful representation of Q_8 which has image in $SL_2(\mathbb{Z}[i])$. $\qquad\square$

Proof of Theorem 2. Corollary 7 gives an upper bound for $p^\circ(GL_n(\mathcal{O}))$. For $n \leq p - 1$, for each $m \leq n$ such that l divides m and m divides $p - 1$, Lemma 9 tells us that $G_1(p, m)$ occurs as a subgroup of $GL_n(\mathcal{O})$, and so by Proposition 1 and Lemma 8, $p^\circ(GL_n(\mathcal{O}))$ is divisible by $p^\circ(G_1(m, p)) = 2m$. This gives the bound for $n \leq p - 1$, and shows that for $n \geq p$, $p^\circ(GL_n(\mathcal{O}))$ is divisible by $2(p - 1)$. Now (for any p) the group $E(p, m)$ is a subgroup of $GL_n(\mathcal{O})$ for each $n \geq 2lp^m$, and has Yagita invariant $2p^m$. This gives the p-part of the bound for $n \geq p$. $\qquad\square$

Proof of Theorem 3. This is similar to the proof of Theorem 2. In the case when $n \leq p - 1$, the l.c.m. occuring in the expression given for $p^\circ(GL_n(\mathcal{O}))$ is clearly equal to the following expression.

$$\text{l.c.m.}\{m \,:\, m \leq n, m = lq^r \text{ for some prime } q, \; q^r | (p - 1)/l\}$$

In other words, we need only consider those m of the form lq^r for some prime q such that q^r divides $(p-1)/l$. If q is an odd prime and q divides l exactly s times, then $G_1(p, q^{r+s})$ is a subgroup of $SL_n(\mathcal{O})$ for $n = lq^r$, and has Yagita invariant $2q^{r+s}$. If 2 divides l exactly s times, and $2^r l$ divides $p - 1$, then $G_1(p, 2^{r+s})$ is a subgroup of $SL_{n+1}(\mathcal{O})$ for $n = l2^r$ and has Yagita invariant 2^{1+r+s}. From these examples it already follows that $p°(SL_n(\mathcal{O}))$ is divisible by $2(p-1)$ for $n \geq p$, and that for $n \leq p - 1$, $p°(SL_n(\mathcal{O}))$ is equal to $p°(GL_n(\mathcal{O}))$ except possibly if n is of the form $2^r l$ and is a factor of $p - 1$, when the Yagita invariant for $SL_n(\mathcal{O})$ might be half the Yagita invariant for $GL_n(\mathcal{O})$. If $\mathcal{O}$ contains a 2^{r+s}th root of -1, or equivalently an nth root of -1 (where $n = l2^r$), then $G_2(p, 2^{r+s})$ is a subgroup of $SL_n(\mathcal{O})$, and has Yagita invariant 2^{1+r+s}, so that in this case too $p°(SL_n(\mathcal{O})) = p°(GL_n(\mathcal{O}))$. $\qquad\square$

Proof of Theorem 5. The groups $GL_n(R)$ and $SL_n(R)$ are subgroups of $GL_n(\mathbb{C})$, so their Yagita invariants are bounded above by $p°(GL_n(\mathbb{C}))$. By the hypothesis on R, the cyclic group C_p admits a faithful representation in $GL_1(R)$. As in Lemma 9 one may use induced representations of the groups $G_i(p, m)$ and $E(p, m)$ to give lower bounds equal to the above upper bounds. We leave the details as an exercise. $\qquad\square$

REMARK. The methods that we use also gives some information concerning the Yagita invariant of the groups $G(\mathcal{O})$ for other algebraic groups G. We hope to address this question in a future publication.

Acknowledgements. The work of the first named author was partially funded by the Centre de Recerca Matemàtica, and the work of the second named author was funded by a DGICYT research fellowship at the Centre de Recerca Matemàtica.

References

[1]   B. Bürgisser, On the p-periodicity of arithmetic subgroups of general linear groups, *Comment. Math. Helv.*, **55** (1980) 499–509.

[2]   B. Bürgisser and B. Eckmann, The p-periodicity of the groups $GL(n, O_S(K))$ and $SL(n, O_S(K))$, *Mathematika*, **31** (1984) 89–97.

[3]   J. H. Conway *et. al.*, *An ATLAS of finite groups*, Oxford University Press, 1985.

[4]   B. Eckmann and G. Mislin, Chern classes of group representations over a number field, *Compositio Math.*, **44** (1981) 41–65.

[5]   H. H. Glover, G. Mislin and Y. Xia, On the Yagita invariant of mapping class groups, *Topology*, **33** (1994) 557–574.

[6]   C. B. Thomas, A model for the classifying space of an extra special p-group, *Mathematika*, **22** (1975) 182–187.

[7]   C. B. Thomas, *Characteristic classes and the cohomology of finite groups*, Cambridge University Press, 1986.

[8]   C. B. Thomas, Free actions by p-groups on products of spheres and Yagita's invariant $po(G)$, *Lecture Notes in Math.*, **1375** (1989) 326–338.

[9]   N. Yagita, On the dimension of spheres whose product admits a free action by a non-abelian group, *Quart. J. Math. Oxford*, **36** (1985) 117–127.

H. H. Glover, Ohio State University, Columbus, Ohio 43210.

I. J. Leary, Centre de Recerca Matemàtica, Institut d'Estudis Catalans, E-08193 Bellaterra.

C. B. Thomas, DPMMS, University of Cambridge, Cambridge CB2 1SB.

Progress in Mathematics, Vol. 136
© 1996 Birkhäuser Verlag Basel/Switzerland

Fibrewise complexes

I. M. James

Abstract. In equivariant homotopy theory the appropriate notion of complex was developed around 1971 by Illman and Matumoto, independently: Chapter II of [3] is a convenient reference for this work. In fibrewise homotopy theory nothing comparable has so far been proposed, as far as I am aware, and the purpose of this note is to try and fill the gap. A more thorough account will be given elsewhere but here we try to convey the basic idea, which is quite simple, rather than strive for maximum generality.

1. Preliminaries

Fibrewise homotopy theory may be regarded as a branch of the homotopy theory of the category Top(2) of pairs of spaces and maps, as discussed by Eckmann and Hilton [5] and others. Unfortunately the term *pair* is potentially confusing in our situation, where it is preferable to describe Top(2) as the category of *spaces over spaces* and *maps over maps*, the latter being classified by *homotopies over homotopies*.

Various treatments of the homotopy theory of the category Top(2) may be found in the literature. Perhaps that of tom Dieck, Kamps and Puppe [4] is the most appropriate for our purposes. However it is convenient to adopt a modification of their terminology and notation, as follows.

The objects of Top(2), of course, are the morphisms of Top. Thus an object consists of a base space B and a space X over B with projection p, say. The morphisms of Top(2) are commutative diagrams of morphisms of Top. Thus if X is a space over B with projection p and X' is a space over B' with projection p' then a morphism from X to X' consists of a map $f : B \to B'$ and a map $F : X \to X'$ such that $p' \circ F = f \circ p$. We may refer to F as a map over f.

Maps over maps are classified by homotopies over homotopies, as follows. Let $f_i : B \to B'$ be a map $(i = 0, 1)$ and let $F_i : X \to X'$ be a map over f_i. Let $f_t : B \to B'$ be a homotopy of f_0 into f_1 and let $F_t : X \to X'$ be a homotopy of F_0 into F_1 such that $p' \circ F_t = f_t \circ p$ for all t. We may refer to F_t as a homotopy over f_t. Homotopy equivalences over homotopy equivalences, etc, are defined in a similar manner.

The category Top_B of spaces over a given base space B may be regarded as contained in the category Top(2) as the subcategory of spaces over B and maps over the identity of B. However the morphisms of Top_B are classified by homotopies over the stationary homotopy of the identity, which is generally a finer classification than that by homotopies over self-homotopies of the identity. In this note it is the former classification which is appropriate.

2.   The basic definition

The notion of CW complex, introduced by J. H. C. Whitehead [10], is covered in all the standard textbooks, for example in Chapter 7 of [9]. To keep this note short we do not consider infinite complexes and so the letters CW (standing for closure finite, weak topology) can be omitted.

Recall that a cellular decomposition of a Hausdorff space B consists, in each dimension n, of a finite collection of maps $\theta : D^n \to B$, satisfying certain conditions. The image of the closed n-ball D^n under the characteristic map θ is called the closed n-cell. That of $D^n - S^{n-1}$ is called the open n-cell, and that of S^{n-1} is called the boundary of the n-cell (the terminology does not refer to the topology of B). The conditions are that θ maps $D^n - S^{n-1}$ homeomorphically onto the open n-cell, and that the boundary of the n-cell is the union of open m-cells for $m < n$. Also the whole collection of open cells forms a decomposition of B, so that every point of B is contained in precisely one open cell. When these conditions are satisfied we describe B as a *complex*.

Choose a cellular decomposition of B, and let K be a fibrewise Hausdorff space over B (see [8] for the terminology of fibrewise topology). A cellular block decomposition of K consists of a decomposition of K into open cellular blocks. Specifically, over each closed n-cell of B, with characteristic map $\theta : D^n \to B$, there exists a finite collection of maps $\Theta : D^n \times T \to K$, over the map θ, where T is compact Hausdorff (the factors T may vary with Θ), satisfying certain conditions. The image of $D^n \times T$ under Θ is called the closed n-cellular block, that of $(D^n - S^{n-1}) \times T$ the open n-cellular block, and that of $S^{n-1} \times T$ the boundary of the n-cellular block (the terminology does not refer to the topology of K). The conditions are that Θ maps $(D^n - S^{n-1}) \times T$ homeomorphically onto the open n-cellular block, and that the boundary of the n-cellular block is the union of open m-cellular blocks for $m < n$. Also the whole collection of open cellular blocks, over the chosen cellular decomposition of B, forms a decomposition of K, so that every point of K is contained in precisely one open cellular block. When these conditions are satisfied we describe K as a *fibrewise complex* over the complex B. There is no requirement for K to be a complex in the ordinary sense.

Clearly any fibre bundle with compact Hausdorff fibre over the complex B can be regarded as a fibrewise complex over B. Examples of fibrewise complexes which are not fibre bundles arise in the theory of transformation groups. Specifically, if K is a G-complex, where G is a compact group, then the orbit space K/G is a complex and K is a fibrewise complex over K/G, as described in [3], the factors of the cellular blocks being orbits of different types.

Thus consider the much-studied family of $O(n)$-manifolds W_k^{2n-1} (see I.7 of [2]) for which the orbit space is the 2-disc D^2. We can construct W_k^{2n-1} by adjoining the 2-cellular block $D^2 \times V_{n,2}$ to S^{n-1} by means of the map $\psi_k : S^1 \times V_{n,2} \to S^{n-1}$, where $\psi_k((\cos\theta, \sin\theta), (u, v)) = (u\cos k\theta, v\sin k\theta)$. Here the factor $V_{n,2}$ is the Stiefel manifold of orthonormal pairs (u, v) in $\mathbb{R}^n$.

Fibrewise complexes in which all the factors in the cellular blocks are complexes in the ordinary sense play a special role in the theory and so we will refer to them as *special* fibrewise complexes. For example the family of $O(n)$-complexes W_k^{2n-1} we have just described are special fibrewise complexes over D^2. Also sphere-bundles over complexes are special. If K is a special fibrewise complex over B the dimension $\dim K$ of K is defined to be the maximum dimension of the cellular blocks in the decomposition of K. For example $2n-1$ is the dimension of W_k^{2n-1}. Also if K is a q-sphere bundle over the complex B then the dimension of K is $q + \dim B$.

3. The basic theory

Let B be a complex and let K be a fibrewise complex over B. We describe a subspace L of K as a *subcomplex* of K if L is the union of open cellular blocks of K subject to the condition that the boundary of each of the cellular blocks of L is also in L. This ensures that L itself is a fibrewise complex over B. If K is special then $\dim(K - L)$ is defined to be the maximum dimension of the cellular blocks in the decomposition of K which do not belong to L. When K is a sphere-bundle over B with section corresponding to a reduction of the structural group then the section forms a subcomplex of K assuming K is regarded as a fibrewise complex in the obvious way.

Returning to the general situation let us denote by K^n, where $n \geq 0$, the subcomplex formed by m-cellular blocks for $m \leq n$. Then $L^n = L \cap K^n$ when L is a subcomplex of K.

It would be convenient if the inclusion of a subcomplex in a fibrewise complex satisfied the condition for a fibrewise cofibration (see §20 of [8] for information about this concept). While this may not be so we can demonstrate a weaker result in this direction which is still useful.

PROPOSITION 3.1 *Let B be a complex and let X be a fibre space over B. Let K be a fibrewise complex over B and let L be a subcomplex of K. Let $f : K \to X$ be a fibrewise map, and let $g_t : L \to X$ be a fibrewise homotopy of $f \mid L$. Then there exists a fibrewise homotopy $h_t : K \to X$ of f such that $g_t = h_t \mid L$.*

Here the term *fibre space*, as distinct from fibrewise space, means that the homotopy lifting property holds. The first step in the proof of (3.1) is to establish the following.

LEMMA 3.2 *Let B be a complex and let $\theta : D^n \to B$ be a map. Let X be a fibre space over B. Let*

$$\phi : (\{0\} \times D^n \cup I \times S^{n-1}) \times T \to X$$

be a map over θ, where T is a complex. Then ϕ can be extended to a map

$$\psi : I \times D^n \times T \to X$$

over θ.

By taking adjoints we obtain from ϕ a map

$$\hat{\phi} : \{0\} \times D^n \cup I \times S^{n-1} \to \text{map}(T, X)$$

over θ, where the codomain is the space of maps with compact-open topology. Since X is a fibre space over B so is $\text{map}(T, X)$. Hence the induced fibre space $\theta^* \text{map}(T, X)$ over D^n is equivalent to the product $D^n \times \text{map}(T, X_0)$, where X_0 is the fibre of X.

Now $\hat{\phi}$ determines a section

$$s : \{0\} \times D^n \cup I \times S^{n-1} \to \theta^* \text{map}(T, X),$$

equivalently a section

$$s' : \{0\} \times D^n \cup I \times S^{n-1} \to D^n \times \text{map}(T, X_0).$$

Consider the second projection

$$s'' : \{0\} \times D^n \cup I \times S^{n-1} \to \text{map}(T, X_0).$$

Since the inclusion $S^{n-1} \to D^n$ is a cofibration we can extend s'' over $I \times D^n$. Therefore s' can be extended to a section over $I \times D^n$, and hence s can be extended to a section over $I \times D^n$. Therefore $\hat{\phi}$ can be extended to a map

$$\hat{\psi} : I \times D^n \to \text{map}(T, X)$$

over θ and finally, taking the adjoint, ϕ can be extended to a map

$$\psi : I \times D^n \times T \to X$$

over θ, as asserted.

Having established this we can now prove (3.1) in the special case where K is obtained from L by adjoining the single n-cellular block $D^n \times T$. All that needs to be done is to precompose with the characteristic map of the block, apply (3.2), and then precompose again with the inverse of the characteristic map. We may then proceed by iteration to the case where K is obtained from L by adjoining a succession of n-cellular blocks, for given n.

In the general case we make an induction on dimension as follows. Assume, for $n \geq 1$, that there exists a fibrewise homotopy $f_t^{n-1} : K^{n-1} \to X$ of $f \mid K^{n-1}$ such that $f_t^{n-1} \mid L^{n-1} = g_t \mid L^{n-1}$, as is clearly true when $n = 1$. Use the special case to extend f_t^{n-1} to a fibrewise homotopy $f_t^n : K^n \to X$ of $f \mid K^n$ such that $f_t^n \mid L^n = g_t \mid L^n$. This deals with the inductive step and so, since $K^n = K$ for sufficiently large n, proves (3.1). In particular, taking $X = \{0\} \times K \cup I \times L$ we obtain

COROLLARY 3.3 *Let B be a complex. Let K be a fibre complex over B and let L be a fibre subcomplex of K. Then the inclusion $L \to K$ is a fibrewise cofibration.*

Here we use the term *fibre complex* to mean a fibre space which is also a fibrewise complex. Next we prove

PROPOSITION 3.4 *Let B be a complex. Let K be a special fibrewise complex over B and let L be a subcomplex of K. Let X be a fibre space over B and let Y be a subspace of X which is also a fibre space over B. Suppose that the pair (X, Y) is d-connected and that $dim(K - L) \leq d$. Then any fibrewise map*

$$f : (K, L) \to (X, Y)$$

is fibrewise homotopic, relative to L, to a fibrewise map of K into Y.

The proof proceeds on similar lines to that of (3.1). The first step is to establish

LEMMA 3.5 *Let X be a fibre space over the complex B and let Y be a subspace of X which is also a fibre space over B. Suppose that the pair (X, Y) is d-connected. Let $\theta : D^n \to B$ be a map and let*

$$\phi : (D^n \times T, S^{n-1} \times T) \to (X, Y)$$

be a map over θ, where T is a complex such that $n + dimT \leq d$. Then relative to $S^{n-1} \times T$, ϕ is homotopic over θ to a map of $D^n \times T$ into Y.

The adjoint of the given map ϕ is a map

$$\hat{\phi} : (D^n, S^{n-1}) \to (\mathrm{map}(T, X), \mathrm{map}(T, Y)).$$

Now $\hat{\phi}$, like ϕ, is over $\theta : D^n \to B$.

As in the proof of (3.2) we see that $\theta^* \mathrm{map}(T, X)$ is equivalent to $D^n \times \mathrm{map}(T, X_0)$ and at the same time $\theta^* \mathrm{map}(T, Y)$ is equivalent to $D^n \times \mathrm{map}(T, Y_0)$, where X_0 and Y_0 are the fibres of X and Y, respectively. Now $\hat{\phi}$ determines a section

$$s : (D^n, S^{n-1}) \to (\theta^* \mathrm{map}(T, X), (\theta \mid S^{n-1})^* \mathrm{map}(T, Y)),$$

equivalently a section

$$s' : (D^n, S^{n-1}) \to (D^n \times \mathrm{map}(T, X_0), S^{n-1} \times \mathrm{map}(T, Y_0)).$$

Consider the second projection

$$s'' : (D^n, S^{n-1}) \to (\mathrm{map}(T, X_0), \mathrm{map}(T, Y_0))$$

of s'. By standard theory (see (7.6.13) of [9], for example) s'' is homotopic rel S^{n-1} to a map of D^n into $\mathrm{map}(T, Y_0)$, since the pair $(\mathrm{map}(T, X_0), \mathrm{map}(T, Y_0))$ is $d - dimT$-connected. Therefore s' is vertically homotopic rel S^{n-1} to a section into $S^{n-1} \times \mathrm{map}(T, Y_0)$, and hence s is vertically homotopic rel S^{n-1} to a section into $(\theta \mid S^{n-1})^* \mathrm{map}(T, Y)$. Finally $\hat{\phi}$ is homotopic, over ψ and relative to S^{n-1}, to a map of D^n into $\operatorname{map}(T, Y)$ and then, taking the adjoint, the original map ϕ is homotopic over θ and relative to $S^{n-1} \times T$, to a map of $D^n \times T$ into Y. This proves the lemma.

Having established this we can now prove the special case of (3.4) where K is obtained from L by adjoining the single n-cellular block $D^n \times T$. All that needs to be done is to precompose with the characteristic map of the block, apply (3.5), and then precompose again with the inverse of the characteristic map. We may then proceed by iteration to the case where K is obtained from L by adjoining a succession of n-cellular blocks, for given n.

In the general case we make an induction on dimension, as follows. Assume, for $n \geq 1$, that there exists a fibrewise homotopy $f_t^{n-1} : K^{n-1} \to X$ of $f \mid K^{n-1}$, relative to L^{n-1}, such that $f_1^{n-1} K^{n-1} \subset Y$, as is clearly true when $n = 1$. Use the fibrewise homotopy extension property, as in (3.1), to extend f_t^{n-1} to a fibrewise homotopy $h_t : K^n \to X$ of $f \mid K^n$ relative to L^n. Using (3.5), since $h_1 K^{n-1} \subset Y$ there exists a fibrewise homotopy $k_t : K^n \to X$ of h_1, relative to $K^{n-1} \cup L^n$, such that $k_1 K^n \subset Y$. By juxtaposition of k_t and h_t we obtain a fibrewise homotopy $f_t^n : K^n \to X$ of $f \mid K^n$, relative to L^n, such that $f_1^n K^n \subset Y$. This deals with the inductive step and so, since $K^n = K$ for sufficiently large n, proves (3.4).

4. Applications

Most of the applications of (3.4) can be derived from special cases of the following

PROPOSITION 4.1 *Let B be a complex and let K be a sectioned special fibrewise complex over B. Let $u : E \to F$ be a k-connected section-preserving fibrewise map, where E and F are sectioned fibre spaces over B. Then the induced function*

$$u_* : \pi_B^B(K, E) \to \pi_B^B(K, F)$$

is injective when $\dim K < k$, surjective when $\dim K \leq k$.

Here the notation $\pi_B^B(K, E)$ means the set of section-preserving fibrewise maps, classified by section-preserving fibrewise homotopy.

By replacing F by the fibrewise mapping cylinder of u we may suppose, without real loss of generality that $E \subset F$. Surjectivity in (4.1) follows at once from (3.4), applied to the pair (K, B), while injectivity follows from (3.4) applied to the pair $(I \times K, \{0\} \times K)$. Of course there is a relative version of this result, proved in the same way.

Having reached this stage we can now improve a number of results in the literature by replacing assumptions that fibrewise spaces are complexes, which is contrary to the spirit of fibrewise homotopy theory, by assumptions that they are special fibrewise complexes. Here we give just one illustration of this out of many possibilities, the fibrewise Freudenthal theorem.

Proofs of this fundamental result, under somewhat different hypotheses, have been given by Becker [1] and myself [6]. However (4.1) enables us to prove the result in the following form.

PROPOSITION 4.2 *Let B be a complex and let K be a sectioned special fibrewise complex over B. Let E be a sectioned fibre space over B with $(m-1)$-connected fibre. Then the fibrewise suspension*

$$\Sigma_* : \pi_B^B(K, E) \to \pi_B^B(\Sigma_B^B K, \Sigma_B^B E)$$

is injective for $dim K < 2m - 1$, surjective for $dim K \leq 2m - 1$.

Here Σ_B^B denotes the reduced fibrewise suspension. To deduce (4.2) from (4.1) we note that $\Sigma_B^B E$ is a fibre space over B, by (6.37) of [6], since E is a fibre space over B, and so the fibrewise loop-space $\Omega_B^B \Sigma_B^B E$ is a fibre space over B, by (6.32) of [7]. Since the fibre of E is $(m-1)$-connected the classical Freudenthal suspension theorem shows that the adjoint

$$u : E \to \Omega_B^B \Sigma_B^B E$$

of the identity is $(2m - 1)$-connected, and so (4.2) follows at once from (4.1). Proceeding in the same way as in [6] we deduce

COROLLARY 4.3 *Let B be a complex. Let K be a sectioned k-sphere bundle and let L be a sectioned l-sphere bundle over B. Then for each sectioned sphere-bundle N over B the fibrewise smash product*

$$N_\# : \pi_B^B(K, L) \to \pi_B^B(N \wedge_B K, N \wedge_B L)$$

is injective when $dim B < 2l - k - 1$, surjective when $dim B \leq 2l - k - 1$.

References

1. J. C. Becker, On the existence of A_k-structures on stable vector bundles, Topology 9 (1970), 367–384.

2. G. Bredon, Introduction to Compact Transformation Groups. Academic Press 1972.

3. T. tom Dieck, Transformation Groups. de Gruyter 1987.

4. T. tom Dieck, K. H. Kamps and D. Puppe, Homotopietheorie. Lecture Notes in Math. no. 57, Springer Verlag 1970.

5. B. Eckmann and P. J. Hilton, Homotopy groups of maps and exact sequences, Comment. Math. Helv. 34 (1960), 271–304.

6. I. M. James, Ex-homotopy theory I, Illinois J. Math. 15 (1971), 324–337.

7. I. M. James, General Topology and Homotopy Theory. Springer Verlag 1985.

8. I. M. James, Fibrewise Topology. Cambridge Univ. Press 1990.

9. E. Spanier, Algebraic Topology. McGraw-Hill 1966.

10. J. H. C. Whitehead, Combinatorial homotopy, Bull. Amer. Math. Soc. 55 (1949), 213–245.

Mathematical Institute, Oxford

Progress in Mathematics, Vol. 136
© 1996 Birkhäuser Verlag Basel/Switzerland

A note on the thick subcategory theorem

Alain Jeanneret, Peter S. Landweber* and Douglas C. Ravenel [†]

1. Introduction

In this paper we will discuss an algebraic version (Theorem 1.6) of the thick subcategory theorem of Hopkins-Smith [HS] (Theorem 1.4). The former is stated as Theorem 3.4.2 in [Rav92], but the proof given there is incorrect. (A list of errata for [Rav92] can be obtained by e-mail from the third author.)

First we recall the nilpotence theorem in its p-local version. Let BP be the Brown-Peterson spectrum at the prime p, which satisfies:

$$\pi_*(BP) \cong BP_* \cong \mathbf{Z}_{(p)}[v_1, v_2, \cdots], \quad |v_i| = 2(p^i - 1).$$

Theorem 1.1 (Nilpotence theorem) *[DHS88]*

   (i) Let R be a p-local ring spectrum. The kernel of the BP Hurewicz homomorphism $BP_* : \pi_*(R) \longrightarrow BP_*(R)$ consists of nilpotent elements.

   (ii) Let $f : F \longrightarrow X$ be a map from a p-local finite spectrum to an arbitrary spectrum. If $BP \wedge f$ is null homotopic, then f is smash nilpotent; i.e. the i-fold smash product $f^{(i)} = f \wedge \cdots \wedge f$ is null for i sufficiently large.

   (iii) Let $\cdots \longrightarrow X_n \xrightarrow{f_n} X_{n+1} \xrightarrow{f_{n+1}} X_{n+2} \longrightarrow \cdots$ be a direct system of p-local spectra with X_n c_n-connected. Suppose that $c_n \geq mn + b$ for some m and b. If $BP_* f_n = 0$ for all n then $\mathrm{hocolim} X_n$ is contractible.

The Baas-Sullivan theory of bordism with singularities allows one to define ring spectra $K(n)$ and $P(n)$ for $0 < n < \infty$ satisfying [Rav86]:

$$\pi_*(K(n)) \cong K(n)_* \cong \mathbf{F}_p[v_n, v_n^{-1}]$$

$$\pi_*(P(n)) \cong P(n)_* \cong \mathbf{F}_p[v_n, v_{n+1}, \cdots]$$

as BP_*-algebras. We also set $P(0) = BP$ and $K(0) = H\mathbf{Q}$, the rational Eilenberg-Mac Lane spectrum. $K(n)$ is known as the nth Morava K-theory at the prime p. The following corollary of the nilpotence theorem will be proved in §2. This is stated in [Rav92] as Corollary 5.1.5, but again the proof given there is incorrect.

---

*Partially supported by the National Science Foundation
[†]Partially supported by the National Science Foundation

COROLLARY 1.2 *Let W, X and Y be p-local finite spectra and $f : X \longrightarrow Y$ a map. Then $W \wedge f^{(k)}$ is null homotopic for $k \gg 0$ if $K(n)_*(W \wedge f) = 0$ for all $n \geq 0$.*

Now let $\mathcal{CP}_0$ be the homotopy category of finite p-local spectra and let $\mathcal{CP}_n \subset \mathcal{CP}_0$ be the full subcategory of $K(n-1)_*$-acyclics. In [Rav84] it was shown that the $\mathcal{CP}_n$ fit into a sequence:

$$\cdots \subset \mathcal{CP}_{n+1} \subset \mathcal{CP}_n \subset \cdots \subset \mathcal{CP}_0.$$

Moreover all the inclusions are strict [Mit85].

DEFINITION 1.3 *A full subcategory $\mathcal{C}$ of $\mathcal{CP}_0$ is* **thick** *if:*

(i) *An object weakly equivalent to an object in $\mathcal{C}$ is in $\mathcal{C}$.*

(ii) *If $X \longrightarrow Y \longrightarrow Z$ is a cofibration in $\mathcal{CP}_0$ and two of $\{X, Y, Z\}$ are in $\mathcal{C}$ then so is the third.*

(iii) *A retract of an object in $\mathcal{C}$ is in $\mathcal{C}$.*

Corollary 1.2 is the form of the nilpotence theorem needed to prove the thick subcategory theorem (see §5.3 of [Rav92]):

THEOREM 1.4 (THICK SUBCATEGORY THEOREM) *If $\mathcal{C}$ is a thick subcategory of $\mathcal{CP}_0$, then there exists an integer k such that $\mathcal{C} = \mathcal{CP}_k$.*

Before we state an algebraic version of Theorem 1.4 let us fix some notation. Let $\mathcal{BP}_0$ be the abelian category of $BP_*(BP)$-comodules finitely presented as BP_*-module [Lan76]. A typical object in $\mathcal{BP}_0$ is $BP_*(X)$ for X in $\mathcal{CP}_0$. We denote by $\mathcal{BP}_k$ the full subcategory of $\mathcal{BP}_0$ whose objects M satisfy $v_{k-1}^{-1}M = 0$ (we set $v_0 = p$). Results of Johnson-Yosimura [JY80] (see also [Lan79] for a more algebraic proof) show that:

$$\cdots \subset \mathcal{BP}_{k+1} \subset \mathcal{BP}_k \subset \cdots \subset \mathcal{BP}_0.$$

DEFINITION 1.5 *Let $\mathcal{A}$ be an abelian category. A full subcategory $\mathcal{C}$ of $\mathcal{A}$ is* **thick** *if it satisfies the following condition:*
If

$$0 \longrightarrow M' \longrightarrow M \longrightarrow M'' \longrightarrow 0$$

is a short exact sequence in $\mathcal{A}$, M belongs to $\mathcal{C}$ if and only if M' and M'' belong to $\mathcal{C}$. (This means that $\mathcal{C}$ is stable under subobjects, quotient objects and extensions.)

The classification of the thick subcategories of $\mathcal{BP}_0$ is now the following; see §3 for the proof.

THEOREM 1.6 (ALGEBRAIC THICK SUBCATEGORY THEOREM) *If $\mathcal{C}$ is a thick subcategory of $\mathcal{BP}_0$, then there exists an integer k such that $\mathcal{C} = \mathcal{BP}_k$.*

Two remarks are in order.

- Theorem 3.4.2 of [Rav92] is the analog of Theorem 1.6 stated in a different category, $C\Gamma$, which is defined in terms of MU rather than BP.

- The BP-homology functor, $BP_*(\cdot) : \mathcal{CP}_0 \longrightarrow \mathcal{BP}_0$ sends the category $\mathcal{CP}_k$ into $\mathcal{BP}_k$. This comes from the fact [Rav84] that if $X \in \mathcal{CP}_0$ then

$$K(n)_*(X) = 0 \iff v_n^{-1}BP_*(X) = 0.$$

Theorem 1.6 can be generalized to the abelian category of $P(n)_*(P(n))$-comodules, finitely presented over $P(n)_*$, which we denote by $\mathcal{P}(n)$. Similarly as for $\mathcal{BP}_0$ we can define the subcategories $\mathcal{P}(n)_k$ and prove the following.

THEOREM 1.7 *If $\mathcal{C}$ is a thick subcategory of $\mathcal{P}(n)$, then there exists an integer $k \geq n$ such that $\mathcal{C} = \mathcal{P}(n)_k$.*

A further generalization of Theorem 1.6 can be obtained in the following setting. Let E_* be a commutative $P(n)_*$-algebra such that $E_* \otimes_{P(n)_*} -$ is an exact functor on $\mathcal{P}(n)$. In [Lan76] the second author gave sufficient conditions for exactness. (The necessity of these conditions was shown by Rudyak in [Rud86].) Define

$$E_*(E) = E_* \otimes_{P(n)_*} P(n)_*(P(n)) \otimes_{P(n)_*} E_*,$$

which can be made into a Hopf algebroid by extending the structure maps for $P(n)_*(P(n))$. Moreover $E_*(E)$ is a flat E_*-module because $P(n)_*(P(n))$ is a flat $P(n)_*$-module and if N is a E_*-module then

$$E_*(E) \otimes_{E_*} N \cong E_* \otimes_{P(n)_*} \left(P(n)_*(P(n)) \otimes_{P(n)_*} N \right).$$

If M is an object of $\mathcal{P}(n)$ then $E_* \otimes_{P(n)_*} M$ is an $E_*(E)$-comodule via the E_*-extension of the composite:

$$M \rightarrow P(n)_*(P(n)) \otimes_{P(n)_*} M \rightarrow E_*(E) \otimes_{P(n)_*} M \rightarrow E_*(E) \otimes_{E_*} \left(E_* \otimes_{P(n)_*} M \right).$$

Let $\mathcal{E}$ be the category whose objects are $E_* \otimes_{P(n)_*} M$ with $M \in \mathcal{P}(n)$ and morphisms $E_* \otimes f : E_* \otimes M_1 \longrightarrow E_* \otimes M_2$ with $f : M_1 \longrightarrow M_2$ in $\mathcal{P}(n)$; then $\mathcal{E}$ is an abelian category equipped with an exact functor:

$$E_* \otimes_{P(n)_*} - : \mathcal{P}(n) \longrightarrow \mathcal{E}.$$

The image of the subcategory $\mathcal{P}(n)_k$, written $\mathcal{E}_k$, satisfies:

$$\cdots \subset \mathcal{E}_{k+1} \subset \mathcal{E}_k \subset \cdots \subset \mathcal{E}_n = \mathcal{E}.$$

We are no longer claiming that the inclusions are strict. The thick subcategories of $\mathcal{E}$ can be described as follows:

THEOREM 1.8 *If C is a thick subcategory of $\mathcal{E}$, then there exists an integer $k \geq n$ such that $C = \mathcal{E}_k$.*

It should be emphasized that under the above assumption on E_*, the functor $E_* \otimes_{P(n)_*} P(n)_*(\cdot)$ is a homology theory [Lan76] taking its values in the category $\mathcal{E}$ as far as finite spectra are concerned.

2. The proof of Corollary 1.2

Let D : $CP_0 \longrightarrow CP_0$ be the anti-equivalence induced by the Spanier-Whitehead duality [Ada74]. If $X \in CP_0$ and Y is any spectrum, the graded group $[X, Y]_*$ is isomorphic to $\pi_*(DX \wedge Y)$. We say that maps $f : \Sigma^n X \longrightarrow Y$ and $\widehat{f} : S^n \longrightarrow DX \wedge Y$ are adjoint if they correspond to each other under the above isomorphism of groups. In particular the adjoint of the identity $X \longrightarrow X$ is a map $e : S^0 \longrightarrow DX \wedge X$. Recall that $X^{(i)}$ is a notation for the i-fold smash product $X \wedge \cdots \wedge X$.

Set $R = DW \wedge W$, a ring spectrum whose unit is e and whose multiplication is the composite

$$R \wedge R = DW \wedge W \wedge DW \wedge W \xrightarrow{DW \wedge De \wedge W} DW \wedge S^0 \wedge W = R.$$

The map $f : X \longrightarrow Y$ is adjoint to $\widehat{f} : S^0 \longrightarrow DX \wedge Y$ and $W \wedge f$ is adjoint to the composite

$$S^0 \xrightarrow{\widehat{f}} DX \wedge Y \xrightarrow{e \wedge DX \wedge Y} R \wedge DX \wedge Y,$$

which we denote by g. Set $F = R \wedge DX \wedge Y$. The map $W \wedge f^{(i)}$ is adjoint to the composite

$$S^0 \xrightarrow{g^{(i)}} F^{(i)} = R^{(i)} \wedge DX^{(i)} \wedge Y^{(i)} \longrightarrow R \wedge DX^{(i)} \wedge Y^{(i)},$$

the latter map being induced by the multiplication in R.

We want to show that $W \wedge f^{(k)}$ is null for large k; by adjointness it suffices to prove that $g^{(k)}$ is null for large k. The second statement of Theorem 1.1 implies that we only need to show that $BP \wedge g^{(i)}$ is null for large i, so we can take k to be an appropriate multiple of i. Let $T_i = R \wedge DX^{(i)} \wedge Y^{(i)}$ and let T be the direct limit of

$$S^0 \xrightarrow{g} T_1 \xrightarrow{T_1 \wedge \widehat{f}} T_2 \xrightarrow{T_2 \wedge \widehat{f}} T_3 \longrightarrow \cdots .$$

The desired conclusion will follow from showing that $BP \wedge T$ is contractible.

At this point we need to use the theory of Bousfield classes. Recall that the Bousfield class of a spectrum X (denoted $\langle X \rangle$) is the collection of spectra Z for which $X \wedge Z$ is not contractible. In [Rav84] it was shown that

$$\langle BP \rangle = \langle K(0) \rangle \vee \langle K(1) \rangle \vee \cdots \vee \langle K(n) \rangle \vee \langle P(n+1) \rangle.$$

By assumption, $K(n) \wedge T$ is contractible for all n. Therefore it suffices to show that $P(m) \wedge T$ is contractible for large m.

Since we are concerned only with finite spectra, we have for large enough m:

$$K(m)_*(W \wedge f) = K(m)_* \otimes_{\mathbf{F}_p} H_*(W \wedge f; \mathbf{F}_p)$$
$$P(m)_*(W \wedge f) = P(m)_* \otimes_{\mathbf{F}_p} H_*(W \wedge f; \mathbf{F}_p).$$

Our hypothesis implies that both of these homomorphisms are trivial, so the smash product $P(m) \wedge T$ is contractible as required.

3.   The proof of Theorem 1.6

The proof of Theorem 1.6 is a consequence of the filtration theorem of the second author, namely

THEOREM 3.1 *[Lan73] Each object* $M \in \mathcal{BP}_0$ *has a filtration*

$$0 = M_s \subset \cdots \subset M_1 \subset M_0 = M$$

in the category $\mathcal{BP}_0$*, so that for* $0 \le i \le s - 1$ *the quotient* M_i/M_{i+1} *is stably isomorphic to* BP_*/I_{n_i} *in* $\mathcal{BP}_0$*, where* $I_{n_i} = (p, v_1, \cdots, v_{n_i-1})$ *are invariant prime ideals of* BP_**. (Stably isomorphic means isomorphic after a dimension shift.)*

For $M \in \mathcal{BP}_0$ define $\mathrm{Spec}(M) = \{m \ge 1 : v_{m-1}^{-1}M = 0\} \bigcup \{0\}$ (set as usual $v_0 = p$). If $M \ne 0$ then $\mathrm{Spec}(M)$ is a finite subset of $\mathbf{N}$ and is of the form:

$$\mathrm{Spec}(M) = \{0, 1, \cdots, N_M\}$$

with $N_M \ge 0$.

Let $\mathcal{C}$ be a thick subcategory of $\mathcal{BP}_0$. Define an integer k by:

$$\bigcap_{M \in \mathcal{C}} \mathrm{Spec}(M) = \{0, 1, \cdots, k\}.$$

From the definition of k, one has $\mathcal{C} \subset \mathcal{BP}_k$ and $\mathcal{C} \not\subset \mathcal{BP}_{k+1}$. Let M in $\mathcal{C}$ be such that

$$v_{k-1}^{-1}M = 0 \text{ and } v_k^{-1}M \ne 0,$$

and let

$$0 = M_s \subset \cdots \subset M_1 \subset M_0 = M$$

be a Landweber filtration of M. As $\mathcal{C}$ is thick and $M \in \mathcal{C}$, all the M_i's belong to $\mathcal{C}$ as well as all the quotients $M_i/M_{i+1} \cong BP_*/I_{n_i}$.

Localization being an exact functor, all the $v_{k-1}^{-1}M_i$ are null and hence $v_{k-1}^{-1}M_i/M_{i+1} \cong v_{k-1}^{-1}BP_*/I_{n_i} = 0$. Therefore

$$n_i \ge k \text{ for } 0 \le i \le s - 1. \tag{3.2}$$

On the other hand, $v_k^{-1}M \neq 0$ implies the existence of a j for which $v_k^{-1}BP_*/I_{n_j} \neq 0$, which forces

$$n_j \leq k \text{ for some } j, \ 0 \leq j \leq s-1. \tag{3.3}$$

From (3.2) and (3.3) we obtain that $n_j = k$ for some j, $0 \leq j \leq s-1$, hence $BP_*/I_k \in \mathcal{C}$. Now it is fairly easy to prove by induction that $BP_*/I_{k+l} \in \mathcal{C}$ for all $l \geq 0$. Consider the exact sequence in $\mathcal{BP}_0$

$$0 \longrightarrow BP_*/I_{k+l} \xrightarrow{v_{k+l}} BP_*/I_{k+l} \longrightarrow BP_*/I_{k+l+1} \longrightarrow 0$$

where the first morphism is multiplication by v_{k+l}. The subcategory $\mathcal{C}$ being thick, $BP_*/I_{k+l} \in \mathcal{C}$ implies $BP_*/I_{k+l+1} \in \mathcal{C}$.

We are now ready to show the inclusion $\mathcal{BP}_k \subset \mathcal{C}$. Let N be an object in $\mathcal{BP}_k$ and $0 = N_s \subset \cdots \subset N_1 \subset N_0 = N$ be a Landweber filtration of N. We have seen that $v_{k-1}^{-1}N = 0$ implies $n_i \geq k$ for all $0 \leq i \leq s-1$ with, as usual, n_i such that $N_i/N_{i+1} \cong BP_*/I_{n_i}$. By downward induction on i we prove that $N_i \in \mathcal{C}$. This works as follows.

First $N_s = 0 \in \mathcal{C}$. Second, the short exact sequence in $\mathcal{BP}_0$

$$0 \longrightarrow N_{i+1} \longrightarrow N_i \longrightarrow BP_*/I_{n_i} \longrightarrow 0$$

is such that $N_{i+1} \in \mathcal{C}$ (by the inductive assumption) and $BP_*/I_{n_i} \in \mathcal{C}$ as $n_i \geq k$. From the thickness of $\mathcal{C}$ we obtain that $N_i \in \mathcal{C}$. For $i = 0$ we have $N \in \mathcal{C}$ and so $\mathcal{BP}_k = \mathcal{C}$, as required.

References

[Ada74] J. F. Adams. *Stable Homotopy and Generalised Homology.* University of Chicago Press, Chicago, 1974.

[DHS88] E. Devinatz, M. J. Hopkins, and J. H. Smith. Nilpotence and stable homotopy theory. *Annals of Mathematics*, 128:207–242, 1988.

[HS] M. J. Hopkins and J. H. Smith. Nilpotence and stable homotopy theory II. To appear in *Annals of Mathematics*.

[JY80] D. C. Johnson and Z. Yosimura. Torsion in Brown-Peterson homology and Hurewicz homomorphisms. *Osaka Journal of Mathematics*, 17:117–136, 1980.

[Lan73] P. S. Landweber. Associated prime ideals and Hopf algebras. *Journal of Pure and Applied Algebra*, 3:175–179, 1973.

[Lan76] P. S. Landweber. Homological properties of comodules over $MU_*(MU)$ and $BP_*(BP)$. *American Journal of Mathematics*, 98:591–610, 1976.

[Lan79] P. S. Landweber. New applications of commutative algebra to Brown-Peterson homology. *Algebraic Topology, Waterloo 1978, Lecture Notes in Mathematics* 741, pages 449–460, Springer-Verlag, New York, 1979.

[Mit85] S. A. Mitchell. Finite complexes with $A(n)$-free cohomology. *Topology*, 24:227–248, 1985.

[Rav84] D. C. Ravenel. Localization with respect to certain periodic homology theories. *American Journal of Mathematics*, 106:351–414, 1984.

[Rav86] D. C. Ravenel. *Complex Cobordism and Stable Homotopy Groups of Spheres.* Academic Press, New York, 1986.

[Rav92] D. C. Ravenel. *Nilpotence and periodicity in stable homotopy theory.* Volume 128 of *Annals of Mathematics Studies*, Princeton University Press, Princeton, 1992.

[Rud86] Yu. B. Rudyak. Exactness theorems for the cohomology theories MU, BP and $P(n)$. *Mat. Zametki*, 40:115–126, 1986. English translation in *Math. Notes* 40:562–569, 1986.

Alain Jeanneret, University of Bern, CH-3012 Bern
jeanner@math-stat.unibe.ch

Peter S. Landweber, Rutgers University
New Brunswick, New Jersey 08903 USA
landwebe@lagrange.rutgers.edu

Douglas C. Ravenel, University of Rochester
Rochester, New York 14627 USA
drav@troi.cc.rochester.edu

Progress in Mathematics, Vol. 136
© 1996 Birkhäuser Verlag Basel/Switzerland

The Morava K-theory Hopf Ring for BP

TAKUJI KASHIWABARA, NEIL STRICKLAND AND PAUL TURNER

1. Introduction

Let K be a p-local complex-oriented homology theory. The K-homology of the even spaces in the Ω-spectrum for BP form a Hopf ring. In [6] Ravenel and Wilson chararacterise this Hopf ring by a purely algebraic universal property, and also prove that the K-homology of each component of each even space is polynomial under the star product. The star-indecomposables in this Hopf ring form an algebra under the circle product.

In this paper we take K to be 2-periodic Morava K-theory, and study the resulting ring R of indecomposables. In propositions 2.2 and 2.3 we give an algebraic universal property which characterises R, and relate this to a better-known description of the stable ring $K_*(BP)$. In theorem 2.11 we nearly provide a splitting of R as a product of indecomposable factors, each of which is isomorphic modulo nilpotents to $K_*(BP)$. In the case $n = 1$, there are no nilpotents and R is the subring of an infinite product of copies of $K_*(BP)$ defined by a certain asymptotic condition; this is proved as theorem 3.3. We give a very simple description of the Dyer-Lashof operation on R in these terms.

In section 4 we again take $n = 1$, replace BP by an arbitrary (-1)-connected spectrum X with only even cells, and obtain some partial results in the same spirit.

It would clearly be desirable to generalise those of our results which are restricted to the case $n = 1$. This will probably be possible using the results of [8] and [2]. However, the statements and proofs will be considerably more elaborate.

All spectra and spaces in this paper will be assumed to be localised at a prime p.

2. Results for Arbitrary Height

First we establish some notation. Fix an integer $n > 0$. We write K for the 2-periodic Morava K-theory spectrum of height n:

$$K = \bigvee_{k=0}^{p^n-2} \Sigma^{2k} K(n) = K(n)[u]/(u^{p^n-1} - v_n)$$

We use the multiplication on $K(n)$ to define a multiplication on K in the obvious way. Thus $K_* = \mathbb{F}_p[u^{\pm 1}]$ as rings. Note that if $n = 1$ then K is just mod p complex K-theory, $K = KU/p$. In sections 3 and 4 we shall specialise to this case.

If X is a spectrum, $\mathbf{X}_i$ will denote the i-th infinite loop space associated to it. If E_* is a homology theory we write

$$\sigma \colon E_m(\mathbf{X}_j) \longrightarrow E_{m+1}(\mathbf{X}_{j+1})$$

for the homology suspension and

$$\sigma^\infty \colon E_m(\mathbf{X}_j) \longrightarrow E_{m-j}(X)$$

for the stabilisation map.

We let y^K denote the usual p-typical orientation class in $K^0(\mathbb{C}P^\infty)$, which gives rise to the p-typical formal group law $F_0(s,t)$ over $\mathbb{F}_p$ with $[p]_{F_0}(s) = s^{p^n}$. We also write $x^K = u^{-1}y^K \in K^2(\mathbb{C}P^\infty)$. This is a strict orientation, in the sense that it restricts to 1 in $K^2(\mathbb{C}P^1) \simeq K^2(S^2) \simeq K^0$.

We let x^{BP} be the usual strict orientation class in $BP^2(\mathbb{C}P^\infty)$, with associated formal group law

$$F(s,t) = \sum_{k,l} a_{kl} s^k t^l.$$

We let $\beta_i \in K_0(\mathbb{C}P^\infty)$ be dual to $(y^K)^i$ and set

$$b_i = x_*^{BP}(\beta_i) \in K_0(\mathbf{BP}_2)$$

$$b(s) = \sum_{i \geq 0} b_i s^i \in K_0(\mathbf{BP}_2)[\![s]\!].$$

For $a \in BP^{2m}(\text{point}) = [S^0, \mathbf{BP}_{-2m}]$, we define

$$[a] = a_*(1) \in K_0(\mathbf{BP}_{-2m}).$$

It is shown in [6] that $K_*(\mathbf{BP}_{2m})$ is concentrated in even degrees, which implies that

$$K_0(\mathbf{BP}_{2m} \times \mathbf{BP}_{2l}) = K_0(\mathbf{BP}_{2m}) \otimes K_0(\mathbf{BP}_{2l})$$

and so on. It follows that the groups $K_0(\mathbf{BP}_{2*})$ form a (singly graded) Hopf ring over $\mathbb{F}_p$. After taking note of the slight difference between our notation and that of [6], we conclude from their results that $K_0(\mathbf{BP}_{2*})$ is the free Hopf ring over $\mathbb{F}_p[BP_*]$ generated by the coalgebra $K_0(\mathbb{C}P^\infty) = \mathbb{F}_p\{b_k \mid k \geq 0\}$ subject to the relations

$$b_0 = [0_2]$$

$$b(s +_{F_0} t) = b(s) +_{[F]} b(t) = \prod_{k,l} [a_{kl}] \circ b(s)^{\circ k} \circ b(t)^{\circ l}$$

Our main object of study will be the ring of indecomposables in this Hopf ring. We shall write Ind instead of the traditional Q for indecomposables, to avoid proliferation of Q's.

DEFINITION 2.1.

$$R_{2*} = \text{Ind}(K_0(\mathbf{BP}_{-2*})) = QK_0(\mathbf{BP}_{-2*})$$

We have changed the sign of the grading to make it compatible with stabilisation. We consider R as a ring with multiplication induced by the circle product in $K_0(\mathbf{BP}_{-2*})$. From now on, identities take place in R (and not in $K_0(\mathbf{BP}_{-2*})$ itself) unless explicitly stated otherwise.

We would like to deduce relations in R from the Ravenel-Wilson relations stated above. First, we consider some generalities. Let A_* be a Hopf ring over $\mathbb{F}_p$ (with grading as for $K_0(\mathbf{BP}_{2*})$). We write

$$\mathrm{Group}(A)_k = \{ \text{ grouplike elements } \} = \{a \in A_k \mid \psi(a) = a \otimes a \text{ and } \epsilon(a) = 1\}$$

$$\mathrm{Ind}(A)_k = \{ \text{ indecomposables in degree } k \}$$

It is not hard to see that $\mathrm{Group}(A)_*$ is a graded ring, with addition given by the star product and multiplication by the circle product. On the other hand, $\mathrm{Ind}(A)_*$ is a graded ring with addition given by ordinary addition and multiplication by the circle product. There is a ring homomorphism $\mathrm{Group}(A) \longrightarrow \mathrm{Ind}(A)$ given by

$$a \mapsto a - [0_k] \qquad a \in \mathrm{Group}(A)_k$$

More generally, we can consider $A[\![s]\!]$ as a Hopf ring over $\mathbb{F}_p[\![s]\!]$. This has to be interpreted in a completed sense — the coproduct is a map

$$\psi \colon A[\![s]\!] \longrightarrow (A \otimes A)[\![s]\!] = A[\![s]\!] \widehat{\otimes}_{\mathbb{F}_p[\![s]\!]} A[\![s]\!]$$

This gives us a ring homomorphism

$$\mathrm{Group}(A[\![s]\!]) \longrightarrow \mathrm{Ind}(A[\![s]\!]) = \mathrm{Ind}(A)[\![s]\!]$$

Similar things apply to $A[\![s,t]\!]$ and so on, of course.

We now take $A_* = K_0(\mathbf{BP}_*)$. For $a \in BP_{2m}$ we have $[a] \in \mathrm{Group}(A)_{-2m}$ and thus $[a] - [0_{2m}] \in \mathrm{Ind}(A)_{-2m} = R_{2m}$. We write $\eta_R(a)$ or just a for this element. We will always consider R as an algebra over BP_* via η_R. We also have $b(s) \in \mathrm{Group}(A[\![s]\!])_2$ and thus

$$\bar{b}(s) = b(s) - [0_2] = \sum_{k>0} b_k s^k \in R[\![s]\!]_{-2}$$

The Ravenel-Wilson relation can be read as a relation in $\mathrm{Group}(A[\![s,t]\!])$. By mapping this ring to $R[\![s,t]\!]_*$, we get

$$\bar{b}(s +_{F_0} t) = \bar{b}(s) +_F \bar{b}(t) \qquad (1)$$

This just means that $\bar{b}(s)$ is a homomorphism of formal group laws $\bar{b} \colon F_0 \longrightarrow F$. The p-series version of this is

$$\bar{b}(s^{p^n}) = [p]_F(\bar{b}(s)) = \sum_{k>0}^{F} v_k \bar{b}(s)^{p^k} \qquad (2)$$

Because the Ravenel-Wilson relation is essentially the only relation in $K_0(\mathbf{BP}_{2*})$, it is not hard to conclude that the above is the only relation in R. Thus, R is generated over BP_*/p by elements b_k for $k > 0$ subject only to the relations got by expanding $\bar{b}(s +_{F_0} t) = \bar{b}(s) +_F \bar{b}(t)$. The following proposition is an immediate consequence:

PROPOSITION 2.2. $(R, \bar{b})$ is the universal example of a BP_*/p-algebra equipped with a homomorphism $\bar{b} \colon F_0 \longrightarrow F$ defined over R.   □

We now write $S_* = K_*(BP)$, so that the stabilisation map σ^∞ is a ring map $R \longrightarrow S$. Note that the composite

$$BP_*/p \xrightarrow{\eta_R} R \xrightarrow{\sigma^\infty} S$$

is just the usual right unit map.

We would like to compare our description of R with a similar description of the stable ring S.

PROPOSITION 2.3. The stabilisation map σ^∞ induces an isomorphism $R[b_1^{-1}] \simeq S$, and thus $(S, \sigma^\infty(\bar{b}))$ is the universal example of an algebra over BP_*/p equipped with an isomorphism $F_0 \longrightarrow F$.

Proof. It is well known (almost by definition) that

$$K_*(BP) = \mathrm{colim}(\mathrm{Ind}(K_*(\mathbf{BP}_{2*})) \xrightarrow{\sigma^2} \mathrm{Ind}(K_{*+2}(\mathbf{BP}_{2*+2})) \xrightarrow{\sigma^2} \dots)$$

To express this more precisely, we write $R_{2k}^{2l} = \mathrm{Ind}(K_{2l}(\mathbf{BP}_{-2k}))$, so that $R_{2k}^0 = R_{2k}$. The following diagram exhibits $S_{2(k+l)} = K_{2(k+l)}(BP)$ as the colimit of the top line

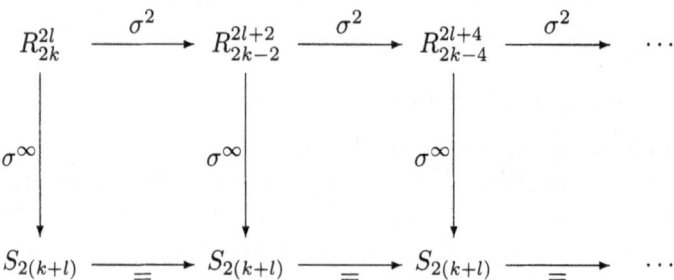

As u is invertible, the following is a colimit diagram in the same sense.

Because $u^{-1}y^K$ is a strict orientation, we have $\sigma^2(x) = u b_1 x$ for $x \in R$ (c.f. [6, Proposition 2.4]). Thus, the horizontal maps in the top line of the above diagram are just multiplication by b_1. It follows immediately that $S = R[b_1^{-1}]$ and that $\sigma^\infty(b_1) = u^{-1}$. Next, observe that a map $f \colon F' \longrightarrow F$ of formal group laws is iso (i.e. invertible under composition) iff $f'(0)$ is a unit. It follows easily that S has the claimed universal property.   □

REMARK 2.4. A description which may be more familiar is that S is the universal example of an algebra over $K_* \otimes BP_* = BP_*[u^{\pm 1}]/p$ equipped with a *strict* isomorphism $c \colon F_0' \longrightarrow F$, where $F_0'(s,t) = u^{-1}F_0(us, ut)$. This is related to the description above by $\sigma^\infty \bar{b}(t) = c(t/u)$. It is not hard to show that the two descriptions are equivalent.

A natural thing to try to do now is to factor a morphism $F_0 \longrightarrow F$ as the projection to the quotient by a subgroup, followed by an isomorphism. However, the theory of subgroups of formal groups is subtle — we shall return to this in later work [8]. For the moment, we shall construct a splitting of R motivated by the above idea, without explaining the connection in detail. Later in this paper, we will take $n = 1$. In this case, the theory of subgroups is easy and we will be able to give a more precise answer.

DEFINITION 2.5. $J_m = (b_1, b_p, \dots, b_{p^m - 1}) \lhd R$

LEMMA 2.6.

$$J_m = (b_i \mid p^m \text{ does not divide } i)$$

Proof. If $f(x)$ is any homomorphism of formal group laws over an $\mathbb{F}_p$-algebra, then there is an integer $n \geq 0$ (the *height* of f) and a power series g such that $f(x) = g(x^{p^n})$ and $g'(0) \neq 0$ (see [5, Appendix A]). We know that $\bar{b}$ is a homomorphism from F_0 to F. Let π be the projection $R \longrightarrow R/J_m$. The homomorphism $\bar{b}$ induces a homomorphism $\pi_* \bar{b} \colon \pi_*(F_0) \longrightarrow \pi_*(F)$. Arguing by induction on k we see that $\pi_* \bar{b}$ has height at least k for each $k \leq m$, in other words that $\pi_* \bar{b}$ has height at least m. This means that $\pi_* \bar{b}(x) = \bar{b}'(x^{p^m})$ for some $\bar{b}'$. The lemma follows easily. $\qquad\square$

We now write $b_{(m)} = b_{p^m}$ and $e_m' = v_n b_{(m)}^{p^n - 1}$.

LEMMA 2.7. e_m' is idempotent modulo J_m. Moreover $b_{(m)}$ is nilpotent modulo (J_m, e_m') and invertible modulo $(J_m, 1 - e_m')$.

Proof. We know that

$$\bar{b}(s^{p^n}) = [p]_F(\bar{b}(s)) = \sum_{k>0}^{F} v_k \bar{b}(s)^{p^k}$$

We work mod J_m, so that

$$\bar{b}(s) = b_{(m)} s^{p^m} + \text{higher terms}$$

For the rest of the proof, we write $b = b_{(m)}$. The lowest order terms on either side of the above relation are

$$b s^{p^{n+m}} + \dots = v_1 b^p s^{p^{m+1}} + \dots$$

If $n = 1$ we have $b = v_1 b^p$. If $n > 1$ we have $v_1 b^p = 0$. If we work mod $\operatorname{ann}(b^p)$ we find that $v_1 = 0$ so the relation gives

$$b s^{p^{n+m}} + \dots = v_2 b^{p^2} s^{p^{m+2}} + \dots$$

If $n = 2$ we have $b = v_2 b^{p^2} \pmod{\mathrm{ann}(b^p)}$, so that $b^{1+p} = v_2 b^{p+p^2}$. If $n > 2$ we have $v_1 = v_2 = 0 \pmod{\mathrm{ann}(b^{p+p^2})}$, and we proceed in the same way. We conclude that for arbitrary n

$$b^{1+p+\dots+p^{n-1}} = v_n b^{p+\dots+p^n} = b^{1+p+\dots+p^{n-1}} v_n b^{p^n-1}$$

Because $p^n - 1 \geq 1 + p + \dots + p^{n-1}$, we conclude that

$$v_n b^{p^n-1} v_n b^{p^n-1} = v_n b^{p^n-1}$$

as claimed. The second statement in the lemma is clear.　　　　$\square$

We next consider the Verschiebung map of the Hopf ring $K_0(\mathbf{BP}_{2*})$. Let us briefly recall the definition. For each m, the ring $A = K_0(\mathbf{BP}_{2m})$ is a bicommutative Hopf algebra over $\mathbb{F}_p$. Write ψ_p for the p-fold coproduct map

$$\psi_p \colon A \longrightarrow (A^{\otimes p})^{\Sigma_p}$$

Let N denote the norm map

$$N \colon (A^{\otimes p})_{\Sigma_p} \longrightarrow (A^{\otimes p})^{\Sigma_p}$$

$$N(a) = \sum_{\sigma \in \Sigma_p} \sigma(a)$$

We also consider the map

$$D \colon A \longrightarrow \mathrm{cok}(N) \qquad a \mapsto a \otimes \dots \otimes a$$

It can be shown that D is iso. The Verschiebung can be defined as the composite

$$A \xrightarrow{\psi_p} (A^{\otimes p})^{\Sigma_p} \longrightarrow \mathrm{cok}(N) \xrightarrow{D^{-1}} A$$

If A were finite-dimensional then this would just be dual to the map $F \colon a \mapsto a^p$ in A^*. We could give a similar description in the general case, but we would have to consider topological vector spaces and continuous duals.

We refer to [7] for formal properties of the Verschiebung map. Here we shall simply note that V is a homomorphism of Hopf rings. From the definition, one can see that

$$V([a]) = [a]$$

$$V(b_k) = \begin{cases} b_{k/p} & \text{if } p \mid k \\ 0 & \text{otherwise} \end{cases}$$

PROPOSITION 2.8. V^m induces isomorphisms $R/J_m \simeq R$ and $R/J_m[b_{(m)}^{-1}] \simeq R[b_1^{-1}] \simeq S$.

Proof. It is visible from the formulae above that $V^m J_m = 0$, and that V is epi, so that V^m induces an epimorphism $R/J_m \longrightarrow R$. In $R/J_m[\![s]\!]$ we have $\bar{b}(s) = \bar{b}'(s^{p^m})$ for some $\bar{b}'$. Because F_0 is defined over $\mathbb{F}_p$, we have

$$(s +_{F_0} t)^{p^m} = s^{p^m} +_{F_0} t^{p^m}$$

It follows that $\bar{b}'$ is also a homomorphism $F_0 \longrightarrow F$. The universal property of R thus gives a map $V' \colon R \longrightarrow R/J_m$ of BP_*-algebras, sending $\bar{b}(s)$ to $\bar{b}'(s)$. It is not hard to show that $V'V^m = 1_{R/J_m}$, and thus that $V^m \colon R/J_m \longrightarrow R$

is mono. This gives the first statement of the proposition. As $V^m b_{(m)} = b_1$, we see that V also induces an isomorphism $R/J_m[b_{(m)}^{-1}] \simeq R[b_1^{-1}]$. This is also isomorphic to S by proposition 2.3. $\square$

We next record some standard facts from commutative algebra.

LEMMA 2.9. *For any ring A and any idempotent $e \in A$ we have*

$$A \simeq A/e \times A/(1-e)$$

$$A/e = A.(1-e) = A[(1-e)^{-1}] = \{x \in A \mid ex = 0\}$$
$$A/(1-e) = A.e = A[e^{-1}] = \{x \in A \mid ex = x\} \quad \square$$

LEMMA 2.10. *If A is a ring and $J \lhd A$ is a nilpotent ideal then $x \in A$ is nilpotent (resp. invertible) iff the same is true of its image in A/J. Moreover, for any idempotent in A/J there is a unique idempotent in A which lifts it.*

Proof. This is easy, except for the lifting of idempotents. That is also well-known, but seems not to appear in the standard references on commutative algebra. We learned the following simple proof from Andrew Ranicki. Suppose $e \in A$ is idempotent mod J. Write $f = 1 - e$, so $ef \in J$ is nilpotent, say $(ef)^n = 0$. Note that

$$e^n + f^n = e + f = 1 \pmod{J}$$

Thus $e^n + f^n$ is invertible mod J, hence invertible in A. Take $e' = e^n/(e^n + f^n)$. This is an idempotent lift of e. If e'' is another such lift, then $e'(1 - e'')$ and $e''(1 - e')$ are both idempotents lying in the nilpotent ideal J, hence are zero. It follows that $e' = e'e'' = e''$. $\square$

THEOREM 2.11. *There are unique idempotents $e_k \in R$ such that for each $m \geq 0$,*

1. $R = R[e_0^{-1}] \times \ldots \times R[e_m^{-1}] \times R/(e_0, \ldots, e_m)$
2. J_m *is nilpotent and $b_{(m)}$ is invertible in $R[e_m^{-1}]$.*
3. J_{m+1} *is nilpotent in $R/(e_0, \ldots, e_m)$.*

Each factor $R[e_k^{-1}]$ is isomorphic to S modulo nilpotents, and thus is indecomposable.

Proof. Suppose we have constructed e_k for $k < m$, with properties as above. Then J_m is nilpotent mod $I_m = (e_0, \ldots, e_{m-1})$. Lemma 2.7 gives us an idempotent $e'_m \in R/J_m$. By the lemma 2.10, there is a unique idempotent $e_m \in R/I_m$ which agrees with e'_m mod J_m. We consider e_m as an idempotent in R via the identification

$$R/I_m = \{x \in R \mid xe_0 = \ldots = xe_{m-1} = 0\}$$

By lemma 2.7 and lemma 2.10 we find that $b_{(m)}$ is nilpotent in $R/(I_m, e_m)$ and invertible in $(R/I_m)[e_m^{-1}]$. By lemma 2.6, we see that J_{m+1} is nilpotent mod $(I_m, e_m) = (e_0, \ldots, e_m) = I_{m+1}$. As $e_k e_m = 0$ for $k < m$ we see that $(R/I)[e_m^{-1}] = R[e_m^{-1}]$. By lemma 2.9 we have

$$R/I_m = R/I_m[e_m^{-1}] \times R/(I_m, e_m) = R[e_m^{-1}] \times R/I_{m+1}$$

This completes the induction step except for the claim that e_m is unique. Suppose that f is another element with the required properties. The decomposition

$$R = R[e_0^{-1}] \times \ldots \times R[e_{m-1}^{-1}] \times R[f^{-1}] \times R/(e_0, \ldots, e_{m-1}, f)$$

implies that $fe_k = 0$ for $k < m$. Let us write $b = b_{(m)}$. The image of f in R/J_m is an idempotent such that b is nilpotent mod f (by property (3)). It follows that $e'_m = v_n b^{p^n - 1}$ is both idempotent and nilpotent mod f, hence is divisible by f. On the other hand, property (2) implies that b is invertible in $(R/J_m)[f^{-1}]$. As f is idempotent, this easily implies that b^k divides f for all $k \geq 0$. On the other hand, we know that b is nilpotent mod e'_m. Thus, for large k we have $e'_m | b^k | f$. As e'_m and f are idempotent and $f | e'_m | f$, we see that $f = e'_m$ in R/J_m. By the uniqueness clause in lemma 2.10, we see that $f = e_m$ in R/I_m. Using the decomposition

$$R = R[e_0^{-1}] \times \ldots \times R[e_{m-1}^{-1}] \times R/I_m$$

we see that $f = e_m$ in R as claimed.

For the last claim, observe that J_m is nilpotent in $R[e_m^{-1}]$ and

$$R[e_m^{-1}]/J_m \simeq R/J_m[(e'_m)^{-1}]$$
$$\simeq R/J_m[b_{(m)}^{-1}]$$
$$\simeq S$$

The first isomorphism follows from the construction of e_m, the second follows from lemma 2.7, and the third is proposition 2.8. On the other hand, $S = K_* BP$ is well-known to be polynomial and therefore indecomposable as claimed. □

REMARK 2.12. Note that $e_0 = e'_0$.

REMARK 2.13. In [10], Wilson constructs a non-canonical splitting of $\mathbf{BP}_{2k}$ as a product of indecomposable spaces. Except for the cases $k = (p^l - 1)/(p - 1)$, this is a splitting as H-spaces. It therefore induces a vector-space splitting of R_{-2k}. It would be interesting to know how this splitting interacts with ours.

3. The Height One Case

From now on we take $n = 1$, so that K is mod p complex K-theory.

LEMMA 3.1. The map

$$\alpha: R \longrightarrow S \times R \qquad\qquad a \mapsto (\sigma^\infty(a), Va)$$

is an isomorphism.

Proof. The idempotent $e_0 = e'_0$ gives a splitting

$$R = R[e_0^{-1}] \times R/e_0$$

We have $e_0 = v_1 b_1^{p-1}$ and $b_1 e_0 = b_1$, which implies that $R[e_0^{-1}] = R[b_1^{-1}]$, so $\sigma^\infty: R[e_0^{-1}] \simeq S$ by proposition 2.3. The same equations give $(e_0) = (b_1) = J_1$. By proposition 2.8 we know that $V: R/J_1 \longrightarrow R$ is iso. Thus

$$R = R[e_0^{-1}] \times R/e_0 = S \times R/J_1 = S \times R$$

as claimed. □

By the Hattori-Stong theorem, the right unit map

$$\eta_R \colon BP_*/p \longrightarrow S_*$$

is injective. We write T_* for the image. This can also be thought of as a subring of R, in which case we have

$$T_* = \{x \in R \mid V(x) = x\}$$

(this follows easily from the formulae for V above).

DEFINITION 3.2. Let R' be the set of sequences $\mathbf{a} = (a_0, a_1, \dots)$ of elements of S such that for $k \gg 0$ we have $a_k = a_{k+1} \in T$. We make this into a ring with coordinatewise operations, and grade it by declaring that $\mathbf{a}$ is homogeneous of degree d iff each a_i is homogeneous of degree d.

THEOREM 3.3. The following map $\tau \colon R_* \longrightarrow R'_*$ is an isomorphism of graded rings.

$$a \mapsto (\sigma^\infty(a), \sigma^\infty(Va), \dots)$$

Proof. By iterating the construction of lemma 3.1, we obtain an isomorphism

$$R \longrightarrow S^m \times R \qquad a \mapsto (\sigma^\infty(a), \sigma^\infty(Va), \dots \sigma^\infty(V^{m-1}a), V^m a)$$

It is not hard to check that this is the same as the decomposition

$$R = R[e_0^{-1}] \times \dots \times R[e_{m-1}^{-1}] \times R/(e_0, \dots, e_{m-1})$$

which we established earlier for general n. We get the desired result by passage to the limit, noting that $V^N(x) \in T_*$ for $N \gg 0$ and that $T_* = \{x \mid V(x) = x\}$. $\qquad\square$

REMARK 3.4. We do not yet understand the analogous limiting condition for $n > 1$.

Given this description, the following operation seems very natural.

DEFINITION 3.5. Define $\tilde{Q} \colon R \to R$ by identifying R with R' and setting

$$\tilde{Q}((a_0, a_1, \dots)) = (0, a_0, a_1, \dots)$$

We show that our $\tilde{Q}$ "agrees" with McClure's Dyer-Lashof operation. Given an infinite loop space X, McClure constructs a map

$$Q \colon KU_0(X; \mathbb{Z}/p^{r+1}) \to KU_0(X; \mathbb{Z}/p^r)$$

This operation is not additive. If we write

$$C(x, y) = \frac{1}{p}(x^p + y^p - (x+y)^p) \in \mathbb{Z}[x, y]$$

then

$$Q(x+y) = Q(x) + Q(y) + C(x, y)$$

It follows easily, however, that Q induces an additive operation on indecomposables. If H_*X is torsion-free then the same is true of $KU_0 X$, and

$$KU_0(X; \mathbb{Z}/p^r) = KU_0(X)/p^r$$

It follows that Q induces an additive map

$$\overline{Q} \colon \operatorname{Ind}(K_0(X)) = \operatorname{Ind}(KU_0(X))/p \longrightarrow \operatorname{Ind}(K_0(X))$$

THEOREM 3.6. $\tilde{Q} = \overline{Q}$ on $\mathrm{Ind}(K_0(\mathbf{BP}_{2*}))$.

We will need following two lemmas whose proofs are postponed until the end of the section. Let $e = 1 - e_0 = \eta_R(1) - \eta_R(v_1) \circ b_1^{\mathrm{op}-1}$.

LEMMA 3.7. $b_1 \circ e = (b_1)^{*p} \in K_0(\mathbf{BP}_2)$.

LEMMA 3.8. Let E be a generalized homology theory such that E_* is concentrated in even degrees. Suppose that X is a space such that $E_*(\Omega^2 X)$ is a polynomial algebra concentrated in even degrees. Then the homology suspension induces a monomorphism $\mathrm{Ind}E_*(\Omega^2 X) \hookrightarrow E_{*+2}(X)$.

Proof of theorem 3.6. Throughout the proof we identify R with R'. We consider the suspension map $u^{-1}\sigma^2 \colon \mathrm{Ind}(K_0(\mathbf{BP}_{2n})) \to K_0(\mathbf{BP}_{2n+2})$ for which we have seen $u^{-1}\sigma^2(xy) = b_1 \circ x \circ y$. It is easy to see that in $\mathrm{Ind}(K_0(\mathbf{BP}_{2n}))$ we have

$$e\tilde{Q}(a) = \tilde{Q}(a). \tag{3}$$

Thus, in $K_0(\mathbf{BP}_{2n+2})$, we have

$$\begin{aligned}
u^{-1}\sigma^2(\tilde{Q}(a)) &= u^{-1}\sigma^2(e\tilde{Q}(a)) \text{ by (3)} \\
&= b_1 \circ e \circ \tilde{Q}(a) \\
&= b_1^{*p} \circ \tilde{Q}(a) \text{ by Lemma 3.7} \\
&= \sum (b_1 \circ a_{i_1}) * \cdots * (b_1 \circ a_{i_p}) \text{ by the distributivity law}
\end{aligned}$$

where

$$\psi(\tilde{Q}(a)) = \sum a_{i_1} \otimes \ldots \otimes a_{i_p}.$$

Notice from the definition of Verschiebung that in $(K_0(\mathbf{BP}_{2n})^{\otimes p})_{\Sigma_p}$ we have

$$\psi(\tilde{Q}(a)) = (V(\tilde{Q}(a)))^{\otimes p}$$

Thus, the right hand side above becomes $(b_1 \circ V(\tilde{Q}(a)))^{*p}$. It is clear that in R we have $V(\tilde{Q}(a))) = a$ so we get

$$u^{-1}\sigma^2(\tilde{Q}(a)) = (b_1 \circ a)^{*p}.$$

According to [4] we also have $u^{-1}\sigma^2(\overline{Q}(a)) = (b_1 \circ a)^{*p}$ so we arrive at

$$\sigma^2(\tilde{Q}(a)) = \sigma^2(\overline{Q}(a)).$$

Since $K_*(\mathbf{BP}_{2n})$ is a polynomial algebra by [9], we can apply Lemma 3.8 to conclude that $\tilde{Q}(a) = \overline{Q}(a)$ in R. □

Now we go back to the proof of the lemmas.

Proof of lemma 3.7. Although this follows from the main relation of [6], we give a proof since it is quite simple. We evaluate the effect of the map induced by the composition:

$$\mathbb{C}P^\infty \xrightarrow{p} \mathbb{C}P^\infty \xrightarrow{x^{BP}} \mathbf{BP}_2$$

and obtain:

$$
\begin{aligned}
(b_1)^{*p} + \eta_R(v_1) \circ (b_1)^{\circ p} &= (px + v_1 x^p)_*(\beta_p) \\
&= (px + v_1 x^p + \cdots)_*(\beta_p) \\
&= ([p]_{BP}(x))_*(\beta_p) \\
&= x_*(p_*(\beta_p)) \\
&= x_*(\beta_1) \\
&= b_1
\end{aligned}
$$

Since $e = \eta_R(1) - \eta_R(v_1) \circ b_1^{\circ p-1}$, we obtain the desired result. $\square$

Proof of lemma 3.8. Since $E_*(\Omega^2 X)$ is a free E_*-module by the assumption, $E_*((\Omega^2 X)^k)$ is isomorphic to $E_*(\Omega^2 X)^{\otimes k}$. Therefore the spectral sequence associated to the bar construction for $\Omega X \simeq B\Omega^2 X$ converging to $E_*(\Omega X)$ has

$$
E_2 \simeq \mathrm{Tor}^{E_*(\Omega^2 X)}(E_*, E_*)
$$

and collapses since E_2 term is generated by the elements of homological degree 1, so that

$$
E_2 \simeq E_\infty \simeq \Lambda(\sigma(\mathrm{Ind}E_*(\Omega^2 X)))
$$

There can be no extension problem, and one has

$$
\sigma \colon \mathrm{Ind}E_*(\Omega^2 X) \simeq \mathrm{Ind}E_{*+1}(\Omega X)
$$

Thus again $E_*(\Omega X)$ is a free E_*-module, so that we have a spectral sequence $\mathrm{Tor}^{E_*(\Omega X)}(E_*, E_*) \Rightarrow E_*(X)$ that collapses for degree reasons and one has

$$
E_2 \simeq E_\infty \simeq \Gamma(\sigma(\mathrm{Ind}E_*(\Omega X)))
$$

where Γ denotes the divided polynomial coalgebra, i.e., the dual of the polynomial algebra generated by the basis elements of $\mathrm{Hom}_{E_*}(\mathrm{Ind}E_*(\Omega X), E_*)$. Thus one has an inclusion

$$
\sigma \colon \mathrm{Ind}E_*(\Omega X) \hookrightarrow E_{*+1}(X)
$$

By composing, one gets the desired result. (Note that this argument also proves that $E^*(X)$ is a formal power series algebra unless E^* is bounded *from below*, in which case it is a polynomaial algebra.) $\square$

4. Spectra with Even Cells

In this section, we again take $n = 1$. Let us say that a p-local spectrum X is *even* iff it is (-1)-connected, and $H_*(X) = H_*(X; \mathbb{Z}_{(p)})$ is a free $\mathbb{Z}_{(p)}$-module concentrated in even degrees. In [1], the first author proves a number of results about $K_0(\mathbf{X}_{2*})$. Later in this section we shall derive some slightly more precise versions of these results; for the moment we simply quote from [1] the fact that $K_1(\mathbf{X}_{2k}) = 0$ for $k \geq 0$.

For such a spectrum X, and an integer $l \leq 0$, we define

$$
M(X)_{2l} = \mathrm{Ind}(K_0(\mathbf{X}_{-2l}))
$$

$$
M'(X)_{2l} = \{\mathbf{a} \in (K_{2l}(X)^{\mathbb{N}} \mid a_k = a_{k+1} \in \mathrm{image}(\pi_{2l}(X) \to K_{2l}(X)) \text{ for } k \gg 0\}
$$

Note that $M(BP) = R$ and $M'(BP) = R'$.

For any space Y with $K_1(Y) = 0$, the vector space $K_0(Y)$ is a coalgebra over $\mathbb{F}_p$, and therefore has a Verschiebung map.

LEMMA 4.1. The image of $\mathbb{F}_p[\pi_0(Y)] \longrightarrow K_0(Y)$ is $\{a \in K_0(Y) \mid Va = a\}$. Moreover, if $a \in K_0(Y)$ then $V^k a = V^{k+1} a$ for $k \gg 0$.

Proof. We may assume that Y is a CW-complex with skeleta Y^k, and that $Y^0 = \pi_0(Y)$. We filter $K^0(Y)$ by $I_k = \ker(K^0(Y) \to K^0(Y^{k-1}))$ and put $R_k = K^0(Y)/I_k \simeq \text{image}(K^0(Y) \to K^0(Y^{k-1}))$. It is well known that $I_k I_l \leq I_{k+l}$, so that the Frobenius induces a map $R_k \to R_{kp}$. Now put

$$C_k = \text{image}(K_0(Y^{k-1}) \to K_0(Y)) = R_k^*$$

By duality, we see that $V C_{kp} \leq C_k$. Moreover, $C_0 = \mathbb{F}_p[\pi_0(Y)]$ and $V = 1$ on C_0. Finally, $K_0(Y) = \bigcup_k C_k$. The lemma follows.   □

For any even spectrum X we can define a map

$$\tau \colon M(X)_{2l} \longrightarrow M'(X)_{2l}$$

$$\tau(a) = (\sigma^\infty(a), \sigma^\infty(Va), \sigma^\infty(V^2 a), \dots)$$

The lemma assures us that $\tau(a) \in M'(A)$ as claimed. It follows directly from proposition 3.3 that τ is iso if X is a wedge of spectra of the form $\Sigma^{2k} BP$.

PROPOSITION 4.2. For $l \leq 0$, τ gives an isomorphism

$$\tau \colon M(X)_{2l} \longrightarrow M'(X)_{2l}$$

We will prove this after some preliminary remarks.

First, we observe that an even spectrum has an essentially functorial skeletal filtration. To be more precise, there is a $2k$-dimensional even spectrum X^{2k} with a map $X^{2k} \to X$ such that $H_*(Y) \to H_*(X)$ is iso up to degree $2k$. Given a map $X \to Y$ of even spectra, and skeleta $X^{2k} \to X$ and $Y^{2k} \to Y$, there is a unique (up to homotopy) compatible map $X^{2k} \to Y^{2k}$. In particular, by considering the identity map of X we see that X^{2k} is determined up to canonical isomorphism. All this follows from elementary obstruction theory. Similarly, X/X^{2k} and X^{2k}/X^{2k-2} are functors of X.

Next, consider two even spectra X and Y, and put $Z = X \vee Y$. Then $\mathbf{Z}_k = \mathbf{X}_k \times \mathbf{Y}_k$ and $K_1(\mathbf{X}_k) = 0 = K_1(\mathbf{Y}_k)$, so

$$K_0(\mathbf{Z}_{2k}) = K_0(\mathbf{X}_{2k}) \otimes K_0(\mathbf{Y}_{2k})$$

$$\text{Ind}(K_0(\mathbf{Z}_{2k})) = \text{Ind}(K_0(\mathbf{X}_{2k})) \oplus \text{Ind}(K_0(\mathbf{Y}_{2k}))$$

$$M(Z) = M(X) \oplus M(Y)$$

In the usual way, this implies that $M \colon [X, Y] \to \text{Hom}(M(X), M(Y))$ is additive (see [3, p193]).

Now consider the map

$$\pi_{2k}(X^{2k}/X^{2k-2}) \times M(S^{2k}) \longrightarrow M(X^{2k}/X^{2k-2}) \qquad (u, a) \mapsto M(u)(a)$$

By the previous remark, this is bilinear. Moreover, $\pi_{2k}(X^{2k}/X^{2k-2})$ is canonically isomorphic to $H_{2k}(X)$. We thus obtain a map

$$H_{2k}(X) \otimes M(S^{2k}) \longrightarrow M(X^{2k}/X^{2k-2})$$

The first author showed in [1] that for $k > 0$ the ring $K_0(\mathbf{X}_{2k})$ is polynomial, and there is a short exact sequence

$$0 \to M(X^{2k-2}) \longrightarrow M(X^{2k}) \longrightarrow M(X^{2k}/X^{2k-2}) \to 0$$

Because we use p-local spectra, $K_0(\mathbf{X}_0)$ may have tensor factors of the form $\mathbb{F}_p[\mathbb{Z}_{(p)}]$. It is not hard to see that $\mathrm{Ind}(\mathbb{F}_p[\mathbb{Z}_{(p)}]) \simeq \mathbb{F}_p$. If we read the proof of proposition 7.1 of [1] in the light of the obove remarks, we find the following:

PROPOSITION 4.3. If X is an even spectrum, then $M(X)$ has a natural filtration, with quotients canonically isomorphic to $\mathrm{H}_{2k}(X) \otimes M(S^{2k})$. $\square$

LEMMA 4.4. If $X \to Y \to Z$ is a cofibration of even spectra then $K_*(X) \to K_*(Y) \to K_*(Z)$ and $M(X) \to M(Y) \to M(Z)$ are short exact.

Proof. From the long exact sequences we see that $\mathrm{H}_*(X) \to \mathrm{H}_*(Y) \to \mathrm{H}_*(Z)$ and $K_*(X) \to K_*(Y) \to K_*(Z)$ are short exact. The sequence $M(X) \to M(Y) \to M(Z)$ is compatible with the filtrations considered above. The associated graded sequence has terms

$$\mathrm{H}_{2k}(X) \otimes M(S^{2k}) \to \mathrm{H}_{2k}(Y) \otimes M(S^{2k}) \to \mathrm{H}_{2k}(Z) \otimes M(S^{2k})$$

It is thus short exact. It follows that $M(X) \to M(Y) \to M(Z)$ is itself short exact. $\square$

We next construct two sequences to which this applies. First, we have the cofibration

$$S \xrightarrow{\eta} BP \xrightarrow{\pi} \overline{BP}$$

Elementary obstruction theory shows that there is a unique map $\alpha \colon \overline{BP} \to BP \wedge BP$ with $\alpha\pi = \eta \wedge 1 - 1 \wedge \eta$, and one can calculate directly that $\mathrm{H}_*(\alpha)$ is a split monomorphism. We write $\beta \colon BP \wedge BP \to Z$ for the cofibre. After smashing these cofibrations with X and applying the lemma we find that the following sequences are exact:

$$0 \to M(X) \longrightarrow M(BP \wedge X) \longrightarrow M(BP \wedge BP \wedge X)$$

$$0 \to K_*(X) \longrightarrow K_*(BP \wedge X) \longrightarrow K_*(BP \wedge BP \wedge X)$$

We next consider the subsequence of the second sequence given by the image of the Hurewicz homomorphism $\pi_* \to K_*$. All the spectra are (-1)-connected, so this is zero in negative degrees, and in degree zero it agrees with

$$0 \to \mathrm{H}_*(X; \mathbb{F}_p) \to \mathrm{H}_*(BP \wedge X; \mathbb{F}_p) \to \mathrm{H}_*(BP \wedge BP \wedge X; \mathbb{F}_p)$$

This is again easily seen to be exact. It follows that the sequence

$$0 \to M'(X)_* \longrightarrow M'(BP \wedge X)_* \longrightarrow M'(BP \wedge BP \wedge X)_*$$

is exact in nonpositive degrees. Finally, we consider the diagram

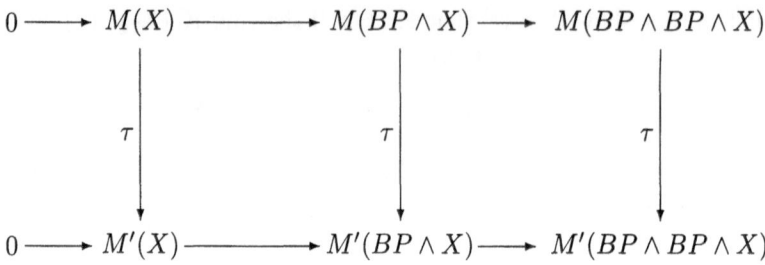

The spectra $BP \wedge X$ and $BP \wedge BP \wedge X$ are wedges of suspended BP's, so the corresponding τ's are iso. In nonpositive degrees both rows are exact, so that $\tau \colon M(X) \to M'(X)$ is also iso. This proves proposition 4.2.   □

References

[1] T. Kashiwabara. On $K(2)$-homology of some infinite loop spaces. To appear in Mathematische Zeitschrift.

[2] T. Kashiwabara. Brown-Peterson cohomology of $\Omega^\infty \Sigma^\infty S^{2n}$. Preprint, 1994.

[3] S. MacLane. *Categories for the Working Mathematician*, volume 5 of *Graduate Texts in Mathematics*. Springer Verlag, 1971.

[4] J. E. McClure. Mod p K-theory of QX. In H_∞ *Ring Spectra and their Applications*, volume 1176 of *Lecture Notes In Mathematics*, pages 291–377. Springer Verlag, 1986.

[5] D. C. Ravenel. *Complex Cobordism and Stable Homotopy Groups of Spheres*. Academic Press, 1986.

[6] D. C. Ravenel and W. S. Wilson. The Hopf ring for complex cobordism. *Journal of Pure and Applied Algebra*, 9:241–280, 1977.

[7] D. C. Ravenel and W. S. Wilson. The Morava K-theories of Eilenberg-MacLane spaces and the Conner-Floyd conjecture. *American Journal of Mathematics*, 102(4):691–748, 1980.

[8] N. P. Strickland. Finite subgroups of formal groups. In preparation, 1994.

[9] W. S. Wilson. The Ω-spectrum for Brown-Peterson cohomology, part I. *Commentarii Mathematici Helvetici*, 48:45–55, 1973.

[10] W. S. Wilson. The Ω-spectrum for Brown-Peterson cohomology, part II. *American Journal of Mathematics*, 97:101–123, 1975.

Takuji Kashiwabara, Laboratoire J. Dieudonné, U.A. au C.N.R.S.N°. 168, Université de Nice Sophia-Antipolis, Parc Valrose, 06034, CEDEX, France
      and
Kyoto University, Department of Mathematics, Kyoto 606 Japan
tk@kusm.kyoto-u.ac.jp
(a J.S.P.S. research fellow, supported by the Ministry of Education of Japan)

Neil Strickland, Department of Mathematics, Massachusetts Institute of Technology, Cambridge MA, 02139 U.S.A.
neil@math.mit.edu
(partially supported by the N.S.F.)

Paul Turner, Mathematisches Institut der Universität
Im Neuenheimer Feld 288, 69120 Heidelberg, Germany
pault@vogon.mathi.uni-heidelberg.de
(supported by the Alexander von Humboldt Stiftung)

Progress in Mathematics, Vol. 136
© 1996 Birkhäuser Verlag Basel/Switzerland

Periodicity of Geometric Dimension for Real Projective Spaces

KEE YUEN LAM AND DUANE RANDALL

ABSTRACT. We investigate periodicity for the geometric dimensions of certain stable vector bundles over RP^n. Sufficient conditions are obtained for periodicity of type 2^L where $L = 1 + [\log_2 n]$. We determine both the geometric dimensions of the bundles $(64s + 32)\xi_{64t+j}$ and the maximum number of desuspensions of their Thom complexes for $0 \leq j \leq 31$ and $s \geq t \geq 1$ with $\binom{s}{t}$ odd.

1 Introduction and Preliminaries

Let $m\xi_n$ denote the m-fold Whitney sum of the Hopf line bundle ξ_n based on real projective n-space RP^n. The geometric dimension of $m\xi_n$, denoted $g.\dim(m\xi_n)$, is the minimum fiber dimension of all vector bundles stably equivalent to $m\xi_n$. The calculation of $g.\dim(m\xi_n)$ is an important and difficult problem with an extensive literature. We extend some results of [11] by determining $g.\dim(64s + 32)\xi_{64t+j}$ for $s \geq t \geq 0$ whenever the binomial coefficient $\binom{s}{t}$ is odd. We introduce a periodicity property for the geometric dimensions of stable vector bundles over real projective spaces. The geometric dimension of $m\xi_n$ satisfies periodicity of type 2^L for some positive integer L whenever certain Whitney sums of $m\xi_n$ with positive multiples of $2^L\xi_n$ possess the same geometric dimension as $m\xi_n$. Existence theorems for nonsingular bilinear maps and numerous sectioning results in the literature are applied to establish sufficient conditions for $g.\dim(m\xi_n)$ to satisfy periodicity of type 2^L with $L = 1 + [\log_2 n]$.

We let $S(m,n)$ denote the maximum number of linearly independent sections of $m\xi_n$. Clearly $S(m,n) + g.\dim(m\xi_n) = m$. We recall that $\widetilde{KO}(RP^n)$ is a cyclic group of order $2^{\sigma(n)}$ where $\sigma(n)$ is the number of positive integers $i \leq n$ such that i is congruent to 0, 1, 2 or 4 mod 8. The elements of $\widetilde{KO}(RP^n)$ are represented by $m\xi_n$ stably for $1 \leq m \leq 2^{\sigma(n)}$. We let $H : RP^m \to QP^n$ denote the Hopf map for $4n \leq m \leq 4n + 3$ so that $H^*\gamma_n = 4\xi_m$ for the quaternionic line bundle γ_n over QP^n. The Thom complex $T(m\xi_n)$ of $m\xi_n$ is homeomorphic to the stunted projective space $RP_m^{m+n} = RP^{m+n}/RP^{m-1}$. The coefficient group for cohomology is understood to be $Z/2$, whenever omitted. Generators for $H^1(RP^n)$ and $H^4(QP^n; Z)$ are denoted by α and δ respectively. The 2-adic valuation $\nu(m)$ of the positive integer m is the exponent of the largest

*)   1991 Subject Classification: 55S40, 55S45, 55S20, 57T20.

*)   Key Words: Geometric dimension, Postnikov resolution, bilinear map

*)   The first author is partially supported by an NSERC grant of Canada.

power of 2 which divides m. Let $\alpha(m)$ denote the number of non-zero entries in the dyadic expansion of the positive integer m. The 2-adic valuation of the binomial coefficient $\binom{m+n}{n}$ is given by $\nu\binom{m+n}{n} = \alpha(n) + \alpha(m) - \alpha(m+n)$. Let $[t]$ denote the greatest integer less than or equal to the real number t.

We let $V_k(R^m)$ denote the Stiefel manifold of orthonormal k-frames in R^m. The antipodal involution on $V_k(R^m)$ sends a k-frame $(u_1, \ldots, u_k)$ to $(-u_1, \ldots, -u_k)$ and thus extends the antipodal involution on $S^m = V_1(R^{m+1})$. A map $f : S^n \circ\!\!\longrightarrow V_k(R^m)$ *equivariant* with respect to the antipodal involutions corresponds to a unique section of the bundle of orthonormal k-frames of $m\xi_n$, and vice versa. Berrick defined a certain subquotient of $\pi_n(V_k(R^m))$, a vector space over $Z/2$, such that the restriction to S^{n-1} of any given equivariant map $f : S^n \circ\!\!\longrightarrow V_k(R^m)$ admits an equivariant extension $g : S^{n+1} \circ\!\!\longrightarrow V_k(R^m)$ if and only if $[f]$ has trivial image in this subquotient by Lemma 6.5 of [2]. For example, this subquotient is given by

$$\pi_n(V_k(R^m))/2\pi_n(V_k(R^m)) \tag{1.1}$$

for n odd and mk even. A fundamental property of equivariant maps formulated first in [6] and used extensively in [9] and [11] affirms that any equivariant map

$$f : S^n \circ\!\!\longrightarrow V_k(R^m) \quad \text{yields an } \textit{equivariant} \text{ map} \tag{1.2}$$
$$g : S^{k-1} \circ\!\!\longrightarrow V_{n+1}(R^m) \quad \text{if } k + 2 \leq 2(m - n) .$$

We utilize a fundamental nonsectioning theorem formulated by Don Davis in [4]. Indeed, Davis extended this nonsectioning result to a nondesuspension theorem for the corresponding Thom complexes in Theorem 1.10 of [4]. The hypothesis involving the binomial coefficient $\binom{m-1}{n}$ can be interpreted as an equal height condition on the 1-dimensional cohomology classes for the map $g : RP^n \to PV_k(R^m)$ induced from an equivariant map $S^n \circ\!\!\longrightarrow V_k(R^m)$ on the double coverings.

THEOREM 1.3 ([3], [4], [5]). Suppose the binomial coefficient $\binom{m-1}{n}$ is odd. Then $g.\dim(m\xi_n) \geq n - 2\nu(m) + \varepsilon(\nu(m), n)$. Here $\varepsilon(\nu(m), n)$ depends only on the mod 4 values of $\nu(m)$ and is given by the following table.

		n mod 4			
		0	1	2	3
$\nu(m)$ mod 4	0	0	1	1	0
	1	1	1	2	1
	2	2	1	2	1
	3	2	1	0	-1

$$\tag{1.4}$$

Equivalently, the maximum number of linearly independent sections in $m\xi_n$ is less than or equal to $m - n + 2\nu(m) - \varepsilon(\nu(m), n)$. Moreover, the maximum number of desuspensions of the Thom complex $T(m\xi_n) = RP^{m+n}_m$ is less than or equal to $m - n + 2\nu(m) - \varepsilon(\nu(m), n)$.

We assume familiarity with modified Postnikov resolutions as employed in [9], [11] and [13]. The calculations of k-invariants require only the evaluation of secondary cohomology operations, defining relations and indeterminacy arguments.

2 Periodicity

We now define the notion of periodicity for the geometric dimension of $m\xi_n$. The case $m \leq n$ must be excluded; however, $g.\dim(m\xi_n) = m$ for $m \leq n$.

DEFINITION 2.1. Let m and n be any integers satisfying $2 \leq n < m < 2^{\sigma(n)}$. The geometric dimension of $m\xi_n$ satisfies periodicity of type 2^L for some positive integer L whenever $m + 2^L < 2^{\sigma(n)}$ and

$$g.\dim(k\xi_n) = g.\dim(m\xi_n) \tag{2.2}$$

for all integers k with $m < k < 2^{\sigma(n)}$ and $k \equiv m \bmod 2^L$.

In other words, periodicity of type 2^L affirms that the bundles $m\xi_n$, $(m + 2^L)\xi_n, \ldots, (m + r2^L)\xi_n$ possess the same geometric dimension where $2^{\sigma(n)} - m - 2^L \leq r2^L < 2^{\sigma(n)} - m$. An immediate consequence of the definition is the fact that $g.\dim(m+2^L)\xi_n$ also satisfies periodicity of type 2^L whenever $g.\dim(m\xi_n)$ satisfies periodicity of type 2^L and $2^{L+1} < 2^{\sigma(n)} - m$. We shall exhibit numerous examples of periodicity of type 2^L for the geometric dimension of $m\xi_n$ where $L = 1 + [\log_2 n]$. This value appears to be minimal, in general, for the periodicity property. For example, we choose values of m and n satisfying $n < m < 2^{\sigma(n)} - n$ and $g.\dim(m\xi_n) < 2^c$ for $c = [\log_2 n]$. Then the geometric dimension of $m\xi_n$ cannot satisfy periodicity of type 2^L for any $L \leq c$, since $g.\dim(m+2^c)\xi_n \geq 2^c$ due to the non-trivial Stiefel-Whitney class of $(m+2^c)\xi_n$ in dimension 2^c. Not every $m\xi_n$ possesses periodicity of type 2^L with respect to geometric dimension. For example, $g.\dim(32\xi_{10}) = 5$ by [11] while $g.\dim(34\xi_{10}) = g.\dim(36\xi_{10}) = 6$, $g.\dim(40\xi_{10}) = 8$ and $g.\dim(48\xi_{10}) = 7$. Thus $g.\dim(32\xi_{10})$ does not possess periodicity of type 2^L for any positive integer L. On the other hand, numerous examples of periodicity for the geometric dimensions of bundles over low-dimensional projective spaces are given by Theorem 3.2 of [11]. For example, $g.\dim(16\xi_{15}) = 7$ satisfies periodicity of type 2^5, and $g.\dim(32\xi_{15}) = 6$ satisfies periodicity of type 2^6.

Interesting examples of periodicity for geometric dimension are provided by the work of Davis, Gitler and Mahowald in [5]. Let e and n be integers satisfying $e \geq 75$ and $n \geq 4e + 16[\log_2(e + 4)] + 42$. The authors of [5] prove that the geometric dimension of any stable vector bundle over RP^n of order 2^e in $\widetilde{KO}(RP^n)$ is given by $2e + \delta(n, e)$ where $-2 \leq \delta(n, e) \leq 2$. We deduce immediately that $g.\dim(2^{\sigma(n)-e}\xi_n)$ satisfies periodicity of type 2^L for $L = \sigma(n) - e + 1$. We also note that the Thom complexes $T(k\xi_n)$ and $T(m\xi_n)$ in (2.2) do not have the same stable homotopy type, in general. Whenever $S(k, n)$ and $S(m, n)$ give the maximum number of desuspensions of the Thom complexes $T(k\xi_n)$ and $T(m\xi_n)$, the periodicity for geometric dimension in (2.2) means that the Thom classes desuspend maximally to classes of the same dimension for all the bundles in (2.2). However, Davis produced bundles $m\xi_n$ in [4] for which $S(m,n)$ is strictly less than the maximum number of desuspensions of $T(m\xi_n)$, the binomial coefficients $\binom{m-1}{n}$ being even integers for these examples.

PROPOSITION 2.3.   Let $m = 8\ell + p$ and $n = 8r + q$ where $\ell > r \geq 1$, $\binom{\ell}{r}$ is odd, and $0 \leq p$, $q \leq 7$. We assume that $m + 2^L < 2^{\sigma(n)}$ where $L = 1 + [\log_2 n]$. Then $g.\dim(m\xi_n)$ satisfies periodicity of type 2^L.

Proof. Now $m + 2^L = 8(\ell + 2^{L-3}) + p$ and $\binom{\ell + 2^{L-3}}{r}$ is also odd. Theorem 1.1 of [9] affirms that

$$S(m,n) = m - n + j \quad \text{and} \quad S(m + 2^L, n) = (m + 2^L) - n + j \qquad (2.4)$$

where j depends only on p and q, and is given by Table I of [9]. We conclude that $g.\dim(m + 2^L)\xi_n = g.\dim(m\xi_n)$. Repetition of this argument gives $g.\dim(m + t2^L)\xi_n = g.\dim(m\xi_n)$ for all $t \geq 1$ such that $m + t2^L < 2^{\sigma(n)}$. $\qquad \square$

PROPOSITION 2.5.   Let $m = 8\ell + p$ and $n = 8r + q$ where $\ell > r > 1$, $\nu\binom{\ell}{r} = 1$, $0 \leq p \leq 6$, and $1 \leq q \leq 7$. We assume that $m + 2^L < 2^{\sigma(n)}$ where $L = 2 + [\log_2 n]$. Then $g.\dim(m\xi_n)$ satisfies periodicity of type 2^L.

Proof. For every $t \geq 1$ such that $m + t2^L < 2^{\sigma(n)}$, we write $m + t2^L = 8(\ell + t2^{L-3}) + p$ and note that $\nu\binom{\ell + t2^{L-3}}{r} = 1$ also. Theorem 3.1 of [9] affirms that $g.\dim(m + t2^L)\xi_n = n - j = g.\dim(m\xi_n)$ where j is given by Table II of [9] and depends only on the values of p and q. Remark 3.4 of [9] gives $j = 6$ for $p = 3$ and $q = 4$. We remark that $\nu\binom{\ell + 2^{c-2}}{r} > 1$ whenever $\binom{m}{2^c}$ is even for $c = [\log_2 n]$. Consequently, periodicity of type 2^{c+1} does not necessarily hold, in contrast to (2.3). $\qquad \square$

We recall an existence theorem for nonsingular bilinear mappings due to Lam [8]. Given any integers $k > h \geq 0$, we write the dyadic expansions $k = \sum_{i \geq 0} a_i 2^i$, $h = \sum_{i \geq 0} b_i 2^i$, and $k - h = \sum_{i \geq 0} c_i 2^i$ where a_i, b_i and c_i belong to $\{0,1\}$. $\tau(k,h)$ is defined to be the cardinality of $\{i \geq 0 \mid c_i = 0 \text{ and } a_i \neq b_i\}$. Theorem 5.5 of [8] affirms the existence of nonsingular bilinear mappings for every $k > h \geq 0$

$$R^{8(k-h)+\tau(k,h)} \times R^{8(h+1)} \to R^{8k} . \qquad (2.6)$$

We select any integers s, t, and r satisfying $s \geq t \geq 1$ with $\binom{s}{t}$ odd and $r \geq 3$. We set $m = s2^{r+1} + 2^r$, $n = t2^{r+1} + r + 4$, and $L = 1 + [\log_2 n]$. For any integer $i \geq 0$, we claim that (2.6) produces nonsingular bilinear mappings

$$R^{n+1} \times R^{(s-t)2^{r+1}+i2^L+2^r} \to R^{m+i2^L} . \qquad (2.7)$$

To obtain the mappings in (2.7) from (2.6), we substitute $k = s2^{r-2} + i2^{L-3} + 2^{r-3}$ and $h = (s-t)2^{r-2} + i2^{L-3} + 2^{r-3} - 1$ so that $k - h = t2^{r-2} + 1$. Now $\tau(s2^{r-2}+2^{r-3}, (s-t)2^{r-2}+2^{r-3}-1) = r-3$ since $\binom{s}{t}$ is odd. Since $L > r+1$, we conclude that $\tau(k, h) \geq r-3$. Thus $8(k-h)+\tau(k, h) \geq 8(t2^{r-2}+1)+r-3 = n+1$ so (2.7) is obtained from (2.6). The nonsingular bilinear mappings in (2.7) show that $(m + i2^L)\xi_n$ admits $(s-t)2^{r+1} + i2^L + 2^r$ independent sections. Since the Stiefel-Whitney class of $(m + i2^L)\xi_n$ in dimension $t2^{r+1}$ is given by $\binom{s}{t}\alpha^{t2^{r+1}} \neq 0$, we deduce that $g \cdot \dim(m + i2^L)\xi_n = t2^{r+1}$. Consequently, we have established the following periodicity result.

PROPOSITION 2.8.   Let s, t and r be any integers satisfying $s \geq t \geq 1$ with $\binom{s}{t}$ odd and $r \geq 3$. We set $m = s2^{r+1} + 2^r$, $n = t2^{r+1} + r + 4$, and $L = 1 + [\log_2 n]$. Whenever $m + 2^L < 2^{\sigma(n)}$, $g. \dim(m\xi_n)$ satisfies periodicity of type 2^L.

The geometric dimensions of odd multiples of $16\xi_n$ produce more examples of periodicity. Our main theorem in §3 produces analogous results for odd multiples of $32\xi_n$.

THEOREM 2.9.   We consider the bundles $(32s + 16)\xi_{32t+j}$ for $0 \leq j \leq 15$ where $s \geq t \geq 0$ with $\binom{s}{t}$ odd. The maximum number of linearly independent sections is $32(s-t) + k$ where the values of k depend only on j and are enumerated by the following list.

j	=	0	1	2	3	4	5	6	7	8	9	10	11	12	13	14	15
k	=	16	16	16	16	16	16	16	16	16	10	9	9	9	9	9	9

The maximum number of desuspensions of the Thom complex of $(32s+16)\xi_{32t+j}$ is also $32(s-t)+k$. Moreover, $g. \dim(m\xi_n)$ satisfies periodicity of type 2^L whenever $m + 2^L < 2^{\sigma(n)}$ where $m = 32s + 16$, $n = 32t + j$, and $L = 1 + [\log_2 n]$.

Proof. Both the geometric dimension and maximum desuspension results for $(32s + 16)\xi_{32t+j}$ for $0 \leq j \leq 8$ and $t > 0$ follow directly from the nonsingular bilinear maps in (2.7) (with $r = 4$ and $i = 0$) together with the non-zero Stiefel-Whitney class in dimension $32t$. We note that the bundles $(32s + 16)\xi_j$ are trivial for $j \leq 8$. The geometric dimension results for $(32s + 16)\xi_{32t+j}$ with $9 \leq j \leq 15$ follow directly from Theorem 3.1 of [9], since $\nu\left(\dfrac{4s + 2}{4t + 1}\right) = \alpha(4t + 1) + \alpha(4s - 4t + 1) - \alpha(4s + 2) = 1$. We now observe that the maximum sectioning results are also maximum desuspension results. For $j = 9$, the second-order k-invariant k_2^1 in [9, p. 101], which obstructs completely the existence of $32(s-t) + 11$ independent sections, is detected by a secondary operation on the Thom class of the bundle. But this operation would vanish for dimensional reasons if the Thom class could desuspend $32(s-t)+11$ times by [7]. The maximum desuspension result for $10 \leq j \leq 15$ follows from the result for $j = 10$ by naturality. A pair of second-order k-invariants obstructs completely the existence of $32(s-t)+10$ independent sections to $(32s+16)\xi_{32t+10}$. Let $\Phi = (\Phi_1, \Phi_2)$ denote a 2-valued stable secondary operation whose component operations are associated to the following relations on integral classes, where δ denotes the integral Bockstein operator.

$$\Phi_1 : Sq^2(\delta Sq^{32t+6}) + Sq^1 Sq^{32t+8} = 0 . \tag{2.10}$$
$$\Phi_2 : Sq^4(\delta Sq^{32t+6}) + Sq^3 Sq^{32t+8} + Sq^{32t+9} Sq^2 = 0 .$$

Now Φ is defined by the Thom class U_ρ of $\rho = (8s+4)\gamma_{8t+2}$, and $\Phi(U_\rho) = (\delta^{8t+2}, 0)$ by [13] since $c_{16t+4}(\rho) = \begin{pmatrix} 8s+4 \\ 8t+2 \end{pmatrix} \delta^{8t+2}$ and $\nu \begin{pmatrix} 8s+4 \\ 8t+2 \end{pmatrix} = 1$. By naturality, $(\alpha^{32(s+t)+24}, 0) \in \Phi(\alpha^{32s+16})$ while the indeterminacy is generated by $(\alpha^{32(s+t)+24}, \alpha^{32(s+t)+26})$. On the other hand, $(0, e \cdot Sq^4 e) \in \Phi(e)$ by [7] where $e = \pi^* \iota$ and $\pi : E \to K(Z, 32t+6)$ is the principal fibration classified by $Sq^2\iota : K(Z, 32t+6) \to K(Z/2, 32t+8)$. If the Thom class α^{32s+16} in $H^* \left(RP_{32s+16}^{32(s+t)+26} \right)$ could desuspend $32(s-t)+10$ times to a class u, then

$$(0,0) = \sum^{32(s-t)+10} (0, u \cdot Sq^4 u) \in \Phi(\alpha^{32s+16}) .$$

However, $\Phi(\alpha^{32s+16})$ must be non-trivial so the maximum sectioning values coincide with the maximum desuspensions of the corresponding Thom complexes. Finally, periodicity for geometric dimension is an immediate consequence. Since $S(m + 2^L, n) = 32(s + 2^{L-5} - t) + k$, we deduce that $g.\dim(m + 2^L)\xi_n = g.\dim(m\xi_n)$. For $j \le 8$, the periodicity property is given by the bilinear maps in (2.7). $\qquad\square$

3 Main Theorem

We determined $g.\dim(64s + 32)\xi_{64s+j}$ for $0 \le j \le 31$ and all $s \ge 0$ in [11]. Moreover, $g.\dim(64s + 32)\xi_j$ was essentially determined in [11] also for $0 \le j \le 31$. The values of $g.\dim(64s + 32)\xi_j$ for $17 \le j \le 31$ follow by applications of (1.2) to results of [11]. For example, there is an equivariant map $S^{17} \circ\!\!\longrightarrow V_{64s+18}(R^{64s+32})$ if and only if there is an equivariant map $S^{64s+17} \circ\!\!\longrightarrow V_{18}(R^{64s+32})$ for every $s \ge 0$ by (1.2). We showed that $g.\dim(64s+32)\xi_j = 5$ for $10 \le j \le 13$ and $s \ge 0$ in section 5 of [11]. Theorem 3.2 of [11] together with (1.3) gives $g.\dim(64s+32)\xi_j = 6$ for $14 \le j \le 15$. Of course, $(64s + 32)\xi_j$ is trivial for $j \le 9$. Numerous results on bundles with low geometric dimension are contained in [1] and [10]. The methods of proof in [11] rely on modified Postnikov towers and knowledge of generators for the 2-primary homotopy of Stiefel manifolds from [12]. We apply here the bilinear maps constructed in [8]. The main result includes the case $t = s$ established in [11], but omits the case $t = 0$ due to the irregular pattern of $g.\dim(64s+32)\xi_j$ for $10 \le j \le 13$.

THEOREM 3.1. We consider the bundles $(64s + 32)\xi_{64t+j}$ for $0 \le j \le 31$ where $s \ge t \ge 1$ with $\binom{s}{t}$ odd. The maximum number of linearly independent sections in $(64s + 32)\xi_{64t+j}$ is given by $64(s-t) + k$ where k depends only on j and is given by the following list.

j	$=$	0	1	2	3	4	5	6	7	8	9	10	1 1	12	13	14	15
k	$=$	32	32	32	32	32	32	32	32	32	32	30	30	29	28	26	26

j	$=$	16	17	18	19	20	21	22	23	24	25	26	27	28	29	30	31
k	$=$	25	18	17	17	17	17	17	17	17	16	14	14	13	12	10	10

The maximum number of desuspensions of the Thom complex of $(64s+32)\xi_{64t+j}$ is exactly $64(s-t) + k$ also. We set $m = 64s+32$, $n = 64t+j$, and $L = 1 + [\log_2 n]$. The $g.\dim(m\xi_n)$ satisfies periodicity of type 2^L whenever $m + 2^L < 2^{\sigma(n)}$.

Proof. The sectioning results for $0 \le j \le 9$ and for $30 \le j \le 31$ follow directly from the nonsingular bilinear maps in (2.7) for $r = 5$ and from Theorem 1.3. The nontriviality of the Stiefel-Whitney class of $(64s + 32)\xi_{64t+j}$ in dimension $64t$ together with Theorem 1.3 shows that the Thom complex of $(64s+32)\xi_{64t+j}$ desuspends maximally $64(s-t) + k$ times for $0 \le j \le 9$ and for $30 \le j \le 31$. Periodicity of geometric dimension is established geometrically through the bilinear maps in (2.7) with $r = 5$ for $0 \le j \le 9$, also a consequence of (2.8). Periodicity for $g.\dim(64s + 32)\xi_{64t+j}$ is a consequence of $S(64s+32+i2^L, 64t+j) = 64(s-t)+10+i2^L$ for $30 \le j \le 31$.

The existence of nonsingular bilinear maps

$$R^{64(s-t)+17} \times R^{64t+24} \to R^{64s+32} \tag{3.2}$$

follows from (2.6) by setting $k = 8s+4$ and $h = 8t+2$ so that $k-h = 8(s-t)+2$. We note that $\tau(k, h) = 1$ since $\binom{s}{t}$ is assumed to be odd. Consequently, $(64s + 32)\xi_{64t+j}$ admits $64(s - t) + 17$ independent sections for $18 \le j \le 23$. These sections produced in (3.2) are maximal by Theorem 3.1 of [9]. Also, Theorem 3.1 of [9] affirms that $(64s + 32)\xi_{64t+17}$ admits $64(s - t) + 18$ independent sections maximally for $s \ge t \ge 0$ with $\binom{s}{t}$ odd. These maximum sectioning results are also maximum desuspension results for the Thom complexes by the arguments in the proof of Theorem 2.9. In other words, the maximum sectioning results for odd multiples of 32ξ obtained from Theorem 3.1 of [9] are also maximum desuspension results for the Thom complex.

We establish the results for $j = 15$ by Postnikov methods and then deduce those for 14 and 25. We claim that a classifying map for the bundle $\rho = (64s + 32)\xi_{64t+15}$ for $s \ge t > 0$ lifts to $BSpin(64t + 6)$ in the Postnikov resolution III of [11]

$$BSpin(64t + 6) \to E_4 \to E_3 \to E_2 \to E_1 \to BSpin \tag{3.3}$$

for the 2-primary homotopy of the fibration $\pi : BSpin(64t+6) \to BSpin$. We enumerate here only the relevant k-invariants for the lifting problem for ρ. The primary k-invariants are given by $(\delta w_{64t+6}, w_{64t+8})$. The higher k-invariants k_j^i belong to $H^{64t+j}(E_i)$ and have defining relations as follows.

$k_8^1 \;\; : Sq^2(\delta w_{64t+6}) + Sq^1 w_{64t+8} = 0.$

$k_{12}^1 : (Sq^4 + \cdot w_4)Sq^1 w_{64t+8} = 0.$

$k_{12}^2 : Sq^1 k_{12}^1 + Sq^2 Sq^3 k_8^1 = 0.$

$\tilde{k}_{15}^2 : (Sq^4 + \cdot w_4)k_{12}^1 + (Sq^6 Sq^2 + w_6 \cdot Sq^2)k_8^1 = 0. \qquad\qquad (3.4)$

$k_{12}^3 : Sq^2 Sq^1 k_{10}^2 + Sq^1 k_{12}^2 = 0.$

$k_{15}^3 : Sq^1 \tilde{k}_{15}^2 + Sq^2 Sq^1 k_{13}^2 + (Sq^4 + \cdot w_4)k_{12}^2 + (Sq^6 + \cdot w_6 + w_4 \cdot Sq^2)k_{10}^2 = 0.$

$k_{14}^4 : Sq^1 k_{14}^3 + Sq^2 Sq^1 k_{12}^3 = 0.$

We first analyze the lifting problems for $\zeta = (16s + 8)\gamma_{16t+4}$ and $i^*\zeta = (16s + 8)\gamma_{16t+3}$. Now $w_{64t+8}(\zeta) = \binom{64s+32}{64t+8}\delta^{16t+2} = 0$ so there is a lifting $f : QP^{16t+4} \to E_1$ of ζ. We claim that both $f^*k_8^1$ and $f^*k_{12}^1$ are trivial so f lifts to E_2 in (3.3). The defining relation for $\tilde{k}_{15}^2$ yields $0 = Sq^4(f^*k_{12}^1)$ in $H^{64t+16}(QP^{16t+4})$ so $f^*k_{12}^1 = 0$. The generating class theorem [14] affirms that $U_\zeta \cdot f^*k_8^1 = \varphi(U_\zeta)$ with zero indeterminacy where φ denotes a secondary operation associated to the relation $Sq^2(\delta Sq^{64t+6}) + Sq^1 Sq^{64t+8} = 0$ on integral classes. The evaluation algorithm of Proposition 3.3 of [13] yields $\varphi(U_\zeta) = U_\zeta \cdot z$ where $2z = c_{32t+4}(\zeta) = \binom{16s+8}{16t+2}\delta^{16t+2}$. Now $\nu\binom{16s+8}{16t+2} = 2$ so $U_\zeta \cdot z = 0$ and thus $f^*k_8^1 = 0$. Let $g : QP^{16t+4} \to E_2$ denote any lifting of f and so of ζ. The defining relation for k_{15}^3 gives $Sq^4(g^*k_{12}^2) = 0$ in $H^{64t+16}(QP^{16t+4})$ and so $g^*k_{12}^2 = 0$. Let $h : QP^{16t+3} \to E_3$ denote any lifting of g restricted to QP^{16t+3} and so of $i^*\zeta$. If $h^*k_{12}^3 \neq 0$, we alter $h \circ H : RP^{64t+15} \to E_3$ through the indeterminacy component based on k_{12}^2 to produce a lifting $\ell : RP^{64t+15} \to E_4$ of ρ. A lifting of ρ to $BSpin(64t + 6)$ in (3.3) is obtained by altering $\ell^*k_{14}^4$ through Sq^1 indeterminacy, if necessary. Consequently, $(64s+32)\xi_{64t+j}$ admits $64(s - t) + 26$ independent sections for $j = 15$ and also for $j = 14$ by restriction. This number gives both the maximum number of linearly independent sections and of desuspensions of the Thom complex by (1.3). The equivariant maps $S^{64(s-t)+15} \!\!\circ\!\!\longrightarrow V_{64t+26}(R^{64s+32})$ just established for $s > t \geq 0$ together with the equivariant maps $S^{64s+15}\!\!\circ\!\!\longrightarrow V_{26}(R^{64s+32})$ of Theorem 4.13 of [11] now produce by (1.2) equivariant maps $S^{64t+25}\!\!\circ\!\!\longrightarrow V_{64(s-t)+16}(R^{64s+32})$ for all $s \geq t \geq 0$ with $\binom{s}{t}$ odd. Again, $64(s - t) + 16$ is the maximum number of independent sections and Thom complex desuspensions for $j = 25$ by (1.3).

We utilize Postnikov resolution II of [11] for producing $64(s-t)+k$ sections to $(64s + 32)\xi_{64t+j}$ for both $j = 27$ and $j = 11$. We proceed to enumerate the higher k-invariants k_j^i in $H^{16t+j}(E_i)$ in a Postnikov resolution for the fibration $\pi : BSpin(16t + 2) \to BSpin$ through dimension $16t + 11$ for $t > 0$. Only the 2-primary homotopy of the fiber is needed for the resolution

$$BSpin(16t + 2) \to E_4 \to E_3 \to E_2 \to E_1 \to BSpin \qquad (3.5)$$

whose primary k-invariants are the triple $\left(\delta w_{16t+2}, w_{16t+4}, w_{16t+8}\right)$.

$$k_4^1 \; : \; Sq^2(\delta w_{16t+2}) + Sq^1 w_{16t+4} = 0.$$
$$k_6^1 \; : \; (Sq^4 + \cdot w_4)\delta w_{16t+2} = 0.$$
$$k_7^1 \; : \; (Sq^4 + \cdot w_4)w_{16t+4} = 0.$$
$$k_8^1 \; : \; (Sq^4 + \cdot w_4)Sq^1 w_{16t+4} + Sq^1 w_{16t+8} = 0.$$
$$k_9^1 \; : \; (Sq^4 + \cdot w_4)Sq^2 w_{16t+4} + Sq^2 w_{16t+8} = 0.$$
$$k_{10}^1 \; : \; (Sq^8 + \cdot w_8)\delta w_{16t+2} + (w_4 \cdot Sq^3 + \cdot w_7)w_{16t+4} = 0.$$
$$k_{11}^1 \; : \; (Sq^8 + \cdot w_8)w_{16t+4} + w_4 \cdot w_{16t+8} = 0.$$
$$k_6^2 \; : \; Sq^2 Sq^1 k_4^1 + Sq^1 k_6^1 = 0. \qquad k_8^2 : Sq^1 k_8^1 + Sq^2 Sq^3 k_4^1 = 0. \qquad (3.6)$$
$$k_9^2 \; : \; (Sq^4 + \cdot w_4)k_6^1 + Sq^2 Sq^1 k_7^1 + Sq^6 k_4^1 + Sq^2(k_4^1 \cdot w_4) = 0.$$
$$k_{10}^2 \; : \; Sq^1 k_{10}^1 + Sq^2 Sq^3 k_6^1 + Sq^1(k_4^1 \cdot w_6) + (w_4 \cdot Sq^1)k_6^1 = 0.$$
$$\tilde{k}_{10}^2 \; : \; (Sq^4 + \cdot w_4)k_7^1 + Sq^3 k_8^1 + Sq^2 k_9^1 = 0.$$
$$k_{11}^2 \; : \; Sq^1 k_{11}^1 + Sq^2 k_{10}^1 + Sq^2 Sq^3 k_7^1 + (Sq^8 + \cdot w_8 + \cdot w_4^2 + w_4 \cdot Sq^4)k_4^1$$
$$\qquad + (Sq^4 Sq^2 + \cdot w_6)k_6^1 + w_4 \cdot k_8^1 = 0.$$
$$k_8^3 \; : \; Sq^2 Sq^1 k_6^2 + Sq^1 k_8^2 = 0. \qquad k_{10}^3 : Sq^1 k_{10}^2 + Sq^2 Sq^3 k_6^2 = 0.$$
$$k_{10}^4 \; : \; Sq^1 k_{10}^3 + Sq^2 Sq^1 k_8^3 = 0.$$

In order to establish Theorem 3.1 for $j = 27$, we let $\zeta : QP^{16t+6} \to BSpin$ classify the bundle $\zeta = (16s+8)\gamma_{16t+6}$ and proceed to evaluate the obstructions to lifting both ζ and $\rho = (64s+32)\xi_{64t+27} = H^*\zeta$ to $BSpin(64t+18)$ in (3.5). (We are replacing t in (3.5) by $4t+1$.) Let $f : QP^{16t+6} \to E_1$ denote any lifting of ζ. Now $U_\zeta \cdot f^* k_4^1 = \varphi(U_\zeta)$ with zero indeterminacy by the generating class theorem of [14] where φ denotes a secondary operation associated to the relation $Sq^2(\delta Sq^{64t+18}) + Sq^1 Sq^{64t+20} = 0$ on integral classes. Evaluation by (3.3) of [13] gives $\varphi(U_\zeta) = U_\zeta \cdot y$ where $2y = c_{32t+10}(\zeta) = \binom{16s+8}{16t+5}\delta^{16t+5}$. But $\nu\binom{16s+8}{16t+5} = 3$ so $\varphi(U_\zeta) = 0$ and thus $f^* k_4^1 = 0$. Similarly, $\Gamma(U_\zeta) = U_\zeta \cdot f^* k_8^1$ with zero indeterminacy by [14] where Γ is a stable secondary operation associated to the relation

$$(Sq^4 Sq^1)Sq^{64t+20} + Sq^1 Sq^{64t+24} + Sq^{64t+23} Sq^2 = 0$$

on integral classes. Proposition 3.4 of [13] affirms that $\Gamma(U_\zeta) = U_\zeta \cdot z + Sq^4(U_\zeta \cdot y)$ where $2y = c_{32t+10}(\zeta)$ and $2z = c_{32t+12}(\zeta)$. We note that $\nu\binom{16s+8}{16t+6} = 2$. We conclude that $\Gamma(U_\zeta)$ is trivial so $f^* k_8^1 = 0$. Given any lifting $g : QP^{16t+6} \to E_2$ of f, the composite map $g \circ H : RP^{64t+27} \to E_2$ annihilates all k-invariants, except possible k_8^2. However, $(0, \alpha^{64t+24}, 0, 0, 0, 0)$ lies in the indeterminacy subgroup of $(k_6^2, k_8^2, k_9^2, k_{10}^2, \tilde{k}_{10}^2, k_{11}^2)(\rho)$. Thus $g \circ H$ can be altered to give a lifting $h : RP^{64t+27} \to E_3$ of ρ. Altering the values of k_8^3, k_{10}^3 and then k_{10}^4 for liftings of ρ through independent Sq^1 indeterminacies in (3.6) suffices to lift ρ to $BSpin(64t+18)$. This number, $64(s-t)+14$, of independent sections to $(64s + 32)\xi_{64t+j}$ for $s \geq t \geq 0$ with $\binom{s}{t}$ odd is maximal for $26 \leq j \leq 27$ by (1.3), and also yields the maximum number of desuspensions of the Thom complex by (1.3). Now (1.2) yields equivariant maps $S^{64t+13} \circ\!\!\!\longrightarrow V_{64(s-t)+28}(R^{64s+32})$ for $s \geq t > 0$ with $\binom{s}{t}$ odd. We apply (1.3) again to conclude that the results are established for $j = 27$, 26 and 13.

We shall deduce the corresponding results for $j = 28$ and $j = 12$ from $j = 27$. The morphism

$$p_* : \pi_{64t+27}\left(V_{64(s-t)+14}(R^{64s+32})\right) \to \pi_{64t+27}\left(V_{64(s-t)+13}(R^{64s+32})\right) \quad (3.7)$$

induced by the projection p is given by multiplication by 2 on $Z/8$, the 2-primary component of each group. The tables of [12] give $p_*\left(i_{64(s-t)+7}[4\nu]\right) = 2i_{64(s-t)+6}[\eta^2]$. Thus p_* in (3.7) maps the homotopy class of any equivariant map to 0 in the subquotient (1.1) defined by Berrick. That is, given any $f : S^{64t+27} \circ\!\!\!\longrightarrow V_{64(s-t)+14}(R^{64s+32})$ with $s \geq t \geq 0$ and $\binom{s}{t}$ odd, there is an equivariant map $S^{64t+28} \circ\!\!\!\longrightarrow V_{64(s-t)+13}(R^{64s+32})$ whose restriction to S^{64t+26} agrees with $p \circ f$. Now $64(s-t)+13$ gives both the maximum number of independent sections and of Thom complex desuspensions for $(64s+32)\xi_{64t+28}$ by (1.3). Applying (1.2) yields equivariant maps $S^{64t+12} \circ\!\!\!\longrightarrow V_{64(s-t)+29}(R^{64s+32})$ for $s \geq t > 0$ with $\binom{s}{t}$ odd. The number $64(s-t)+29$ is again maximal for both sectioning and desuspending by (1.3).

The result for $j = 11$ is established in a manner completely analogous to $j = 27$ by lifting a classifying map for $\rho = (64s+32)\xi_{64t+11}$ to $BSpin(64t+2)$ in (3.5) for $s \geq t > 0$ with $\binom{s}{t}$ odd. The details are too similar to justify repeating. The number, $64(s-t)+30$, of independent sections to $(64s+32)\xi_{64t+j}$ for $10 \leq j \leq 11$ is maximal for both the sectioning problem and desuspension problem by (1.3). We obtain equivariant maps $S^{64t+29} \circ\!\!\!\longrightarrow V_{64(s-t)+12}(R^{64s+32})$ for all $s > t \geq 0$ with $\binom{s}{t}$ odd by (1.2). Applying also Theorem 4.14 of [11] gives $64(s-t)+12$ independent sections to $(64s+32)\xi_{64t+29}$ for all $s \geq t \geq 0$ with $\binom{s}{t}$ odd. Again (1.3) affirms that $64(s-t)+12$ maximizes both the sectioning and desuspension problems for $(64s+32)\xi_{64t+29}$.

It remains to establish Theorem 3.1 only for $j = 24$ and $j = 16$. We utilize Postnikov resolution III of [13, p. 148] in order to lift a classifying map for $\rho = (64s+32)\xi_{64t+24}$ to $BSpin(64t+15)$ whenever $s \geq t \geq 0$ with $\binom{s}{t}$ odd. Let $f : QP^{16t+6} \to E_1$ denote a lifting for $\zeta = (16s+8)\gamma_{16t+6}$ in the resolution for $\pi : BSpin(64t+15) \to BSpin$ in [13] through dimension $64t+24$. The defining relation for k_3^2 gives $Sq^4(f^*k_2^1) = 0$ so $f^*k_2^1 = 0$ in $H^{64t+20}(QP^{16t+6})$. The generating class theorem [14] affirms that $U_\zeta \cdot f^*k_4^1 = \varphi(U_\zeta)$ where φ is a stable secondary opeation associated to the relation

$$(Sq^8 Sq^1)Sq^{64t+16} + Sq^{64t+23}Sq^2 + Sq^1\left(Sq^{64t+20}Sq^4\right) = 0$$

on integral classes. Proposition 3.5 of [13] affirms that $\varphi(U_\zeta) = Sq^8(U_\zeta \cdot y)$ where $2y = c_{32t+8}(\zeta) = \binom{16s+8}{16t+4}\delta^{16t+4}$. Now $\nu\binom{16s+8}{16t+4} = 1$ so y is an odd multiple of δ^{16t+4}. However, $Sq^8(U_\zeta \cdot \delta^{16t+4}) = 0$ so $f^*k_4^1 = 0$ in

$H^{64t+24}(QP^{16t+6})$. We take any lifting $g : QP^{16t+6} \to E_2$ of ζ and consider the composite lifting $g \circ H : RP^{64t+24} \to E_2$ of ρ. The indeterminacy subgroup of $\left(k_1^2, k_2^2, k_3^2, k_4^2, k_5^2\right)(\rho)$ is generated by $(0, \alpha^{64t+20}, 0, 0, 0)$, $(0, 0, 0, \alpha^{64t+24}, 0)$ and $(0, 0, 0, 0, \alpha^{64t+24})$. Consequently, $g \circ H$ can be altered, if necessary, through indeterminacies to produce a lifting $RP^{64t+24} \to E_3$ for ρ. Altering values of k_1^3, k_2^3 and k_1^4 through independent Sq^1 indeterminacies produces a lifting of ρ to $BSpin(64t + 15)$. Now $64(s - t) + 17$ gives the maximum number of independent sections to $(64s + 32)\xi_{64t+24}$ and also the maximum number of desuspensions for the Thom complex by (1.3). Applying (1.2) and Theorem 4.6 of [11] produce $64(s - t) + 25$ independent sections to $(64s + 32)\xi_{64t+16}$ for all $s \geq t \geq 0$ with $\binom{s}{t}$ odd. Again, this number is maximal for both sections and Thom complex desuspensions by (1.3).

We have established the values k for $0 \leq j \leq 31$ in Theorem 3.1. Knowledge of geometric dimension shows that $g.\dim(m + 2^L)\xi_n = g.\dim(m\xi_n)$ so periodicity of type 2^L is a direct consequence. $\square$

References

[1]   J. F. Adams, Geometric dimension of vector bundles over RP^n, Proc. Int. Conf. on Prospects of Math., Kyoto (1973), 1–17.

[2]   A. J. Berrick, The Smale invariants of an immersed projective space, Math. Proc. Camb. Phil. Soc. 86 (1979), 401–411.

[3]   D. M. Davis, Generalized homology and the vector field problem, Quart. J. Math. Oxford 25 (1974), 169–193.

[4]   D. M. Davis, Desuspensions of stunted projective spaces, Pac. J. Math. 113 (1984), 35–49.

[5]   D. M. Davis, S. Gitler, M. Mahowald, The stable geometric dimension of vector bundles over real projective spaces, Trans. Amer. Math. Soc. Oxford 268 (1981), 39–61; Corrections 280 (1983), 841–843.

[6]   S. Gitler, The projective Stiefel manifolds II. Applications, Topology 7 (1968), 47–53.

[7]   A. Hughes, E. Thomas, A note on certain secondary cohomology operations, Bol. Soc. Mat. Mexicana 13 (1968), 1–17.

[8]   K. Y. Lam, Construction of some bilinear maps, Bol. Soc. Mat. Mexicana 13 (1968), 88–94.

[9]   K. Y. Lam, Sectioning vector bundles over real projective spaces, Quart. J. Math. Oxford 23 (1972), 97–106.

[10] K. Y. Lam, D. Randall, Low dimensional spinor representations, Adams maps and geometric dimension, Adams Memorial Symposium on Algebraic Topology, London Math. Soc. Lecture Note Series 175 (1992), 89–102.

[11] K. Y. Lam, D. Randall, Geometric dimension of bundles over real projective spaces, Contemp. Math., Amer. Math. Soc., to appear.

[12] Y. Nomura, Some homotopy groups of real Stiefel manifolds in the metastable range I–VI, Science reports. Coll. Edu. Osaka Univ. 17 (1978), 1–31; 27 (1978), 55-97; 28 (1979), 1–26; 28 (1979), 35–60; 29 (1980), 159–183; 30 (1981), 11–57.

[13] A. D. Randall, Some immersion theorems for projective spaces, Trans. Amer. Math. Soc. 147 (1970), 135–151.

[14] E. Thomas, The index of a tangent 2-field, Comment. Math. Helv. 42 (1967), 86–110.

Kee Yuen Lam, Mathematics Department, University of British Columbia
Vancouver, B.C., V6t 1Z2, Canada
e-mail: lam@math.ubc.ca

Duane Randall, Department of Mathematics and Computer Science
Loyola University, New Orleans, LA 70118 USA
e-mail: ffmsadr@music.loyno.edu

Progress in Mathematics, Vol. 136
© 1996 Birkhäuser Verlag Basel/Switzerland

Further Structure in $K(1)_* \Omega^k S^{2n+1}$

LISA LANGSETMO

ABSTRACT. We determine the coalgebra structure for $K(1)_*(\Omega^k S^{2n+1})$ and thus complete the description as Hopf algebras. We also study the higher order K-theory Bockstein structure for $K(1)_*(\Omega^k S^{2n+1})$. To do this we use the p-local K-theory equivalences given by Bousfield and the author in [1] and [4]. We then may use the results of McClure, who determines the K-theory Bockstein spectral sequence for $\Omega^\infty \Sigma^\infty X$. As an application we use the existence of elements of order p^r in K-theory to determine a lower bound for the suspension order of the stable summands of $\Omega^{2k} S^{2n+1}$. Using the upper bound results of Silberbush this gives the suspension order at odd primes.

1. Introduction

One result of the author's paper [4] was to compute $K(1)_*(\Omega^k S^{2n+1})$ as algebras. In this paper we will compute some further structure, namely the associated coalgebra structure and the structure of the K-theory Bockstein spectral sequence.

Throughout this paper we will consider a fixed prime p. We let E, P and Γ denote an exterior algebra, a polynomial algebra and a divided polynomial algebra, respectively. T^m is a polynomial algebra truncated at height m. We will use $K(1)_*$ to denote mod p complex K-theory, $q = 2p - 2$. Let B denote $B\Sigma_p$ localised at p, B^{qn} denote the qn-skeleton of B and $B_{(n+1)q-1}$ denote the cofiber of the inclusion $B^{qn} \longrightarrow B$. Let F^{2n+1} denote the fiber of the Snaith map $QS^{2n+1} \longrightarrow Q\Sigma^{2n+1} B_{(n+1)q-1}$. The first result follows from [1] and [4].

PROPOSITION 1. *a)* $K(1)_*(\Omega^{2n+1} S_0^{2n+1}) \cong P(y_1, \dots, y_n) \otimes E(z_1, z_2, \dots, z_n)$.
b) If $k < 2n + 1$ then

$$K(1)_*(\Omega^k S^{2n+1}) \cong \begin{cases} E(w_0, w_1, \dots, w_l) \otimes T^{p^l}(x_1, x_2, \dots) & \text{if } k = 2l \\ P(y_0, y_1, \dots, y_l) \otimes E(z_1, z_2, \dots, z_l) & \text{if } k = 2l + 1 \end{cases}$$

as algebras over $K(1)_*$ for $p > 2$ or k odd and as graded vector spaces otherwise.

Part a) is proved by Bousfield in [1] and relies on calculations done by Mahowald and Thompson. In particular we have that the Snaith map

$$\Omega^{2n+1} S_0^{2n+1} \xrightarrow{\ s_n\ } QB^{qn}$$

is both an isomorphism in mod p v_1-periodic homotopy and mod p K-theory. Part b) was proved in [4] with the exception of $k = 2n$. The crux of this argument is to use $\Omega^k F^{2n+1}$ as a model for $\Omega^k S^{2n+1}$ and then to calculate the mod p K-theory of the former. Now we consider the case where $k = 2n$. Let

Research partially supported by NSF and Wayne State University.

$X\langle 1 \rangle$ denote the 1-connected cover of X. The Snaith map, s_n, deloops to a map

$$\Omega^{2n} S^{2n+1}\langle 1 \rangle \longrightarrow Q\Sigma B^{qn}$$

which is also an isomorphism in mod p K-theory. To see that it is we may repeat the argument used for s_n or we may invoke the bar spectral sequence. We have a map of fibrations

$$
\begin{array}{ccccc}
\Omega^{2n+1} S_0^{2n+1} & \longrightarrow & * & \longrightarrow & \Omega^{2n} S^{2n+1}\langle 1 \rangle \\
\downarrow & & \downarrow & & \downarrow \\
QB^{qn} & \longrightarrow & * & \longrightarrow & Q\Sigma B^{qn}
\end{array}
$$

Thus we have a map of bar spectral sequences in which the maps of fibers and total spaces induce isomorphisms. We conclude that the map of base spaces is an isomorphism as well. $K(1)_* Q\Sigma B^{qn}$ is given by results of [6]. To recover $K(1)_*(\Omega^{2n} S^{2n+1})$ we note that the fibration

$$\Omega^{2n} S^{2n+1}\langle 1 \rangle \longrightarrow \Omega^{2n} S^{2n+1} \longrightarrow S^1$$

is split and then apply the Künneth formula for $K(1)_*$.

2. $K(1)$-cohomology of $\Omega^k S^{2n+1}$

THEOREM 2. *a)* $K(1)^*(\Omega^{2n+1} S_0^{2n+1}) \cong T^{p^n}(c_1, c_2, \dots) \otimes E(d_1, d_2, \dots, d_n)$.
b) If $k < 2n+1$ *then*

$$
K(1)^*(\Omega^k S^{2n+1}) \cong
\begin{cases}
E(a_0, a_1, \dots, a_l) \otimes P(b_1, b_2, \dots, b_l) & \text{if } k = 2l \\
T^{p^{l+1}}(c_0, c_1, \dots) \otimes E(d_1, d_2, \dots, d_l) & \text{if } k = 2l+1
\end{cases}
$$

as $K(1)^*$ *algebras. In this statement we are taking the polynomial algebra to be a polynomial power series algebra.*

Proof. The cases $k = 0, 1, 2$ are known and the latter appears in [8]. To prove the remaining cases we will study the appropriate spectral sequences.

   First let us consider the case $k = 2l$ with $0 < l \leq n$. We have a principal fibration

$$\Omega^{2l+1} S^{2n+1} \longrightarrow * \longrightarrow \Omega^{2l} S^{2n+1}.$$

Since we know the $K(1)$-homology of both the left and right side it is easy to see what happens in the bar spectral sequence. However, the bar spectral sequence is also a spectral sequence of coalgebras following the construction given in [10] and generalized to K-theory in [2]. Further discussion may be found in [5] and [9]. We wish to examine to examine the coalgebra structure more carefully. We have that

$$E_2 \cong E(u_0, u_1, \dots, u_l) \otimes \Gamma(v_1, v_2, \dots, v_l)$$

as Hopf algebras where Γ denotes a divided polynomial Hopf algebra. As an algebra this falls apart mod p, i.e. $\Gamma(x) \cong T^p(x, \gamma_p(x), \gamma_{p^2}(x), \dots)$. The natural coalgebra structure is given by

$$\psi(\gamma_m(v)) = \sum_{i=0}^{m} \gamma_i(v) \otimes \gamma_{m-i}(v).$$

The spectral sequence collapses since all the odd generators are known to be non-trivial permanent cycles.

By Theorem 1 we see that there are algebra extensions. Note that the generator x_1 is primitive and consequently so is $x_1^{p^j}$. We can choose classes so that $x_1^{p^j}$ is represented by v_{j+1}. This may be seen by analyzing how the suspension behaves in the spectral sequence as done in [7]. In particular we have the relations

$$v_{j+1} = \sigma z_{j+1} = \sigma Q^j z_1 = (\sigma z_1)^{p^j} = (v_1)^{p^j}$$

where σ denotes the homology suspension and Q^j is the Dyer-Lashof operation iterated j times.

The algebra structure of $K(1)^*$ follows from the coalgebra structure of $K(1)_*$ which can be read from the E_∞ term of the spectral sequence. Here the $K(1)^*$ generators are dual to to the $K(1)_*$ primitives, namely u_i and v_i. Finally, we note that for $k = 2n$ it suffices to consider the bar spectral sequence converging to its 1-connective cover. The argument is similar to that above except for a missing generator on both sides. To recover the answer we again use the split fibration. The added exterior generator is primitive and there are no extensions.

If we look at the analogous fibration for $k = 2l + 1$, we see that there are differentials. This is a complicated spectral sequence and we leave the analysis to the interested reader. We instead look at the cobar spectral sequence for the fibration

$$\Omega^{2l+1} S^{2n+1} \longrightarrow * \longrightarrow \Omega^{2l} S^{2n+1}.$$

By [13] we know this is a spectral sequence of Hopf algebras and the E_∞ term is an associated graded to $K(1)_*$ of the fiber. As input to this spectral sequence we need to know the coalgebra structure of $K(1)_*(\Omega^{2l} S^{2n+1})$. We now use the identification of Cotor over a coalgebra with Ext over the dual algebra and the previous result to get that

$$E_2 \cong P(s_0, s_1, \dots, s_l) \otimes E(t_1, t_2, \dots, t_l).$$

The polynomial generators cannot be sources of non-trivial differentials since they must hit odd primitives of different homological filtration. The Hopf-algebra structure of the spectral sequence gives that the only possible sources of non-trivial differentials are the t_i. By comparison with Theorem 1 (as algebras) we see that the spectral sequence collapses after E_2 since the number of odd primitive classes must be l in the E_∞ term as well.

It remains to show that there are non-trivial coalgebra extensions giving the desired answer. To see this we look at the map

$$j : \Omega^{2l+1} S^{2n+1} \longrightarrow QS^{2n-2l}$$

where j is the iterated loops of the the inclusion $i : S^{2n+1} \longrightarrow QS^{2n+1}$. Observe that j factors as follows:

$$\Omega^{2l+1} S^{2n+1} \longrightarrow \Omega^{2l+1} F^{2n+1} \longrightarrow QS^{2n-2l}.$$

As shown in [4], the left map induces an isomorphism and the right map maps the polynomial generators over non-trivially. We can now use the coalgebra structure of $K(1)_*(QS^{2n-2l})$ to deduce the extensions. This structure follows from results in [3] and [12]. In particular we note that only the first polynomial generator in $K(1)_*\Omega^{2l+1}S^{2n+1}$ is primitive.

In the above arguments we used only information from the bar and cobar spectral sequence for the fibration

$$\Omega^{2l+1}S^{2n+1} \longrightarrow * \longrightarrow \Omega^{2l}S^{2n+1}.$$

Both these spectral sequences collapse and the only question was to show that there were non-trivial extensions. In contrast the bar and cobar spectral sequence for the fibration

$$\Omega^{2l}S^{2n+1} \longrightarrow * \longrightarrow \Omega^{2l-1}S^{2n+1}$$

both have non-trivial differentials. Thus any attempt to compute the $K(1)$ homology iteratively must determine the differential pattern at every other step. This is in contrast to ordinary homology where the spectral sequence collapses conveniently after the E_2 term. Using the rich Hopf algebra structure might be a way to determine the general pattern of diffferentials in the cobar spectral sequence for the non-trivial cases. Such a description may also illuminate the differential structure in the $K(n)$ cobar spectral sequence as well.

3. The Bockstein Spectral Sequence

THEOREM 3. *Let $1 \le m < \infty$, then the mth stage of the Bockstein spectral sequence, E_*^m, is additively isomorphic to*

$$E_*^m(\Omega^{2n+1}S_0^{2n+1}) \cong \begin{cases} P(y_1,\dots,y_{n-m+1},y_{n-m+2}^p,\dots,y_l^{p^{m-1}}) \\ \otimes E(z_1,\dots,z_{n-m+1},\beta_{n-m+2},\dots,\beta_l) \end{cases}$$

$$E_*^m(\Omega^{2l}S^{2n+1}) \cong \begin{cases} E(w_0,w_1,\dots,w_{l-m+1},\alpha_{l-m+2},\dots,\alpha_l) \\ \otimes_{i=t}^\infty T^{p^k}(x_i) \end{cases}$$

where $l \le n$, $t = \max(1, m-l+1)$ and $k = \min(l, l+i-m)$

$$E_*^m(\Omega^{2l+1}S^{2n+1}) \cong \begin{cases} P(y_0,y_1,\dots,y_{l-m+1},y_{l-m+2}^p,\dots,y_l^{p^{m-1}}) \\ \otimes E(z_1,\dots,z_{l-m+1},\beta_{l-m+2},\dots,\beta_l) \end{cases}$$

where $l < n$.

Proof. In the case $n = 0,1$ the generators are clearly permanent cycles. The case $k = 2n+1$ follows from the fact that the Snaith map $s_n : \Omega^{2n+1}S_0^{2n+1} \longrightarrow QB^{qn}$ is a rational equivalence, hence it is also a $K_*(\ ,\mathbf{Z}_{(p)})$ equivalence. From this we know that the corresponding Bockstein spectral sequences are the same. The Bockstein spectral sequence for QX is described in [6]. The case $k = 2n$ is similar. As noted in the first section the Snaith map deloops to a map

$$\Omega^{2n}S^{2n+1}\langle 1\rangle \longrightarrow Q\Sigma B^{qn}.$$

This map is also a rational equivalence, and thus a $K_*(\ ,\mathbf{Z}_{(p)})$ equivalence. The extra exterior generator in $\Omega^{2n}S^{2n+1}$ comes from S^1 and hence is a permanent cycle in the Bockstein spectral sequence. The Bockstein spectral sequence for the rest corresponds to that of $Q\Sigma B^{qn}$.

For the remaining cases we consider F^{2n+1}, the fiber of the Snaith map. By [4] we have that the map $\Omega^k S^{2n+1} \longrightarrow \Omega^k F^{2n+1}$ is both a mod p and a p-local K-theory equivalence so it suffices to consider the Bockstein spectral sequence of the latter. For $k = 2l$ we have a map of principle fibrations

$$
\begin{array}{ccccc}
\Omega^k F^{2n+1} & \longrightarrow & QS^{2n+1-k} & \overset{\Omega^k s}{\longrightarrow} & Q\Sigma^m B_{(n+1)q-1} \\
\downarrow \scriptstyle{f} & & \downarrow \scriptstyle{s} & & \downarrow \scriptstyle{=} \\
Q\Sigma^m B^{nq}_{(n+1-l)q-1} & \overset{Q(i)}{\longrightarrow} & Q\Sigma^m B_{(n+1-l)q-1} & \overset{Q(p)}{\longrightarrow} & Q\Sigma^m B_{(n+1)q-1}
\end{array}
$$

where $m = 2n + 1 - k$.

The right hand square commutes, hence there is a lift f which makes the diagram commute. All the vertical maps are rational equivalences and f is an epimorphism in $K(1)_*$. The kernel of this map consists of the ideal generated by the bottom class. The bottom class is a permanent cycle hence we can determine the Bockstein spectral sequence of $\Omega^k F^{2n+1}$ from that of $Q\Sigma^{2n+1-k} B^{nq}_{(n+1-l)q-1}$. The case $k = 2l + 1$ is done by looping the whole diagram back and repeating the argument.

4. Application to Stable Order

We start with some definitions. We say that a space X has suspension order r if the identity map of ΣX to itself has order r in the group $[\Sigma X, \Sigma X]$. We then define X to have stable order r where r is the suspension order of $\Sigma^m X$ and m is sufficiently large. Now let $D_{j,l}$ denote the p^j piece in the splitting of $\Omega^{2l} S^{2n+1}$.

PROPOSITION 4. [11] *For $p > 2$ and $j > 0$, the stable order of $D_{j,l}$ divides p^{j+l-1}.*

COROLLARY 5. *For $p > 2$ and $j > 0$, the stable order of $D_{j,l}$ is p^{j+l-1}.*

To prove this we will use the results of the previous section together with results from [11]. In particular we will use the observation that if X has stable order dividing r then $rh_*(\Sigma X) = 0$ for any reduced cohomology theory h_*. Therefore, if we can show there exist an element that survives to the mth stage of the Bockstein spectral sequence then the stable order must be greater or equal to p^m.

LEMMA 6. *There exists an element in $K(1)_*(D_{j,l})$ which survives to the $(j + l - 1)$ stage of the Bockstein spectral sequence.*

Proof. From examination of Theorem 3 we see that the elements we want are the generators x_j. Thus it remains to show that the element x_j lives on the appropriate piece of the stable splitting, i.e. $D_{j,l}$. This is certainly true for $l = 1$. Unfortunately, the inclusion map takes all these generators to zero so we are unable to get more information from this. If we examine the Bockstein structure for $l = 1$ we note that w_1 and x_1 are connected by a primary Bockstein and in general $w_1 x_1^{p-1} \ldots x_{j-1}^{p-1}$ and x_j are connected by a j-th order Bockstein. Using weight filtration we note that the weight of x_j is the same as the weight of $w_1 x_1^{p-1} \ldots x_{j-1}^{p-1}$, i.e. they are on the same piece of the stable splitting. Now the

weight of a homology product is the sum of the weights, hence the weight of
x_j is p times the weight of x_{j-1}.

Now consider the case where $l > 1$. Now w_1 and x_1 are connected by an
l-th order Bockstein and w_2 and x_1^p are connected $(l-1)$-st order Bockstein.
Now the only possible target for $w_1 x_1^{p-1}$ is still x_2 and they are connected by
a $(l+1)$-st order Bockstein. In general the primitives w_i are connected by an
$(l-i+1)$-st order Bockstein to the primitives $x_1^{p^i}$. This in turn forces the classes
$w_1 x_1^{p-1} \ldots x_{j-1}^{p-1}$ to be the source of a $(l+j-1)$-st order Bockstein which hits
the class x_j. As in the case $l = 1$ the weight of x_j is then p times the weight of
x_{j-1}. Since x_1 is on the piece $D_{1,l}$ the lemma follows by induction.

References

[1] A. K. Bousfield, *Localization and Periodization in Unstable Homotopy Theory*,
    preprint.
[2] L. Hodgkin, *On the K-theory of Lie groups*, Topology **6** (1967), 1–36.
[3] L. Hodgkin, *Dyer-Lashof operations in K-theory*, London Math. Soc. Lecture
    Notes **11** (1974).
[4] L. Langsetmo, *The K-theory Localization of Loops on an Odd Sphere and Appli-
    cations*, Topology **32** (1993), 577–591.
[5] J. P. May, *Classification spaces and fibrations* Am. Math. Soc. Memoirs **155**
    (1975).
[6] J. McClure, *The mod p K-theory of QX*, H_∞ Ring Spectra and Their Applica-
    tions, SLNS **1176** (1986), 291–383.
[7] H. Miller, V. Snaith, *On $K_*(QRP^n, \mathbf{Z}/2)$*, Can. Math. Soc. Conf. Proc. **2** part 1
    (1982), 233–245.
[8] D. Ravenel, *The Homology and Morava K-theory of $\Omega^2 SU(n)$*, to appear in Fo-
    rum Mathematicum.
[9] D. Ravenel, S. Wilson, *The Morava K-theories of Eilenberg-MacLane spaces and
    the Conner-Floyd Conjecture*, Am. Jour. Math. **102** (1980), 691–748.
[10] M. Rothenberg, N. Steenrod, *The cohomology of classifying spaces of H-spaces*,
    Bull. Am. Math. Soc. **71** (1965), 872–875, and mimeographed notes.
[11] P. Silberbush, *Order of the Identity of the Stable Summands of $\Omega^{2k} S^{2n+1}$*, Thesis,
    University of Rochester.
[12] V. Snaith, *Dyer-Lashof Operations in K-theory*, SLNS **496** (1976), 103–294.
[13] D. Tamaki, *A Dual Rothenberg-Steenrod Spectral Sequence*, Topology **33** (1994),
    631–662.

Wayne State University and University of Minnesota
lisa@math.wayne.edu

Progress in Mathematics, Vol. 136
© 1996 Birkhäuser Verlag Basel/Switzerland

Tor et Ext-dimensions des H^*V-A-modules instables qui sont de type fini comme H^*V-modules

Jean Lannes et Saïd Zarati

1 Introduction

Soient V un 2-groupe abélien élémentaire (en d'autres termes, un $\mathbb{F}_2$-espace vectoriel de dimension finie) et A l'algèbre de Steenrod modulo 2 ; H^*V désigne la cohomologie modulo 2 de V. Un H^*V-A-module instable est un A-module instable M muni d'une structure de H^*V-module définie par une application $H^*V \otimes M \to M$ qui est A-linéaire (l'exemple type d'un tel objet est la cohomologie modulo 2 équivariante d'un V-espace). Cet article poursuit l'étude, commencée dans [DW] [LZ], de la catégorie des H^*V-A-modules instables qui sont de type fini comme H^*V-modules.

Nous associons à tout H^*V-A-module instable M deux entiers (en fait deux éléments de $\mathbb{N} \cup \{+\infty\}$) :
– Le premier, que nous appelons Tor-dimension et que nous notons $d_V M$, est défini par :

$$d_V M = \operatorname{Inf} \left\{ n \in \mathbb{N} \; ; \; (\operatorname{Tor}_s^{H^*V}(\mathbb{F}_2, M))^t = 0 \text{ pour } t - s > n \right\} .$$

Ci-dessus $(\operatorname{Tor}_s^{H^*V}(\mathbb{F}_2, M))^t$ désigne le sous-groupe de $\operatorname{Tor}_s^{H^*V}(\mathbb{F}_2, M)$ formé des éléments de degré t.
Il est à remarquer que $d_V M$ ne dépend que de la structure de H^*V-module de M.
– Le second, que nous appelons Ext-dimension et que nous notons $e_V M$, est défini par :

$$e_V M = \operatorname{Inf} \left\{ n \in \mathbb{N} \; ; \; (\operatorname{Ext}_{H^*V-\mathcal{U}}^s(\Sigma^t(-), M) = 0 \text{ pour } s + t > n \right\} .$$

Ci-dessus $H^*V - \mathcal{U}$ désigne la catégorie des H^*V-A-modules instables et Σ^t le t-ième itéré du foncteur suspension (que l'on considère comme un endofoncteur de $H^*V - \mathcal{U}$).
Notre résultat principal est que ces deux entiers coïncident si M est de type fini comme H^*V-module (dans ce cas la Tor-dimension de M est finie).
La méthode de démonstration utilise notre classification des H^*V-$\mathcal{U}$-injectifs (il s'agit là des injectifs de la catégorie H^*V-$\mathcal{U}$ [LZ], une définition équivalente de la Ext-dimension en termes de H^*V-$\mathcal{U}$-résolutions injectives et quelques-unes des propriétés des foncteurs Fix du premier auteur [L] [LZ]. Le résultat énoncé ci-dessus implique notamment qu'un H^*V-A-module instable M qui est

de type fini comme H^*V-module admet une H^*V-$\mathcal{U}$-résolution injective d'un type bien précis (voir 2.1.3.2) : elle est en particulier de longueur inférieure à la Tor-dimension de M et la complexité de chaque terme est aussi contrôlée par cette Tor-dimension.

Indiquons pour terminer cette introduction un contexte dans lequel la Tor-dimension apparaît naturellement.

Soit X un V-CW-complexe fini. B. Oliver, H.-W. Henn et le premier auteur ont montré l'inégalité suivante :

$$d_V H_V^* X \leq (2^{\dim V} - 1) \dim X \ ,$$

$H_V^* X$ désignant la cohomologie modulo 2 équivariante de X.

Il résulte d'autre part des travaux de J. Duflot [D] que dans le cas où X est une variété compacte (éventuellement à bord) on a l'inégalité suivante :

$$d_V H_V^* X \leq \dim X \ .$$

En conséquence, si un $V - CW$-complexe fini X admet un plongement V-équivariant dans une V-variété de dimension n alors on a nécessairement :

$$d_V H_V^* X \leq n \ .$$

2 Enoncé du résultat principal

2.1. Définitions et notations

2.1.1. Soit V un 2-groupe abélien élémentaire. On désigne par BV un classifiant du groupe V et on pose $H^*V = H^*(BV; \mathbb{F}_2)$. On note A l'algèbre de Steenrod modulo 2, $\mathcal{U}$ la catégorie des A-modules instables et H^*V-$\mathcal{U}$ la catégorie dont les objets sont les H^*V-A-modules instables c'est-à-dire les A-modules instables M munis d'une structure de H^*V-module définie par une application $H^*V \otimes M \to M$ qui est A-linéaire. Les morphismes de $H^*V - \mathcal{U}$ sont les applications de degré zéro à la fois A-linéaires et H^*V-linéaires.

La catégorie $H^*V - \mathcal{U}$ est une catégorie abélienne qui possède assez de projectifs et d'injectifs.

Tout H^*V-A-module instable projectif est en fait isomorphe à une somme directe de H^*V-A-modules instables de la forme $H^*V \otimes F(n)$, $F(n)$ désignant le A-module instable librement engendré par un élément de degré n.

La classification des objets injectifs de la catégorie $H^*V - \mathcal{U}$ (en abrégé : $H^*V - \mathcal{U}$-injectifs) est la suivante [LZ].

– Soit $\mathcal{L}$ un système de représentants pour les classes de $\mathcal{U}$-isomorphismes des facteurs directs indécomposables de $H^*((\mathbb{Z}/2)^m)$, m parcourant $\mathbb{N}$ (chacune de ces classes est donc représentée dans L une et une seule fois).

– Soit $\mathcal{W}$ l'ensemble des sous-groupes de V.

– Soit $J_V(n)$, $n \in \mathbb{N}$, le H^*V-A-module instable caractérisé, à isomorphisme près, par la bijection fonctorielle en le H^*V-A-module instable M :

$$\mathrm{Hom}_{H^*V - \mathcal{U}}(M, J_V(n)) \cong \mathrm{Hom}_{\mathbb{F}_2}(M^n, \mathbb{F}_2) \ .$$

– Soit (W, n) un élément de $\mathcal{W} \times \mathbb{N}$, on pose

$$E(V, W, n) = H^*V \otimes_{H^*V/W} J_{V/W}(n)$$

(dans cette formule H^*V est un H^*V/W-module *via* l'application induite en cohomologie modulo 2 par la surjection canonique $V \to V/W$).

Théorème 2.1.1.— *Soit I un H^*V-$\mathcal{U}$-injectif. Il existe une unique famille de cardinaux $(a_{L,W,n})_{(L,W,n)\in\mathcal{L}\times\mathcal{W}\times\mathbb{N}}$ telle que I est isomorphe à la somme directe $\oplus_{(L,W,n)}(L\otimes E(V,W,n))^{\oplus a_{L,W,n}}$.*
Réciproquement tout H^*V-A-module instable de cette forme est injectif.

(Dans cette formule la notation $(-)^{\oplus a}$ désigne la somme directe de a copies de $(-)$.)

2.1.2. Dans cet article nous aurons fréquemment à faire l'hypothèse qu'un H^*V-A-module M est de type fini comme H^*V-module, c'est-à-dire engendré comme H^*V-module par un nombre fini d'éléments. En d'autres termes, M est de type fini comme H^*V-module si le A-module instable $\mathbb{F}_2\otimes_{H^*V}M = M\,/\,\widetilde{H}^*V\cdot M$ est fini ; $\mathbb{F}_2$ est un H^*V-A-module instable *via* l'augmentation de H^*V, $\widetilde{H}^*V$ désigne l'idéal d'augmentation et $\widetilde{H}^*V\cdot M$ l'image de l'application $\widetilde{H}^*V\otimes M\to M$. Par exemple, la cohomologie modulo 2 équivariante d'un V-CW-complexe fini est un H^*V-A-module de type fini comme H^*V-module.
Nous abrégerons souvent "H^*V-A-module de type fini comme H^*V-module" en "$(H^*V)_{tf}$-A-module".
Le théorème 2.1.1 implique facilement le suivant (voir encore [LZ]) :

Théorème 2.1.2.— *L'enveloppe injective d'un H^*V-A-module instable qui est de type fini comme H^*V-module est isomorphe à une somme directe finie de $E(V,W,n)$.*

2.1.3. Ext-dimension d'un H^*V-A-module instable.

2.1.3.1. Soit $\Sigma^t, t\in\mathbb{N}$, le t-ième itéré du foncteur suspension $\Sigma : H^*V$-$\mathcal{U}\to H^*V$-$\mathcal{U}$. Nous définissons la Ext-dimension d'un H^*V-A-module instable M, que nous notons $e_V M$, par :

$$e_V M = \text{Inf } \{n\in\mathbb{N};\ (\text{Ext}^s_{H^*V-\mathcal{U}}(\Sigma^t(-),M) = 0 \text{ pour } s+t > n\} \ ;$$

e_V prend ses valeurs dans $\mathbb{N}\cup\{+\infty\}$.
Nous reformulons cette définition en termes de H^*V-$\mathcal{U}$-résolutions injectives dans la proposition ci-dessous ; étant donné les applications que nous avons en vue nous nous bornons au cas des H^*V-A-modules instables qui sont de type fini comme H^*V-modules.

Proposition 2.1.3.2.— *Soient M un H^*V-A-module instable qui est de type fini comme H^*V-module et n un entier. Les propriétés suivantes sont équivalentes :*
  (i) $e_V M \leq n$;
  (ii) *M admet une H^*V-$\mathcal{U}$-résolution injective*

$$0\to M\to I_0\to I_1\to\cdots\to I_m\to\cdots$$

  vérifiant :
  – $I_m = 0$ pour $m > n$,
  – *pour $m\leq n$, I_m est isomorphe à une somme directe finie de $E(V,W,k)$ avec $k\leq n-m$.*

La démonstration (facile) de cette proposition est reportée à 5.1.

Scholie 2.1.3.3.— *Si la Ext-dimension d'un $(H^*V)_{tf}$-A-module instable est inférieure ou égale à un entier n alors son enveloppe injective est une somme directe finie de $E(V, W, k)$ avec $k \leq n$.*

2.1.4. Tor-dimension d'un H^*V-A-module instable.

2.1.4.1. Soit s un entier, nous notons $\text{Tor}_s^V - = \text{Tor}_s^{H^*V}(\mathbb{F}_2, -)$ le s-ième foncteur dérivé du foncteur $\mathbb{F}_2 \otimes_{H^*V} - : H^*V\text{-}\mathcal{U} \to \mathcal{U}$. Nous définissons la Tor-dimension d'un H^*V-A-module instable M, que nous notons $d_V M$, par :

$$d_V M = \text{Inf}\ \{n \in \mathbb{N};\ (\text{Tor}_s^V M)^t = 0 \text{ pour } t - s > n\}\ .$$

La notation $(\text{Tor}_s^V M)^t$ désigne le sous-groupe de $\text{Tor}_s^V M$ formé des éléments de degré t ; d_V prend ses valeurs dans $\mathbb{N} \cup \{+\infty\}$.

2.1.4.2. *Remarques*
(a) La Tor-dimension ne dépend que de la structure de H^*V-module de M.
(b) Soit $P_\bullet \to \mathbb{F}_2 \to 0$ une H^*V-$\mathcal{U}$-résolution de $\mathbb{F}_2$ telle que $P_\bullet$ est libre comme H^*V-module. Alors on a un $\mathcal{U}$-isomorphisme canonique $\text{Tor}_s^V M \cong H_s(P_\bullet \otimes_{H^*V} M)$.
(c) Notons τM le sous-A-module de M formé des éléments x annulés par $\tilde{H}^*V$; nous appelons τM la partie triviale de M. On vérifie que le A-module instable $\text{Tor}_{\dim V}^V(M)$ s'identifie à $\Sigma^{\dim V}\tau M$.
(d) Si M est de type fini comme H^*V-module alors $d_V M$ est fini.
(e) Introduisons la notation suivante : soit E un $\mathbb{F}_2$-espace vectoriel $\mathbb{N}$-gradué, nous posons $||E|| = \text{Sup}\ \{n \in \mathbb{N}; E^n \neq 0\}; ||E||$ est un élément de $\mathbb{N} \cup \{-\infty, +\infty\}$.
    Si M est non nul et si $||M||$ est fini alors on a $d_V M = ||M||$. L'inégalité $d_V M \leq ||M||$ est claire, l'égalité résulte de la remarque (c) ci-dessus.
(f) Supposons $M = H^*V \otimes_{H^*V/W} N$ avec N un H^*V/W-A-module instable. Alors on a $d_V M = d_{V/W} N$ puisque $\text{Tor}_s^V M$ et $\text{Tor}_s^{V/W} N$ sont isomorphes (en fait comme A-modules instables).

2.1.5. Le résultat principal de cet article, qui sera démontré en 5.2, est le suivant :

Théorème 2.1.5.— *Soit M un H^*V-A-module instable qui est de type fini comme H^*V-module. Alors les Ext et Tor-dimensions de M sont égales :*

$$e_V M = d_V M\ .$$

2.1.6. *Remarque.* Définissons comme précédemment la Ext-dimension d'un A-module instable N (il s'agit en fait du cas $V = 0$!) :

$$e(N) = \text{Inf}\ \{n \in \mathbb{N}\ ;\ (\text{Ext}_\mathcal{U}^s(\Sigma^t(-), N) = 0 \text{ pour } s + t > n\}$$

(ci-dessus Σ^t désigne cette fois le t-ième itéré du foncteur suspension $\Sigma : \mathcal{U} \to \mathcal{U}$). Soit maintenant M un H^*V-A-module instable. On se convainc facilement que l'on a l'inégalité $e(M) \leq e_V M$. Par contre on n'a pas l'égalité en général, même si l'on suppose M de type fini comme H^*V-module. Voici un exemple. Posons $H = H^*(\mathbb{Z}/2), \widetilde{H} = \widetilde{H}^*(\mathbb{Z}/2), r = \dim V$, écrivons $H^*V = H^{\otimes r}$ et prenons $M = \widetilde{H}^{\otimes r}$; cet H^*V-A-module instable est bien de type fini comme H^*V-module et l'on a $e(M) = 0, d_V M = r$ et donc $e_V M = r$.

La démonstration du théorème 2.1.5 se fait par récurrence sur la dimension du $\mathbb{F}_2$-espace vectoriel V. Cette récurrence utilise de façon cruciale des propriétés des foncteurs Fix introduits dans [L] (voir aussi [LZ]) que nous allons commencer par établir.

3 Les foncteurs Fix

3.1. Soit V' un sous-groupe de V. Nous notons $\mathrm{Fix}_{(V,V')} : H^*V\text{-}\mathcal{U} \to H^*V/V'\text{-}\mathcal{U}$ l'adjoint à gauche du foncteur $H^*V \otimes_{H^*V/V'} - : H^*V/V'\text{-}\mathcal{U} \to H^*V\text{-}\mathcal{U}$. Nous avons donc la bijection fonctorielle :

$$\mathrm{Hom}_{H^*V-\mathcal{U}}(M, H^*V \otimes_{H^*V/V'} N) \cong \mathrm{Hom}_{H^*V/V'-\mathcal{U}}(\mathrm{Fix}_{(V,V')}M, N)$$

pour tout H^*V-A-module instable M et tout H^*V/V'-A-module instable N.
Il est clair que $\mathrm{Fix}_{(V,0)} = \mathrm{id}_{H^*V-\mathcal{U}}$ et que $\mathrm{Fix}_{(V,V)}$ (resp. $\mathrm{Fix}_{(V'\oplus V'',V')}$) est le foncteur noté Fix_V (resp. $\mathrm{Fix}_{V',V''}$) dans [LZ]. La relation que l'on y trouve entre foncteurs $\mathrm{Fix}_{V',V''}$ et $T_{V'}$ se reformule comme suit. Le foncteur $T_{V'} : \mathcal{U} \to \mathcal{U}$ induit un foncteur $H^*V\text{-}\mathcal{U} \to (T_{V'}H^*V)-\mathcal{U}$ que l'on note encore $T_{V'}$ et l'on introduit le $\mathcal{K}$-morphisme $\delta : T_{V'}H^*V \to H^*V$ adjoint du $\mathcal{K}$-morphisme $H^*V \to H^*V' \otimes H^*V \cong H^*(V' \oplus V)$ induit par l'homomorphisme $(v', v) \mapsto v' + v$.

Proposition 3.1.— *Soit M un H^*V-A-module instable. On a un isomorphisme naturel de H^*V-A-modules instables :*

$$H^*V \otimes_{T_{V'}H^*V} T_{V'}M \cong H^*V \otimes_{H^*V/V'} \mathrm{Fix}_{(V,V')}M$$

H^*V étant vu comme un $(T_{V'}H^*V)$-module via δ.

3.2. Les foncteurs $\mathrm{Fix}_{(V,V')}$ possèdent les propriétés suivantes.

Proposition 3.2.1.— *Le foncteur $\mathrm{Fix}_{(V,V')} : H^*V\text{-}\mathcal{U} \to H^*V/V'\text{-}\mathcal{U}$ est exact.*

(Rappelons que c'est une conséquence de la proposition 3.1 et de l'exactitude du foncteur $T_{V'}$.)
Soient M un H^*V-module et α un élément non nul de H^*V, nous notons $M[\alpha^{-1}]$ le localisé de M par rapport à la partie multiplicativement stable de H^*V engendrée par l'élément α.

Proposition 3.2.2.— *Soient M un H^*V-A-module instable et u un élément de $H^1V - \{0\}$ (que l'on peut voir comme une forme linéaire non nulle sur V). Les propriétés suivantes sont équivalentes :*
  (i) $M[u^{-1}] = 0$;
  (ii) $\mathrm{Fix}_{(V,V')}M = 0$ pour tout sous-groupe V' de V de dimension un tel que $u(V')$ est non nul.

Démonstration. Le H^*V-A-module instable M est aussi un $H^*(V/\ker(u))$-A-module instable, nous pouvons donc considérer le A-module instable $\mathrm{Fix}_{V/\ker(u)}M$. Dans [LZ] on montre qu'il y a équivalence entre les propriétés suivantes :

(i) $M[u^{-1}] = 0$;

(ii) $\mathrm{Fix}_{V/\ker(u)}M = 0$;

(iii) $\mathrm{Fix}_{s(V/\ker(u)),\ker(u)}M = 0$ pour toute section (linéaire) s de la projection $V \to V/\ker(u)$.

Il est évident que la propriété (iii) est équivalente à la suivante :

(iii-bis) $\mathrm{Fix}_{V',\ker(u)}M = 0$ pour tout sous-groupe V' de dimension un de V tel que $u(V')$ est non nul.

La proposition 3.2.2 découle de l'isomorphisme $\mathrm{Fix}_{V',\ker(u)}M \cong \mathrm{Fix}_{(V,V')}M$.

Lemme 3.2.3.— *Soit M un H^*V-A-module instable vérifiant :*

(a) $\mathrm{Fix}_{(V,V')}M = 0$, pour tout sous-espace V' de dimension un de V ;

(b) la partie triviale τM de M est nulle.

Alors un tel M est nul.

Démonstration. Ecrivons $H^*V = \mathbb{F}_2[u_1, u_2, \cdots, u_r]$ avec $u_i, 1 \le i \le r$, de degré un. Il résulte de la proposition 3.2.2 que $M[u_i^{-1}]$ est nul pour tout i. Soit x un élément de M, il existe donc, pour tout i, un entier ℓ_i tel que $u_i^{\ell_i}x$ est nul. Considérons maintenant l'ensemble $E(x)$ des entiers n tels que $u_1^{m_1}u_2^{m_2}\cdots u_r^{m_r}x$ est nul pour tout r-uplet d'entiers $(m_1, m_2, \cdots, m_r)$ vérifiant $\Sigma_{1 \le i \le r}\, m_i \ge n$. L'ensemble $E(x)$ satisfait les deux propriétés suivantes :

$- \Sigma_{1 \le i \le r}\, \ell_i \in E(x)$;

$- n+1 \in E \Longrightarrow n \in E(x)$ (utiliser le fait que τM est nul).

Ceci prouve $E(x) = \mathbb{N}$ et donc $x = 0$.

3.3. Foncteurs Fix et enveloppes injectives

Proposition 3.3.1.— *Soient V' un sous-groupe de V et I un H^*V-$\mathcal{U}$-injectif qui est de type fini comme H^*V-module. Alors $\mathrm{Fix}_{(V,V')}I$ est un H^*V/V'-$\mathcal{U}$-injectif qui est de type fini comme H^*V/V'-module.*

Démonstration. La classification des H^*V-$\mathcal{U}$-injectifs (voir 2.1.1) montre que tout H^*V-$\mathcal{U}$-injectif qui est de type fini comme H^*V-module est isomorphe à une somme directe finie de $E(V, W, n)$. Il suffit donc de montrer que $\mathrm{Fix}_{(V,V')}E(V, W, n)$ est un H^*V/V'-$\mathcal{U}$-injectif qui est de type fini comme H^*V/V'-module.

Soit M un H^*V/W-A-module instable fini, on vérifie que l'on a :

$$\mathrm{Fix}_{(V,V')}(H^*V \otimes_{H^*V/W} M) \cong \begin{cases} H^*V/V' \otimes_{H^*V/W} M, & \text{si } V' \subset W, \\ 0, & \text{sinon} . \end{cases}$$

En particulier, pour $M = J_{V/W}(n)$, on obtient :

$$\mathrm{Fix}_{(V,V')}E(V, W, n) \cong \begin{cases} E(V/V', W/V', n) & \text{, si } V' \subset W, \\ 0 & \text{, sinon} . \end{cases}$$

Remarque 3.3.2. On vérifie également que l'application naturelle

$$\eta_{(V,V')} : E(V,W,n) \to H^*V \otimes_{H^* \cdot V/V'} E(V,W,n)$$

est un isomorphisme si $V' \subset W$ et est triviale sinon.

3.3.3. Soit M un H^*V-A-module instable. Rappelons que l'enveloppe injective de M est un couple (E, i) où E est un H^*V-$\mathcal{U}$-injectif et $i : M \to E$ est une application H^*V-A-linéaire essentielle (c'est-à-dire qu'un sous-H^*V-A-module P de E est trivial si et seulement si $i^{-1}(P)$ l'est). En particulier i est injective.

Proposition 3.3.3.— *Soient M un $(H^*V)_{tf}$-A-module instable et E son enveloppe injective. Alors pour tout sous-groupe V' de V, $\mathrm{Fix}_{(V,V')}E$ est l'enveloppe injective de $\mathrm{Fix}_{(V,V')}M$.*

Démonstration. Soient M un $(H^*V)_{tf}$-A-module instable et E son enveloppe injective. On a le diagramme commutatif suivant :

$$
\begin{array}{ccc}
M & \xrightarrow{\eta_{(V,V')}} & H^*V \otimes_{H^* \cdot V/V'} \mathrm{Fix}_{(V,V')}M \\
\uparrow & & \uparrow \\
0 \to \ker \eta_{(V,V')} \to E & \xrightarrow{\eta_{(V,V')}} & H^*V \otimes_{H^* \cdot V/V'} \mathrm{Fix}_{(V,V')}E \to 0
\end{array}
$$

dont la ligne du bas est une suite exacte scindable d'après la remarque 3.3.2. Soit P un sous-H^*V-A-module de $H^*V \otimes_{H^* \cdot V/V'} \mathrm{Fix}_{(V,V')}E$ avec tel que l'intersection $P \cap (H^*V \otimes_{H^* \cdot V/V'} \mathrm{Fix}_{(V,V')}M)$ est nul. Comme la suite exacte du bas est scindable, il existe un sous-H^*V-A-module Q de E vérifiant $\eta_{(V,V')}Q = P$ et $Q \cap \ker \eta_{(V,V')} = 0$. On a $Q \cap M = 0$ puisque $Q \cap M$ est inclus dans $\ker \eta_{(V,V')}$; on a $Q = 0$ puisque E est l'enveloppe injective de M. Il en résulte que $H^*V \otimes_{H^* \cdot V/V'} \mathrm{Fix}_{(V,V')}E$ est l'enveloppe injective de $H^*V \otimes_{H^* \cdot V/V'} \mathrm{Fix}_{(V,V')}M$. Ceci implique aisément que $\mathrm{Fix}_{(V,V')}E$ (qui est H^*V/V'-$\mathcal{U}$-injectif d'après 3.3.1) est l'enveloppe injective de $\mathrm{Fix}_{(V,V')}M$.

3.4. Foncteurs $\mathrm{Fix}_{(V,V')}$ et H^*V-A-modules instables à Tor-dimension finie
Soient P un $H^*V - A$-module instable et V' un sous-espace de V. On note $\bar{\eta}_{(V,V')} : \mathbb{F}_2 \otimes_{H^* \cdot V/V'} P \to \mathbb{F}_2 \otimes_{H^*V/V'} (H^*V \otimes_{H^* \cdot V/V'} \mathrm{Fix}_{(V,V')}P) = H^*V' \otimes (\mathbb{F}_2 \otimes_{H^* \cdot V/V'} \mathrm{Fix}_{(V,V')}P)$ le H^*V'-$\mathcal{U}$-morphisme induit par $\eta_{(V,V')} : P \to H^*V \otimes_{H^* \cdot V/V'} \mathrm{Fix}_{(V,V')}P$.
On vérifie que l'adjointe de $\bar{\eta}_{(V,V')}$:

$$\mathrm{Fix}_{V'}(\mathbb{F}_2 \otimes_{H^* \cdot V/V'} P) \to \mathbb{F}_2 \otimes_{H^*V/V'} \mathrm{Fix}_{(V,V')}P$$

est un $\mathcal{U}$-isomorphisme. Plus généralement :

Lemme 3.4.1.— *Soient M un H^*V-A-module instable et V' un sous-groupe de V. Alors, pour tout s, on a un $\mathcal{U}$-isomorphisme naturel :*

$$\mathrm{Fix}_{V'}(\mathrm{Tor}_s^{V/V'}(M)) \to \mathrm{Tor}_s^{V/V'}(\mathrm{Fix}_{(V,V')}M) \ .$$

Démonstration. Soit $P_\bullet \to M \to 0$ une H^*V-$\mathcal{U}$-résolution libre de M ; c'est aussi une H^*V/V'-$\mathcal{U}$-résolution de M qui est H^*V/V'-libre. Le lemme résulte des points suivants :

– d'après ce qui précède on a un isomorphisme naturel de $\mathcal{U}$-complexes $\mathrm{Fix}_{V'}(\mathbb{F}_2 \otimes_{H^*V/V'} P_\bullet) \to \mathbb{F}_2 \otimes_{H^*V/V'} \mathrm{Fix}_{(V,V')} P_\bullet$;

– le foncteur $\mathrm{Fix}_{(V,V')}$ transforme un H^*V-$\mathcal{U}$-objet libre en un H^*V/V'-$\mathcal{U}$-objet libre ;

– le foncteur $\mathrm{Fix}_{(V,V')}$ est exact.

Lemme 3.4.2.— *Soient M un H^*V-A-module instable et V' un sous-groupe de V de dimension un. Alors, pour tout s, on a une injection naturelle :*

$$\mathrm{Tor}_0^{V'}(\mathrm{Tor}_s^{V/V'} M) \hookrightarrow \mathrm{Tor}_s^V M .$$

Démonstration (à la main !). Soient M un H^*V-A-module instable et $P_\bullet \to M \to 0$ une H^*V-$\mathcal{U}$-résolution libre de M. Posons $H = H^*V'$; H est une algèbre polynômiale sur un générateur u de degré un. On a $\mathrm{Tor}_s^V M = H_s(\mathbb{F}_2 \otimes_H C_\bullet)$ avec $C_\bullet = \mathbb{F}_2 \otimes_{H^*V/V'} P_\bullet$. Comme chaque terme du complexe $C_\bullet$ est H-libre on a une suite exacte de complexes $0 \to uC_\bullet \to C_\bullet \to \mathbb{F}_2 \otimes_H C_\bullet \to 0$ qui donne une longue suite exacte

$$\cdots \to u\,\mathrm{Tor}_s^{V/V'} M \to \mathrm{Tor}_s^{V/V'} M \to \mathrm{Tor}_s^V M \to \cdots .$$

D'où en particulier une injection naturelle $\mathrm{Tor}_0^{V'}(\mathrm{Tor}_s^{V/V'} M) \hookrightarrow \mathrm{Tor}_s^V M$.

Lemme 3.4.3.— *Soit M un H^*V-A-module instable. Soit k un entier ; on suppose $\|\mathrm{Tor}_0^V M\| \leq k$. Alors :*

(a) *On a également $\|\mathrm{Fix}_V M\| \leq k$.*

(b) *L'homomorphisme $\bar{\eta}_V : \mathrm{Tor}_0^V M \to \mathrm{Fix}_V M$ induit par $\eta_V : M \to H^*V \otimes \mathrm{Fix}_V M$ est surjectif en degré k.*

(La notation $\| - \|$ est introduite en 2.1.4.2 (e).)

Démonstration. Il est clair que (b) implique (a). Le point (b) peut être reformulé ainsi : l'homomorphisme canonique

$$\mathrm{Hom}_{H^*V - \mathcal{U}}(M, H^*V \otimes J(k)) \to \mathrm{Hom}_{H^*V - \mathcal{U}}(M, (H^*V \otimes J(k))/(\widetilde{H}^*V \otimes J(k)))$$
$$= \mathrm{Hom}_{\mathcal{U}}(\mathrm{Tor}_0^V M, J(k))$$

est injectif. (Rappelons que $J(k)$ est le A-module instable caractérisé, à isomorphisme près, par la bijection fonctorielle en le A-module instable $L : \mathrm{Hom}_{\mathcal{U}}(L, J(k)) \cong \mathrm{Hom}_{\mathbb{F}_2}(L^k, \mathbb{F}_2)$.). Il suffit donc de montrer que $\mathrm{Hom}_{H^*V - \mathcal{U}}(M, \widetilde{H}^*V \otimes J(k))$ est trivial. Pour cela on considère la filtration

$$\widetilde{H}^*V \otimes J(k) = N_1 \supset N_2 \supset \cdots \supset N_m \supset \cdots$$

définie par $N_m = (H^*V)^{\geq m} \otimes J(k)$, $(H^*V)^{\geq m}$ désignant l'idéal de H^*V formé des éléments de degré supérieur ou égal à m (la puissance m-ième de l'idéal d'augmentation). Le quotient N_m/N_{m+1} est isomorphe à $\Sigma^m H^m V \otimes J(k)$ et se plonge dans $H^m V \otimes J(k+m)$; ici les A-modules instables $\Sigma^m H^m V \otimes J(k)$ et $H^m V \otimes J(k+m)$ sont considérés comme des H^*V-A-modules instables *via* l'augmentation de H^*V. On en déduit que l'homomorphisme

$$\mathrm{Hom}_{H^*V - \mathcal{U}}(M, N_{m+1}) \to \mathrm{Hom}_{H^*V - \mathcal{U}}(M, N_m)$$

est surjectif pour tout $m \geq 1$ et que $\mathrm{Hom}_{H^*V - \mathcal{U}}(M, N_1)$ est bien trivial.

Proposition 3.4.4.— *Soient M un $(H^*V)_{tf}$-A-module instable et V' un sous-espace de V de dimension un. Alors on a l'inégalité :*

$$d_{V/V'} \, \mathrm{Fix}_{(V,V')} M \le d_V M \ .$$

Démonstration. On pose $d_V M = n$, il s'agit de prouver :

$$\|\mathrm{Tor}_s^{V/V'} \, \mathrm{Fix}_{(V,V')} M\| \le s + n \ .$$

Pour cela, on montre que le morphisme $\mathrm{Tor}_s^V M \to \mathrm{Tor}_s^{V/V'} \, \mathrm{Fix}_{(V,V')} M$ induit par l'application naturelle $\eta_{(V,V')} : M \to H^*V \otimes_{H^*V/V'} \mathrm{Fix}_{(V,V')} M$ est surjectif en degré supérieur ou égal à $s + n$. Ceci résulte du diagramme commutatif

$$
\begin{array}{ccc}
\mathrm{Tor}_s^V M & \longrightarrow & \mathrm{Tor}_s^{V/V'} \, \mathrm{Fix}_{(V,V')} M \\[2mm]
\big\uparrow & & \big\uparrow {\scriptstyle \cong} \\[2mm]
\mathrm{Tor}_0^{V'} \, \mathrm{Tor}_s^{V/V'} M & \longrightarrow & \mathrm{Fix}_{V'} \, \mathrm{Tor}_s^{V/V'} M
\end{array}
$$

dans lequel :
- la flèche verticale de droite est l'isomorphisme du lemme 3.4.1 ;
- la flèche verticale de gauche est le monomorphisme du lemme 3.4.2 ;
- la flèche horizontale du bas est l'homomorphisme $\bar\eta$ du lemme 3.4.3 (V étant remplacé par V' et M par $\mathrm{Tor}_s^{V/V'} M$).

4 Sur les H^*V-A-modules instables à Tor-dimension finie

Dans ce paragraphe nous donnons quelques propriétés des H^*V-A- modules instables à Tor-dimension finie. Nous étudions en particulier leurs enveloppes injectives. La proposition 4.2.3 est le point technique clé de notre article.

4.1. Commençons par rappeler ce qui se passe dans le cas $V = 0$.
Soient $\Sigma : \mathcal{U} \to \mathcal{U}, M \mapsto \Sigma M$, le foncteur suspension et $\widetilde{\Sigma} : \mathcal{U} \to \mathcal{U}, N \mapsto \widetilde{\Sigma} N$ son adjoint à droite ; $\Sigma \widetilde{\Sigma} N$ est simplement le sous-A-module de N formé des éléments x tels que $Sq^{|x|}x$ est nul. On a donc pour tous A-modules instables M et N :
$$\mathrm{Hom}_{\mathcal{U}} \, (\Sigma M, N) \cong \mathrm{Hom}_{\mathcal{U}} \, (M, \widetilde{\Sigma} N) \ .$$

On vérifie sans difficultés l'énoncé suivant (on note $\widetilde{\Sigma}^k$ le k-ième itéré du foncteur $\widetilde{\Sigma}$) :

Proposition 4.1.— *Soit n un entier. Soit M un A-module instable vérifiant $\|M\| \le n$ et E son enveloppe injective. Alors :*
(a) $\widetilde{\Sigma}^{n+1} M = 0$;
(b) $\widetilde{\Sigma}^n (E/M) = 0$;
(c) $\widetilde{\Sigma}^{n+1} E = 0$;
(d) $\widetilde{\Sigma}^n M$ est isomorphe à $\widetilde{\Sigma}^n E$.

4.2. Comme nous l'avons déjà remarqué, le $\mathcal{U}$-endofoncteur Σ induit un H^*V-$\mathcal{U}$-endofoncteur que l'on note toujours Σ ; il en est de même pour $\widetilde{\Sigma}$. De plus l'isomorphisme $\widetilde{\Sigma}(H^*V \otimes N) \cong H^*V \otimes \widetilde{\Sigma}N$ (qui est évident "pour $p = 2$" parce que l'élévation au carré est injective dans H^*V) montre que $\widetilde{\Sigma}$ est encore l'adjoint à droite de Σ dans la catégorie H^*V-$\mathcal{U}$.

Ce foncteur $\widetilde{\Sigma}$ vérifie les propriétés suivantes dont les démonstrations sont laissées au lecteur.

Lemme 4.2.1.—
(a) *Le foncteur* $\widetilde{\Sigma}$: H^*V-$\mathcal{U} \to H^*V$-$\mathcal{U}$ *est exact à gauche.*
(b) *Il transforme* H^*V-$\mathcal{U}$-*injectif en* H^*V-$\mathcal{U}$-*injectif. On a de plus :*

$$\widetilde{\Sigma}E(V,W,n) \cong E(V,W,n-1).$$

(c) *Soient* M *un* H^*V-A-*module instable et* E *son enveloppe injective. Alors* $\widetilde{\Sigma}E$ *est l'enveloppe injective de* $\widetilde{\Sigma}M$.

Soient M un H^*V-A-module instable et V' un sous-espace de V. Considérons l'application naturelle $\eta_{(V,V')}$: $M \to H^*V \otimes_{H^*V/V'} \mathrm{Fix}_{(V,V')}M$; $\widetilde{\Sigma}\eta_{(V,V')}$ s'identifie à un H^*V-$\mathcal{U}$-morphisme $\widetilde{\Sigma}M \to H^*V \otimes_{H^*V/V'} \widetilde{\Sigma}\,\mathrm{Fix}_{(V,V')}M$ dont l'adjoint est un H^*V/V'-$\mathcal{U}$-morphisme σ : $\mathrm{Fix}_{(V,V')}\widetilde{\Sigma}M \to \widetilde{\Sigma}\,\mathrm{Fix}_{(V,V')}M$.

Proposition 4.2.2.— *Soit* M *un* H^*V-A-*module instable. L'homomorphisme naturel* σ : $\mathrm{Fix}_{(V,V')}\widetilde{\Sigma}M \to \widetilde{\Sigma}\,\mathrm{Fix}_{(V,V')}M$ *est un isomorphisme.*

Démonstration. Le calcul de $\mathrm{Fix}_{(V,V')}E(V,W,n)$ évoqué dans la démonstration de 3.3.1 et l'isomorphisme de 4.2.1 (b) montrent que σ est un isomorphisme pour $M = E(V,W,n)$. On en déduit qu'il en est de même pour $M = L_{\ t}\otimes E(V,W,n)$ (notations de 2.1.1 et de [LZ, 1.2]) ; on utilise ici que l'on a un H^*V/V'-$\mathcal{U}$-isomorphisme naturel

$$\mathrm{Fix}_{(V,V')}(L_{\ t}\otimes M) \cong T_{V'}L_{\ t}\otimes \mathrm{Fix}_{(V,V')}M$$

(ci-dessus L et M sont respectivement un A-module instable et un H^*V-A-module instable arbitraires) ce dont on peut se convaincre avec 3.1. On conclut par un argument standard de résolutions injectives.

Voici maintenant l'analogue dans la catégorie H^*V-$\mathcal{U}$ de la proposition 4.1. Sa démonstration est reportée à 4.2.7 ; on observera cependant que (c) et (d) sont conséquences immédiates de (a) et (b) !

Proposition 4.2.3.— *Soit* n *un entier. Soit* M *un* A-*module instable vérifiant* $d_V M \leq n$ *et* E *son enveloppe injective. Alors :*
(a) $\widetilde{\Sigma}^{n+1}M = 0$;
(b) $\widetilde{\Sigma}^n(E/M) = 0$;
(c) $\widetilde{\Sigma}^{n+1}E = 0$;
(d) $\widetilde{\Sigma}^n M$ *est isomorphe à* $\widetilde{\Sigma}^n E$.

Le théorème 2.1.2 et l'isomorphisme de 4.2.1 (b) impliquent alors :

Scholie 4.2.4.— *Si la Tor-dimension d'un $(H^*V)_{tf}$-A-module instable est inférieure ou égale à un entier n alors son enveloppe injective est une somme directe finie de $E(V, W, k)$ avec $k \leq n$.*

Avant de prouver la proposition 4.2.3 nous avons besoin de dégager certaines propriétés du foncteur "partie triviale", $\tau : H^*V\text{-}\mathcal{U} \to \mathcal{U}$ (voir 2.1.4.2 (c)). Le lemme suivant est immédiat :

Lemme 4.2.5.—
(a) *Le foncteur $\tau : H^*V\text{-}\mathcal{U} \to \mathcal{U}$ est exact à gauche.*
(b) *Le foncteur $\tau : H^*V\text{-}\mathcal{U} \to \mathcal{U}$ transforme $H^*V\text{-}\mathcal{U}$-injectif en $\mathcal{U}$-injectif.*
(c) *Soient M un H^*V-A-module instable et E son enveloppe injective, alors τE est l'enveloppe injective de τM.*

Lemme 4.2.6.— *Soient M un H^*V-A-module instable et E son enveloppe injective. On a les inégalités :*
(a) $\|\tau M\| \leq d_V M$;
(b) $\|\tau(E/M)\| \leq d_V M - 1$.

Démonstration. On pose $n = d_V M$ (il est clair que l'on peut supposer cette Tor-dimension finie) et on note r la dimension du $\mathbb{F}_2$-espace vectoriel V. On a $\|\tau M\| \leq n$ puisque le A-module instable $\Sigma^r \tau M$ est isomorphe à $\mathrm{Tor}_r^V M$ et que l'on a par définition $\|\mathrm{Tor}_r^V M\| \leq r + n$. Le point (c) de 4.2.5 et la proposition 4.1.1 montrent en outre que $(\tau M)^n$ est isomorphe à $(\tau E)^n$; l'inégalité (b) découle alors de la suite exacte

$$0 \to \Sigma^r \tau M \to \Sigma^r \tau E \to \Sigma^r \tau(E/M) \to \mathrm{Tor}_{r-1}^V M \to \cdots$$

et de l'inégalité $\|\mathrm{Tor}_{r-1}^V M\| \leq r - 1 + n$.

4.2.7. Démonstration de la proposition 4.2.3.
La démonstration se fait par récurrence sur la dimension du $\mathbb{F}_2$-espace vectoriel V. Le cas $\dim V = 0$ est réglé par la proposition 4.1.1. Montrons donc que si la proposition est vérifiée pour $\dim V = r - 1$ elle l'est encore pour $\dim V = r$. Soient V un $\mathbb{F}_2$-espace vectoriel de dimension r et V' un sous-espace de dimension un. D'après la proposition 3.4.4 nous avons $d_{V/V'} \mathrm{Fix}_{(V,V')} M \leq n$. Les propositions 3.2.1, 4.2.2, 3.3.3 et l'hypothèse de récurrence impliquent :
(a') $\mathrm{Fix}_{(V,V')}(\widetilde{\Sigma}^{n+1} M) = 0$;
(b') $\mathrm{Fix}_{(V,V')}(\widetilde{\Sigma}^n (E/M)) = 0$.
Pour montrer que $\widetilde{\Sigma}^{n+1} M$ et $\widetilde{\Sigma}^n (E/M)$ sont nuls il suffit, d'après 3.2.3, de montrer que les parties triviales $\tau(\widetilde{\Sigma}^{n+1} M)$ et $\tau(\widetilde{\Sigma}^n (E/M))$ le sont.
Puisque $\Sigma^{n+1} \tau(\widetilde{\Sigma}^{n+1} M) = \tau(\Sigma^{n+1} \widetilde{\Sigma}^{n+1} M)$ est un sous-module de τM on a d'une part $\|\tau(\widetilde{\Sigma}^{n+1} M)\| \leq \|\tau M\| - n - 1$. On a d'autre part $\|\tau M\| \leq n$ d'après 4.2.6 (a). D'où $\tau(\widetilde{\Sigma}^{n+1} M) = 0$ (avec nos conventions $\|0\| = -\infty$!). En utilisant 4.2.6 (b) on montre pareillement $\tau(\widetilde{\Sigma}^n (E/M)) = 0$.
Nous terminons ce paragraphe par le lemme ci-dessous. Le point (a) affirme que 2.1.5 est vrai pour $M = E(V, W, n)$; (c) est un point technique qui sera utilisé au paragraphe suivant.

Lemme 4.2.8.— *Soient W un sous-groupe de V et n un entier. Alors :*

(a) $d_V E(V, W, n) = n$;

(b) $d_V(E(V, W, n)/\Sigma^n \widetilde{\Sigma}^n E(V, W, n)) = n - 1$ *pour* $n > 0$;

(c) *l'homomorphisme* $\operatorname{Tor}_s^V(\Sigma^n \widetilde{\Sigma}^n E(V, W, n)) \to \operatorname{Tor}_s^V E(V, W, n)$ *est surjectif en degré* $s + n$, *pour tout* s.

Démonstration. Le point (c) est conséquence du point (b). Les points (a) et (b) sont conséquences des points (f) et (e) de 2.1.4.2 et des formules :

$$\|J_{V/W}(n)\| = n, \|J_{V/W}(n)/\Sigma^n \widetilde{\Sigma}^n J_{V/W}(n)\| = n - 1 .$$

5 Démonstrations de 2.1.3.2 et 2.1.5

5.1. Démonstration de la proposition 2.1.3.2.

Seule l'implication (i) $\Rightarrow$ (ii) mérite une démonstration.

Soit M un $(H^*V)_{tf}$-A-module instable avec $e_V M \leq n$. Il suffit de montrer que l'enveloppe injective E de M est une somme directe finie de $E(V, W, k)$ avec $k \leq n$ et que l'on a $e_V(E/M) \leq n - 1$, pour $n > 0$, ou $E/M = 0$, pour $n = 0$. L'hypothèse $e_V M \leq n$ implique en particulier que le foncteur $\operatorname{Hom}_{H^*V - \mathcal{U}}(\Sigma^t(-), M)$ est nul pour $t = n + 1$ et exact pour $t = n$; en d'autres termes que le H^*V-A-module instable $\widetilde{\Sigma}^t M$ est nul pour $t = n + 1$ et injectif pour $t = n$. On a donc d'après 4.2.1 (c) :

(a) $\widetilde{\Sigma}^{n+1} E$ est nul ;

(b) $\widetilde{\Sigma}^n M \to \widetilde{\Sigma}^n E$ est un isomorphisme.

Le point (a), le théorème 2.1.2 et l'isomorphisme de 4.2.1 (b) impliquent bien que E est une somme directe finie de $E(V, W, k)$ avec $k \leq n$.

Le point (b) dit que E/M est nul pour $n = 0$ et en général que la transformation naturelle $\operatorname{Hom}_{H^*V - \mathcal{U}}(\Sigma^n(-), M) \to \operatorname{Hom}_{H^*V - \mathcal{U}}(\Sigma^n(-), E)$ est un isomorphisme ; on voit alors en considérant les longues suites exactes de foncteurs Ext associées à la suite exacte $0 \to M \to E \to E/M \to 0$ que l'on a bien $e_V(E/M) \leq n - 1$ pour $n > 0$.

5.2. Démonstration du théorème 2.1.5

L'inégalité $d_V M \leq e_V M$ est facile à établir ; par contre l'inégalité $e_V M \leq d_V M$ est plus délicate (mais tout le travail a déjà été fait dans les paragraphes précédents !).

Soient E l'enveloppe injective de M et n un entier.

5.2.1. Démonstration de l'implication $e_V M \leq n \Rightarrow d_V M \leq n$.

Nous raisonnons par récurrence sur n. D'après 2.1.3.3 et 4.2.8 (a) on a $d_V E \leq n$. Pour $n = 0$, M est isomorphe à E. Pour $n > 0$, on a $e_V(E/M) \leq n - 1$ d'après 2.1.3.2, donc $d_V(E/M) \leq n - 1$ par hypothèse de récurrence, ce qui implique $d_V M \leq n$ (considérer la longue suite exacte de foncteurs Tor associée à la suite exacte $0 \to M \to E \to E/M \to 0$).

5.2.2. Démonstration de l'implication $d_V M \leq n \Rightarrow e_V M \leq n$.

Nous raisonnons toujours par récurrence sur n. D'après 4.2.4 on a $e_V E \leq n$. Pour $n = 0$, M est isomorphe à E d'après 4.2.3. Pour $n > 0$, compte tenu comme précédemment de l'hypothèse de récurrence, il suffit de montrer $d_V(E/M) \leq n-1$. Pour cela on considère le diagramme commutatif suivant :

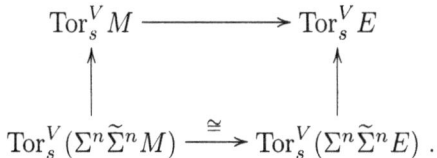

Celui-ci montre que l'homomorphisme $\mathrm{Tor}_s^V M \to \mathrm{Tor}_s^V E$ est surjectif en degré $s+n$; en effet, d'après 4.2.8 (c), il en est déjà ainsi pour l'homomorphisme $\mathrm{Tor}_s^V(\Sigma^n \widetilde{\Sigma}^n E) \to \mathrm{Tor}_s^V E$ et, d'après 4.2.3, $\widetilde{\Sigma}^n M$ est isomorphe à $\widetilde{\Sigma}^n E$. On voit alors en considérant encore la longue suite exacte de foncteurs Tor associée à la suite exacte $0 \to M \to E \to E/M \to 0$ que l'on a bien $d_V(E/M) \leq n-1$.

Références

[D] J. Duflot, Smooth toral actions, *Topology*, **22**, 1983, 253–265.

[DW] W. G. Dwyer, et C. W. Wilkerson, Smith theory and the functor T, *Comm. Math. Helv.*, **66**, 1991, 1–17.

[L] J. Lannes, Sur les espaces fonctionnels dont la source est le classifiant d'un p-groupe abélien élémentaire, *Pub. Sci., I. H. E. S.*, **75**, 1992, 135–244.

[LZ] J. Lannes et S. Zarati, Théorie de Smith algébrique et classification des H^*V-$\mathcal{U}$-injectifs, Bull. Soc. Math. France, **123**, 1995, 101–135.

Jean Lannes, URA du CNRS D0169, Centre de Mathématiques
Ecole Polytechnique, F-91128 Palaiseau Cedex, France

Saïd Zarati, Université de Tunis II
Le Campus Universitaire, TN-1060 Tunis, Tunisie

Progress in Mathematics, Vol. 136
© 1996 Birkhäuser Verlag Basel/Switzerland

A bound on the exponent of the cohomology of BC-bundles

I. J. LEARY

We give a lower bound for the exponent of certain elements in the integral cohomology of the total spaces of principal BC-bundles for C a finite cyclic group. We are mainly interested in the case when the total space is BG for some discrete group G having a central subgroup isomorphic to C. As applications we give a proof of the theorem of A. Adem and H.-W. Henn that a p-group is elementary abelian if and only if its integral cohomology has exponent p, and we exhibit some infinite groups of finite virtual cohomological dimension whose Tate-Farrell cohomology contains torsion of order greater than the l.c.m. of the orders of their finite subgroups. Our examples include a class of groups having similar properties discovered by Adem and J. Carlson. As a third application, we examine the integral cohomology of a class of p-groups expressible as central extensions with cyclic kernel and quotient abelian of p-rank two. For each such G we determine the minimal n such that almost all (i.e. all but possibly finitely many) of the groups $H^i(BG)$ have exponent dividing p^n. The lemma we use to give an upper bound for the exponents of almost all of the groups $H^i(BG)$ applies to any p-group and may be of independent interest. Here, and throughout the paper, the coefficients for cohomology are to be the integers when not otherwise stated, and we write $\mathbb{Z}_n$ for the integers modulo n. The author gratefully acknowledges that this work was funded by the DGICYT.

PROPOSITION 1. *Let C be a cyclic group of order n, and let E be a principal BC-bundle over a connected space X, classified by $\xi \in H^2(X;C)$ of order m. Then for any $i \geq 0$, any element of $H^{2i}(E)$ restricting to the fibre as a generator for $H^{2i}(BC)$ has order divisible by mn.*

REMARK. Note that we do not claim that such elements always exist, nor do we rule out the possibility that they have infinite order.

Proof. In [4] Cartan and Eilenberg computed the ring $H^*(BC;R)$ for any coefficient ring R. Recall that we have the following ring isomorphisms:

$$H^*(BC) \cong \mathbb{Z}[z]/(nz), \qquad H^*(BC;\mathbb{Z}_n) \cong \mathbb{Z}_n[x,y]/(ny, nx, y^2 - ex),$$

where $e = 0$ if n is odd and $e = n/2$ if n is even, and y has degree 1 while x and z have degree two. The natural map from integral to mod-n cohomology sends z to x, and if we let β stand for the Bockstein for the coefficient sequence

$$0 \to \mathbb{Z} \to \mathbb{Z} \to \mathbb{Z}_n \to 0,$$

then it is easy to see that $\beta(y) = z$, and that therefore $\beta(yx^i) = z^{i+1}$.

Now consider the spectral sequence for the given fibration with coefficients in $\mathbb{Z}_n$. By assumption the fundamental group of X acts trivially on the cohomology of BC, and so

$$E_2^{i,j} \cong H^i(X;\mathbb{Z}_n) \otimes H^j(BC;\mathbb{Z}_n).$$

Now $1 \otimes yx^j$ represents a generator for $E_2^{0,2j+1}$ and $1 \otimes x^j$ represents a generator for $E_2^{0,2j}$. Comparing this spectral sequence with the spectral sequence for the path-loop fibration over an Eilenberg-MacLane space $K(C,2)$ it is easy to see that $d_2(1 \otimes y) = \xi$ and $d_2(1 \otimes x) = 0$. (In fact, $d_3(1 \otimes x) = \xi' \otimes 1$, where ξ' is the image of $\beta(\xi)$ under the map from $H^3(X)$ to $H^3(X;\mathbb{Z}_n)$, and d_4 may be described using the argument given in [8], but we do not need this here.) Now $d_2(1 \otimes x^j y) = \xi \otimes x^j$ and $d_2(1 \otimes x^j) = 0$, from which it follows that $E_3^{0,2j}$ is generated by $1 \otimes x^j$ and $E_3^{0,2j+1}$ by $m(1 \otimes yx^j)$. The map from $H^*(E;\mathbb{Z}_n)$ to $H^*(BC;\mathbb{Z}_n)$ factors through $E_\infty^{0,*}$, which is a subgroup of $E_3^{0,*}$, and so we see that the image of $H^{2j+1}(E;\mathbb{Z}_n)$ in $H^{2j+1}(BC;\mathbb{Z}_n)$ must be contained in the subgroup generated by myx^j.

Now recall that the image of the Bockstein β defined above is exactly the elements of integral cohomology of order dividing n. Let $f : BC \to E$ be the inclusion of the fibre of the above fibration. Now let χ be an element of $H^*(E)$ such that $f^*(\chi) = z^{j+1}$ for some j. If χ has infinite order then there is nothing to prove. Otherwise, the order of χ must be a multiple of n (the order of z^{j+1}), say $m'n$, and it remains to show that m divides m'. Now $m'\chi$ has order n, so there exists $\chi' \in H^{2j+1}(E;\mathbb{Z}_n)$ such that $\beta(\chi') = m'\chi$. However, the spectral sequence argument shows that $f^*(\chi')$ is in the subgroup of $H^{2j+1}(BC;\mathbb{Z}_n)$ generated by myx^j and hence $\beta f^*(\chi')$ is in the subgroup of $H^{2j+2}(BC)$ generated by mz^{j+1}, but $\beta f^*(\chi') = f^*\beta(\chi') = f^*(\chi) = m'z^{j+1}$. □

COROLLARY 1. *Let C be a cyclic subgroup of order n of a group G. If there exists an element of $H^*(BG)$ of order n whose image in $H^*(BC)$ is a generator for $H^{2i}(BC)$ for some i, then C is a direct factor of its centraliser in G.*

Proof. This is just Proposition 1 applied to the principal BC-bundle with total space the classifying space of the centraliser of C.   □

COROLLARY 2. *Let G be a discrete group expressible as a central extension with kernel C cyclic of order n. Let Q be the quotient G/C, and let the extension class of G in $H^2(BQ;C)$ have order m. If G has a normal subgroup N of finite index whose intersection with C is trivial (for example, if G is finite or residually finite), then for infinitely many i, $H^{2i}(BG)$ contains elements of order mn.*

REMARK. The condition that the extension class of G has order m may be rephrased as follows: If D is the smallest subgroup of C such that G/D is isomorphic to $(C/D) \times Q$, then D has order m.

Proof. Let G' be the quotient G/N, and let C' be the image of C in G'. Then C' is isomorphic to C and G' is finite. By either Evens' argument using the Norm map from $H^*(BC)$ to $H^*(BG)$ [5,6] or Venkov's argument using Chern classes of a representation of G' restricting faithfully to C' [10], we see that for infinitely many i there exists $\chi' \in H^{2i}(BG')$ whose image in $H^{2i}(BC')$ is a generator. If χ is the image of χ' in $H^*(BG)$, then χ has finite order (dividing the order of G') and its image in $H^{2i}(BC)$ is a generator. Hence by Proposition 1, some multiple of χ has order exactly mn.   □

The first example of a group whose Tate-Farrell cohomology contains elements of order greater than the l.c.m. of the orders of its finite subgroups is due to Adem [2]. The following application of Corollary 2 is more closely related to some other examples due to Adem and Carlson [3]. In particular, Corollary 3 may be compared with Theorem 3.1 of [3], which gives stronger cohomological information about a smaller class of groups.

COROLLARY 3. *With notation and hypotheses as in Corollary 2, assume also that Q has finite cohomological dimension (or equivalently, assume that there is a finite-dimensional CW-complex BQ). Then*

   a) *G has finite virtual cohomological dimension and hence the Tate-Farrell cohomology groups $\hat{H}^i(G)$ are defined,*
   b) *C consists of all the elements of G of finite order, and*
   c) *$\hat{H}^i(G)$ contains elements of order mn for infinitely many i.*

Proof. The subgroup N of G has finite index and is isomorphic to a subgroup of Q, so has cohomological dimension less than or equal to that of Q. Hence G has finite vcd. The group Q is torsion-free, and so any element of $G - C$ has infinite order because its image in Q does. If i is greater than vcd$G$ then $\hat{H}^i(G)$ is isomorphic to $H^i(BG)$, and so the third claim follows from Corollary 2.   □

The following Corollary is due to Adem [1] and Henn [7].

COROLLARY 4. *Let G be a finite p-group. Then G is not elementary abelian if and only if $H^i(BG)$ contains elements of order p^2 for some i if and only if $H^i(BG)$ contains elements of order p^2 for infinitely many i.*

Proof. If G is elementary abelian (i.e. is isomorphic to a product of cyclic groups of order p) then $H^i(G)$ has exponent p for $i > 0$ by the Künneth theorem. Conversely, if G is not elementary abelian then G contains a central subgroup of order p which is not a direct factor, or equivalently, C of order p such that the extension class of G in $H^2(BG/C; C)$ has order p. The result now follows by applying Corollary 2.   □

The following application of Proposition 1 is new.

PROPOSITION 2. *For positive integers α, β, γ, δ satisfying the inequalities $0 \leq \gamma - \delta \leq \min\{\alpha, \beta\}$, let $G = G(\alpha, \beta, \gamma, \delta)$ be a p-group with the following presentation.*

$$G = \langle a, b, c \mid [a, c] = [b, c] = 1 = a^{p^{\alpha}} = b^{p^{\beta}} = c^{p^{\gamma}}, \quad [a, b] = c^{p^{\delta}} \rangle$$

Now let ϵ be $\max\{\alpha, \beta, 2\gamma - \delta\}$. Then for infinitely many i, $H^i(BG)$ has exponent p^{ϵ}, and at most finitely many of the groups $H^i(BG)$ have higher exponent.

REMARK. It is easy to see that any group having a presentation of the above form for arbitrary $(\alpha, \beta, \gamma, \delta)$ also has a presentation of the above form in which the inequalities are satisfied: If γ is less than δ, then $c^{p^{\delta}} = c^{p^{\gamma}} = 1$, and so in this case $G(\alpha, \beta, \gamma, \delta)$ is isomorphic to $G(\alpha, \beta, \gamma, \gamma)$. On the other hand, the order of $[a, b] = c^{p^{\delta}}$ is bounded by the orders of a and b given that c is central, and so the order of c is bounded by $p^{\alpha+\delta}$ and $p^{\beta+\delta}$. Thus given a presentation as above but not satisfying the second inequality we could replace γ by $\gamma' = \min\{\alpha+\delta, \beta+\delta\}$ and obtain another presentation of the same group.

Proof. First we recall that for any G and any split surjection from G onto Q, $H^*(BQ)$ occurs as a direct summand of $H^*(BG)$. Now the above group G may be expressed as a split extension with kernel $\langle a, c \rangle$ and quotient $\langle b \rangle \cong \mathbb{Z}/p^{\beta}$, or as a split extension with kernel $\langle b, c \rangle$ and quotient $\langle a \rangle \cong \mathbb{Z}/p^{\alpha}$. Hence we deduce that $H^{2i}(BG)$ has elements of exponents p^{α} and p^{β} for all $i > 0$.

G may also be viewed as a central extension with kernel $\langle c \rangle$ which is isomorphic to $\mathbb{Z}/p^{\gamma}$, and quotient isomorphic to $\mathbb{Z}/p^{\alpha} \oplus \mathbb{Z}/p^{\beta}$ generated by the images of a and b. The extension class of this extension is easily seen to have order $p^{\gamma-\delta}$, and so it follows from Corollary 1 that for infinitely many i, $H^{2i}(BG)$ contains elements of order $p^{2\gamma-\delta}$.

For the partial converse, note that G has subgroups $\langle a, c \rangle$, $\langle b, c \rangle$, and $\langle a, b^{p^{\gamma-\delta}} \rangle$ of index p^{α}, p^{β} and $p^{2\gamma-\delta}$ respectively whose intersection is trivial, and then apply the following Lemma.

LEMMA 1. *Let G be a (finite) p-group, let $H_1, \ldots, H_k$ be a family of subgroups of G such that the index $|G : H_j|$ of each H_j is less than or equal to p^n, and suppose that the intersection*

$$\bigcap_{g \in G, 1 \leq j \leq k} H_j^g$$

of the conjugates of the subgroups H_j is trivial. Then $H^i(BG)$ has exponent dividing p^n for all but finitely many i.

Proof. Let Σ_m be the symmetric group on m symbols and let G_n be the Sylow p-subgroup of Σ_{p^n}. Since the index of $(\Sigma_m)^p$ in Σ_{mp} divides exactly once by p an easy induction argument using the transfer shows that for all $i > 0$ and all n, $H^i(BG_n)$ has exponent dividing p^n. If H is a subgroup of G, then the kernel of the permutation representation of G on the cosets of H is the intersection of the conjugates of H. Hence if G has subgroups $H_1, \ldots, H_k$ as in the statement then G occurs as a subgroup of a product of k symmetric groups on at most p^n symbols, and hence as a subgroup of $(G_n)^k$. The result now follows from the observation due to Adem [1] that for any group G' and any subgroup G, the finite generation of $H^*(BG)$ as an $H^*(BG')$-module implies that at most finitely many of the groups $H^i(BG)$ can have higher exponent than the reduced cohomology $\tilde{H}^*(BG')$. $\square\,\square$

REMARK. The bound given by Lemma 1 for the exponent of almost all of the integral cohomology groups of a p-group is attained for many groups. For example, Proposition 2 shows that the bound is attained for the groups $G(\alpha, \beta, \gamma, \delta)$. We were tempted to conjecture that the bound is always attained, but have recently found a group of order 128 whose index four subgroups intersect non-trivially and whose integral cohomology has exponent four [9]. Adem has conjectured that for G a finite group, if $H^i(BG)$ contains elements of order p^n for some i, then it does so for infinitely many i [1], and Henn has asked if this is the case [7]. We do not know if this holds for the groups $G(\alpha, \beta, \gamma, \delta)$.

References

[1]   A. Adem, Cohomological exponents of $\mathbf{Z}G$-lattices, J. Pure and Appl. Alg. **58** (1989), 1–5.

[2]   A. Adem, On the exponent of the cohomology of discrete groups, Bull. London Math. Soc. **21** (1989), 585–590.

[3]   A. Adem and J. F. Carlson, Discrete groups with large exponents in cohomology, J. Pure and Appl. Alg. **66** (1990), 111–120.

[4]   H. Cartan and S. Eilenberg, Homological Algebra, Princeton Univ. Press (1956).

[5]   L. Evens, The cohomology ring of a finite group, Trans. Amer. Math. Soc. **101** (1961), 224–239.

[6]   L. Evens, A generalization of the transfer map in the cohomology of groups, Trans. Amer. Math. Soc. **108** (1963), 54–65.

[7]   H.-W. Henn, Classifying spaces with injective mod-p cohomology, Comment. Math. Helvetici **64** (1989), 200–206.

[8]   I. J. Leary, A differential in the Lyndon-Hochschild-Serre spectral sequence, J. Pure and Appl. Alg. **88** (1993), 155–168.

[9]   I. J. Leary, Integral cohomology of some wreath products, in preparation.

[10]   B. B. Venkov, Cohomology algebras for some classifying spaces, Dokl. Akad. Nauk SSSR **127** (1959), 943–944 (in Russian).

Centre de Recerca Matematica, Institut d'Estudis Catalans,
Apartat 50, E-08193 Bellaterra.

Progress in Mathematics, Vol. 136
© 1996 Birkhäuser Verlag Basel/Switzerland

A Counter-example to a Conjecture of Cohen

RAN LEVI

ABSTRACT. Let G be a finite p-superperfect group. A conjecture of F. Cohen suggests that $\Omega BG_p^\wedge$ is resolvable by finitely many fibrations over spheres and iterated loop spaces on spheres, where $(-)_p^\wedge$ denotes the p-completion functor of Bousfield and Kan. We produce a counter-example to this conjecture and discuss some related aspects of the homotopy type of $\Omega BG_p^\wedge$.

1. Introduction

Let p be a prime number. Recall that a group G is said to be p-perfect if $H_1(BG; \mathbb{F}_p) = 0$ and p-superperfect if, in addition $H_2(BG; \mathbb{F}_p) = 0$. The group G is said to be perfect (superperfect) if it is p-perfect (p-superperfect) with respect to any prime p. A conjecture of F. Cohen [2] suggests that if G is a finite superperfect group then ΩBG^+ is spherically resolvable of finite weight, where $(-)^+$ denotes the Quillen "plus" construction.

A simple observation due to Bousfield and Kan [1] shows that for any finite group G, the space BG^+ is homotopy equivalent to the product (or wedge) of the p-completed classifying spaces $BG_p^\wedge$ taken over all primes p dividing the order of G. Thus a more general version of Cohen's conjecture appears in [5], in which the notion of "superperfect" is replaced by p-superperfect and the "plus" construction by the p-completion functor of Bousfield and Kan [1]. A considerable number of examples for the conjecture are given in [2, 5].

As we observe below, Cohen's conjecture is related to questions of classical interest in homotopy theory. Unfortunately the conjecture turns out to be false. Indeed the purpose of this note is to prove the following

THEOREM 1.1. *Let r be a positive integer and let $p \geq 13$ be a prime such that 4 divides $p - 1$. Then there exists a finite p-superperfect group $D_2(p^r)$ such that $\Omega BD_2(p^r)_p^\wedge$ is not spherically resolvable of finite weight.*

The study of spaces of the form $BG_p^\wedge$ for G finite and p-perfect appears to be related to various aspects of classical homotopy theory. As we shall observe, any example of a finite p-perfect group G, for which Cohen's conjecture holds has the property that $BG_p^\wedge$ has a global exponent in homotopy groups. This in turn produces a family of finite elliptic complexes, which satisfy the Moore finite exponents conjecture [4]. On the other hand an example of a finite p-perfect group G for which $BG_p^\wedge$ admits no global homotopy exponent would produce a counter-example to Moore's conjecture.

1991 *Mathematics Subject Classification.* Primary 55R35, Secondary 55R40, 55Q52.
The author is supported by a DFG grant.

Our counter-example has a feature which appear mildly unusual. The groups G which are shown to fail Cohen's conjecture turn out to have the property that the single loop space on a certain mod-p Moore space is a retract of $\Omega BG_p^\wedge$. This observation is crucial in showing that those groups are indeed counter-examples for Cohen's conjecture and in addition raises the question how much of the homotopy theory of Moore spaces can be retrieved by studying spaces of the form $\Omega BG_p^\wedge$ for finite p-perfect groups G. Notice that this stands in contrast to the fact that there are no essential maps from $BG_p^\wedge$ to an iterated loop space on a finite complex by the Sullivan conjecture. We also remark that our example shows that in general spaces of the form $\Omega BG_p^\wedge$ do not have an H-space exponent as the same is true for the single loop space on a Moore space. Note however that in view of the remark above concerning the Moore conjecture, one might like to believe that the homotopy groups of $BG_p^\wedge$ do have an exponent for every finite p-perfect group G. It might be the case that for every finite p-perfect group G, some iterated loop space $\Omega^k BG_p^\wedge$ has an H-space exponent.

In previous study of spaces of the form $\Omega BG_p^\wedge$ [2, 5], several computational examples led to the question whether the mod-p loop space homology of $BG_p^\wedge$ is a commutative algebra (or at least Lie nilpotent). Our example provides a negative answer to this question.

Throughout this article all space are assumed simply connected, p-complete and to have the homotopy type of a CW-complex. By $H_*(-)$ we shall always mean homology with coefficients in the prime field $\mathbb{F}_p$.

2. Homological rate of growth

A space X is said to be spherically resolvable of weight $\leq r$ if there exists a tower of principal fibrations:

$$X_r \longrightarrow X_{r-1} \longrightarrow \cdots \longrightarrow X_2 \longrightarrow X_1 = X$$

such that:

1. For each $i \geq 1$ the fibration, $X_{i+1} \longrightarrow X_i$, is induced from the path loop fibration over $\Omega^{n_i} S^{n_i+k_i}$, $n_i \geq 0$, $k_i > 0$ via a map $\rho_i : X_i \longrightarrow \Omega^{n_i} S^{n_i+k_i}$.
2. $X_r \simeq \Omega^{n_r} S^{n_r+k_r}$ for some $n_r \geq 0$ and $k_r > 0$.

The main observation needed to produce a counter-example for Cohen's conjecture is stated below and is proven in [6].

THEOREM 2.1. *Let X be a space which is finitely resolvable by fibrations over spheres and iterated loop spaces on spheres. Let $\theta(d)$ denote the coefficient of t^d in the poincaré series for $H_*(X)$. Then for every real number $\lambda > 1$,*

$$\lim_{d \to \infty} \frac{\theta(d)}{\lambda^d} = 0.$$

The theorem is proven by first considering iterated loop spaces on sphere, for which the homological structure is well known [3], and showing that the theorem holds for those spaces. The general statement follows by using the

Serre spectral sequence to observe that exponential growth for the coefficients $\theta(d)$ cannot be obtained for a space which is finitely resolvable.

As a consequence of theorem 2.1, it suffices to find an example of a finite p-perfect group G such that $H_*(\Omega BG_p^\wedge)$ grows exponentially in order to disprove Cohen's conjecture.

For a finite group G, consider faithful representations of G in some unitary group $U(n)$. Any such representation ρ gives rise to a fibration

$$U(n)/\rho \longrightarrow BG \longrightarrow BU(n) \tag{1}$$

with a simply-connected base space, where $U(n)/\rho$ denotes the orbit space of $U(n)$ by the G-action via ρ. Notice that p-completion respects fibrations with simply-connected base space.

Now consider the case where G is p-superperfect and let a faithful representation $G \xrightarrow{\rho} U(n)$ be given. Then $B\rho_p^\wedge$ can be lifted to a map $BG_p^\wedge \xrightarrow{B\tilde{\rho}_p^\wedge} BSU(n)_p^\wedge$, whose fibre which we denote $(SU(n)/\rho)_p^\wedge$ is the 1-connected cover of $(U(n)/\rho)_p^\wedge$. Thus we get a sequence

$$\Omega SU(n)_p^\wedge \longrightarrow \Omega(SU(n)/\rho)_p^\wedge \longrightarrow \Omega BG_p^\wedge \longrightarrow SU(n)_p^\wedge, \tag{2}$$

in which every consecutive pair of maps is a fibration sequence up to homotopy.

LEMMA 2.2. *Let G be a finite p-superperfect group. Then for every unitary faithful representation $\rho : G \longrightarrow U(n)$*

1. $\Omega BG_p^\wedge$ *is spherically resolvable of finite weight if and only if $\Omega(SU(n)/\rho)_p^\wedge$ is.*
2. $H_*(\Omega BG_p^\wedge)$ *grows exponentially if and only if $H_*(\Omega(SU(n)/\rho)_p^\wedge)$ does.*
3. $\pi_* BG_p^\wedge$ *has an exponent if and only if $\pi_*(SU(n)/\rho)_p^\wedge$ does.*

Proof. The first statement follows at once from [5, II.3.0.2]; the second follows by inspection of the Serre spectral sequences of the two fibrations in 2 above; the third statement follows by considering the long exact homotopy sequence for 2, taking into account the fact that $SU(n)_p^\wedge$ is spherically resolvable of finite weight and thus admits a homotopy exponent. □

Notice that $(SU(n)/\rho)_p^\wedge$ has the homotopy type of a finite elliptic complex. This justifies the remark made in the introduction about the Moore conjecture.

3. The groups $D_k(p^r)$ and a related representation

We assume all through that p is a prime such that 2^k divides $p-1$. In that case the group $\mathbb{Z}/2^{k+1}\mathbb{Z}$ operates on a free $\mathbb{Z}/p^r\mathbb{Z}$-module T_r of rank 2 as follows. Let ζ denote an element of multiplicative order 2^k in the group of units $(\mathbb{Z}/p^r\mathbb{Z})^*$. Let a and b denote a choice of generators for T_r. Let σ denote a generator for $\mathbb{Z}/2^{k+1}\mathbb{Z}$. Define an action of σ on T_r by $\sigma(a) = b$ and $\sigma(b) = \zeta a$. Define $D_k(p^r)$ to be the semidirect product of T_r with $\mathbb{Z}/2^{k+1}\mathbb{Z}$ with respect to the action given above. A presentation of $D_k(p^r)$ is given by

$$D_k(p^r) = \langle a, b, \sigma \mid a^{p^r} = b^{p^r} = \sigma^{2^{k+1}} = [a,b] = 1; \sigma a \sigma^{-1} = b; \sigma b \sigma^{-1} = a^g \rangle,$$

where g is an integer such that $g^{2^k} \equiv 1 \bmod p^r$.

Define a unitary representation $\rho_r : D_k(p^r) \longrightarrow U(2^{k+1})$ as follows. Let θ denote a complex primitive root of 1 of order p^r. Define $\rho_r(\sigma)$ to be the permutation matrix whose i-th row is the standard unit vector e_{i+1} for $1 \leq i \leq 2^{k+1} - 1$ and whose 2^{k+1}-st row is e_1. Define

$$\rho_r(a) = diag(\theta, 1, \theta^g, 1 \cdots, \theta^{g^{2^k}-1}, 1),$$

$$\rho_r(b) = diag(1, \theta^g, 1, \theta^{g^2}, \cdots, \theta^{g^{2^k}-1}, 1, \theta).$$

One easily verifies that the relations are satisfied and ρ_r is evidently faithful. Let T denote a 2^{k+1}-fold product of the 1-sphere S^1 and let $\psi : T \longrightarrow U(2^{k+1})$ denote the canonical inclusion. Fix the values of p, r and k and let G denote $D_k(p^r)$. Let $\phi : T_r \longrightarrow G$ denote the inclusion. Then the restriction ρ'_r of ρ_r to T_r factors through T and we have $\psi\rho'_r = \rho_r\phi$. Using this factorization one easily computes the Chern classes of ρ_r. To fix our notation let

$$H^*(BT) = P[u_1, \cdots, u_{2^{k+1}}], \ |u_j| = 2,$$
$$H^*(BT_r) = P[v_1, v_2] \otimes E[x_1, x_2], \ |v_j| = 2, \ |x_j| = 1.$$

The following lemma is an easy exercise.

LEMMA 3.1. *The total Chern class of ρ_r restricts to*

$$1 - (v_1^{2^k} + v_2^{2^k}) + (v_1 v_2)^{2^k}$$

in $H^*(BT_r)$. Thus $w_{2^k}(\rho_r) = -(v_1^{2^k} + v_2^{2^k})$, $w_{2^{k+1}}(\rho_r) = (v_1 v_2)^{2^k}$ and $w_j(\rho_r) = 0$ otherwise.

COROLLARY 3.2. *Let $U(2^{k+1})/\rho_r$ denote the orbit space of $U(2^{k+1})$ by $D_k(p^r)$ with respect to the representation ρ_r. Then there is an isomorphism of algebras*

$$H^*(U(2^{k+1})/\rho_r) \cong H^*(BD_k(p^r))/(w_{2^k}, w_{2^{k+1}}) \otimes E,$$

where E is an exterior algebra on $2^{k+1} - 2$ generators, corresponding to the zero Chern classes of ρ_r.

Proof. This is an immediate consequence of the big collapse theorem of L. Smith [7].   □

Next if the prime p in the definition of $D_k(p^r)$ is sufficiently large, then there is an obvious map

$$f : (U(2^{k+1})/\rho_r)_p^{\wedge} \longrightarrow \prod_{i \neq 2^{k-1}-1, 2^k-1} (S^{2i+1})_p^{\wedge},$$

realizing the factor E in $H^*((U(2^{k+1})/\rho_r)_p^{\wedge})$. Indeed one obtains this map by defining it component by component, starting with a suitable skeleton and than extending, using the fact that possible obstructions vanish for large primes. Notice that the dimension of $(U(2^{k+1})/\rho_r)_p^{\wedge}$ is independent of p.

Let $X_k(p^r)$ denote the homotopy fibre of f. The following lemma is obvious by inspection of the Eilenberg-Moore spectral sequence for the map f.

LEMMA 3.3. *Let p and k be chosen so that the map f defined above exists. Then there is an isomorphism of algebras*

$$H^*(X_k(p^r)) \cong H^*(BD_k(p^r))/(w_{2^k}, w_{2^{k+1}}).$$

We remark that the assumption that the prime p is sufficiently large makes calculations easier but seems not to play an important role otherwise. In the next section we produce our counter-example based on the calculations carried out here. We find it suitable to conjecture that results similar to those given below can be obtained for all $D_k(p^r)$.

4. The counter-example

We specialize to the case $k = 2$ and calculate the cohomology algebra of $X_2(p^r)$. Notice that $X_2(p^r)$ can be constructed as above if $p \geq 13$.

PROPOSITION 4.1. *There is an isomorphism of algebras*

$$H^*(BD_2(p^r))/(w_4, w_8) \cong P[a_6, a_6', b_7, b_7', d_7, d_7', t_8, s_{15}, s_{15}', q_{16}]/R,$$

where R is the set of relations given by $at = d'b'$; $ds = aq = s'b = d'b't = a't^2$; $t^3 = 0$; all other possible products of generators except for those given above and in addition t^2, $b't$ and $d't$ vanish. Thus $H^*(X_2(p^r))$ is given in the cell diagram below.

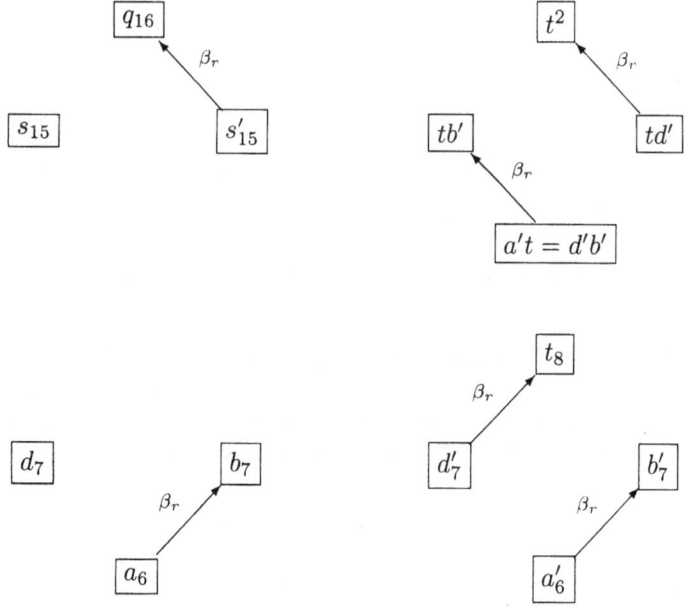

$$1$$

Proof. Let $T_r < D_2(p^r)$ denote the Sylow p-subgroup and write

$$H^*(BT_r) \cong P[v_1, v_2] \otimes E[x_1, x_2].$$

Let $\zeta \in \mathbb{F}_p$ denote a primitive root of unity of order 4. Computing the algebra structure modulo the Chern classes w_4 and w_8, whose restrictions to $H^*(BT_r)$ are given by $-(v_1^4 + v_2^4)$ and $v_1^4 v_2^4$ respectively, one observes that the resulting algebra is 22-dimensional. In fact obtaining an $\mathbb{F}_p$ vector space basis is easy by routine invariant calculation.

Let $res : H^*(BD_2(p^r)) \longrightarrow H^*(BT_r)$ denote the restriction. We conclude the proof by spelling out the restrictions of the specified generators and leave it for the reader to verify that all the promised relations hold modulo the ideal (w_4, w_8).

(1) $res(a_6) = (v_1^2 - \zeta v_2^2) x_1 x_2$.

(2) $res(a_6') = -v_1 v_2 x_1 x_2$.

(3) $res(b_7) = (v_1^3 x_2 + \zeta v_2^3 x_1) - (v_1^2 v_2 x_1 + \zeta v_1 v_2^2 x_2)$.

(4) $res(b_7') = v_1 v_2^2 x_1 - v_1^2 v_2 x_2$.

(5) $res(d_7) = (v_1^3 x_1 + v_2^3 x_2) + \zeta(v_1^2 v_2 x_2 - v_1 v_2^2 x_1)$.

(6) $res(d_7') = v_1 v_2^2 x_2 - \zeta v_1^2 v_2 x_1$.

(7) $res(t_8) = v_1 v_2^3 - \zeta v_1^3 v_2$.

(8) $res(s_{15}) = v_1^3 v_2^4 x_1 + v_1^4 v_2^3 x_2$.

(9) $res(s_{15}') = \zeta v_1^7 x_2 - v_2^7 x_1$.

(10) $res(q_{16}) = \zeta v_1^7 v_2 - v_1 v_2^7$. $\qquad\qquad\qquad\qquad\qquad\qquad\qquad\square$

PROPOSITION 4.2. *There is a retract*

$$P^{16}(p^r) \bigvee S^{15} \xrightarrow{\ g\ } X_2(p^r) \xrightarrow{\ f\ } P^{16}(p^r) \bigvee S^{15},$$

where $P^{16}(p^r)$ is the 16-dimensional mod-p^r Moore space.

Proof. Let X denote $X_2(p^r)$ and recall our convention that all spaces are p-complete. First observe that there is a map on the 16- skeleton

$$f' : X^{(16)} \longrightarrow P^{16}(p^r) \bigvee S^{15}$$

given by pinching down the 8-skeleton together with the cells corresponding to the products on the right hand side of the diagram above. An obstruction to extending f' to $f : X \longrightarrow P^{16}(p^r) \bigvee S^{15}$ might exist in $\pi_{21} P^{16}(p^r) \bigvee S^{15}$, which vanishes for the primes under consideration.

Next notice that
$$X^{(8)} \simeq P^7(p^r) \bigvee P^7(p^r) \bigvee P^8(p^r) \bigvee S^7.$$

If the prime p is sufficiently large, computing $\pi_i X^{(8)}$ for $i = 14$ and 15 is straight forward by using the Hilton-Milnor theorem [8]. In particular one observes that the homotopy in those dimensions is generated by primary Whitehead products and thus a non-trivial attaching map results in "attaching" a decomposable cohomology class. Since s, s' and q are indecomposable, we conclude that the corresponding attaching maps are trivial and thus the desired map
$$g : P^{16}(p^r) \bigvee S^{15} \longrightarrow X$$

is obtained.

Finally notice that the composite $f \circ g$ induces an isomorphism on mod-p cohomology and is thus a homotopy equivalence. $\square$

COROLLARY 4.3. *The homology algebra $H_*(\Omega X_2(p^r))$ contains a tensor algebra on three generators. In particular $H_*(\Omega X_2(p^r))$ grows exponentially and hence is not spherically resolvable of finite weight.*

We now consider $\Omega B D_2(p^r)_p^\wedge$. Through the end of this paper let X denote as before $X_2(p^r)$ and let G denote $D_2(p^r)$. Then the Moore space $P^{16}(p^r)$ is a retract of X. Moreover, Let $\pi : X \longrightarrow BG_p^\wedge$ denote the map obtained by the composite
$$X \longrightarrow (U(8)/\rho_2)_p^\wedge \longrightarrow BG_p^\wedge.$$

One readily verifies that the homotopy fibre of π is equivalent to $(S^7 \times S^{15})_p^\wedge$ and that the fibre inclusion map into X induces the zero map on mod-p cohomology. This together with the assumption that the prime p is sufficiently large implies that the composite
$$S^7 \times S^{15} \longrightarrow X \longrightarrow P^{16}(p^r)$$

is null homotopic and thus yields a homotopy commutative diagram

$$
\begin{array}{ccccccc}
\Omega X & \longrightarrow & \Omega BG_p^\wedge & \longrightarrow & S^7 \times S^{15} & \longrightarrow & X \\
\downarrow & & \downarrow & & \downarrow & & \downarrow \\
\Omega P^{16}(p^r) & \overset{=}{\longrightarrow} & \Omega P^{16}(p^r) & \longrightarrow & * & \longrightarrow & P^{16}(p^r)
\end{array}
$$

Note that the map $\Omega BG_p^\wedge \longrightarrow \Omega P^{16}(p^r)$ is not multiplicative in general, however the composite
$$\Omega P^{16}(p^r) \longrightarrow \Omega X \longrightarrow \Omega BG_p^\wedge \longrightarrow \Omega P^{16}(p^r)$$

is evidently homotopic to the identity. Thus we have proven

PROPOSITION 4.4. *The space $\Omega P^{16}(p^r)$ is a retract of $\Omega BG_p^\wedge$ for $p \geq 13$.*

COROLLARY 4.5. *For $p \geq 13$, the algebra $H_*(\Omega BG_p^\wedge)$ contains a tensor algebra on two generators and thus grows exponentially*

Corollary 4.5 combined with theorem 2.1 completes the proof of theorem 1.1. In addition we have

COROLLARY 4.6. *For $p \geq 13$, any power map on $\Omega BG_p^{\wedge}$ is essential.*

Proof. By [4] any power map on a single loop space on a Moore space is essential. The result follows. □

5. Speculations

If G is a finite p-superperfect group than $\Omega BG_p^{\wedge}$ does not satisfy Cohen's conjecture unless possibly if the loop space homology $H_*(\Omega BG_p^{\wedge})$ does not grow exponentially. In view of the fact that the groups $D_2(p^r)$ are by no means "pathological", it seems reasonable to wonder what group theoretic properties of G would imply that the loop space homology of BG grows polynomialy or at least subexponentially. We refer the reader to [2, 5] to inspect that, in fact, all the examples of groups G known to satisfy Cohen's conjecture have the property that $H_*(\Omega BG_p^{\wedge})$ grows polynomialy.

Although theorem 1.1 shows that Cohen's conjecture is false as stated in [5], it is still conceivable by results in [2, 4] that the conjecture holds in general if one drops the finiteness requirement on the length of a resolution.

Another point to be emphasized is the significance of proposition 4.4. This result suggests that understanding the homotopy type of spaces of the from $\Omega BG_p^{\wedge}$ might possibly shed new light on objects of classical interest in homotopy theory. One might wonder for example whether it is possible to obtain $\Omega P^n(p^r)$ for any given values of n and p as a retract of $\Omega BG_p^{\wedge}$ for some finite p-perfect group G.

Many interesting p-perfect groups G are not p-superperfect but in this case the 1-connected cover of $BG_p^{\wedge}$ is $B\tilde{G}_p^{\wedge}$, where $\tilde{G}$ is the p-universal central extension of G and is finite and p-superperfect. It is easy to observe that if $H_*(\Omega BG_p^{\wedge})$ grow exponentially then so does $H_*(\Omega B\tilde{G}_p^{\wedge})$. For instance $BD_1(3)_3^{\wedge} \simeq BSL_2(\mathbb{F}_9)_3^{\wedge}$ and is 3-perfect but not 3-superperfect. In this case the corresponding space $X_1(3)$ can be obtained at the prime 3 and is 10-dimensional. In addition there is a fibration

$$Y \longrightarrow \Omega BD_1(3)_3^{\wedge} \longrightarrow X_1(3),$$

where Y is spherically resolvable of weight 2. However the existence of a non-zero P^1 in $H^*(X_1(3))$ implies that a result corresponding to prop 4.2 for $X_1(3)$ fails to hold. It would be interesting to know whether or not the 1-connected cover of $\Omega BD_1(p^r)_p^{\wedge}$ is spherically resolvable of finite weight as this could in some sense provide a minimal counter-example (of order $4p^2$).

Finally, the referee has suggested that the loop space on the space $X_1(3)$ might have the loop space of a 4-cell complex as a retract, where the 4-cell complex supports a non-trivial P^1. We remark that such a retract, if it exists cannot be multiplicative, i.e. it does not exist before looping due to the existence of products in $H^*(X_1(3))$. The question whether or not it exists after looping once remains unsolved.

References

[1] A. K. Bousfield and D. M. Kan; *Homotopy Limits Completions and Localizations*; LNM 304, (1972), Springer-Verlag.

[2] F. R. Cohen; *Remarks on the Homotopy Theory associated to Finite Perfect Groups*; LNM 1509 (1992) Springer-Verlag.

[3] F. R. Cohen, T. J. Lada and J. P. May; *The Homology of Iterated Loop Spaces*; LNM 533, (1976) Springer-Verlag.

[4] F. R. Cohen, J. C. Moore and J. Neisendorfer; *Exponents in Homotopy Theory*; Ann. of Math. Studies, 133 (1987), 3–34.

[5] R. Levi; *On Finite Groups and Homotopy Theory*; to appear in the Memoirs of the A.M.S.

[6] R. Levi; *On Homological Rate of Growth and the Homotopy Type of $\Omega BG_p^\wedge$*; to appear,

[7] L. Smith; *Homological Algebra and the Eilenberg-Moore Spectral Sequence*; AMS Translatl. 129, (1967), 58–93.

[8] G. Whitehead; *Elements of Homotopy Theory*; Springer-Verlag (1978).

Mathematisches Institut, Universität Heidelberg,
Im Neuenheimer Feld 288, 69120 Heidelberg, Germany
rlvi@vogon.mathi.uni-heidelberg.de

Progress in Mathematics, Vol. 136
© 1996 Birkhäuser Verlag Basel/Switzerland

The complete Steenrod algebra and the generalized Dickson algebra

IRENE LLERENA AND NGUYÊN H.V. HU'NG*

ABSTRACT. We prove that the mod 2 complete Steenrod algebra $\hat{A}$ is closely related to the Dickson algebra, the invariant algebra of $GL(k, \mathbb{Z}/2)$. More precisely, $\hat{A}$ is dual to $D_\infty^{\vee}$, the generalized Dickson algebra on infinitely many generators, as a $\mathbb{Z}[\frac{1}{2}]$-graded algebra. We also show that the generalized operations in $\hat{A}$ are derived from the generalized Dickson invariants in a similar way as the operations in A are derived from the Dickson invariants (see Mùi [5], Madsen–Milgram [9], Lomonaco [7]).

0 Introduction and background

Throughout this paper homology and cohomology are taken with coefficients in $\mathbb{Z}/2$.

In the last two decades, since the classical paper by Mùi [4], a lot of work has been done to analyze the cohomology operations and the structure of the Steenrod algebra by means of the Dickson invariants of the general linear group $GL(k, \mathbb{Z}/2)$.

In [5] Mùi showed that the cohomology operations derived from the Dickson invariants are exactly the ones dualizing Milnor monomials in A_*, the dual of the Steenrod algebra A. (See also Madsen–Milgram [9] and Lomonaco [7] for related results.)

In [2] Bullett and Macdonald pointed out that the Adem relations actually stand behind the Dickson invariants of $GL(2, \mathbb{Z}/2)$ and they gave an elegant proof of these relations.

From another point of view, Singer also recognized the connection between the Adem relations and the invariants of $GL(2, \mathbb{Z}/2)$. This observation is one of the main ingredients of his beautiful modular-invariant theoretic description of the Lambda algebra Λ, the E_1-term in the Adams spectral sequence for the spheres [17].

Lannes and Zarati [6] applied the Dickson invariants to compute the derived functors of the destabilization in some interesting cases.

Many authors studied the action of the Steenrod algebra A on the Dickson algebra D_k (see the relevant discussion and the references in [12]). Hai and

1991 *Mathematics Subject Classification.* Primary 55S05, 55S10.

Key words and phrases. Steenrod algebra, Dickson algebra.

*) The first-named author was partially supported by the DGICYT, PB 91-0467.

The second-named author was supported by the DGU through the CRM (Barcelona).

Hu'ng have given an explicit formula for this action [11], [12]. Since D_k is dual to the coalgebra of Dyer–Lashof operations of length k [8], the Hai–Hu'ng formula can be thought of as a version of the Nishida relations.

Hu'ng and Peterson studied a minimal set of $\mathcal{A}$-generators for the Dickson algebra in [14]. This has been applied by the second-named author of the present paper to investigate spherical classes in $Q_0 S^0$ [13].

Recently, Arnon contributed a new view on the action of $\mathcal{A}$ on D_k in his thesis [1]. He extended this action to an action of the complete Steenrod algebra $\hat{\mathcal{A}}$ on the generalized Dickson algebra $D_k^\vee$, which is also a ring of invariants of $GL(k, \mathbb{Z}/2)$.

In this paper, we develop Arnon's idea to get a deeper relationship between the Steenrod algebra and the Dickson algebra.

The paper is organized as follows. In Section 1, we recall the definition of the complete Steenrod algebra $\hat{\mathcal{A}}$, give a precise definition of its Hopf algebra structure and describe explicitly the product (the Adem relations) and the coproduct. In Section 2, we construct an exact functor, the root functor, from the category of $\mathcal{A}$-algebras to the category of $\hat{\mathcal{A}}$-algebras. In particular, $\hat{\mathcal{A}}$ acts naturally on the root closure of the cohomology of any topological space X. Our main result, proved in Section 3, is $\hat{\mathcal{A}}_* \cong D_\infty^\vee$ as $\mathbb{Z}[\frac{1}{2}]$-graded algebras, where $\hat{\mathcal{A}}_*$ denotes the graded dual of $\hat{\mathcal{A}}$. The key point is that $D_k^\vee$ is canonically embedded into $D_{k+1}^\vee$, so that $D_\infty^\vee = \varinjlim D_k^\vee$ is well defined, while D_k cannot be embedded into D_{k+1}. In Section 4, we use the generalized Steenrod total power to derive operations in $\hat{\mathcal{A}}$ from the generalized Dickson invariants. We obtain an analogue of the formula of Mùi [5], Madsen–Milgram [9] and Lomonaco [7].

1 The complete Steenrod algebra as a Hopf algebra

In this section we recall Arnon's definition of the complete Steenrod algebra [1]. For later use, we give a precise definition of its Hopf algebra structure and describe explicitly its product and coproduct.

The following is a variation of a definition in [1]. See Remark 1.5 for the reason of making this variation.

DEFINITION 1.1.

(1) A *large* $\mathbb{Z}[\frac{1}{2}]$-*graded algebra* A is an algebra $A \subset \prod_{n \in \mathbb{Z}[\frac{1}{2}]} A_n$, whose multiplication is defined by maps $A_m \otimes A_n \to A_{m+n}$, for $m, n \in \mathbb{Z}[\frac{1}{2}]$. If additionally $A = \oplus_{n \in \mathbb{Z}[\frac{1}{2}]} A_n$, then A is called a $\mathbb{Z}[\frac{1}{2}]$-*graded algebra*

(2) A *(large)* $\mathbb{Z}[\frac{1}{2}]$-*graded Hopf algebra* A is a Hopf algebra which is a (large) $\mathbb{Z}[\frac{1}{2}]$-graded algebra with respect to its multiplication, and its comultiplication is given by maps $A_m \to A_n \otimes A_{m-n}$ for $m, n \in \mathbb{Z}[\frac{1}{2}]$.

(3) A $\mathbb{Z}[\frac{1}{2}]$-*graded module* M over a (large) $\mathbb{Z}[\frac{1}{2}]$-graded algebra A is an A-module of the form $M = \oplus_{n \in \mathbb{Z}[\frac{1}{2}]} M_n$ where the action of A on M is given by maps $A_m \otimes M_n \to M_{m+n}$, for $m, n \in \mathbb{Z}[\frac{1}{2}]$.

Let M be a (large) $\mathbb{Z}[\frac{1}{2}]$-graded object (i.e. algebra, Hopf algebra or module). We define $2^t M$ to be the (large) $\mathbb{Z}[\frac{1}{2}]$-graded object which is isomorphic to M as ungraded object and has $(2^t M)_n = M_{2^{-t}n}$, for any $t \in \mathbb{Z}$.

Let us consider the Steenrod algebra $\mathcal{A}$ as a $\mathbb{Z}[\frac{1}{2}]$-graded Hopf algebra which is zero in negative and fractional degrees. The *halving homomorphism* $d\colon \frac{1}{2}\mathcal{A} \to \mathcal{A}$, defined by $d(Sq^{2n}) = Sq^n$ and $d(Sq^{2n+1}) = 0$, is a degree preserving homomorphism of Hopf algebras.

DEFINITION 1.2 [1]. The *complete Steenrod algebra* is the inverse limit

$$\hat{\mathcal{A}} = \varprojlim \{ \cdots \xrightarrow{d} \frac{1}{2^{t+1}}\mathcal{A} \xrightarrow{d} \frac{1}{2^t}\mathcal{A} \xrightarrow{d} \cdots \}.$$

The increasing sequence of ideals $I_t = \ker(\hat{\mathcal{A}} \to 2^t \mathcal{A})$ filters $\hat{\mathcal{A}}$,

$$\cdots \subset I_{-1} \subset I_0 \subset I_1 \subset \cdots \subset \hat{\mathcal{A}}.$$

Moreover, $\mathcal{A} \cong \hat{\mathcal{A}}/I_0$.

Let $\hat{d}\colon \frac{1}{2}\hat{\mathcal{A}} \to \hat{\mathcal{A}}$ denote the degree preserving homomorphism induced by the halving homomorphism. We will also consider $\hat{d}$ as a map from $\hat{\mathcal{A}}$ to $\hat{\mathcal{A}}$ that halves degrees. Thus, for instance, $\hat{d}(I_t) = I_{t-1}$.

To equip $\hat{\mathcal{A}}$ with a Hopf algebra structure one needs to define $\hat{\mathcal{A}} \otimes \hat{\mathcal{A}}$. Note that the classical way of definition

$$(\hat{\mathcal{A}} \otimes \hat{\mathcal{A}})_m = \bigoplus_{p+q=m} \hat{\mathcal{A}}_p \otimes \hat{\mathcal{A}}_q,$$

for $m, p, q \in \mathbb{Z}[\frac{1}{2}]$, does not work because the right hand side does not contain certain sums of infinitely many terms in $\hat{\mathcal{A}} \otimes \hat{\mathcal{A}}$ as the ones appearing in Proposition 1.7.

DEFINITION 1.3. $\hat{\mathcal{A}} \otimes \hat{\mathcal{A}} = \varprojlim_t \left(\frac{1}{2^t}\mathcal{A} \otimes \frac{1}{2^t}\mathcal{A} \right).$

Observe that $\varprojlim_t \left(\frac{1}{2^t}\mathcal{A} \otimes \frac{1}{2^t}\mathcal{A} \right) = \varprojlim_{i,j} \left(\frac{1}{2^i}\mathcal{A} \otimes \frac{1}{2^j}\mathcal{A} \right).$

Let $\varphi_t\colon \frac{1}{2^t}\mathcal{A} \otimes \frac{1}{2^t}\mathcal{A} \to \frac{1}{2^t}\mathcal{A}$ and $\psi_t\colon \frac{1}{2^t}\mathcal{A} \to \frac{1}{2^t}\mathcal{A} \otimes \frac{1}{2^t}\mathcal{A}$ be respectively the product and the coproduct induced by those of $\mathcal{A}$. Since the halving homomorphism d is compatible with φ_t and ψ_t, we can define a product and a coproduct in $\hat{\mathcal{A}}$ as follows.

DEFINITION 1.4. $\varphi = \varprojlim \varphi_t \colon \hat{\mathcal{A}} \otimes \hat{\mathcal{A}} \to \hat{\mathcal{A}}$ and $\psi = \varprojlim \psi_t \colon \hat{\mathcal{A}} \to \hat{\mathcal{A}} \otimes \hat{\mathcal{A}}$.

They equip $\hat{\mathcal{A}}$ with a large $\mathbb{Z}[\frac{1}{2}]$-graded Hopf algebra structure.

We denote by $Sq^{r/2^t} \in \hat{\mathcal{A}}$, $r, t \in \mathbb{Z}$, the element represented by $\{ Sq^{2^n r} \in \frac{1}{2^{t+n}}\mathcal{A} \mid n \in \mathbb{N} \}$. Here and from now on $\mathbb{N}$ denotes the set of non-negative integers. Since $d(Sq^{2r}) = Sq^r$, we get $Sq^{r/2^t} = Sq^{2r/2^{t+1}}$. Therefore, the notation is not confusing. If r is odd, then $Sq^{r/2^t} \in I_{-n}$ for any $n < t$. We have $\deg(Sq^{r/2t}) = r/2t$.

The operation Sq^I, with I a finite sequence of elements in $\mathbb{N}[\frac{1}{2}]$, is defined in the usual way.

Remark 1.5.

(1) It should be noted that $\hat{A}$ is *not* of finite type. For example, the operations $Sq^{\frac{2^n-1}{2^n}} Sq^{\frac{1}{2^n}}$, $n > 1$, are of degree 1 and are linearly independent.

(2) $\hat{A}$ is a large $\mathbb{Z}[\frac{1}{2}]$-graded Hopf algebra but not a $\mathbb{Z}[\frac{1}{2}]$-graded Hopf algebra. Indeed, $\alpha = \sum_{n \in \mathbb{N}} Sq^{\frac{1}{2^n}} \in \hat{A}$ but $\alpha \notin \oplus_{n \in \mathbb{Z}[\frac{1}{2}]} \hat{A}_n$. Furthermore, $\hat{A} \neq \prod_{n \in \mathbb{Z}[\frac{1}{2}]} \hat{A}_n$. In fact, $\beta = \sum_{n \in \mathbb{N}} Sq^n \notin \hat{A}$ but $\beta \in \prod_{n \in \mathbb{Z}[\frac{1}{2}]} \hat{A}_n$. It is easy to check that a sum $\sum_{I \in \mathcal{I}} Sq^I$ belongs to $\hat{A}$ if and only if, for each $t \in \mathbb{Z}$, the number of sequences $I = (i_1, \ldots, i_m) \in \mathcal{I}$ such that $2^t i_q \in \mathbb{N}$ for all i_q, $q = 1, \ldots, m$, is finite.

DEFINITION 1.6 [1]. For $a, b \in \mathbb{N}[\frac{1}{2}]$, define the binomial coefficient $\binom{a}{b}$ to be the residue class mod 2 of $\binom{2^N a}{2^N b}$, where N is a big enough integer such that $2^N a, 2^N b \in \mathbb{Z}$.

This definition makes sense since it does not depend on N.

The following proposition describes the product and coproduct in the Hopf algebra $\hat{A}$.

PROPOSITION 1.7.

(1) *(Generalized) Adem relations*

$$Sq^a Sq^b = \sum_{\substack{0 \leq i \leq \frac{a}{2} \\ i \in \mathbb{N}[\frac{1}{2}]}} \binom{b - i - 2^{-N(a,b,i)}}{a - 2i} Sq^{a+b-i} Sq^i$$

for $0 < a < 2b$. Here $N = N(a, b, i)$ is a big enough integer so that $2^N a, 2^N b, 2^N i \in \mathbb{Z}$.

(2) $$\psi(Sq^a) = \sum_{\substack{b+c=a \\ b,c \in \mathbb{N}[\frac{1}{2}]}} Sq^b \otimes Sq^c$$

for any $a \in \mathbb{N}[\frac{1}{2}]$.

Proof. First let us check that the binomial coefficient does not depend on N. If $A = a_0 + a_1 2 + \cdots + a_n 2^n$ and $B = b_0 + b_1 2 + \cdots + b_n 2^n$ are the dyadic expansions of A and B respectively, then $\binom{A}{B} = \prod_i \binom{a_i}{b_i}$ mod 2 (see e.g. [18, I.2.6]). Hence

$$\binom{2A - 1}{2B} = \binom{2A - 2}{2B}\binom{1}{0} = \binom{A - 1}{B}.$$

Thus,
$$\binom{b - i - 2^{-(N+1)}}{a - 2i} = \binom{2(2^N b - 2^N i) - 1}{2(2^N a - 2 \cdot 2^N i)}$$
$$= \binom{2^N b - 2^N i - 1}{2^N a - 2 \cdot 2^N i} = \binom{b - i - 2^{-N}}{a - 2i}.$$

Now, let $\pi_t \colon \hat{\mathcal{A}} \to \frac{1}{2^t}\mathcal{A}$ denote the projection. To prove (1), it suffices to show that, for any t big enough, the images by π_t of both sides are identical. Take t such that $2^t a, 2^t b \in \mathbb{Z}$. We have

$$\pi_t(Sq^i) = \begin{cases} Sq^{2^t i} & \text{if } 2^t i \in \mathbb{N}, \\ 0 & \text{otherwise.} \end{cases}$$

Applying π_t to both sides of the formula in (1), we get

$$Sq^{2^t a} Sq^{2^t b} = \sum_{\substack{0 \le i \le \frac{a}{2} \\ 2^t i \in \mathbb{N}}} \binom{b - i - 2^{-t}}{a - 2i} Sq^{2^t a + 2^t b - 2^t i} Sq^{2^t i}.$$

Note that the sum becomes now a finite sum. Taking $a' = 2^t a$, $b' = 2^t b$, $i' = 2^t i$, we get the classical Adem relations

$$Sq^{a'} Sq^{b'} = \sum_{i'=0}^{[\frac{a'}{2}]} \binom{b' - i' - 1}{a' - 2i'} Sq^{a' + b' - i'} Sq^{i'}$$

for any $a', b', i' \in \mathbb{N}$ and $0 < a' < 2b'$. This proves (1).

Part (2) can be similarly proved by using π_t and the definition of the coproduct in $\hat{\mathcal{A}}$. $\square$

Admissible operations Sq^I, for $I = (i_1, \dots, i_m)$ with $i_j \in \mathbb{N}[\frac{1}{2}]$, $m > 0$, are defined in the usual way: $i_q \ge 2i_{q+1}$, for all q, and $i_m \ne 0$. They are linearly independent, since their images by π_t, for every t big enough, are admissible operations in $\frac{1}{2^t}\mathcal{A}$. However, they are not enough to generate $\hat{\mathcal{A}}$ as a module.

2 Root $\hat{\mathcal{A}}$-algebras

An $\mathcal{A}$-module M can be viewed as an $\hat{\mathcal{A}}$-module via the projection $\hat{\mathcal{A}} \to \mathcal{A}$. However, the new structure brings nothing new, since all fractional degree operations in $\hat{\mathcal{A}}$ act trivially on M. If we suppose additionally that M is an $\mathcal{A}$-algebra, then we can define the root closure of M and equip it with an $\hat{\mathcal{A}}$-algebra structure.

DEFINITION 2.1 [1]. A *root algebra* B is a $\mathbb{Z}[\frac{1}{2}]$-graded commutative $\mathbb{Z}/2$-algebra such that the degree preserving homomorphism $\delta \colon 2B \to B$ defined by $\delta(x) = x^2$ is an isomorphism.

In other words, B is a root algebra if any element $x \in B$ has a unique square root $y = \sqrt{x}$, i.e. the element y with $\delta(y) = x$. In general, set $\sqrt[2^t]{x} := \sqrt{\sqrt[2^{t-1}]{x}}$.

EXAMPLE 2.2 [1]. Let $P(x_1, \dots, x_k)$ be the polynomial algebra over $\mathbb{Z}/2$ generated by $x_1, \dots, x_k$ with some prescribed degrees (usually, $\deg(x_i) = 1$). Then

$$R(x_1, \dots, x_k) = \varinjlim_{\delta} \frac{1}{2^t} P(x_1, \dots, x_k)$$

is the *free root algebra* generated by $x_1, \dots, x_k$. Here $\delta(x_i) = x_i^2$. This example also works for the case of infinitely many generators $x_1, x_2 \dots$.

DEFINITION 2.3.

(1) Given a $\mathbb{Z}$-graded algebra M, the *root closure* of M is the $\mathbb{Z}[\frac{1}{2}]$-graded algebra

$$M^{\checkmark} = \varinjlim_{\delta} \left(\frac{1}{2^t} M \right),$$

where $\delta \colon \frac{1}{2^t} M \to \frac{1}{2^{t+1}} M$ is the homomorphism given by $\delta(x) = x^2$. Clearly, $M^{\checkmark}$ is a root algebra.

(2) Let $f \colon M \to N$ be a homomorphism of $\mathbb{Z}$-graded algebras. Since f commutes with the homomorphism δ, it induces a homomorphism of algebras

$$f^{\checkmark} \colon M^{\checkmark} \to N^{\checkmark}.$$

In particular, we have $f^{\checkmark}(\sqrt{u}) = \sqrt{f^{\checkmark}(u)}$, $u \in M^{\checkmark}$.

Remark 2.4.

(1) If B is a root algebra, then it is obviously the root closure of its *integral part* defined as the $\mathbb{Z}$-graded algebra $B_{\mathbb{Z}} = \oplus_{n \in \mathbb{Z}} B_n$. Every homomorphism of root algebras is induced by a homomorphism of their integral parts.

(2) A $\mathbb{Z}$-graded algebra M is canonically embedded into its root closure. Moreover, for any $u \in M^{\checkmark}$, there exists $z \in M$ such that $u = \sqrt[2^t]{z}$ for some $t \geq 0$. Indeed, if u is represented by $\{z^{2^i} \in \frac{1}{2^{t+i}} M \mid i \geq 0\}$, for some $t \geq 0$, the element $z \in M$ represented by $\{z^{2^i} \in \frac{1}{2^t} M \mid i \geq 0\}$ satisfies $u^{2^t} = z$, or equivalently $u = \sqrt[2^t]{z}$.

We want now to provide the root closure $M^{\checkmark}$ of a $\mathbb{Z}$-graded $\mathcal{A}$-algebra M with an $\hat{\mathcal{A}}$-module structure. Note first that, if $\theta \in \frac{1}{2^{t+1}} \mathcal{A}$ and $x \in \frac{1}{2^t} M$, then

$$\theta \, \delta x = \delta(d\theta \, x).$$

Indeed, we only need to check it for $\theta = Sq^k$, k positive integer. If k is odd, $d(Sq^k) = 0$ and $Sq^k \, \delta x = Sq^k \, x^2 = 0$. If $k = 2h$, $Sq^k(\delta x) = Sq^{2h} x^2 = (Sq^h x)^2 = \delta(d(Sq^{2h}) x)$.

Hence, the following extension of the action of $\mathcal{A}$ on M to an action of $\hat{\mathcal{A}}$ on $M^{\checkmark}$ does not depend on the representative.

DEFINITION 2.5. Let M be an $\mathcal{A}$-algebra, $\xi \in \hat{A}$ and $u \in M^{\vee}$. Then ξu is the element in $M^{\vee}$ represented by $(\pi_t \xi)\, x$, where $x \in \frac{1}{2^t} M$ is a representative of u.

In particular, we have $\sqrt{\xi u} = \hat{d}(\xi)\sqrt{u}$, \qquad for $\xi \in \hat{A}, u \in M^{\vee}$.

The *total square in* $\hat{A}$ is defined as follows $Sq = \sum\limits_{i \in \mathbb{N}[\frac{1}{2}]} Sq^i$.

It is only a formal notation and is not an element of $\hat{A}$. It is "represented" in each $\frac{1}{2^t}\mathcal{A}$ by the usual total square. We emphasize that *our Sq is different from that of* [1]. We choose this definition for Sq in order to get Proposition 2.7 below.

We say that the *condition on finiteness* holds for an $\mathcal{A}$-module M (respectively, $\hat{A}$-module $\mathcal{M}$) if for each $x \in M$ (respectively, $x \in \mathcal{M}$), $Sq^i(x) \neq 0$ for only finitely many i's, $i \in \mathbb{N}$ (respectively, $i \in \mathbb{N}[\frac{1}{2}]$). Under this condition, $Sq(x)$ is well defined for any $x \in M$ (respectively, $x \in \mathcal{M}$), where Sq is the usual total square (respectively, the total square in $\hat{A}$). If M satisfies the condition on finiteness, then so does $M^{\vee}$.

DEFINITION 2.6. A $\mathbb{Z}[\frac{1}{2}]$-graded $\hat{A}$-module $\mathcal{M}$ is called an $\hat{A}$-*algebra* if it is an algebra, satisfies the condition on finiteness and

$$Sq(uv) = Sq(u)Sq(v)$$

for any $u, v \in \mathcal{M}$, where Sq is the total square in $\hat{A}$.

The above equality is equivalent to the fact that the multiplication $\mathcal{M} \otimes \mathcal{M} \to \mathcal{M}$ is a homomorphism of $\hat{A}$-modules, where $\mathcal{M} \otimes \mathcal{M}$ is endowed with an $\hat{A}$-module structure via the coproduct of $\hat{A}$ and under the condition on finiteness.

The following proposition is a direct consequence of Proposition 1.7.

PROPOSITION 2.7 *(Cartan formula). Let M be an $\mathcal{A}$-algebra satisfying the condition on finiteness. Then*

$$Sq(uv) = Sq(u)Sq(v),$$

for every $u, v \in M^{\vee}$, or equivalently

$$Sq^a(uv) = \sum_{\substack{b+c=a \\ b,c \in \mathbb{N}[\frac{1}{2}]}} Sq^b(u)Sq^c(v),$$

for every $a \in \mathbb{N}[\frac{1}{2}]$. In other words, $M^{\vee}$ is an $\hat{A}$-algebra. $\square$

COROLLARY 2.8. *If M is an $\mathcal{A}$-algebra satisfying the condition on finiteness, then*

$$Sq(\sqrt[2^t]{u}) = \sqrt[2^t]{Sq(u)}.$$

for any $u \in M^{\vee}$ and $t \in \mathbb{Z}$. $\square$

If $f: M \to N$ is a homomorphism of $\mathcal{A}$-algebras, $f^{\vee}: M^{\vee} \to N^{\vee}$ is a homomorphism of $\hat{\mathcal{A}}$-algebras. Therefore, we obtain a functor, called the root functor, from the category of $\mathcal{A}$-algebras to the category of $\hat{\mathcal{A}}$-algebras, which sends M to $M^{\vee}$ and f to $f^{\vee}$. One can easily check that if f is either a monomorphism or an epimorphism, then so is $f^{\vee}$. In other words, the root functor is exact.

In particular, if X is a topological space, we denote by $H^{\vee}(X)$ the root closure of $H^*(X)$. Every continuous map between spaces $f: X \to Y$ induces a homomorphism of $\mathbb{Z}[\frac{1}{2}]$-graded $\hat{\mathcal{A}}$-algebras

$$f^{\vee}: H^{\vee}(Y) \longrightarrow H^{\vee}(X).$$

By unstability, the condition on finiteness holds for the root closure of the cohomology of any topological space.

3 The generalized Dickson algebra and the complete Steenrod algebra

Let $GL_k = GL(k, \mathbb{Z}/2)$ act as usual on the polynomial algebra $P_k = P(x_1, \ldots, x_k)$ with $\deg(x_i) = 1$. The Dickson algebra D_k is defined to be the algebra of invariants $P_k^{GL_k}$. Dickson [3] proved that D_k is also a polynomial algebra on k generators

$$D_k = P(Q_{k,0}, Q_{k,1}, \ldots, Q_{k,k-1}),$$

where $Q_{k,s}$ is the Dickson invariant of degree $2^k - 2^s$. An inductive definition of $Q_{k,s}$ is

$$Q_{k,s} = Q_{k-1,s-1}^2 + V_k Q_{k-1,s},$$

where, by convention, $Q_{k,k} = 1$, $Q_{k,s} = 0$ if $s < 0$ and

$$V_k = \prod_{\lambda_i \in \mathbb{Z}/2} (\lambda_1 x_1 + \cdots + \lambda_{k-1} x_{k-1} + x_k).$$

Arnon [1] called $D_k^{\vee}$ the *generalized Dickson algebra* and observed that

$$D_k^{\vee} := \left(P_k^{GL_k} \right)^{\vee} = \left(P_k^{\vee} \right)^{GL_k} = R(x_1, \ldots, x_k)^{GL_k}.$$

To describe $D_k^{\vee}$ explicitly he used the following elements defined by Peterson [15]:

NOTATION 3.1. $_k w_n = \displaystyle\sum_{\substack{s_1 + \cdots + s_k = n \\ s_i = 0 \text{ or } 2^{r_i}, r_i \in \mathbb{Z}}} x_1^{s_1} x_2^{s_2} \cdots x_k^{s_k} \in D_k^{\vee},$

for $n \in \mathbb{N}[\frac{1}{2}]$. Note that this sum is always finite.

PROPOSITION 3.2 [15]. *Let $\alpha(n)$ be the number of 1's in the dyadic expansion of n. For all $k, n \in \mathbb{N}$, $_k w_n \in D_k$ if and only if $2^{k-\alpha(n)} | n$. In particular,*

$$Q_{k,s} = {}_k w_{2^k - 2^s}.$$

PROPOSITION 3.3 [1].

(1) For any $k \in \mathbb{N}$ and $n \in \mathbb{N}[\frac{1}{2}]$, $_kw_n$ is GL_k-invariant.

(2) $D_k^{\checkmark}$ is the free root algebra generated by $\{_kw_{2^i-1}\}_{i=1}^k$.

Note that $_kw_n^2 = {}_kw_{2n}$. Thus, in $D_k^{\checkmark}$, $\quad _kw_{2^i-1} = {}^{2^{k-i}}\!\!\sqrt{Q_{k,k-i}}$.

DEFINITION 3.4. Let $i_k: D_k^{\checkmark} \to D_{k+1}^{\checkmark}$ be the degree preserving monomorphism of algebras given by $i_k({}_kw_{2^i-1}) = {}_{k+1}w_{2^i-1}$, for $1 \leq i \leq k$. Then we define

$$D_\infty^{\checkmark} = \varinjlim_k D_k^{\checkmark} \quad \text{and} \quad w_{2^i-1} = \varinjlim_k {}_kw_{2^i-1}.$$

$D_\infty^{\checkmark}$ is the free root algebra generated by $\{w_{2^i-1}\}_{i=1}^\infty$, with $\deg(w_{2^i-1}) = 2^i - 1$.

Remark 3.5.

(1) By [1, Thm. 2.11], which expresses $_kw_n$ as a polynomial on the generators $\{_kw_{2^i-1}\}_{i=1}^k$,

$$i_k({}_kw_n) = {}_{k+1}w_n \quad \text{for any } n \in \mathbb{N}[\tfrac{1}{2}].$$

(2) i_k is *not* a homomorphism of $\hat{\mathcal{A}}$-algebras. Indeed, we have

$$i_k(Q_{k,s}^2) = i_k\big({}_kw_{2^k-s-1}^{2^{s+1}}\big) = {}_{k+1}w_{2^k-s-1}^{2^{s+1}} = Q_{k+1,s+1}.$$

However, by [12, Thm. B], $Sq^1 Q_{k+1,1} = Q_{k+1,0}$, while $Sq^1 Q_{k,0}^2 = 0$ (by the Cartan formula).

The main theorem relates the "dual" of the complete Steenrod algebra $\hat{\mathcal{A}}$ to invariants of the general linear groups. Since $\hat{\mathcal{A}}$ is *not* of finite type, we need to give a precise definition of its dual, $\hat{\mathcal{A}}_*$.

DEFINITION 3.6. Let $d^*: \frac{1}{2^{t+1}}\mathcal{A}_* \leftarrow \frac{1}{2^t}\mathcal{A}_*$ be the dual of the halving homomorphism $d: \frac{1}{2^{t+1}}\mathcal{A} \to \frac{1}{2^t}\mathcal{A}$. The *graded dual* of $\hat{\mathcal{A}}$ is the direct limit

$$\hat{\mathcal{A}}_* = \varinjlim_{d^*} \big(\frac{1}{2^t}\mathcal{A}_*\big),$$

where $\mathcal{A}_*$ is the graded dual of the finite type algebra $\mathcal{A}$ (see [10]).

THEOREM 3.7. *There is an isomorphism of* $\mathbb{Z}[\frac{1}{2}]$*-graded algebras*

$$\hat{\mathcal{A}}_* \cong D_\infty^{\checkmark}.$$

Proof. According to Milnor [10], there is an isomorphism of $\mathbb{Z}$-graded algebras

$$\mathcal{A}_* \cong P(\xi_1, \xi_2, \dots).$$

Here $\xi_i = (Sq^{2^{i-1}} \dots Sq^2 Sq^1)^*$ is the dual of $(Sq^{2^{i-1}} \dots Sq^2 Sq^1)$ with respect to the basis of admissible operations in $\mathcal{A}$. Its degree is $2^i - 1$.

Let $d^* : \frac{1}{2^{t+1}}A_* \leftarrow \frac{1}{2^t}A_*$ be as in Definition 3.6. We want to show that

$$d^*(\xi_i) = \xi_i^2$$

for every i. By the definition of d

$$d(Sq^I) = \begin{cases} Sq^J & \text{if } I = 2J, \\ 0 & \text{otherwise}. \end{cases}$$

Here I and J are integral multi-indices. In particular,

$$d(Sq^{2^i} \ldots Sq^4 Sq^2) = Sq^{2^{i-1}} \ldots Sq^2 Sq^1.$$

So, passing to the dual,

$$d^*(\xi_i) = d^*(Sq^{2^{i-1}} \ldots Sq^2 Sq^1)^* = (Sq^{2^i} \ldots Sq^4 Sq^2)^*.$$

By [18, II.3.4],

$$\langle \xi^2, Sq^I \rangle = \begin{cases} \langle \xi, Sq^J \rangle & \text{if } I = 2J, \\ 0 & \text{otherwise}, \end{cases}$$

for $\xi \in A_*$ and I, J integral. Therefore, for all i,

$$\xi_i^2 = (Sq^{2^i} \ldots Sq^4 Sq^2)^* = d^*(\xi_i).$$

Now, by the Milnor isomorphism and Example 2.2,

$$\hat{A}_* = \varinjlim_{d^*} \left(\frac{1}{2^t} A_* \right) \cong \varinjlim_{d^*} \frac{1}{2^t} P(\xi_1, \xi_2, \ldots) = R(\xi_1, \xi_2, \ldots).$$

On the other hand, $D_\infty^{\checkmark}$ is also a free root algebra generated by the elements w_{2^i-1} with $\deg(w_{2^i-1}) = 2^i - 1$. Therefore

$$\begin{array}{ccc} R(\xi_1, \xi_2, \ldots) & \cong & D_\infty^{\checkmark} \\ \xi_i & \longmapsto & w_{2^i-1}. \end{array}$$

is an isomorphism of $\mathbb{Z}[\frac{1}{2}]$-graded algebras. $\square$

The next corollary follows immediately from the proof of Theorem 3.7.

COROLLARY 3.8.

$$\langle \sqrt[2^m]{\xi_i}, Sq^I \rangle = 1 \quad \text{if and only if} \quad I = (2^{i-m-1}, \ldots, 2^{1-m}, 2^{-m}). \quad \square$$

4 The generalized Steenrod total power

The aim of this section is to derive generalized cohomology operations in $\hat{A}$ from the generalized Dickson invariants in a similar way as the cohomology operations in A are derived from the classical Dickson invariants (see [5], [7], [9]).

Let E^n denote an elementary abelian 2-group of rank n and Σ_{2^n} the symmetric group on 2^n letters. For each topological space X, consider the homomorphism

$$d_n^* P_n : H^*(X) \xrightarrow{P_n} H^*\left(E\Sigma_{2^n} \times_{\Sigma_{2^n}} X^{2^n}\right) \xrightarrow{d_n^*} H^*(BE^n) \otimes H^*(X),$$

where P_n is the Steenrod power map, which sends u to $1 \otimes u^{2^n}$ at cochain level and d_n^* is induced by the regular permutation representation $E^n \subset \Sigma_{2^n}$ and the diagonal map of X.

Steenrod [18] used $d_1^* P_1$ to define the squares Sq^i as follows

$$d_1^* P_1(u) = \sum_{i=0}^{q} x_1^{q-i} \otimes Sq^i(u),$$

where $u \in H^q(X)$ and x_1 is the generator of the algebra $H^*(BE^1) \cong P(x_1)$.

In the general case, since the Weyl group of E^n in Σ_{2^n} is GL_n,

$$\mathrm{Im}(d_n^* P_n) \subset H^*(BE^n)^{GL_n} \otimes H^*(X).$$

Mùi [5] proved that, if $u \in H^q(x)$,

$$d_n^* P_n(u) = \sum_{\substack{R=(r_1,\ldots,r_n) \\ r_i \in \mathbb{N}}} Q_{n,0}^{q-(r_1+\cdots+r_n)} Q_{n,1}^{r_1} \cdots Q_{n,n-1}^{r_{n-1}} \otimes St^R u,$$

where St^R, $R = (r_1,\ldots,r_n)$, denotes the dual element of $\xi_1^{r_1} \cdots \xi_n^{r_n}$ with respect to the Milnor basis of $\mathcal{A}_*$ consisting of all monomials in the ξ_i's. (See also Madsen–Milgram [9] for a related result.)

In [7] Lomonaco considered the following stable version of $d_n^* P_n$. Let

$$\Phi_n = H^*(BE^n)[Q_{n,0}^{-1}].$$

For $n = 1$, $\Phi_1 \cong P(x_1, x_1^{-1})$. There is a unique way to extend the $\mathcal{A}$-action on $H^*(E^n)$ to an $\mathcal{A}$-action on Φ_n (see [19]). In particular, for any $k \in \mathbb{Z}$, we have

$$Sq^j(x_1^k) = \binom{k}{j} x_1^{k+j},$$

where $\binom{k}{j}$ stands for the coefficient of x^j in the expansion of $(1+x)^k$. Now, define

$$S_1 : H^*(X) \longrightarrow \Phi_1 \otimes H^*(X) \quad \text{by} \quad S_1(u) = \sum_{i \geq 0} x_1^{-1} \otimes Sq^i(u),$$

for any $u \in H^*(X)$. Again, $\Phi_1 \otimes H^*(X)$ is an $\mathcal{A}$-module by means of the diagonal action (Cartan formula). So we can iterate S_1 and define

$$S_n(u) = \sum_{i_j \geq 0} (x_1^{-i_1} \otimes Sq^{i_1})(x_2^{-i_2} \otimes Sq^{i_2}) \ldots (x_n^{-i_n} \otimes Sq^{i_n}(u)).$$

Lomonaco proved that, for every $u \in H^*(X)$,

$$S_n(u) \in \Phi_n^{GL_n} \otimes H^*(X),$$

and obtained the following normalized version of Mùi's formula

$$S_n(u) = \sum_{\substack{R=(r_1,\ldots,r_n) \\ r_i \in \mathbb{N}}} Q_{n,0}^{-(r_1+\cdots+r_n)} Q_{n,1}^{r_1} \ldots Q_{n,n-1}^{r_{n-1}} \otimes St^R u.$$

$S_n: H^*(X) \to \Phi_n^{GL_n} \otimes H^*(X)$, as well as $d_n^* P_n$, is a homomorphism of algebras. In this section we extend S_n to root closures of cohomology.

DEFINITION 4.1. Let $\mathcal{M}$ be an $\hat{A}$-module satisfying the condition on finiteness. We define $\mathcal{S}_1 : \mathcal{M} \to \Phi_1^{\vee} \otimes \mathcal{M}$ by $\mathcal{S}_1(u) = \sum_{i \in \mathbb{N}[\frac{1}{2}]} x_1^{-1} \otimes Sq^i(u), \quad u \in \mathcal{M}.$

The diagonal action (Cartan formula) endows $\Phi_1^{\vee} \otimes \mathcal{M}$ with an $\hat{A}$-module structure. We inductively define $\mathcal{S}_n(u)$ as follows

$$\mathcal{S}_n(u) = \mathcal{S}_1(\mathcal{S}_{n-1}(u)) = \mathcal{S}_1 \circ \mathcal{S}_1 \circ \cdots \circ \mathcal{S}_1(u)$$

$$= \sum_{i_j \in \mathbb{N}[\frac{1}{2}]} (x_1^{-i_1} \otimes Sq^{i_1})(x_2^{-i_2} \otimes Sq^{i_2}) \ldots (x_n^{-i_n} \otimes Sq^{i_n}(u)).$$

$\mathcal{S}_n$ is an $\hat{A}$-module homomorphism.

LEMMA 4.2. If X is a topological space, then $\mathcal{S}_n : H^{\vee}(X) \to (\Phi_n^{GL_n})^{\vee} \otimes H^{\vee}(X)$ is an algebra homomorphism and $\mathcal{S}_n = S_n^{\vee}$.

Proof. It suffices to prove the lemma for $n = 1$. By Remark 2.4 (2), for every $u \in H^{\vee}(X)$, there exists $z \in H^*(X)$ such that $u = \sqrt[2^t]{z}$ for some t. Moreover, by the definition of the action of $\hat{A}$ on $H^{\vee}(X)$, $Sq^i(\sqrt[2^t]{z}) = \sqrt[2^t]{Sq^{2^t \cdot i}(z)} = 0$ whenever $2^t i \notin \mathbb{N}$. Therefore,

$$\mathcal{S}_1(u) = \sum_{i \in \mathbb{N}[\frac{1}{2}]} x_1^{-i} \otimes Sq^i(\sqrt[2^t]{z}) = \sum_{2^t \cdot i \in \mathbb{N}} x_1^{-i} \otimes \sqrt[2^t]{Sq^{2^t \cdot i}(z)}$$

$$= \sqrt[2^t]{\sum_{2^t \cdot i \in \mathbb{N}} x_1^{-2^t \cdot i} \otimes Sq^{2^t \cdot i}(z)} = \sqrt[2^t]{S_1(z)} = S_1^{\vee}(\sqrt[2^t]{z}) = S_1^{\vee}(u). \quad \square$$

LEMMA 4.3. Let $d : \frac{1}{2^{t+1}}A \to \frac{1}{2^t}A$ be the halving homomorphism defined in Section 1. Then

$$d(St^{(r_1,\ldots,r_n)}) = \begin{cases} St^{(r_1/2,\ldots,r_n/2)} & \text{if every } r_i \text{ is even,} \\ 0 & \text{otherwise.} \end{cases}$$

Proof. Consider $d^* : \frac{1}{2^t}\mathcal{A}_* \to \frac{1}{2^{t+1}}\mathcal{A}_*$, the dual homomorphism of d. In the proof of Theorem 3.7 we showed that $d^*(\xi_i) = \xi_i^2$. Thus, since d^* is a homomorphism of algebras,

$$d^*(\xi_1^{t_1} \ldots \xi_n^{t_n}) = \xi_1^{2t_1} \ldots \xi_n^{2t_n}.$$

By passing to the dual, the lemma follows. $\square$

If $R = (r_1, \ldots, r_n)$ with $r_i \in \mathbb{N}[\frac{1}{2}]$ we denote by St^R the element of $\hat{\mathcal{A}}$ represented by $\pi_t(St^R) = St^{2^t R} \in \frac{1}{2^t}\mathcal{A}$, where t is such that $2^t R := (2^t r_1, \ldots, 2^t r_n) \in \mathbb{Z}^n$. This is compatible with the notation $Sq^{\frac{r}{2^t}}$ introduced in Section 1, as $St^{(r)} = Sq^r$.

COROLLARY 4.4.
$$\langle \xi^S, St^R \rangle = \delta_S^R,$$
for $R = (r_1, \ldots, r_n), S = (s_1, \ldots, s_m)$ with $r_i \in \mathbb{N}[\frac{1}{2}], s_j \in \mathbb{N}[\frac{1}{2}]$. $\square$

COROLLARY 4.5. *Let $\hat{d} \colon \frac{1}{2}\hat{\mathcal{A}} \to \hat{\mathcal{A}}$ be the homomorphism defined in Section 1. Then*
$$\hat{d}(St^R) = St^{\frac{R}{2}},$$
with $R = (r_1, \ldots, r_n)$ and $r_i \in \mathbb{N}[\frac{1}{2}]$ for all i. $\square$

The next proposition is a generalization of the formula mentioned above by Mùi [5], Madsen–Milgram [9], Lomonaco [7].

PROPOSITION 4.6. *For every $u \in H^{\vee}(X)$,*
$$\mathcal{S}_n(u) = \sum_{\substack{R=(r_1,\ldots,r_n) \\ r_i \in \mathbb{N}[\frac{1}{2}]}} {}_n w_{2^n-1}^{-(r_1+\cdots+r_n)} \, {}_n w_{2^{n-1}-1}^{2r_1} \cdots {}_n w_1^{2^{n-1}r_{n-1}} \otimes St^R(u).$$

Proof. Observe that, for every u, this is a finite sum. Now given $u \in H^{\vee}(X)$, take $z \in H^*(X)$ such that $u = \sqrt[2^t]{z}$ for some t (see Remark 2.4). By Lemma 4.2,

$$\mathcal{S}_n(u) = \mathcal{S}_n(\sqrt[2^t]{z}) = \sqrt[2^t]{\mathcal{S}_n(z)}$$

$$= \sqrt[2^t]{\sum_{\substack{R'=(r_1',\ldots,r_n') \\ r_i' \in \mathbb{N}}} Q_{n,0}^{-(r_1'+\cdots+r_n')} Q_{n,1}^{r_1'} \cdots Q_{n,n-1}^{r_{n-1}'} \otimes St^{R'}(z)}$$

$$= \sum_{\substack{R=(r_1,\ldots,r_n) \\ r_i = \frac{r_i'}{2^t},\, r_i' \in \mathbb{N}}} {}_n w_{2^n-1}^{-(r_1+\cdots+r_n)} \, {}_n w_{2^{n-1}-1}^{2r_1} \cdots {}_n w_1^{2^{n-1}r_{n-1}} \otimes \sqrt[2^t]{St^{2^t R}(z)}.$$

Moreover, by definition of the $\hat{\mathcal{A}}$-action on $H^{\vee}(X)$ and Corollary 4.5,

$$\sqrt[2^t]{St^{2^t R}(z)} = \hat{d}^t(St^{2^t R})(\sqrt[2^t]{z}) = St^R(\sqrt[2^t]{z}). \qquad \square$$

ACKNOWLEDGEMENTS. *The authors thank Ian Leary for helpful discussions on $\hat{\mathcal{A}} \otimes \hat{\mathcal{A}}$. The second-named author warmly thanks Manuel Castellet and all the colleagues at the CRM (Barcelona) for their hospitality.*

References

1. D. Arnon, *Generalized Dickson invariants*, Ph. D. Thesis, MIT, 1994.
2. S. R. Bullett and I. G. Macdonald, *On the Adem relations*, Topology **21** (1982), 329–332.
3. L. E. Dickson, *A fundamental system of invariants of the general modular linear group with a solution of the form problem*, Trans. Amer. Math. Soc. **12** (1911), 75–98.
4. Huỳnh Mùi, *Modular invariant theory and the cohomology algebras of symmetric groups*, J. Fac. Sci. Univ. Tokyo Sec. IA Math. **22** (1975), 319–369.
5. Huỳnh Mùi, *Dickson invariants and Milnor basis of the Steenrod algebra*, Eger Internat. Colloq. Topology (1983), 345–355.
6. J. Lannes and S. Zarati, *Sur les foncteurs dérivés de la déstabilisation*, Math. Zeit. **194** (1987), 25–59.
7. L. Lomonaco, *The iterated total squaring operation*, Proc. Amer. Math. Soc. **115** (1992), 1149–1155.
8. I. Madsen, *On the action of the Dyer-Lashof algebra in $H_*(G)$*, Pacific J. Math. **60** (1975), 235–275.
9. I. Madsen and J. Milgram, The classifying spaces for surgery and cobordism of manifolds, Ann. of Math. Studies, No. 92, Princeton Univ. Press, 1979.
10. J. Milnor, *The Steenrod algebra and its dual*, Ann. of Math. **67** (1958), 150–171.
11. Nguyên N. Hai and Nguyên H. V. Hu'ng, *Steenrod operations on mod 2 homology of the iterated loop space*, Acta Math. Vietnam. **13** (1988), 113–126.
12. Nguyên H. V. Hu'ng, *The action of the Steenrod squares on the modular invariants of linear groups*, Proc. Amer. Math. Soc. **113** (1991), 1097–1104.
13. Nguyên H. V. Hu'ng, *Spherical classes and the algebraic transfer*, CRM Preprint.
14. Nguyên H. V. Hu'ng and F. P. Peterson, *A-generators for the Dickson algebra*, Trans. Amer. Math. Soc. (to appear).
15. F. P. Peterson, *Private communication*.
16. W. M. Singer, *A new chain complex for the homology of the Steenrod algebra*, Proc. Cambridge Philos. Soc. **90** (1981), 279–292.
17. W. M. Singer, *Invariant theory and the lambda algebra*, Trans. Amer. Math. Soc. **280** (1983), 673–693.
18. N. E. Steenrod and D. B. A. Epstein, Cohomology operations, Ann. of Math. Studies, No. 50, Princeton Univ. Press, 1962.
19. C. Wilkerson, *Classifying spaces, Steenrod operations and algebraic closure*, Topology **16** (1977), 227–237.

Irene Llerena: Departament d'Àlgebra i Geometria, Fac. de Matemàtiques, Universitat de Barcelona, Gran Via 585, 08007 Barcelona, Spain
llerena@cerber.ub.es

Nguyên H.V. Hu'ng: Department of Mathematics, University of Hanôi, 90 Nguyên Trãi Street, Hanôi, Viêtnam

Current address: Centre de Recerca Matemàtica, Institut d'Estudis Catalans, Apartat 50, E-08193 Bellaterra, Barcelona, Spain
hung@cadi.crm.es

Progress in Mathematics, Vol. 136
© 1996 Birkhäuser Verlag Basel/Switzerland

On the localization genus of a space

C. A. McGibbon

Let X be a connected nilpotent CW-complex of finite type over $\mathbf{Z}$. If Y is another nilpotent CW-complex such that the localizations $Y_{(p)}$ and $X_{(p)}$ are homotopy equivalent for each prime p, it does *not* necessarily follow that Y is homotopy equivalent to X; nor does it follow that Y must have finite type. Define the *extended genus* of X to be the set of homotopy types $[Y]$ of nilpotent CW-spaces Y which are locally homotopy equivalent to X at each prime.[1] Define the Mislin genus of X, denoted $\mathcal{G}(X)$, to be the subset of the extended genus consisting of those $[Y]$s of finite type. This paper deals with the extended genus and the Mislin genus of a space. In many interesting cases the Mislin genus is finite as the following result of Wilkerson, [16], shows.

THEOREM 1 *Let X be a 1-connected CW-complex of finite type over $\mathbf{Z}$. If $H_n(X;\mathbf{Z}) = 0$ for n sufficiently large or if $\pi_n(X) = 0$ for n sufficiently large, then the Mislin genus $\mathcal{G}(X)$ is finite.*

In contrast it will be shown here that the extended genus of X is often huge!

THEOREM 2 *Let X satisfy the hypotheses of Theorem 1. Then the extended genus of X has just one element if and only if X has the rational homotopy type of a point. Otherwise the extended genus of X is uncountably large.*

Some special cases of Theorem 2 — that of an Eilenberg-MacLane space and a sphere — were noted in [6] and [11]. The general case follows from the next result which is proved using the Postnikov system of X and certain non-cyclic subgroups of the rational numbers $\mathbf{Q}$. The subgroups in question are not finitely generated but are locally equivalent to $\mathbf{Z}$ at each prime p. A good family of examples to keep in mind is the following: for $n = 1, 2, \ldots$ let A_n denote the additive subgroup of $\mathbf{Q}$ generated by the fractions $1/p^n$, where p runs through all the prime numbers. Thus, for example, A_1 is the subgroup of rationals with square-free denominators. Although these subgroups are not finitely generated, they are locally isomorphic to $\mathbf{Z}$ at each prime. According to [4] or [5], there are up to isomorphism, uncountably many subgroups of $\mathbf{Q}$ with this property. This fact, together with the following construction, accounts for the large cardinality of the extended genus.

THEOREM 3 *Let X be a rationally non-trivial space which satisfies the hypotheses of Theorem 1 and let κ be the smallest integer k such that $\pi_k(X) \otimes \mathbf{Q} \neq 0$.*

---

[1]The notion of a genus set in homotopy theory is due to Mislin, [14]. Hilton began the study of the extended genus of a nilpotent group in [5] and that of a nilpotent space in [6].

For each abelian group A, such that $A_{(p)} \approx \mathbf{Z}_{(p)}$ for every prime p, there exists a space X_A, which is locally equivalent to X at each prime, and also has the property that, modulo torsion, $\pi_\kappa(X_A)$ is isomorphic to the direct sum of r copies of A, where r is the rank of $\pi_\kappa(X) \otimes \mathbf{Q}$. If B is another abelian group locally equivalent to $\mathbf{Z}$ at each prime, then the spaces X_A and X_B are homotopy equivalent only if the groups A and B are isomorphic.

The construction just mentioned is not functorial; there are choices involved in the higher homotopy groups of X_A when, for example, X is a product of two rationally non-trivial spaces with different rational connectivities. The following example shows that the higher homology groups of X_A are more complicated than one might first suspect. In this example $X = \mathbf{CP}^\infty = K(\mathbf{Z}, 2)$ and A is the group A_1 described earlier; in other words, X_A is the Eilenberg-MacLane space $K(A_1, 2)$.

Observe that A_1 can be obtained as a direct limit of the sequence

$$\mathbf{Z} \xrightarrow{\;2\;} \mathbf{Z} \xrightarrow{\;3\;} \mathbf{Z} \xrightarrow{\;5\;} \cdots,$$

where each group is $\mathbf{Z}$ and the nth homomorphism is multiplication by the nth prime. Since homotopy commutes with direct limits, the space $K(A_1, 2)$ can be obtained as an infinite telescope using $\mathbf{CP}^\infty$ and the corresponding power maps. Using the ring structure on $H^*(\mathbf{CP}^\infty; \mathbf{Z})$, it is easy to determine how these power maps behave in homology. Since homology also commutes with direct limits it follows that the integral homology of $K(A_1, 2)$ is isomorphic to A_n in dimension $2n$.

The rest of this paper deals with Zabrodsky's presentation of the Mislin genus $\mathcal{G}(X)$. This is a powerful tool which for certain spaces X reduces the calculation of $\mathcal{G}(X)$ to a problem of classifying, up to rational homology, certain self maps of X. Assume for the moment that X is a 1-connected H-space, of finite type, and with only a finite number of nonzero homotopy groups. Zabrodsky's theorem involves the following sequence,

$$\mathcal{E}_t(X) \xrightarrow{\quad d \quad} (\mathbf{Z}_t^* / \pm 1)^\ell \xrightarrow{\quad \xi \quad} \mathcal{G}(X).$$

In this sequence t is a certain natural number that depends on the space X. In the middle term, $\mathbf{Z}_t^*$ denotes the group of units in the ring of integers modulo t. The exponent ℓ is the number of integers k, for which $\pi_k(X) \otimes \mathbf{Q} \neq 0$. If there is more than one such k, order them $k_1 < \cdots < k_\ell$. The first term in the sequence, $\mathcal{E}_t(X)$, denotes the monoid (under composition) of homotopy classes of those self-maps of X which are local equivalences at each prime divisor of t. The function d assigns to each such map a sequence of determinants — or rather the image of such a sequence in the middle group. The jth determinant here is that of the linear transformation on $\pi_{k_j}(X) \otimes \mathbf{Q}$ induced by f. [2]

---

[2] Zabrodsky defined this map in a slightly different way. Instead of using the rational homotopy groups of X he used the quotient module $QH^*(X; \mathbf{Z})$/torsion. Since X is an H_o-space this module is rationally isomorphic to $\pi_*(X) \otimes \mathbf{Q}$ and there is sufficient naturality here to see that the two sequences of determinants are the same.

The map ξ is given by the following pullback construction. Let

$$K(F,\bar{n}) = K(F_1,k_1) \times \cdots \times K(F_\ell,k_\ell)$$

where F_i is the abelian group isomorphic to the free part of $\pi_{k_i}(X)$. Given $u_i \in \mathbf{Z}_t^*$, let f_i be a self map of $K(F_i,k_i)$ such that $d(f_i) = u_i$. The space $K(F,\bar{n})$ has the same rational homotopy type as X; indeed, the two spaces are equivalent away from t. Fix a rational equivalence $h : X \to K(F,\bar{n})$ which induces an isomorphism on $QH^*(-;\mathbf{Z})/$torsion. Then the function ξ sends a sequence $(u_1,\ldots,u_\ell)$ to the homotopy type of the pullback of the diagram

$$
\begin{array}{ccc}
 & & X \\
 & & \downarrow{\scriptstyle h} \\
K(F,\bar{n}) & \xrightarrow{\ f_1 \times \cdots \times f_\ell\ } & K(F,\bar{n})
\end{array}
$$

In [17], Zabrodsky showed that the map ξ is an epimorphism. In [18] he improved this result in two ways. First, he established exactness at the middle term and hence obtained an abelian group structure on $\mathcal{G}(X)$. Second, he enlarged the class of spaces for which this sequence is exact from H-spaces to H_o-spaces. More precisely, he showed that the above sequence is exact when X is a 1-connected H_o-space (again of finite type and with only a finite number of nonzero homotopy groups). Recall that an H_o-space is a nilpotent space whose rationalization is an H-space. Obviously every H-space is an H_o-space. Perhaps the simplest counterexample to the converse is the sphere S^5. Other nontrivial examples of H_o-spaces include the complex and quaternionic Stiefel manifolds. I will show that Zabrodsky's result also holds in the non-simply connected case.

THEOREM 4 *Zabrodsky's presentation of $\mathcal{G}(X)$ is valid for connected nilpotent, finite type H_o-spaces with at most a finite number of nonzero homotopy groups.*

An important special case of Theorem 4 is when X is a $K(\pi,1)$. For a finitely generated nilpotent group N, the Mislin genus of N is defined to be the set of isomorphism classes of finitely generated nilpotent groups M such that $N_{(p)} \approx M_{(p)}$ for all primes p. Clearly, for such a group N, the Mislin genus of the group N is isomorphic to the Mislin genus of a the space $K(N,1)$. In [7], Hilton and Mislin established a similar presentation for the Mislin genus of a finitely generated nilpotent group whose commutator subgroup is finite. Under these conditions, N is rationally equivalent to its abelianization and so it is evident that the space $K(N,1)$ is an H_o-space. Thus Theorem 4 yields the Hilton-Mislin result as a special case.

There is another version of Zabrodsky's presentation which is dual to the previous ones in the sense of Eckmann and Hilton. Recall that a co-H_o-space is a space which has the rational homotopy type of a bouquet of spheres.

THEOREM 5 *Given a 1-connected finite co-H_o-space X, there is a sequence*

$$\mathcal{E}_t(X) \xrightarrow{\ d\ } (\mathbf{Z}_t^*/\pm 1)^\ell \xrightarrow{\ \xi\ } \mathcal{G}(X)$$

which is exact in the middle and surjective on the right. In this sequence, the determinant map d is defined in terms of endomorphisms of $H_*(X; \mathbf{Q})$ and the map ξ involves a pushout diagram with a bouquet a spheres.

Davis proved a result like this one for finite complexes in the stable range in [3]. Subsequently, a shorter proof of this stable result was given in [10]. The later paper also contains a classification, up to homology, of the stable self maps of complex projective space $\mathbf{C}P^n$. It turns out that every self map of $\Sigma^k \mathbf{C}P^n$ appears, up to homology, to be a linear combination of self maps of $\mathbf{C}P^n$. In particular, the homology classification of such maps does not change once $k \geq 1$. Therefore the calculation of the stable genus of $\mathbf{C}P^n$ in [10], together with Theorem 5, implies the following result.

EXAMPLE A *The Mislin genus of $\Sigma \mathbf{C}P^n$ is trivial for $n \leq 2$. For $n \geq 3$ it is isomorphic to* $\prod_{m=3}^{n} (\mathbf{Z}_{m!}^* / \pm 1)$.

Here is one application of Theorem 5.

COROLLARY 5.1 *If X is a 1-connected finite co-H_o-space and Y is in the Mislin genus of X, then there is a finite bouquet of spheres S such that $X \vee S$ and $Y \vee S$ are homotopy equivalent. Thus X is a co-H-space if and only if Y is.*

In other words, among 1-connected finite complexes the existence of a a co-H-structure is a generic property. I do not know if this remains true if one drops the finite dimensional requirement. The following example shows that, in general, having the homotopy type of a suspension is **not** a generic property.

EXAMPLE B *The Mislin genus of $\Sigma \mathbf{C}P^\infty$ is uncountably large and yet only one of its members has the homotopy type of a suspension.*

Are there presentations of $\mathcal{G}(X)$ in terms of the self maps of X which are valid for a class of spaces X beyond those just considered? While this seems to be an open question, the following example shows that the most naive attempts to extend Zabrodsky's presentation further are doomed to fail.

EXAMPLE C *Let $X = S^3 \vee Sp(2)$. Then for any natural number t, the determinant map $\delta : \mathcal{E}_t(X) \to (\mathbf{Z}_t^* / \pm 1)^2$ is surjective. However, the cardinality of $\mathcal{G}(X)$ is at least 2.*

The determinant map in this example is defined using the indecomposables in cohomology. Notice that in this example X does not have the rational homotopy type of an H-space nor of a co-H-space.

Proofs

All spaces and maps considered here are pointed. All spaces here are also assumed to have the homotopy type of CW-complexes. The localization of a space X at a prime p will be denoted $X_{(p)}$; its rationalization will be denoted X_o. The proof of Theorem 3 will be given first and then Theorem 2 will be derived as a consequence.

PROOF OF THEOREM 3: Let $\{X^{(n)} \mid n = 1, 2, \ldots\}$ denote the Postnikov decomposition of X. Assume that there exists a natural number N such that $\pi_n(X) = 0$ for all $n \geq N$. Let S be the set of primes p for which there is no p-torsion in either $\pi_*(X)$ or in $H_*(X^{(n)}; \mathbf{Z})$ in dimensions $\leq N$ for $n = 1, 2, \ldots$ The finite type hypothesis on X ensures that the set of primes not in S is finite. For the moment assume that X has been localized at S. Then $X^{(\kappa)} \simeq K(G, \kappa)$ where G is a free $\mathbf{Z}_{(S)}$-module of rank $r > 0$. Let $X_A^{(\kappa)} = K(H, \kappa)$ where H is the direct sum of r copies of $A_{(S)}$.

The induction step amounts to solving the following problem. Let $Y = X^{(n-2)}$ where $n < N$ and assume that M is a torsion-free finitely generated $\mathbf{Z}_{(S)}$-module. Given a map

$$k : Y \longrightarrow K(M, n)$$

and a space Y' in the extended genus of Y, find an abelian group M' and a map

$$k' : Y' \longrightarrow K(M', n)$$

such that the maps k and k' are locally equivalent at every prime in S. More precisely we require, for every prime p not in S, a commutative diagram

$$
\begin{array}{ccc}
Y_{(p)} & \xrightarrow{\ k_{(p)}\ } & K(M_{(p)}, n) \\
\downarrow & & \downarrow \\
Y'_{(p)} & \xrightarrow{\ k'_{(p)}\ } & K(M'_{(p)}, n)
\end{array}
$$

whose columns are homotopy equivalences. From this it will follow, of course, that the fibers of k and k' will be in the same extended genus.

The class k corresponds uniquely to a homomorphism of the same name,

$$k : H_n(Y; \mathbf{Z}) \longrightarrow M,$$

by the universal coefficient theorem for cohomology and the absence of torsion after localizing at S. The search for the right k-invariant for Y' thus amounts to finding the appropriate homomorphism $k' : H_n(Y'; \mathbf{Z}) \longrightarrow M'$. The homomorphism k' and its target M' will be constructed as pullbacks.

For each $p \in S$ there is a rational equivalence $r_p : Y_{(p)} \longrightarrow Y_o$ such that Y is homotopy equivalent to the pullback of the diagram

$$\prod_p Y_{(p)}$$
$$\Big\downarrow \prod_p r_p$$
$$Y_o \xrightarrow{\quad \Delta \quad} \prod_p Y_o.$$

Now Y' is in the extended genus of Y and it too can be reconstructed from the pieces Y_o and $\{Y_{(p)} \mid p \in S\}$; it is only necessary to change the maps. There are rational equivalences $r'_p : Y_{(p)} \longrightarrow Y_o$ such that Y' is homotopy equivalent to the pullback of the diagram obtained by replacing the maps r_p by r'_p in the diagram just displayed. To prove this one uses the Mayer-Vietoris sequence for homotopy groups, just as in the proof of Theorem 5.7 in [8], page 86. The key ingredient here is that the homotopy groups of Y and Y' can be pulled back from their localizations. The same is true for the homology groups of these spaces in dimensions $\leq N$. Indeed,

$$H_n(Y'; \mathbf{Z}) \approx (r')_* H_n(Y'; \mathbf{Z}) = \bigcap_{p \in S} (r'_p)_* H_n(Y_{(p)}; \mathbf{Z})$$

Here $r' : Y' \to Y_o$ is a rational equivalence with the property that, for every $p \in S$, $r' = r'_p e_p$ where e_p is a p-equivalence. In this way we can identify $H_n(Y'; \mathbf{Z})_{(p)}$ with $H_n(Y_{(p)}; \mathbf{Z})$. For each prime p choose rational equivalences $\varphi_p : M_{(p)} \to M_o$ such that the following diagram commutes

$$
\begin{array}{ccc}
H_n(Y_{(p)}; \mathbf{Z}) & \xrightarrow{\quad k_{(p)} \quad} & M_{(p)} \\
\Big\downarrow{\scriptstyle (r'_p)_*} & & \Big\downarrow{\scriptstyle \varphi_p} \\
H_n(Y_o; \mathbf{Z}) & \xrightarrow{\quad k_o \quad} & M_o
\end{array}
$$

In other words, the homomorphisms φ_p become isomorphisms when rationalized. It is clear that such maps exist; it is also evident that they are unique only when k_o is surjective. Now define M' to be the intersection $\cap \varphi_p(M_{(p)})$. The homomorphism k' can be obtained as $(r')_* : H_n(Y'; \mathbf{Z}) \to H_n(Y_o, \mathbf{Z})$ followed by the restriction of k_o to $\cap (r'_p)_* (H_n(Y_{(p)}; \mathbf{Z}))$.

The construction of $(X_A)_{(S)}$ now follows by a finite induction. To obtain a global version of X_A, define it as the pullback of the diagram

$$(X_A)_{(S)}$$
$$\Big\downarrow$$
$$X_{(T)} \xrightarrow{\qquad} X_o$$

where T denotes the complement of S in the set of all primes, and the two maps are rational equivalences. Again there are choices involved with the maps, but it is clear that the resulting pullback will be in the extended genus of X. Since A is locally equivalent to $\mathbf{Z}$, the group A can be constructed as a direct limit

$$\mathbf{Z} \xrightarrow{p_1} \mathbf{Z} \xrightarrow{p_2} \mathbf{Z} \xrightarrow{p_3} \cdots$$

wherein multiplication by any one prime occurs at most a finite number of times. As usual, one can ignore the first N terms and the limit remains unchanged. In particular, this means that A is isomorphic to the pullback of the diagram

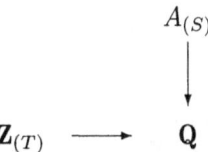

where the two maps are rational equivalences. It follows that the homotopy group $\pi_\kappa(X_A)$ has the stated form.

To complete the proof of Theorem 3, suppose that $\pi_q(X) \otimes \mathbf{Q} \neq 0$ for some q. If X_A and X_B are homotopy equivalent, then it follows that $A^r \approx B^r$ for some $r \geq 1$. Rank 1 abelian groups, like A and B are classified in the following manner. Take a nonzero element x in A and consider the height sequence $(h_2(x), h_3(x), \ldots, h_p(x), \ldots)$ where $h_p(x)$ denotes the highest power of p that divides x. If y is another nonzero element of A, then $y = rx$ for some rational number r, since A has rank 1, and hence $h_p(x) = h_p(y)$ for almost all primes p. The notion of two height sequences being eventually equal defines an equivalence relation and it turns out that A and B are isomorphic if and only if their corresponding height sequences are equivalent, [4]. Notice that the height sequence of any nonzero element in A^r is likewise equivalent to the height sequence of A. Thus if $A^r \approx B^r$, it follows that A and B must be isomorphic.

The case where the homology of X is nonzero in only a finite number of dimensions follows easily from the one just considered. One works with skeletons of Postnikov approximations of the spaces involved.

Theorem 2 follows easily from Theorem 3. Of course, if X is rationally a point, then $X \simeq \vee_p X_{(p)}$ and so its genus consists of just $[X]$.

While on the subject of the extended genus, let me correct a remark I made in [11]. There I claimed in the first section, that if one constructs the Moore space $M(A,n)$ as an infinite telescope using the sphere and its self-maps, then the inclusion of S^n into the left end of the telescope would be a p-equivalence at each prime p. This is nonsense, because a map which is a p-equivalence for every prime is itself an equivalence. On the algebra level it *is* true that for each finite family of primes there is an embedding of $\mathbf{Z}$ into A which is a p-equivalence for each prime in the family, but there is no one embedding which is a p-equivalence for all primes. To get a map of S^n into $M(A,n)$ which is an equivalence at say $p = 7$, one should embed it at a stage in the telescope after which multiplication by 7 does not occur.

Proof of Theorem 4

Recall that Theorem 4 deals with connected H_o-spaces which are nilpotent, of finite type, and have only finitely many nonzero homotopy groups. The following proof is an adaptation of the one Zabrodsky gave in [18] for the simply connected case. The proof depends upon a number of results about maps between such spaces — self maps, maps between members of the same genus, and maps involving Eilenberg-MacLane spaces in both the target and the domain. These preliminary results have applications beyond computations of $\mathcal{G}(X)$; see for example [12]. The proofs often involve Postnikov decompositions; Alex was certainly a master of such calculations. In some places I have (hopefully) streamlined his arguments; in others, such as the following result and its corollary, more detail has been added.

PROPOSITION 4.1 *Let X satisfy the hypotheses of Theorem 4. Then for each prime p there exists a self map of X, say f_p, such that*

  i) *f_p is an equivalence at each prime $q \neq p$, and*

  ii) *f_p induces multiplication by some positive power of p on each homotopy group of X.*

REMARK: This result is false in general, even among 1-connected finite CW-complexes. For example, in [13], Mimura and Toda construct a nontrivial simply connected finite complex every self map of which induces either the identity or the trivial map in integral homology.

Assume for the moment that Proposition 4.1 is true. The next result points out the connection between self maps of X and maps between members of $\mathcal{G}(X)$.

COROLLARY 4.2 *Let X satisfy the hypotheses of Theorem 4. Then for any $[Y] \in \mathcal{G}(X)$ and for any prime p, there exists a p-equivalence $g_p : Y \to X$.*

PROOF OF 4.2: If Y is in the genus of X it is clear there exists a p-equivalence, say $g' : Y \to X_{(p)}$. Now the localization $X_{(p)}$ can be constructed as the infinite mapping telescope using the family of maps $\mathcal{F} = \{f_q \mid (q,p) = 1\}$ described in Proposition 4.1. Let Y_N denote the N-skeleton of Y where N is large with respect to those q's for which $\pi_q(X) \neq 0$. Since Y is nilpotent and of finite type, Y_N can be taken to be a finite complex. By compactness, the restriction of g' to Y_N factors through the nth stage of the telescope (for some finite n), as indicated in the following diagram

$$
\begin{array}{ccc}
Y & \xrightarrow{\;g'\;} & X_{(p)} \simeq Tel_\infty(X, \mathcal{F}) \\
\uparrow & & \uparrow \\
Y_N & \xrightarrow{\;g''\;} & Tel_n(X, \mathcal{F}) \xrightarrow{\;\pi_n\;} X
\end{array}
$$

The vertical maps here are the obvious inclusions. The map π_n projects the n-stage telescope onto its right end. All the maps here are p-equivalences in dimensions $< N$. The map $g_p : Y \to X$ can then be taken to be a Postnikov approximation of the bottom row up to dimension $N - 2$. $\square$

The following group theory result will be used in the first step of the proof of Proposition 4.1.

LEMMA 4.3 *Suppose that G is a finitely generated nilpotent group with a finite commutator subgroup. Then there exists an integer $N = N(G) \geq 2$ such that the power map $g \mapsto g^\lambda$ is a homomorphism (of G to itself) if and only if*

$$\lambda(\lambda - 1) \equiv 0 \bmod N.$$

PROOF: For $g \mapsto g^\lambda$ to be a homomorphism it is necessary and sufficient that

$$(xy)^\lambda = x^\lambda y^\lambda$$

for all x, y in G. Now in any group, one has the identity

$$x^m y^m = (x^{m-1}[x, y^{m-1}]y^{-m+1})(x^{m-2}[x, y^{m-2}]y^{-m+2}) \cdots (x[x, y]y^{-1})(xy)^m$$

Write this identity as

$$x^m y^m = \Phi_m(x, y)(xy)^m.$$

Let M be the smallest natural number m such that g^m is central for any $g \in G$. Clearly M exists and equals the exponent of the finite quotient $G/Z(G)$. Notice that for all $x, y \in G$,

$$\begin{aligned}
\Phi_M(x, y) &= x^M y^M (xy)^{-M} &= x^M y^M xy(xy)^{-1}(xy)^{-M} \\
&= x^{M+1} y^{M+1}(xy)^{-M-1} &= \Phi_{M+1}(x, y).
\end{aligned}$$

A simple exercise shows that if $r \equiv s \bmod M$ then $x^r[x, y^r]y^{-r} = x^s[x, y^s]y^s$. Hence the terms $x^r[x, y^r]y^{-r}, 1 \leq r \leq M$ recur periodically in the expression for $\Phi_m(x, y)$. Indeed if $m = eM + r$, then $\Phi_m = \Phi_r(\Phi_M)^e$. Let $\| \Phi_M \|$ denote the maximum order of the elements $\Phi_M(x, y)$ in $[G, G]$ and set $N = M \cdot \| \Phi_M \|$. Since $\Phi_M = \Phi_{M+1}$, it follows that the function

$$\Phi_\lambda = 1 \quad \text{if} \quad \lambda \equiv 0 \text{ or } 1 \bmod N.$$

If N is a power of some prime p, then the condition that $\lambda \equiv 0$ or $1 \bmod N$ is equivalent to saying that $\lambda(\lambda - 1) \equiv 0 \bmod N$. Of course, this is what happens when one localizes the group G at p. Assume now that this has been done so that the resulting numbers M_p and N_p are powers of p.

To see that $\Phi_\lambda = 1$ implies that $\lambda(\lambda - 1) \equiv 0 \bmod N_p$ suppose that $\Phi_\lambda(x, y) = 1$ for all $x, y \in G_{(p)}$. Thus $(xy)^\lambda = x^\lambda y^\lambda$ and $(yx)^\lambda = y^\lambda x^\lambda$. Therefore,

$$x^{\lambda-1} y^\lambda = x^{-1}(xy)^\lambda = (yx)^\lambda x^{-1} = y^\lambda x^{\lambda-1}.$$

Either λ or $\lambda - 1$ is relatively prime to p; assume the latter holds. Since x^{M_p} and $x^{\lambda-1}$ both centralize y^λ it follows that x also centralizes y^λ. Since x was arbitrary, it follows that $y^\lambda \in Z(G)$. Since y was arbitrary it follows from the definition of M_p that $\lambda = aM_p$ for some integer a. Thus $1 = \Phi_\lambda = (\Phi_{M_p})^a$ and hence $a \equiv 0 \bmod \| M_p \|$. To put it another way, $\lambda \equiv 0 \bmod N_p$. Similarly if one assumes that $(\lambda, p) = 1$, one arrives at the conclusion that $\lambda - 1 \equiv 0 \bmod N_p$. Hence for each prime p, the solutions to $\Phi_\lambda = 1$ coincide with the set of solutions to the congruence $\lambda(\lambda - 1) \equiv 0 \bmod N_p$. By the Chinese Remainder theorem it then follows that $\Phi_\lambda = 1$ if and only if $\lambda(\lambda - 1) \equiv 0 \bmod N$.   □

PROOF OF PROPOSITION 4.1: Since X is nilpotent, the Postnikov decomposition of X admits a principal refinement at each stage, see e.g. [8], page 65. In other words, for each n, the projection $X^{(n)} \to X^{(n-1)}$ can be factored as a sequence of principal fibrations,

$$X^{(n)} = X_c^n \to X_{c-1}^n \to \cdots \to X_1^n \to X_0^n = X^{(n-1)},$$

where each $X_{i+1}^n \to X_i^n$ is induced by a map $k_i^n : X_i^n \to K(G_i, n+1)$. Here G_i is the quotient Γ^i / Γ^{i+1} where Γ^i is the ith term in the lower central series of the $\pi_1(X)$-action on $\pi_n(X)$.

Since X is an H_o-space, each of the k-invariants, k_i^n, has finite order. The proof of 4.1 will go by induction up this principal refinement.

To begin the induction, note that, for any prime p, some positive power of it is a solution to the congruence $\lambda(\lambda - 1) \equiv 0 \bmod N$. Thus if G is the fundamental group of X, there is, by Lemma 4.3, a self map f_p of $K(G, 1)$ induced by the power map on G of power $p^\alpha > 1$. In the induction step assume that the required map f_p exists on X_i^n. It then suffices to find a map g so that the following diagram commutes

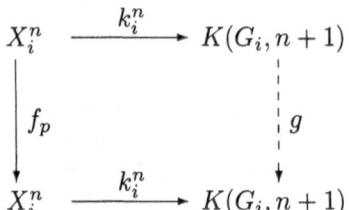

Let T denote the torsion subgroup of $[X_i^n, K(G_i, n+1)]$. This is a finite abelian group which contains the k-invariant k_i^n. Write $T = T_p \oplus T_{(1/p)}$ where the first summand represents the p-primary part of T. The map f_p induces an endomorphism of T and, by replacing f_p with some iterate of itself, if necessary, one may assume that f_p induces the zero homomorphism on T_p and the identity on the other summand. Multiplication by p^r on G_i likewise induces an endomorphism of T and, by replacing r with some larger number, if necessary, one may assume that $(p^r)_* = f^*$ on T, and (more importantly) on k_i^n. It follows that the map g may be taken to be the one induced by $p^r : G_i \to G_i$. The resulting self map on the fiber X_{i+1}^n is the desired map. Since X has only finitely many nonzero homotopy groups, this is a finite induction and the result follows.

PROPOSITION 4.4 *Let X satisfy the hypotheses of Theorem 4 and let A and B be disjoint sets of primes. Then there exists a map $\mu : X \times X \to X$ which* *restricts to an A-equivalence on the first factor and to a B-equivalence on the second.*

PROOF: This proof also uses the principal refinement of the Postnikov decomposition of X. Once again G denotes $\pi_1(X)$. The first step in the induction is the construction of a map

$$K(G,1) \times K(G,1) \longrightarrow K(G,1)$$

which restricts to an A-equivalence on the first factor and to a B-equivalence on the second factor. This is equivalent to finding a homomorphism $G \times G \to G$ with the same property. Such a homomorphism is given by the formulae

$$(x,y) \mapsto x^a y^b, \quad \text{where} \quad \begin{array}{l} (a,A) = (b,B) = 1. \\ (xy)^a = x^a y^a \text{ and } (xy)^b = x^b y^b \text{ for all } x,y \in G \\ x^a y^b = y^b x^a \text{ for all } x,y \in G \end{array}$$

The first condition is saying that no member of A divides a, etc. I leave it to the reader to verify that the function $(x,y) \mapsto x^a y^b$ is a homomorphism if the second and third conditions hold. It remains to show that one can find integers, a and b, with these properties.

Let $TZ(G)$ denote the torsion subgroup of the center of G and recall that the *free center* of G is defined to be

$$FZ(G) = \{x \in Z(G) \mid x = y^n \text{ for some } y \in Z(G) \text{ where } n = \exp(TZ(G))\}.$$

Then $FZ(G)$ is a free abelian group of finite index in G, [15]. For the moment, call this index κ. If $\kappa = 1$, then G is abelian and step 1 is essentially done; so assume otherwise. Let $\kappa = \alpha\beta$ where α and β are relatively prime with the prime divisors of α not in A and the prime divisors of β not in B. Let $a = \alpha^n$ where n is chosen so that a is a solution to the congruence $\lambda(\lambda - 1) \equiv 0 \bmod N(G)$. Define b similarly. Notice that for any $g \in G$, the order of the image of g^a in $G/FZ(G)$ divides b. Interchange the roles of a and b and the statement still holds. Since elements of relatively prime order commute in a nilpotent group it follows that for any $x,y \in G$, the commutator $[x^a, y^b] \in FZ(G)$. This commutator must be trivial since it is an element of finite order in a free abelian group. This completes the first step in the proof.

The induction step amounts to solving the lifting problem in the following diagram

$$\xymatrix{ & X_{i+1}^n \ar[d] \\ X \times X \ar@{-->}[ur] \ar[r]^{\mu_i^n} & X_i^n \ar[r]^{k_i^n} & K(G_i, n+1). }$$

The obstruction to a lift is a class of finite order in $H^{n+1}(X \times X; G_i)$, since the map k_i^n has finite order. When restricted to the wedge $X \vee X$ this obstruction vanishes and so it comes from a class, say ϑ, of finite order in $H^*(X \wedge X; G_i)$.

If u is a class whose order is some power of p in $H^*(X; G)$, it is easy to check that u will be killed by some iterate of the map f_p, described in *4.1*. More generally, one can choose composites, say

$$f_1 = f_{p_1} \cdots f_{p_r} \quad \text{and} \quad f_2 = f_{q_1} \cdots f_{q_s},$$

so that $(f_1 \wedge f_2)^* \vartheta = 0$. These choices can be made so that no p_i is in A and no q_j is in B. Having done this, replace the map μ_i^n in the given lifting problem by the composition

$$X \times X \xrightarrow{\ f_1 \times f_2\ } X \times X \xrightarrow{\ \mu_i^n\ } X_i^n$$

This replacement lifts; let μ_{i+1}^n denote a lift of $\mu_i^n(f_1 \times f_2)$ to X_{i+1}^n. Since this process needs to be repeated at most a finite number of times, the proof follows.

COROLLARY 4.5 *Let X satisfy the hypotheses of Theorem 4. For any finite set of primes C and any $Y \in \mathcal{G}(X)$, there is a C-equivalence $f : Y \to X$.*

PROOF: Apply Zabrodsky's proof of Proposition 1.5 in [18], using Propositions 4.1 and 4.4. □

Recall that $K(F, \bar{n})$ denotes a generalized Eilenberg-MacLane space with finitely generated torsion free homotopy groups and with the same rational homotopy type as X. In the set of all rational equivalences from X to $K(F, \bar{n})$ there are some maps which are in a sense maximal. These are the ones which come as close as possible to inducing epimorphisms in integral homotopy and in integral cohomology. To be more precise, let $\bar{\Delta} : X \to X \wedge X$ denote the reduced diagonal map. Define the module of *integral primitives*, denoted $PH_*(X; \mathbf{Z})$, to be the kernel induced by this map in integral homology and define the module of *integral indecomposables*, $QH^*(X; \mathbf{Z})$, to be the cokernel induced by $\bar{\Delta}$ in integral cohomology.

DEFINITION: A *maximal map* $h : X \to K(F, \bar{n})$ is one that induces an isomorphism on $QH^*(\ ; \mathbf{Z})/$torsion.

Thus a maximal map amounts to a set of classes in $H^*(X; \mathbf{Z})$ which project to a basis for $QH^*(X; \mathbf{Z})/$torsion. Of course, as graded groups $F \approx QH^*(X; \mathbf{Z})/$torsion. Notice that a maximal map will be surjective on fundamental groups. However, it will not necessarily induce an epimorphism on the higher homotopy groups [3] (consider the case where X is a Postnikov approximation of BU). The kernel, in $\pi_*(X)$, of the homomorphism induced by a maximal map is obviously the torsion subgroup of $\pi_*(X)$.

Recall the number $t = t(X)$ mentioned in the introduction. This number can be regarded as a measure of the difference between X and $K(F, \bar{n})$. Its definition involves certain exponents associated with the Hurewicz homomorphism for X. Given a finite abelian group G, let $\exp(G)$ denote its exponent, i.e., the smallest integer $n \geq 1$ such that $ng = 0$ for all $g \in G$. Let $\sigma_n : \pi_n(X) \to PH_n(X; \mathbf{Z})/$torsion denote the obvious quotient of the Hurewicz homomorphism. Since X is an H_o-space, the map σ_n will have a finite kernel

---

[3]Nor will it necessarily induce an epimorphism in cohomology; e.g., consider maps between ΩS^3 and $\mathbf{C}P^\infty$.

and a finite cokernel for each n. The kernel of σ_n is, of course, the torsion subgroup of $\pi_n(X)$. Consider the lower central series for the $\pi_1(X)$-action on this subgroup ([8], page 34). Denote this series by

$$* = \Gamma_n^c \subseteq \cdots \subseteq \Gamma_n^2 \subseteq \Gamma_n^1 \;\; = \;\; \text{torsion in } \pi_n(X).$$

and define

$$t_n(X) \;\; = \;\; \exp(\text{coker } \sigma_{n+1}) \prod_{i \geq 1} \exp(\Gamma_n^i / \Gamma_n^{i+1}).$$

DEFINITION: The number $t = t(X)$ is the least common multiple of the product $\Pi_{n \geq 1} t_n(X)$ and those primes which divide the exponent of the torsion in $QH^k(X; \mathbf{Z})$ where $k = k_1, k_2, \ldots, k_\ell$.

One connection between the numbers just introduced and maximal maps is the following. Suppose that $h : X \to K(F, \bar{n})$ is a maximal map and let $\mathcal{F}_h$ denote its homotopy fiber. This fiber is a connected nilpotent space with finite homotopy groups. Indeed, for each n there is a short exact sequence of $\pi_1(\mathcal{F}_h)$-modules,

$$0 \to \text{coker } \pi_{n+1}(h) \longrightarrow \pi_n(\mathcal{F}_h) \longrightarrow \ker \pi_n(h) \to 0.$$

Since the fibration $\Omega K(F, \bar{n}) \to \mathcal{F}_h \to X$, is principal, the $\pi_1(\mathcal{F}_h)$-action on the cokernel of $\pi_{n+1}(h)$ is trivial. Clearly, $\ker \pi_n(h) \approx \ker \sigma_n(X)$ and, since σ_* is an isomorphism on $K(F, \bar{n})$, it follows that $\text{coker } \pi_{n+1}(h) \approx \text{coker } \sigma_{n+1}(X)$. Therefore if one takes the lower central series for the $\pi_1(\mathcal{F}_h)$-action on $\pi_n(\mathcal{F}_h)$ and takes the product of the exponents on the associated graded group, it follows that this product will divide $t_n(X)$. This is a key point in the proof of the following result.

PROPOSITION 4.6 *Let X satisfy the hypotheses of Theorem 4. Given a maximal map $h : X \to K(F, \bar{n})$ there exists an action $\eta : X \times K(F, \bar{n}) \to X$ which restricts to the identity on the first factor and fits in the following commutative diagram*

$$
\begin{array}{ccc}
X \times K(F, \bar{n}) & \xrightarrow{\;\;\eta\;\;} & X \\
\downarrow{\scriptstyle h \times \lambda} & & \downarrow{\scriptstyle h} \\
K(F, \bar{n}) \times K(F, \bar{n}) & \xrightarrow{\;\;\mu\;\;} & K(F, \bar{n})
\end{array}
$$

where μ is the standard multiplication on $K(F, \bar{n})$ and λ denotes the power map of power $t = t(X)$.

PROOF: To simplify notation, let $\mathcal{K} = K(F, \bar{n})$ in this proof. The Postnikov decomposition of the map $h : X \to \mathcal{K}$ admits a principal refinement since both source and target are nilpotent and since h induces a surjection on fundamental groups ([1], chapter II). A typical term in the principal refinement of h fits into

the commutative diagram

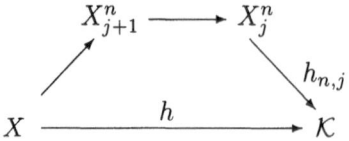

wherein the map on the left side is $(n-1)$-connected and the top row is part of a fibration sequence

$$X_{j+1}^n \longrightarrow X_j^n \longrightarrow K(\Gamma_n^j/\Gamma_n^{j+1}, n+1),$$

where the Γ_n^j are terms in the lower central series for the $\pi_1(\mathcal{F}_h)$-action on $\pi_n(\mathcal{F}_h)$. Once again $\mathcal{F}_h$ denotes the fiber of h. Refine this lower central series further so that each quotient of successive terms, say $G_{n,i}$, has exponent a prime p. Now consider the following two diagrams,

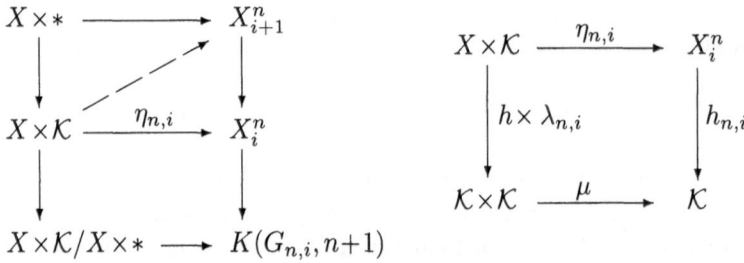

Since $X_0^0 = \mathcal{K}$ both diagrams exist and commute when $n = i = 0$. In the induction step assume, in the left hand diagram, that the group $G_{n,i}$ is an elementary abelian p-group. The mod p cohomology of $\mathcal{K}$ is primitively generated with respect to the standard (product) multiplication. Therefore if 1 denotes the identity on X and p denotes the p-th power map on $\mathcal{K}$, it follows that $1 \times p$ induces the trivial homomorphism on the mod p cohomology of $X \times \mathcal{K}/X \times *$. Therefore the composite $\eta_{n,i}(1 \times p)$ lifts to X_{i+1}^n in the left diagram. Then by replacing $\eta_{n,i}$ by $\eta_{n,i}(1 \times p)$ and $\lambda_{n,i}$ by $p\lambda_{n,i}$ in the right hand diagram, it follows that the same two diagrams exist and commute for a larger value of i. The proof follows by a finite induction. It is apparent, from the remarks made just before the statement of 4.6, that the power of λ needed in this result is no larger than t and that it can be taken to equal t.

PROPOSITION 4.7 *Let X satisfy the hypotheses of Theorem 4. Given a maximal map $h : X \to K(F, \bar{n})$ and $[Y] \in \mathcal{G}(X)$, there is a commutative homotopy pullback diagram*

$$
\begin{array}{ccc}
Y & \xrightarrow{\ \ f'\ \ } & X \\
\downarrow{\scriptstyle h'} & & \downarrow{\scriptstyle h} \\
K(F, \bar{n}) & \xrightarrow{\ \ f\ \ } & K(F, \bar{n})
\end{array}
$$

where f and f' are t-equivalences and where h' is also a maximal map of Y. Moreover f can be taken to be a product map of the form $f = f_1 \times \cdots \times f_\ell$ where each f_i is a self map of $K(F_i, n_i)$.

PROOF: Let $g : Y \to X$ be a t-equivalence, which exists by Corollary 4.5. Using the action described in Proposition 4.6, we will modify the map g and choose the components of the maps f and the projections of the map h' starting in the bottom dimension and working up, one dimension at a time.

Suppose that the bottom dimension is 1. No changes to g are necessary here. The map h induces an epimorphism from $\pi_1(X)$ to F_1. Choose the projection of h' onto $K(F_1, 1)$ to also induce an epimorphism from $\pi_1(Y)$ to F_1. Then let f_1 be the self map of $K(F_1, 1)$ that assigns $h'_*(\alpha) \mapsto h_* g_*(\alpha)$ for any $\alpha \in \pi_1(Y)$. Clearly, this is a well-defined homomorphism and f_1 is a t-equivalence since g is.

Let $k = k_2$, the next dimension in which $\pi_*(X) \otimes \mathbf{Q}$ has nonzero rank, say r. Let $\iota_1, \ldots, \iota_r$ be a basis for $H^k(K(F_2, k); \mathbf{Z})$ and let $x_j = h^*(\iota_j)$ in $H^*(X; \mathbf{Z})$. Since g is a t-equivalence, we can write

$$g^*(x_i) \; = \; \lambda_i y_i + v_i$$

where each λ_i is relatively prime to t, and where the set $\{y_1, \ldots, y_r\}$ projects to a basis for $QH^k(Y; \mathbf{Z})/\text{torsion}$ and where each v_i is in the kernel of $H^*(Y; \mathbf{Z}) \to QH^*(Y; \mathbf{Z})/\text{torsion}$. Let a_i denote the finite order of v_i in $QH^*(Y; \mathbf{Z})$. Thus

$$a_i v_i \; = \; w_i$$

for some decomposable $w_i \in H^k(Y; \mathbf{Z})$. The coefficient a_i divides some power of t and so $(a_i t, \lambda_i) = 1$. For each i choose integers b_i and c_i so that $1 + a_i b_i t = c_i \lambda_i$. Let $\varphi : Y \to K(F, \bar{n})$ be the map whose projections are trivial except in the case of $K(F_2, k)$ where

$$\varphi^*(\iota_i) \; = \; b_i w_i \quad \text{for} \quad i = 1, \ldots, r$$

Now take the composition

$$Y \xrightarrow{\;\;\Delta\;\;} Y \times Y \xrightarrow{\;\;g \times \varphi\;\;} X \times K(F, \bar{n}) \xrightarrow{\;\;\eta\;\;} X.$$

Let g_2 denote this composition. It follows from Proposition 4.6 that, for each i,

$$
\begin{aligned}
g_2^*(x_i) \; &= \; \lambda_i y_i + v_i + a_i b_i t v_i \\
&= \; \lambda_i (y_i + c_i v_i).
\end{aligned}
$$

Let the projection of h' onto $K(F_2, k)$ be the map which sends each ι_i to $y_i + c_i v_i$. Of course, the restriction of the product map f to $K(F_2, k)$ should send each ι_j to $\lambda_j \iota_j$.

It is clear that this process can be repeated in dimension k_3, if necessary, without effecting the behavior of the three maps f, h' and g_2, in lower dimensions. Continuing on, one obtains f, h', and a sequence of maps $g = g_1, g_2, g_3, \ldots$

that fulfill the requirements of this proposition through a growing range of dimensions. Set $f' = g_\ell$ and the proof follows.

To prove Theorem 4 now, one fixes a maximal map $h : X \to K(F, \bar{n})$ and considers pullbacks of the form

$$
\begin{array}{ccc}
 & & X \\
 & & \downarrow h \\
K(F, \bar{n}) & \xrightarrow{\ f_1 \times \cdots \times f_\ell\ } & K(F, \bar{n})
\end{array}
$$

where each f_i is a t-equivalence of $K(F_i, k_i)$. Such a pullback is connected and nilpotent by [8], pages 97 and 98. Since h is a $1/t$-equivalence this pullback will be in the Mislin genus of X.

Let $\mathcal{M}_t(\mathbf{Z}, r)$ denote the monoid of $r \times r$-matrices with integer entries and with determinant relatively prime to t. Let $r_i =$ the rank of F_i and choose a basis for each F_i. The pullback operation just described induces a function

$$
\xi : \mathcal{M}_t(\mathbf{Z}, r_1) \times \cdots \times \mathcal{M}_t(\mathbf{Z}, r_\ell) \longrightarrow \mathcal{G}(X)
$$

and Proposition 4.7 implies that this function is surjective. There is the obvious left action of $GL(r_1, \mathbf{Z}) \times \cdots \times GL(r_\ell, \mathbf{Z})$ on $\mathcal{M}_t(\mathbf{Z}, r_1) \times \cdots \times \mathcal{M}_t(\mathbf{Z}, r_\ell)$ and it follows easily that ξ assigns to each orbit of this action a single value. In other words, ξ only depends upon the absolute values of the determinants of the matrices involved. From Proposition 4.6 it follows that these absolute values can be taken mod t. In this way one obtains a surjection, of the same name,

$$
(\mathbf{Z}_t^* / \pm 1)^\ell \xrightarrow{\ \xi\ } \mathcal{G}(X).
$$

It is easy to see that the kernel of this map contains the image $d(\mathcal{E}_t(X))$. Zabrodsky's Proposition 2.6 works equally well here to show this image equals the kernel. This completes the proof of Theorem 4.

Proof of Theorem 5

Throughout this section X will denote a space which satisfies the hypotheses of Theorem 5; that is, X will be a 1-connected finite CW-complex with the rational homotopy type of a bouquet of spheres. The proof follows the outline of the proof of Theorem 4 and so the results leading up to it are numbered accordingly. Thus Statement $5.n$ can be regarded as the Eckmann-Hilton dual of Statement $4.n$.

PROPOSITION 5.1 *Let X satisfy the hypotheses of Theorem 5. Then for each prime p there exists a self map of X, say f_p, such that*

i) f_p is an equivalence at each prime $q \neq p$, and

ii) f_p induces multiplication by some positive power of p on each reduced integral homology group of X.

PROOF: Let $* = X_1 \subseteq X_2 \subseteq \cdots$ denote a homology decomposition of X. Thus for each n, there is a cofibration sequence,

$$M(G,n) \xrightarrow{k_n} X_n \longrightarrow X_{n+1},$$

where $M(G,n)$ denotes a Moore space with reduced integral homology G concentrated in dimension n and where the inclusion $X_n \subseteq X$ induces an isomorphism in integral homology in dimensions $\leq n$. The first nontrivial piece X_n in this decomposition is a suspension and so it has self maps of the required form. Since X has the rational homotopy type of a suspension. the attaching maps, k_n, in this decomposition have finite order. The argument used in the induction step of the proof of 4.1 can be used here to show that for any prime p there are self-maps, say f and g, of $M(G,n)$ and X_n, respectively, which induce multiplication by the same positive power of p in reduced integral homology and such that $k_n f = g k_n$. Thus, the cofiber X_{n+1} also has such self-maps and the proof follows by induction.

COROLLARY 5.2 *Let X satisfy the hypotheses of Theorem 5. Then for any $[Y] \in \mathcal{G}(X)$ and for any prime p, there exists a p-equivalence $g_p : X \to Y$.*

PROOF: The proof of Corollary 4.2 works just as well here.

PROPOSITION 5.4 *Let X satisfy the hypotheses of Theorem 5 and let A and B be disjoint sets of primes. Then there exists a map $\nu : X \to X \vee X$ which projects to an A-equivalence on the first factor and to a B-equivalence on the second factor.*

PROOF: Use the homology decomposition of X and the notation introduced in the proof of Proposition 5.1. Since the first nontrivial X_n in this decomposition is a co-H-space, use the comultiplication on it to obtain the desired map $\nu : X_n \to X \vee X$. The finite order of the attaching maps, k_n, is exploited once again. Choose f and g in the following composition,

$$M(G,n) \xrightarrow{k_n} X_n \xrightarrow{\nu_n} X \vee X \xrightarrow{f \vee g} X \vee X$$

to be composites of the self maps in Proposition 5.1, chosen so that f is an A-equivalence which kills the p-primary localization of $\nu_n k_n$ for all primes not in A, and g is a B-equivalence which kills the A-primary part of $\nu_n k_n$. Such choices exist because maps of nilpotent spaces which localize integral homology must also localize integral homotopy. Take ν_{n+1} to be the extension to X_{n+1} of $(f \vee g)\nu_n$. The proof then follows by induction.

COROLLARY 5.5 *Let X satisfy the hypotheses of Theorem 5. For any finite set of primes C and for any $[Y] \in \mathcal{G}(X)$, there exists a C-equivalence $f : X \to Y$.*

PROOF: Zabrodsky's proof of Proposition 1.5 dualizes easily to handle this result.

Now let F denote the graded free abelian group $\tilde{H}_*(X;\mathbf{Z})/$torsion and let $M(F,\bar{n})$, or just $\mathcal{M}$, denote the bouquet of spheres whose reduced integral homology is isomorphic to F. In other words, $\mathcal{M}$ is a bouquet of spheres with the same rational homotopy type as X. It will play the same role in this proof that the generalized Eilenberg-MacLane space $K(F, \bar{n})$ played in the proof of Theorem 4.

A maximal map from $\mathcal{M}$ to X is one which induces an isomorphism on a certain quotient of $\pi_*(\ )$. To be more precise, for any space Y, let $Q\pi_*(Y)$ denote the quotient of $\pi_*(Y)$ by the (graded) subgroup generated by all Whitehead products in $\pi_*(Y)$.

DEFINITION A maximal map $h : \mathcal{M} \to X$ is one that induces an isomorphism in $Q\pi_*(\ )/\text{torsion}$.

It follows from the Hilton-Milnor theorem that $Q\pi_*(\mathcal{M})/\text{torsion} \approx F \approx Q\pi_*(X)/\text{torsion}$. Thus a maximal map for X corresponds to a minimal set of generators in $\pi_*(X)$ which projects to a basis for $Q\pi_*(X)/\text{torsion}$.

The number $t = t(X)$ depends upon the Hurewicz homomorphism on X and the exponent of the torsion in $Q\pi_k(X)$ in dimensions $k = k_1, \ldots k_\ell$. To be more precise, let e_n denote the exponent of the cokernel of the Hurewicz homomorphism $\pi_n(X) \to H_n(X; \mathbf{Z})$. Notice that if $h : \mathcal{M} \to X$ is a maximal map with mapping cone C_h, then e_n is also the exponent of $H_n(C_h; \mathbf{Z})$. Next consider the fibration given by projection onto the first factor,

$$\mathcal{F}_\pi \longrightarrow X \vee \mathcal{M} \xrightarrow{\pi} X.$$

For each $n \geq 1$, let λ_n denote the smallest positive integer λ such that the power map on $\mathcal{M}$ of power λ induces the zero morphism on $[M(\mathbf{Z}/e_{n+1}, n), \mathcal{F}_\pi]$. Since $\Omega \mathcal{F}_\pi \simeq \Omega \mathcal{M} \times \Omega\Sigma(\Omega X \wedge \Omega \mathcal{M})$ it is easy to see that λ_n exists. Finally let $t(X)$ be the least common multiple of the product $\lambda_1 \lambda_2 \cdots$ and those primes which divide the exponent of the torsion in $Q\pi_k(X)$ in dimensions $k_1, \ldots, k_\ell$.

PROPOSITION 5.6 *Let X satisfy the hypotheses of Theorem 5. Given a maximal map $h : \mathcal{M} \to X$ there exists a co-action $\zeta : X \longrightarrow X \vee \mathcal{M}$ which projects to the identity on the first factor and which fits into the following commutative diagram*

$$
\begin{array}{ccc}
X & \xrightarrow{\ \zeta\ } & X \vee \mathcal{M} \\
\uparrow{\scriptstyle h} & & \uparrow{\scriptstyle h \vee \lambda} \\
\mathcal{M} & \xrightarrow{\ \nu\ } & \mathcal{M} \vee \mathcal{M}
\end{array}
$$

where ν is the standard multiplication on $\mathcal{M}$ and λ denotes the power map of power $t = t(X)$.

PROOF: The Moore decomposition of the map $h : \mathcal{M} \to X$ ([19], page 44) has the form

$$\mathcal{M} = X_0 \xrightarrow{\ h_0\ } X_1 \xrightarrow{\ h_1\ } X_2 \xrightarrow{\ h_2\ } \quad \cdots \quad \xrightarrow{\ h_N\ } X_{N+1} \simeq X$$

where, for each $n \geq 1$, there is a cofiber sequence

$$M(G_n, n) \xrightarrow{\ k_n\ } X_n \xrightarrow{\ h_n\ } X_{n+1}$$

Here $G_n \approx H_{n+1}(C_h; \mathbf{Z})$. Since h is a rational equivalence, each group G_n in this decomposition is finite. Now consider the following two diagrams

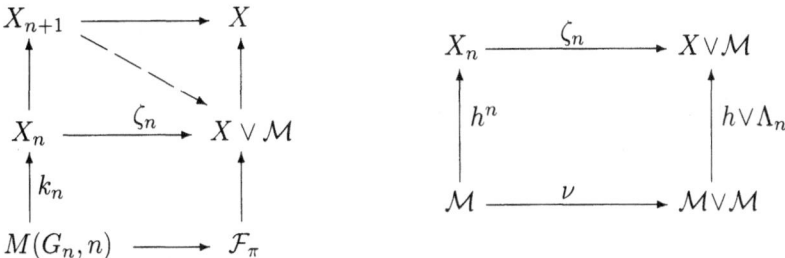

In the first diagram the left column is the cofiber sequence just introduced and the right column is the fiber sequence mentioned in the discussion of $t(X)$. Of course, ζ_n is a Moore approximation to ζ. In the right hand diagram h^n is the identity when $n = 0$ and the composition $h_{n-1} \cdots h_0$ otherwise. The map Λ_n is also the identity when $n = 0$ and is the power map of power $\lambda_1 \cdots \lambda_{n-1}$ otherwise. When $n = 0$ both diagrams exist since X_0 is the co-H-space $\mathcal{M}$. Assume then that both diagrams exist for some $n \geq 1$. If, in the left diagram, the composition $\zeta_n k_n$ is not null homotopic, replace ζ_n by $(1 \vee \lambda_n)\zeta_n$. It follows from the description of $t(X)$ that this replacement for ζ_n will extend to X_{n+1}. The proof then follows by a finite induction.

PROPOSITION 5.7 *Let X satisfy the hypotheses of Theorem 5. Given a maximal map $h : \mathcal{M} \to X$ and $[Y] \in \mathcal{G}(X)$, there is a homotopy pushout diagram*

$$
\begin{array}{ccc}
\mathcal{M} & \xrightarrow{\ f\ } & \mathcal{M} \\
{\scriptstyle h}\downarrow & & \downarrow{\scriptstyle h'} \\
X & \xrightarrow{\ f'\ } & Y
\end{array}
$$

where f and f' are t-equivalences and where h' is also a maximal map of Y. Moreover f can be taken to be a map of the form $f = f_1 \vee \cdots \vee f_\ell$ where each f_i is a self map of $M(F_i, n_i)$.

PROOF: The proof is dual to the one given for Proposition 4.7. Start with a t equivalence $g : X \to Y$, which exists by Corollary 5.5. Use Proposition 5.6 to alter this map, if necessary, by adding to it the appropriate decomposable terms. This involves compositions of the form

$$
X \xrightarrow{\ \zeta\ } X \vee \mathcal{M} \xrightarrow{\ g \vee \varphi\ } Y \vee Y \xrightarrow{\ \text{fold}\ } Y.
$$

The remaining details are left to the reader.

All of the ingredients for the proof of Theorem 5 are now in place. One fixes a maximal map $h : \mathcal{M} \to X$ and considers pushouts of the diagram

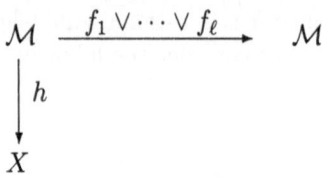

where each f_i is a t-equivalence. Each such pushout is easily seen to be in the Mislin genus of X and by Proposition 5.7, every member of $\mathcal{G}(X)$ can be obtained in this way. The rest of the proof for Theorem 5 follows the outline given for Theorem 4.

The Examples

PROOF OF EXAMPLE B: It is not difficult to see that $\mathcal{G}(\Sigma \mathbf{CP}^{\infty})$ contains the inverse limit of the tower $\mathcal{G}(\Sigma \mathbf{CP}^1) \leftarrow \mathcal{G}(\Sigma \mathbf{CP}^2) \leftarrow \cdots$. The maps in this tower are surjections, [10], and so its inverse limit is uncountably large. Whether or not this inverse limit is all of $\mathcal{G}(\Sigma \mathbf{CP}^{\infty})$ is still an open question.

Suppose that $\Sigma X \in \mathcal{G}(\Sigma \mathbf{CP}^{\infty})$. Then X must be a connected space whose integral cohomology groups are additively isomorphic to those of $\mathbf{CP}^{\infty}$. Let $f : X \to \mathbf{CP}^{\infty}$ be a map which induces an integral cohomology isomorphism in dimension 2. I claim that this map must be a homology equivalence (and hence must suspend to a homotopy equivalence). To see this, choose a generator $x \in H^2(X; \mathbf{Z})$ and choose a prime p. Let $\bar{x}$ denote the image of x in mod p cohomology and note that $\bar{x}^p = \mathcal{P}^1 \bar{x} \neq 0$. The equality follows from the unstable axiom for the Steenrod algebra and inequality follows because ΣX is p-equivalent to $\Sigma \mathbf{CP}^{\infty}$. Consequently, p cannot divide x^m for $m \leq p$. In mod p cohomology the ring homomorphism f^* must therefore be an isomorphism in dimensions $\leq 2p$. Recall that as a module over the Steenrod algebra, $\bar{H}^*(\mathbf{CP}^{\infty}; \mathbf{Z}/p)$ has $p-1$ indecomposable summands with the kth summand starting in degree $2k$. The same must be true of $\bar{H}^*(X; \mathbf{Z}/p)$. From the facts just cited it follows that f must induce an isomorphism in mod p cohomology in all dimensions. Since this is true for every prime p, it follows that f must induce an isomorphism in integral homology as well.

PROOF OF EXAMPLE C: To see that the determinant map (from cohomology) is surjective, it suffices to consider self maps of the form $f \vee g$ where f is a self map of S^3, of degree, say, λ and g is a power map on $Sp(2)$ of power, say, μ. Given t, the image of δ in $\mathbf{Z}_t^* \oplus \mathbf{Z}_t^*$ will consist of pairs $(\lambda\mu, \mu)$. There are no restrictions on the numbers λ and μ, other than being relatively prime to t, and so it follows that δ is surjective.

If $E_{5\omega}$ denotes the nontrivial member of $\mathcal{G}(Sp(2))$, i.e., the Hilton-Roitberg criminal, [9], then it is clear that $Y = S^3 \vee E_{5\omega}$ is in the genus of X. Suppose that there was a homotopy equivalence

$$\psi : X \longrightarrow Y.$$

Let v generate the free part of $\pi_7(X)$. This generator can be regarded as coming from the homotopy of $Sp(2)$. The Whitehead product on $\pi_*(X)$ then gives rise to a homomorphism

$$\pi_3(X) \xrightarrow{\ g \mapsto [g, v]\ } \pi_9(X)$$

whose kernel is 1-dimensional. Let u generate this kernel. Since Whitehead products are trivial in an H-space this provides a way of locating a generator in $\pi_3(X)$ which also comes from $\pi_*(Sp(2))$. Whitehead products are natural and so it follows that the image $\psi_*(u)$ must generate the kernel of the analogous pairing on $\pi_*(Y)$. Identify $H_*(Sp(2))$ as a retract of $H_*(X)$ and likewise $H_*(E_{5\omega})$ as a retract of $H_*(Y)$, in an obvious natural way. The induced isomorphism ψ_* must then take the first retract onto the second. It follows, using homology, that there exists an inclusion i of $Sp(2)$ and a retraction r of $E_{5\omega}$ such that the composition

$$Sp(2) \longrightarrow X \xrightarrow{\ \psi\ } S^3 \vee E_{5\omega} \xrightarrow{\ r\ } E_{5\omega}$$

is a homotopy equivalence. This is a contradiction, of course, and hence the equivalence ψ cannot exist.

Acknowledgement. I would like to thank George Peschke and the referee for spotting a serious error in an earlier version of Theorem 3. I would also like to thank the topologists in Barcelona for making BCAT-94 such a wonderful event.

References

[1] A. K. Bousfield and D. M. Kan, *Homotopy limits, completions and localizations*, Springer Lecture Notes in Math. **304** (1972)

[2] E. H. Brown and A. H. Copeland, *A homology analogue of Postnikov systems*, Michigan Math. Jour. **6** (1959) 313–330

[3] D. Davis, *BP-operations and mappings of stunted complex projective spaces*, Springer Lecture Notes in Math. **741** (1979) 373–393

[4] L. Fuchs, *Infinite Abelian Groups*, Pure and Applied Math. Series, 36-I, Academic Press, New York, 1970

[5] P. Hilton, *On groups of pseudo-integers*, Acta Math. Sinica **4** (1988) 189–192

[6] P. Hilton, *On the extended genus*, Acta Math. Sinica **4** (1988) 372–382

[7] P. Hilton and G. Mislin, *On the genus of a nilpotent group with finite commutator subgroup*, Math. Zeit. **146** (1976) 201–211

[8] P. Hilton, G. Mislin, and J. Roitberg, *Localization of Nilpotent Groups and Spaces*, Math. Studies **15**, North-Holland, Amsterdam, 1975

[9] P. Hilton and J. Roitberg, *On principal S^3-bundles over spheres*, Annals of Math. **90** (1969) 91–107

[10] C. A. McGibbon, *Self maps of projective spaces*, Trans. Amer. Math. Soc. **271** (1982) 325–346

[11] C. A. McGibbon, *The Mislin genus of a space*, Centre de Recherches Mathématiques, CRM Proceedings and Lecture Notes **6** (1994) 75–102

[12] C. A. McGibbon, *Loop spaces and phantom maps*, Contemp. Math. **148** (1993) 297–308

[13] M. Mimura and H. Toda, *On p-equivalences and p-universal spaces*, Comment. Math. Helv. **46** (1971) 87–97

[14] G. Mislin, *The genus of an H-space*, Springer Lecture Notes in Math. **249** (1971) 75–83

[15] G. Mislin, *Nilpotent groups with finite commutator subgroups*, Springer Lecture Notes in Math. **418** (1974) 103–120

[16] C. W. Wilkerson, *Applications of minimal simplicial groups*, Topology **15** (1976) 111–130

[17] A. Zabrodsky, *On the genus of finite CW H-spaces*, Comment. Math. Helv. **49** (1974) 48–64

[18] A. Zabrodsky, *p-equivalences and homotopy type*, Springer Lecture Notes in Math. **418** (1974) 160–171

[19] A. Zabrodsky, *Hopf Spaces*, Math. Studies **22**, North-Holland, Amsterdam, 1976

Charles A. McGibbon, Department of Mathematics
Wayne State University, Detroit, Michigan, 48202

mcgibbon@math.wayne.edu

Progress in Mathematics, Vol. 136

Extensions of p-compact Groups

JESPER MICHAEL MØLLER

ABSTRACT. The classification of short exact sequences of p-compact groups and of rational isomorphisms of not necessarily connected p-compact groups is discussed.

1. Introduction

The concept of a p-compact group was introduced by Dwyer and Wilkerson [8] as a homotopy theoretic version of a compact Lie group. In a subsequent paper [7], they showed that the center of any p-compact group agrees with the centralizer of the identity map. That result is the starting point of this note.

For given p-compact groups, X and Y, let $\mathrm{Ext}(X, Y)$ denote the set of equivalence classes of short exact sequences

$$Y \to G \to X$$

of p-compact groups. Two such extensions of X by Y are declared equivalent if there exists a homomorphism over X and under Y between them.

The discussion of $\mathrm{Ext}(X, Y)$ proceeds along two parallel tracks. One track is concerned with the case where Y is a (completely reducible [10]) connected p-compact group while the other track deals with the case where $Y = Z$ is an abelian p-compact (toral) group. For fixed homotopy actions ρ and ζ of $\pi_0(X)$ on Y and Z let $\mathrm{Ext}_\rho(X, Y) \subseteq \mathrm{Ext}(X, Y)$ and $\mathrm{Ext}_\zeta(X, Z) \subseteq \mathrm{Ext}(X, Z)$ denote the subsets of extensions realizing the actions ρ and ζ, respectively. As is quickly seen, $\mathrm{Ext}_\zeta(X, Z)$ is an abelian group and it turns out [Theorem 3.4] that $\mathrm{Ext}_\rho(X, Y)$ is an affine group with $\mathrm{Ext}_{Z\rho}(X, Z(Y))$ as group of operators. Here Z denotes the conjugation action of the group of self-homotopy equivalences of BY on the classifying space $BZ(Y) = \mathrm{map}(BY, BY)_{B1}$ [7, Theorem 1.3] of the center $Z(Y)$ of Y [13, 7].

The abelian group $\mathrm{Ext}_\zeta(X, Z)$ enjoys nice bifunctorial properties. The affine group $\mathrm{Ext}_\rho(X, Y)$ is functorial in X by pull back but only restricted functorial in Y: Any equivariant rational isomorphism $g \colon Y \to Y'$, where Y' is a connected p-compact group locally isomorphic to Y equipped with a homotopy action ρ' by $\pi_0(X)$, induces a push forward map

$$g_* \colon \mathrm{Ext}_\rho(X, Y) \to \mathrm{Ext}_{\rho'}(X, Y')$$

which is affine [Lemma 3.8]. Given also a homomorphism $h \colon X \to X'$, pull back and push forward of extension classes provide obstructions to the existence of

---

1991 *Mathematics Subject Classification.* 55P35, 55S37.
Key words and phrases. Universal fibration, center, extension.

homomorphisms

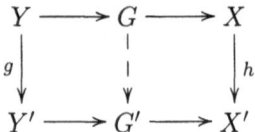

under g and over h. Indeed, such a homomorphism exists [Theorem 3.9] if and only if $g_*(G) = h^*(G')$ in $\mathrm{Ext}(X, Y')$.

Fibrewise discrete approximations to fibered abelian p-compact groups are briefly discussed in Section 4. Lemma 4.1–4.3 show that provided the identity component X_0 of X is simply connected, there is a group isomorphism

$$\mathrm{Ext}_\zeta(X, Z) \cong H_\zeta^2(\pi_0(X); \check{Z})$$

where $\check{Z}$ is the discrete approximation to Z.

The above concepts are exploited in the final section for the classification of rational automorphisms of not necessarily connected p-compact groups. When combined with [11, Theorem 4.3] [10, Theorem 3.5], the short exact sequence of Theorem 5.2 could potentially lead to a fairly explicit classification of rational automorphisms of any given p-compact group.

2. Universal fibrations

Thanks to the homotopy equivalence [7, Theorem 1.3] between the center and the centralizer of the identity map of a p-compact group, the classification of fibrations with p-compact group classifying spaces as fibres is surprisingly manageable.

Let's first fix some notation. For any two p-compact groups, X and X', put $\mathrm{Hom}(X, X') = [BX, *; BX']$, the set of based homotopy classes of maps, and $\mathrm{Rep}(X, X') = [BX, BX']$, the set of unbased homotopy classes of maps. A homomorphism $h \in \mathrm{Hom}(X, X')$ is said to be a rational isomorphism if [11, Definition 2.1] the map

$$H^*(Bh_0; \mathbb{Z}_p) \otimes_{\mathbb{Z}_p} \mathbb{Q}_p \colon H^*(BX_0'; \mathbb{Z}_p) \otimes_{\mathbb{Z}_p} \mathbb{Q}_p \to H^*(BX_0; \mathbb{Z}_p) \otimes_{\mathbb{Z}_p} \mathbb{Q}_p,$$

induced by the restriction $Bh_0 \colon BX_0 \to BX_0'$ of h to the identity components, is an isomorphism. Let $\varepsilon_{\mathbb{Q}}(X, X') \subseteq \mathrm{Hom}(X, X')$ denote the subset of rational isomorphisms.

If $X = X'$, $\mathrm{End}(X) = \mathrm{Hom}(X, X)$ is a monoid (under composition) containing $\varepsilon_{\mathbb{Q}}(X) = \varepsilon_{\mathbb{Q}}(X, X)$ as a submonoid and having $\mathrm{Aut}(X)$ as its group of invertible elements. $\mathrm{Out}(X)$ denotes the invertible elements of the monoid $\mathrm{Rep}(X, X)$. If X is connected or abelian, BX is simple so there is no difference between the based or unbased case: $\mathrm{End}(X) = \mathrm{Rep}(X, X)$ and $\mathrm{Aut}(X) = \mathrm{Out}(X)$.

Turning to classifying fibrations, let Y be a p-compact group with center [13, 7] $Z(Y)$ and adjoint form $P(Y) = Y/Z(Y)$. Then $Z(Y)$ is an abelian p-compact toral group and there exists a fibration

$$BZ(Y) \to BY \to BP(Y)$$

of classifying spaces. Using a Borel construction as in the proof of [8, Proposition 8.3] this fibration may be extended one step further to the right to give a fibration

$$BY \to BP(Y) \xrightarrow{Bk} B^2 Z(Y) \qquad (1)$$

which is universal for fibrations with fibre BY over simply connected base spaces [7, Remark 1.11].

Assume from now on that Y is *connected* and let $g\colon Y \to Y'$ be a rational isomorphism into another connected p-compact group Y' locally isomorphic [11, Definition 2.7] to Y. Then g induces [11, Corollary 3.2, Theorem 3.3] a fibre map

$$
\begin{array}{ccccc}
BZ(Y) & \longrightarrow & BY & \longrightarrow & BP(Y) \\
\scriptstyle BZ(g) \downarrow & & \scriptstyle Bg \downarrow & & \scriptstyle BP(g) \downarrow \\
BZ(Y') & \longrightarrow & BY' & \longrightarrow & BP(Y')
\end{array}
$$

which also extends one step to the right.

LEMMA 2.1. *Any rational automorphism g of Y extends to a fibre self map*

$$
\begin{array}{ccccc}
BY & \longrightarrow & BP(Y) & \xrightarrow{Bk} & B^2 Z(Y) \\
\scriptstyle Bg \downarrow & & \scriptstyle BP(g) \downarrow & & \scriptstyle B^2 Z(g) \downarrow \\
BY' & \longrightarrow & BP(Y') & \xrightarrow{Bk} & B^2 Z(Y')
\end{array}
$$

of the universal fibration (1).

Proof. The claim is that $B^2 Z(g) \circ Bk$ and $Bk \circ BP(g)$ are homotopic. Since looping provides a bijection $\Omega\colon [BP(Y), B^2 Z(Y')] \to [P(Y), BZ(Y')]$, this follows from the extension one step to the left of the fibre map $(BZ(g), Bg, BP(g))$ shown above. $\qquad\square$

The fibration which is universal for fibrations with fibre BY over arbitrary base spaces has the form

$$BY \to BP(Y)_{\mathrm{hOut}(Y)} \xrightarrow{Bk_{\mathrm{hOut}(Y)}} B^2 Z(Y)_{\mathrm{hOut}(Y)} \qquad (2)$$

where $\mathrm{Out}(Y) = \pi_0 \operatorname{aut}(BY, *) = \pi_0 \operatorname{aut}(BY)$ is the group of homotopy classes of homotopy self-equivalences of BY and the homotopy orbit space $BP(Y)_{\mathrm{hOut}(Y)}$ $(B^2 Z(Y)_{\mathrm{hOut}(Y)})$ denotes the classifying space of the group-like topological monoid $\operatorname{aut}(BY, *)$ $(\operatorname{aut}(BY))$ of based (free) homotopy self-equivalences of BY. The monodromy action associated to the homotopy orbit space $B^2 Z(Y)_{\mathrm{hOut}(Y)}$ is induced from the conjugation action of $\mathrm{Out}(Y)$ on $BZ(Y) \simeq \operatorname{map}(BY, BY)_{B1}$, i.e. from the action $Z\colon \mathrm{Out}(Y) \to \mathrm{Out}(Z(Y))$ of [11, Corollary 3.2].

Suppose now that the locally isomorphic p-compact groups, Y and Y', are equipped with homotopy actions, $\rho\colon \pi \to \mathrm{Out}(Y)$ and $\rho'\colon \pi' \to \mathrm{Out}(Y')$, by discrete groups, π and π'.

Pulling back the universal fibration (2) along the maps

$$Bρ: Bπ → \mathrm{BOut}(Y), Bρ': Bπ' → \mathrm{BOut}(Y')$$

produces fibrations

$$
\begin{array}{ccccc}
BY & \longrightarrow & BP(Y)_{hρπ} & \xrightarrow{Bk_{hρπ}} & B^2Z(Y)_{hρπ} \\
\Big\downarrow{\scriptstyle Bg} & & \Big\downarrow & & \Big\downarrow \\
BY' & \longrightarrow & BP(Y')_{hρ'π'} & \xrightarrow{Bk_{hρ'π'}} & B^2Z(Y')_{hρ'π'}
\end{array}
$$

that are universal for fibrations with fibre BY and with monodromy action restricting to $ρ$, $ρ'$. The projection map $Bk_{hρπ}$, $Bk_{hρ'π'}$ is a map over $Bπ$, $Bπ'$ since the universal projection map $Bk_{\mathrm{hOut}(Y)}$ is a map over $\mathrm{BOut}(Y)$. Thus the first obstruction to extending Bg to a fibre map $Bk_{hρπ} → Bk_{hρ'π'}$ is that g be $χ$-equivariant, i.e. $g \cdot ρ(γ) = ρ'(χ(γ)) \cdot g$ in $ε_{\mathbb{Q}}(Y,Y')$ for all $γ \in π$, for some group homomorphism $χ: π → π'$. Provided the mapping space $\mathrm{map}(BP(Y), BP(Y'))_{BP(g)}$ is contractible, as is this case if Y and Y' are *completely reducible* [10, Definition 3.10] p-compact groups, this is in fact the only obstruction to extending.

LEMMA 2.2. *Suppose that* $g \in ε_{\mathbb{Q}}(Y,Y')$ *is a $χ$-equivariant rational isomorphism between the locally isomorphic completely reducible p-compact groups Y and Y'.*

1. *There exists, up to vertical homotopy, exactly one extension*

$$BP(g)_{hχ}: BP(Y)_{hρπ} → BP(Y')_{hρ'π'}$$

of $BP(g): BP(Y) → BP(Y')$ to a map over $Bχ$.

2. *There exists, up to vertical homotopy, exactly one extension*

$$B^2Z(g)_{hχ}: B^2Z(Y)_{hρπ} → B^2Z(Y')_{hρ'π'}$$

of $B^2Z(g): B^2Z(Y) → B^2Z(Y')$ to a map over $Bχ$ such that

$$
\begin{array}{ccc}
BP(Y)_{hρπ} & \xrightarrow{Bk_{hρπ}} & B^2Z(Y)_{hρπ} \\
\Big\downarrow{\scriptstyle BP(g)_{hχ}} & & \Big\downarrow{\scriptstyle B^2Z(g)_{hχ}} \\
BP(Y')_{hρ'π'} & \xrightarrow[Bk_{hρ'π'}]{} & B^2Z(Y')_{hρ'π'}
\end{array}
$$

commutes up to vertical homotopy.

The proof is based on the fibred mapping space construction occuring e.g. in [2, 3]:

Let $p: U → A$ and $q: V → B$ be fibrations over connected and pointed base spaces. Suppose that $g: p^{-1}(*) → q^{-1}(*)$ is a map between the fibres and $h: (A, *) → (B, *)$ a map between the base spaces such that the pair (g, h) respects the monodromy action in the sense that $g \cdot ζ = π_1(h)(ζ) \cdot g$ holds in $[p^{-1}(*), q^{-1}(*)]$ for all $ζ \in π_1(A, *)$. The question of whether (g, h) comes from a fibre map can be turned into a section problem.

Define the set

$$\text{fibmap}(U, V)_h^g = \coprod_{a \in A} \text{map}(p^{-1}(a), q^{-1}(h(a)))_{g_a}$$

where $g_a \in [p^{-1}(a), q^{-1}(h(a))]$ is the homotopy class making

$$
\begin{array}{ccc}
p^{-1}(*) & \xrightarrow{\;\;g\;\;} & q^{-1}(*) \\
{\scriptstyle \varsigma}\downarrow & & \downarrow{\scriptstyle h(\varsigma)} \\
p^{-1}(a) & \xrightarrow[\;\;g_a\;\;]{} & q^{-1}(h(a))
\end{array}
$$

homotopy commutative for any path ς from the base point $*$ to $a \in A$. Using the topology of [3], we obtain a fibration

$$\text{map}(p^{-1}(*), q^{-1}(*))_g \longrightarrow \text{fibmap}(U, V)_h^g \longrightarrow A$$

whose based section space, by the fibrewise exponential law [2, Theorem 1], is homeomorphic to the space of maps of U into V under g and over h. Of course, $\text{fibmap}(U, V)_h^g = \text{fibmap}(U, h^*V)_1^g$ where h^*V is the pull back of V along h and 1 denotes the identity map of A.

Proof of Lemma 2.2. Composition with the maps $Bk_{h\rho\pi}$ and $Bk_{h\rho'\pi'}$ induces, since $Bk \circ BP(g) \simeq B^2Z(g) \circ Bk$ by Lemma 2.1, fibre maps

$$
\begin{array}{ccccc}
\text{map}(BP(Y),BP(Y'))_{BP(g)} & \longrightarrow & \text{fibmap}(BP(Y)_{h\rho\pi},BP(Y')_{h\rho'\pi'})_{BX}^{BP(g)} & \longrightarrow & B\pi \\
{\scriptstyle \underline{Bk}}\downarrow & & \downarrow & & \parallel \\
\text{map}(BP(Y),B^2Z(Y'))_{B(kP(g))} & \longrightarrow & \text{fibmap}(BP(Y)_{h\rho\pi},B^2Z(Y')_{h\rho'\pi'})_{BX}^{B(kP(g))} & \longrightarrow & B\pi \\
{\scriptstyle \overline{Bk}}\uparrow {\scriptstyle\simeq} & & \uparrow & & \parallel \\
\text{map}(B^2Z(Y),B^2Z(Y'))_{B^2Z(g)} & \longrightarrow & \text{fibmap}(B^2Z(Y)_{h\rho\pi},B^2Z(Y')_{h\rho'\pi'})_{BX}^{B^2Z(g)} & \longrightarrow & B\pi
\end{array}
$$

of fibred mapping spaces. The map $\overline{Bk}$ is easily seen to be a homotopy equivalence and the fibre $\text{map}(BP(Y), BP(Y'))_{BP(g)}$ is contractible [7, Theorem 1.3] [10, Theorem 3.11] since Y and Y' and with them their adjoint forms $P(Y)$ and $P(Y')$ are completely reducible. Thus there exists up to vertical homotopy exactly one section of the upper fibration inducing a corresponding section of the lower fibration. $\square$

Note that the fibre map $(BP(g)_{h\pi}, B^2Z(g)_{h\pi})$ of point (ii) of Lemma 2.2 is an extension of the fibre map $(BP(g), B^2Z(g))$ of Lemma 2.1 and thus restricts to the map $Bg \colon BY \to BY'$ on the fibres.

The above constructions pertaining to the *connected* p-compact groups can also be carried out for *abelian* p-discrete or p-compact toral groups [8, Definition 6.3, Definition 6.5].

Let $\check{Z}$ be an abelian p-discrete toral group and Z its closure [8, Definition 6.6]. The group $\text{Aut}(\check{Z})$ of abelian group automorphisms of $\check{Z}$ acts by based homeomorphisms on $B^2\check{Z}$ so we may apply the Borel construction to the path fibration $PB^2\check{Z} \to B^2\check{Z}$ to obtain the fibration

$$\Omega B^2\check{Z} \to (PB^2\check{Z})_{\mathrm{hAut}(\check{Z})} \xrightarrow{\check{\sigma}_0} (B^2\check{Z})_{\mathrm{hAut}(\check{z})} \tag{3}$$

which is universal for fibrations with $B\check{Z}$ as fibre. Note that both the total space and the base space are spaces over and under $\mathrm{BAut}(\check{Z})$ and that the projection map $\check{\sigma}_0$ is a map over and under $\mathrm{BAut}(\check{Z})$. Since p-completion induces isomorphisms $\mathrm{End}(\check{Z}) \to \mathrm{End}(Z)$ and $\mathrm{Aut}(\check{Z}) \to \mathrm{Out}(Z)$ [13, Proposition 3.2], fibrewise completion of (3) results in the fibration

$$BZ \to \mathrm{BOut}(Z) \xrightarrow{\sigma_0} (B^2Z)_{\mathrm{hOut}(Z)} \tag{4}$$

which is universal for fibrations with BZ as fibre. The projection map σ_0 is a map of spaces over and under $\mathrm{BOut}(Z)$.

The abelian group structure on $\check{Z}$ induces on $B^2\check{Z}$ the structure of an abelian topological group. Let $\check{\nabla}\colon B^2\check{Z} \times B^2\check{Z} \to B^2\check{Z}$ be the addition map and $\check{\nu}\colon B^2\check{Z} \to B^2\check{Z}$ the inversion map such that

$$\check{\nabla} \circ \tau = \check{\nabla}, \quad \check{\nabla} \circ (\check{\nu} \times 1) \circ \Delta = 0$$

where τ is the switch and Δ the diagonal map. The p-completions of $\check{\nabla}$ and $\check{\nu}$, $\nabla\colon B^2Z \times B^2Z \to B^2Z$ and $\nu\colon B^2Z \to B^2Z$, promote B^2Z to an abelian group-like space. Moreover, since $\mathrm{Aut}(\check{Z})$ acts on $B^2\check{Z}$ through group isomorphisms, $\check{\nabla}$ and $\check{\nu}$ extend to maps over and under $\mathrm{BAut}(\check{Z})$

$$\check{\nabla}\colon \Delta^*(B^2\check{Z}_{\mathrm{hAut}(\check{z})} \times B^2\check{Z}_{\mathrm{hAut}(\check{z})}) \to B^2\check{Z}_{\mathrm{hAut}(\check{z})} \tag{5}$$

$$\check{\nu}\colon B^2\check{Z}_{\mathrm{hAut}(\check{z})} \to B^2\check{Z}_{\mathrm{hAut}(\check{z})} \tag{6}$$

where Δ is the diagonal on $\mathrm{BAut}(\check{Z})$. The fibrewise p-completion of these maps are maps over and under $\mathrm{BOut}(Z)$

$$\nabla\colon \Delta^*(B^2Z_{\mathrm{hOut}(Z)} \times B^2Z_{\mathrm{hOut}(Z)}) \to B^2Z_{\mathrm{hOut}(Z)} \tag{7}$$

$$\nu\colon B^2Z \to B^2Z \tag{8}$$

extending the structure maps ∇ and ν on B^2Z.

Suppose now that $\check{Z}'$ is another p-discrete toral group and that $\check{Z}$ and $\check{Z}'$ support group actions $\zeta\colon \pi \to \mathrm{Aut}(\check{Z})$, $\zeta'\colon \pi' \to \mathrm{Aut}(\check{Z}')$. Any χ-equivariant abelian group homomorphism $\check{j}\colon \check{Z} \to \check{Z}'$ extends to a topological group homomorphism $B^2\check{j}\colon B^2\check{Z} \to B^2\check{Z}'$ and thus to a map

$$B^2\check{j}_{h\chi}\colon B^2\check{Z}_{h\zeta\pi} \to B^2\check{Z}'_{h\zeta'\pi'} \tag{9}$$

over and under $B\chi$ such that

$$\check{\nabla}' \circ \Delta^*(B^2\check{j}_{h\chi} \times B^2\check{j}_{h\chi}) = B^2\check{j}_{h\chi} \circ \check{\nabla}, \quad \check{\nu}' \circ B^2\check{j}_{h\chi} = B^2\check{j}_{h\chi} \circ \check{\nu}$$

where $\check{\nabla}'$ and $\check{\nu}'$ are the structure maps for $B^2\check{Z}'$.

Let Z' denote the ablian p-compact toral group which is the closure of $\check{Z}'$. Fibrewise p-completion of $B^2\check{j}_{h\chi}$ is a map

$$B^2j_{h\chi}\colon B^2Z_{h\zeta\pi} \to B^2Z'_{h\zeta'\pi'} \tag{10}$$

over and under $B\chi$ such that

$$\nabla' \circ \Delta^*(B^2 j_{h\chi} \times B^2 j_{h\chi}) = B^2 j_{h\chi} \circ \nabla, \quad \nu' \circ B^2 j_{h\chi} = B^2 j_{h\chi} \circ \nu$$

$$(11)$$

where ∇' and ν' are the structure maps on $B^2 Z'$.

3. Short exact sequences

This section contains information about fibrations of p-compact group classi-fying spaces.

Let X and Y be p-compact groups with classifying spaces BY and BX and let $\mathrm{cd}_{\mathbb{F}_p}(-)$ denote mod p cohomological dimension [8, Definition 6.13].

LEMMA 3.1. *Let* $BY \to BG \to BX$ *be a fibration sequence. Then* G *is a* p-*compact group and* $\mathrm{cd}_{\mathbb{F}_p}(G) = \mathrm{cd}_{\mathbb{F}_p}(X) + \mathrm{cd}_{\mathbb{F}_p}(Y)$.

Proof. As the base space as well as the fibre are p-complete spaces, the Fibre lemma [5, II.5.1–5.2] implies that also the total space BG is p-complete.

Let Y_0 denote the identity component of Y. By pulling back the fibration $G \to X \to BY$ to the universal covering space BY_0 we obtain a fibration denoted $G \to X|BY_0 \to BY_0$. Extending this fibration one step to the left gives the fibration $Y_0 \to G \to X|BY_0$ with connected fibre. The action of the fundamental group of any component of the base on $H_i(Y_0; \mathbb{F}_p)$, $i \geq 0$, is nilpotent because it factors through the finite p-group $\pi_0(Y)$ (acting on $H_i(Y; \mathbb{F}_p)$). Hence [8, Lemma 6.16] the corresponding Serre spectral sequence is concentrated in a rectangle of dimensions $\mathrm{cd}_{\mathbb{F}_p}(X)$ by $\mathrm{cd}_{\mathbb{F}_p}(Y)$ and the group in the upper right corner is nontrivial. The fact that G is $\mathbb{F}_p$-finite and the formula for its mod p cohomological dimension now follows as in the proof of [8, Proposition 6.14].

This shows [8, Lemma 2.1, Definition 2.2] that G is a p-compact group. $\square$

It is a consequence of Lemma 3.1 that the composition of two epimor-phisms [8, 3.2] is an epimorphism.

DEFINITION 3.2. *An* extension *of* X *by* Y *is a fibration of based maps*

$$BY \to BG \to BX$$

over BX *with fibre* BY. *Two extensions are* equivalent *if there exists a fibre map of the form*

$$
\begin{array}{ccccc}
BY & \longrightarrow & BG & \longrightarrow & BX \\
\| & & \downarrow & & \| \\
BY & \longrightarrow & BH & \longrightarrow & BX
\end{array}
$$

between them. $\mathrm{Ext}(X, Y)$ *denotes the set of all equivalence classes of extensions of* X *by* Y.

Since the total space BG is the classifying space of a p-compact group [Lemma 3.1], any extension of X by Y is a short exact sequence of p-compact groups [8, 3.2]. The extension $BY \to BG \to BX$ is often referred to simply as

$Y \to G \to X$. In this notation, two extensions are equivalent if there exists a homomorphism of the form

$$
\begin{array}{ccc}
Y \longrightarrow & G \longrightarrow & X \\
\| & \downarrow & \| \\
Y \longrightarrow & H \longrightarrow & X
\end{array}
$$

between them.

Associated to the short exact sequence $Y \to G \to X$ is a homotopy action $\rho \colon \pi_0(X) \to \mathrm{Out}(Y)$. Observe that this monodromy action is an invariant of the equivalence class so that it makes sense to let $\mathrm{Ext}_\rho(X, Y)$ denote the subset of $\mathrm{Ext}(X, Y)$ represented by all short exact sequences realizing the action ρ.

Assume from now on that Y is a *connected p*-compact group.

Let $[BX, B^2 Z(Y)_{\mathrm{hOut}(Y)}]_{B\rho}$ denote the set of vertical homotopy classes of lifts

$$
\begin{array}{ccc}
 & & B^2 Z(Y)_{\mathrm{hOut}(Y)} \\
 & \nearrow & \downarrow \\
BX \xrightarrow{\ B\pi_0\ } B\pi_0(X) & \xrightarrow{\ B\rho\ } & B\mathrm{Out}(Y)
\end{array}
$$

of the map $B\rho \circ B\pi_0$.

Similarly, if Z is an abelian *p*-compact group with discrete approximation $\check{Z} \to Z$ and $\zeta \colon \pi_0(X) \to \mathrm{Aut}(\check{Z}) = \mathrm{Out}(Z)$ an action, let $[BX, B^2 \check{Z}_{\mathrm{hAut}(\check{Z})}]_{B\zeta}$ and $[BX, B^2 Z_{\mathrm{hOut}(Z)}]_{B\zeta}$ denote the sets of vertical homotopy classes of lifts of $B\zeta \circ B\pi_0$.

Define $\mathrm{Ext}_\zeta(X, \check{Z})$ to be the set of equivalence classes (with respect to fibre homotopy equivalences under $B\check{Z}$ and over BX) of fibrations $B\check{Z} \to BG \to BX$ with monodromy action ζ.

LEMMA 3.3. *Let Y be a connected and X any p-compact group. Then there are bijections*

$$
[BX, B^2 Z(Y)_{\mathrm{hOut}(Y)}]_{B\rho} \to \mathrm{Ext}_\rho(X, Y)
$$
$$
[BX, B^2 \check{Z}_{\mathrm{hAut}(\check{Z})}]_{B\zeta} \to \mathrm{Ext}_\zeta(X, \check{Z})
$$
$$
[BX, B^2 Z_{\mathrm{hOut}(Z)}]_{B\zeta} \to \mathrm{Ext}_\zeta(X, Z)
$$

defined by pulling back the universal fibrations (2), (3), and (4), respectively.

Proof. The base space of the fibration of based mapping spaces

$$
\mathrm{map}_*(BX, B^2 Z(Y)_{\mathrm{hOut}(Y)}) \to \mathrm{map}_*(BX, B\mathrm{Out}(Y))
$$

is homotopically discrete [12, Lemma 2.2]. Therefore, the total space is homotopically equivalent to the disjoint union over all homomorphisms $\rho \colon \pi_0(X) \to \mathrm{Out}(Y)$ of the spaces of based lifts of $B\pi_0 \circ B\rho$.

By classification theory, pull back of the universal bundle (2) provides a bijection

$$
\pi_0(\mathrm{map}_*(BX, B^2 Z(Y)_{\mathrm{hOut}(Y)})) \to \mathrm{Ext}(X, Y)
$$

which by the above remarks restricts to a bijection between the based and vertical homotopy classes of lifts of $B\pi_0 \circ B\rho$ and $\text{Ext}_\rho(X, Y)$. However, since the fibre $B^2 Z(Y)$ is simply connected, the clause that the lifts be based is superfluous.

Similar arguments apply in the remaining two cases. $\square$

The chosen equivalence relation [Definition 3.2] on BY-fibrations over BX (assumed to have a nondegenerate base point) corresponds by Allaud [1] to *based* homotopy classes of maps of BX into the classifying space $B^2 Z(Y)_{\text{hOut}(Y)}$. See [2] for an account of the relationship between the free and the based case.

For the following, assume that $Y \to G \to X$ and $Y \to H \to X$ are two short exact sequences realizing the same homotopy action $\rho \colon \pi_0(X) \to \text{Out}(Y)$. Choose [Lemma 3.3] based lifts (also denoted) $G, H \colon BX \to B^2 Z(Y)_{\text{hOut}(Y)}$ of $B\rho \circ B\pi_0$ classifying the two fibrations. Define

$$\Omega B^2 Z(Y) \to B\Delta(H, G) \to BX$$

to be the fibration whose fibre over any point $b \in BX$ is the space of vertical (i.e. having constant projection in $\text{BOut}(Y)$) paths in $B^2 Z(Y)_{\text{hOut}(Y)}$ from $G(b)$ to $H(b)$. This fibration represents an element in $\text{Ext}_{Z\rho}(X, Z(Y))$.

THEOREM 3.4. *Let* $G \in \text{Ext}_\rho(X, Y)$ *where* Y *is a connected and* X *an arbitrary p-compact group. Then the map*

$$\Delta(-, G) \colon \text{Ext}_\rho(X, Y) \to \text{Ext}_{Z\rho}(X, Z(Y))$$

is a bijection.

Proof. Pulling back to BX the two fibrations shown as downward pointing arrows in the diagram

provides two sectioned fibrations, $B^2 Z(Y)_{h\rho\pi_0 X} \to BX$ and $B^2 Z(Y)_{hZ\rho\pi_0 X} \to BX$, with fibre $B^2 Z(Y)$. (See (4) for the section σ_0.) These two spaces over and under BX are equivalent in the sense that there exists up to homotopy over and under BX exactly one extension of the identity map of $B^2 Z(Y)$ to a map

$$u_G \colon B^2 Z(Y)_{h\rho\pi_0 X} \to B^2 Z(Y)_{hZ\rho\pi_0 X}$$

over and under BX. This follows from the fact that the fibre of the *based* fibred mapping space

$$\text{map}_*(B^2 Z(Y), B^2 Z(Y))_{B^2 1} \longrightarrow \text{fibmap}_*(B^2 Z(Y)_{h\rho\pi_0 X}, B^2 Z(Y)_{hZ\rho\pi_0 X})_{B1}^{B^2 1} \longrightarrow BX$$

is contractible. Composition with u_G induces a bijection

$(u_G)_* \colon \text{Ext}_\rho(X, Y) \xrightarrow{\cong} \text{Ext}_{Z\rho}(X, Z(Y))$ identical to the map $\Delta(-, G)$. $\square$

It is a consequence of Theorem 3.4 that $\mathrm{Ext}_\rho(X, Y)$ is an affine group [4, §9, no 1].

To see this, note that the structure maps (7) and (8) make $\mathrm{Ext}_\zeta(X, Z)$ into an abelian group. The neutral element of this group is represented by the short exact sequence $Z \to Z \rtimes_\zeta X \to X$ classified by the map $\sigma_0 \circ B\zeta \circ B\pi_0$. The sum of two short exact sequences $Z \to A \to X$ and $Z \to B \to X$, with classifying maps $A, B \colon BX \to B^2 Z_{\mathrm{hOut}(Z)}$, is the short exact sequence classified by the lift $\nabla \circ (A \times B) \circ \Delta$ of $B\zeta \circ B\pi_0$. The inverse $-A$ is classified by $\nu \circ A$.

Note that the projection $B(Z \rtimes_\zeta X) \to BX$ admits a based section, i.e. that the short exact sequence

$$Z \longrightarrow Z \rtimes_\zeta X \rightleftarrows X \tag{12}$$

is a split short exact sequence, and that there exist homomorphisms

$$
\begin{array}{ccc}
Z \times Z \longrightarrow \Delta^*(A \times B) \longrightarrow X \\
\Big\downarrow{\scriptstyle\nabla} \qquad \Big\downarrow \qquad \Big\| \\
Z \longrightarrow A + B \longrightarrow X
\end{array}
\tag{13}
$$

$$
\begin{array}{ccc}
Z \longrightarrow A \longrightarrow X \\
\Big\downarrow{\scriptstyle\nu} \qquad \Big\downarrow \qquad \Big\| \\
Z \longrightarrow -A \longrightarrow X
\end{array}
\tag{14}
$$

of short exact sequences. These properties are characterizing.

LEMMA 3.5. *Suppose that* $Z \to C \to X$ *is a short exact sequence representing an element of* $\mathrm{Ext}_\zeta(X, Z)$.

1. *If there exists a splitting*

$$Z \longrightarrow C \rightleftarrows X$$

   then $C = 0$ *in* $\mathrm{Ext}_\zeta(X, Z)$.
2. *If there exists a short exact sequence homomorphism*

$$
\begin{array}{ccc}
Z \times Z \longrightarrow \Delta^*(A \times B) \longrightarrow X \\
\Big\downarrow{\scriptstyle\nabla} \qquad \Big\downarrow \qquad \Big\| \\
Z \longrightarrow C \longrightarrow X
\end{array}
$$

   then $C = A + B$ *in* $\mathrm{Ext}_\zeta(X, Z)$.
3. *If there exists a short exact sequence homomorphism*

$$
\begin{array}{ccc}
Z \longrightarrow A \longrightarrow X \\
\Big\downarrow{\scriptstyle\nu} \qquad \Big\downarrow \qquad \Big\| \\
Z \longrightarrow C \longrightarrow X
\end{array}
$$

   then $C = -A$ *in* $\mathrm{Ext}_\zeta(X, Z)$.

Proof. (i) The *based* fibred mapping space

$$\text{map}_*(BZ, BZ)_{B1} \to \text{fibmap}_*(B(Z \rtimes_\zeta X), BC)_{B1}^{B1} \to BX$$

admits a section because its fibre is contractible.

(ii) Precomposition with the short exact sequence homomorphism (13) determines a fibre map

$$\begin{array}{ccccc}
\text{map}(BZ, BZ)_{B1} & \longrightarrow & \text{fibmap}(B(A+B), BC)_{B1}^{B1} & \longrightarrow & BX \\
\Big\downarrow \overline{\nabla} \simeq & & \Big\downarrow & & \Big\| \\
\text{map}(BZ \times BZ, BZ)_\nabla & \longrightarrow & \text{fibmap}(B\Delta^*(A \times B), BC)_{B1}^\nabla & \longrightarrow & BX
\end{array}$$

which is a fibre homotopy equivalence since, Z being abelian, $\overline{\nabla}$ is a homotopy equivalence. As the lower fibration admits a section, so does the upper one.

(iii) Similar to (ii). $\qquad\square$

In case $Z = Z(Y)$ is the center of the connected p-compact group Y, the difference map Δ from Theorem 3.4 and the additive structure in $\text{Ext}_{Z\rho}(X, Z(Y))$ are nicely related.

LEMMA 3.6. *Let $G, H, K \in \text{Ext}_\rho(X, Y)$. Then $\Delta(K, G) = \Delta(K, H) + \Delta(H, G)$ in $\text{Ext}_{Z\rho}(X, Z(Y))$.*

Proof. Since composition of paths defines a map

$$\Delta^*(B\Delta(H, G) \times B\Delta(K, H)) \to B\Delta(K, G)$$

over BX and under the H-space structure on $\Omega B^2 Z$, this formula follows from Lemma 3.5. $\qquad\square$

The formula of Lemma 3.6 implies that

$$\Delta(G, G) = Z(Y) \rtimes_{Z\rho} X, \quad -\Delta(H, G) = \Delta(G, H)$$

for all $G, H \in \text{Ext}_\rho(X, Y)$. More formally

COROLLARY 3.7. $\text{Ext}_\rho(X, Y)$ *is an affine group with the abelian group* $\text{Ext}_{Z\rho}(X, Z(Y))$ *as its group of operators.*

Let's now look at functorial properties of the Ext-affine groups.

Let $Y' \to G' \to X'$ be another short exact sequence of p-compact groups with associated homotopy action $\rho' \colon \pi_0(X') \to \text{Out}(Y')$. Any p-compact group homomorphism $h \colon X \to X'$ induces a map

$$h^* \colon \text{Ext}_{\rho'}(X', Y') \to \text{Ext}_{\rho' \pi_0(h)}(X, Y') \tag{15}$$

defined by pull back. Note that $h^*(G')$ is indeed a p-compact group by Lemma 3.1 and that h extends to a morphism

$$\begin{array}{ccccc}
Y' & \longrightarrow & h^*(G') & \longrightarrow & X \\
\Big\| & & \Big\downarrow & & \Big\downarrow h \\
Y' & \longrightarrow & G' & \longrightarrow & X'
\end{array} \tag{16}$$

of short exact sequences.

As to functorial properties in the second variable, assume now that Y' is connected, completely reducible, locally isomorphic to Y, and that $g: Y \to Y'$ is a rational isomorphism which is χ-equivariant for some group homomorphism $\chi: \pi_0(X) \to \pi_0(X')$. Let

$$g_*: \mathrm{Ext}_\rho(X, Y) \to \mathrm{Ext}_{\rho'\chi}(X, Y') \tag{17}$$

be the map induced by composing classifying maps with the essentially uniquely determined map

$$B^2 Z(g)_{h\chi}: B^2 Z(Y)_{h\rho\pi_0(X)} \to B^2 Z(Y')_{h\rho'\chi\pi_0(X)}$$

from Lemma 2.2. Note that the rational isomorphism g extends to a short exact sequence homomorphism

$$
\begin{array}{ccc}
Y \longrightarrow G \longrightarrow X \\
g \downarrow \qquad \downarrow \qquad \| \\
Y' \longrightarrow g_*(G) \longrightarrow X
\end{array}
\tag{18}
$$

where the middle arrow is induced from $BP(g)_{h\chi}$.

There are similar functorial properties in the abelian case. Let Z and Z' be abelian p-compact toral groups equipped with homotopy actions $\zeta: \pi_0(X) \to \mathrm{Out}(Z)$, $\zeta': \pi_0(X') \to \mathrm{Out}(Z')$. Pull back along the map $Bh: BX \to BX'$ induces a map

$$h^*: \mathrm{Ext}_{\zeta'}(X', Z') \to \mathrm{Ext}_{\zeta'\pi_0(h)}(X, Z')$$

which clearly is a group homomorphism. Also, if $j: Z \to Z'$ is a χ-equivariant homomorphism, composition with the map

$$B^2 j_{h\chi}: B^2 Z_{h\zeta\pi_0(X)} \to B^2 Z'_{h\zeta'\chi\pi_0(X)}$$

over and under $B\pi_0(X)$ from (10) induces

$$j_*: \mathrm{Ext}_\zeta(X, Z) \to \mathrm{Ext}_{\zeta'\chi}(X, Z')$$

which is a group homomorphism by the identities (11).

The χ-equivariant rational isomorphism $g: Y \to Y'$ induces [11, Corollary 4.2] a χ-equivariant rational isomorphism $Z(g): Z(Y) \to Z(Y')$.

LEMMA 3.8. *Let* $h: X \to X'$ *be a homomorphism and* $g: Y \to Y'$ *a χ-equivariant rational isomorphism.*

1. *The pull back (15) along h is an affine map with*

   $$h^*: \mathrm{Ext}_{Z\rho'}(X', Z(Y')) \to \mathrm{Ext}_{Z\rho'\pi_0(h)}(X, Z(Y'))$$

   as its corresponding operator group homomorphism.
2. *The push forward (17) along g is an affine map with*

   $$Z(g)_*: \mathrm{Ext}_{Z\rho}(X, Z(Y)) \to \mathrm{Ext}_{Z\rho'\chi}(X, Z(Y'))$$

   as its corresponding operator group homomorphism.

Proof. (i) It is immediate that $\Delta(h_*G', h^*H') = h^*\Delta(G', H')$ for all $G', H' \in \mathrm{Ext}_{\rho'}(X', Y')$.

(ii) In the diagram

$$
\begin{array}{ccc}
B^2Z(Y)_{h\rho\pi_0 X} & \xrightarrow{\;u_G\;} & B^2Z(Y)_{Z\rho\pi_0 X} \\
{\scriptstyle B^2Z(g)_{hX}}\big\downarrow & & \big\downarrow{\scriptstyle B^2Z(g)_{hX}} \\
B^2(Y')_{h\rho'\chi\pi_0 X} & \xrightarrow[\;u_{g_*(G)}\;]{} & B^2Z(Y')_{hZ\rho'\chi\pi_0 X}
\end{array}
$$

the left vertical map is the one defined in Lemma 2.2 and the right vertical map is, despite the notational coincidence, the one defined in formula (10). However, all maps in this diagram are maps over and under BX and as such maps are essentially unique, cfr. the proof of Theorem 3.4, $B^2Z(g)_{hX} \circ u_G$ and $u_{g_*G} \circ B^2Z(g)_{hX}$ are homotopic over and under BX. Hence $Z(g)_*\Delta(-,G) = \Delta(g_*(-), g_*G)$. $\qquad\square$

For the final result of this section, suppose that the rational isomorphism $g\colon Y \to Y'$ is $\pi_0(h)$-equivariant such that push forward along g and pull back along h

$$
\mathrm{Ext}_\rho(X, Y) \xrightarrow{\;g_*\;} \mathrm{Ext}_{\rho'\pi_0(h)}(X, Y') \xleftarrow{\;h^*\;} \mathrm{Ext}_{\rho'}(X', Y')
$$

have the same target.

THEOREM 3.9. *Assume that Y and Y' are locally isomorphic connected, completely reducible p-compact groups. Then there exists an extension homomorphism of the form*

$$
\begin{array}{ccc}
Y & \longrightarrow\; G \;\longrightarrow & X \\
{\scriptstyle g}\big\downarrow & \big\downarrow & \big\downarrow{\scriptstyle h} \\
Y' & \longrightarrow\; G' \;\longrightarrow & X'
\end{array}
$$

if and only if $g_*(G) = h^*(G')$ in $\mathrm{Ext}_{\rho'\pi_0(h)}(X, Y')$.

Proof. Precomposition with the map $BG \to B(g_*G)$ under Bg and over BX and postcomposition with the map $B(h^*(G')) \to BG'$ under BY' and over Bh induce a fibre map

$$
\begin{array}{ccccc}
\mathrm{map}(BY', BY')_{B1} & \longrightarrow & \mathrm{fibmap}(B(g_*G), B(h^*(G'))_{B1}^{B1} & \longrightarrow & BX \\
{\scriptstyle \overline{Bg}}\big\downarrow{\scriptstyle \simeq} & & \big\downarrow & & \big\| \\
\mathrm{map}(BY, BY')_{Bg} & \longrightarrow & \mathrm{fibmap}(BG, BG')_{Bh}^{Bg} & \longrightarrow & BX
\end{array}
$$

of fibred mapping spaces. Since Y and Y' are completely reducible, this is a fibre homotopy equivalence [10, Theorem 3.11]. Hence one of the two fibrations admits a section if and only if the other one does. $\qquad\square$

A Lie group version of the material contained in this section can be found in Notbohm [14].

4. Approximations

In this section obstruction theory is used to equate Ext-affine groups in certain advantageous situations.

As in the previous sections, X is any p-compact group, Y is a connected p-compact group, Z is an abelian p-compact (toral) group with discrete approximation $\check{Z}$, and $\rho\colon \pi_0(X) \to \mathrm{Out}(Y)$ and $\zeta\colon \pi_0(X) \to \mathrm{Aut}(\check{Z}) = \mathrm{Out}(Z)$ are homotopy actions.

LEMMA 4.1. *Suppose that the identity component X_0 of X is simply connected. Then the component homomorphism $\pi_0\colon X \to \pi_0(X)$ induces bijections*

$$\pi_0^*\colon \mathrm{Ext}_\rho(\pi_0(X), Y) \to \mathrm{Ext}_\rho(X, Y)$$
$$\pi_0^*\colon \mathrm{Ext}_\zeta(\pi_0(X), Z) \to \mathrm{Ext}_\rho(X, Z)$$

of equivalence classes of extensions.

Proof. Since the composite map $BX_0 \to BX \to B\pi_0(X)$ is nonessential, the homotopy orbit space $B^2Z(Y)_{hX_0} \simeq B^2Z(Y) \times BX_0$ and the homotopy fixed point space $B^2Z(Y)^{hX_0} \simeq \mathrm{map}(BX_0, B^2Z(Y)) \simeq B^2Z(Y)$ because BX_0 is 3-connected by Browder [6] [13, Corollary 5.6]. Hence [8, Lemma 10.5, Remark 10.8]

$$B^2Z(Y)^{hX} \simeq (B^2Z(Y)^{hX_0})^{h\rho\pi_0(X)} \simeq B^2Z(Y)^{h\rho\pi_0(X)}$$

and these homotopy equivalences induce bijections

$$\mathrm{Ext}_\rho(X, Y) = \pi_0(B^2Z(Y)^{h\rho\pi_0 X}) = \pi_0(B^2Z(Y)^{h\rho\pi_0(X)}) = \mathrm{Ext}_\rho(\pi_0(X), Y)$$

of Ext-sets. This proves the lemma for extensions of X by Y; extensions of X by Z are handled similarly.   □

The referee pointed out the following

COROLLARY 4.2. *Suppose that X is a connected and simply connected p-compact group. Then every extension of X by Y is equivalent to the trivial extension $Y \to Y \times X \to X$.*

The structure maps (5) and (6) on $B^2\check{Z}_{\mathrm{hAut}(\check{Z})}$ make $\mathrm{Ext}_\zeta(X, \check{Z})$ into an abelian group and the map

$$e_*\colon \mathrm{Ext}_\zeta(X, \check{Z}) \to \mathrm{Ext}_\zeta(X, Z),$$

induced by fibrewise completion $e\colon B^2\check{Z}_{\mathrm{hAut}(\check{Z})} \to B^2Z_{\mathrm{hOut}(Z)}$, into an abelian group homomorphism.

The next lemma shows that extensions of X by Z have unique fibrewise discrete approximations if the identity component of X is semisimple.

A connected p-compact group is said to be semisimple if its fundamental group or, equivalently [13, Theorem 5.3], its center is finite.

LEMMA 4.3. *The above group homomorphism e_* is surjective and also injective provided the identity component X_0 of X is semisimple.*

Proof. The sets $\text{Ext}_\zeta(X, \check{Z})$ and $\text{Ext}_\zeta(X, Z)$ correspond to vertical homotopy classes of the lifts indicated by dashed arrows in the diagram

$$B\pi_0(X) \longleftarrow B^2 Z_{h\zeta\pi_0(X)} \xleftarrow{\ e\ } B^2 \check{Z}_{h\zeta\pi_0(X)}$$

$$B\pi_0 \uparrow$$

$$BX$$

where the two spaces to the right are total spaces for the pull backs of the classifying fibrations (3) and (4) along $B\zeta \colon B\pi_0(X) \to \text{BAut}(\check{Z}) = \text{BOut}(Z)$.

The obstruction to lifting a map $BX \to B^2 Z_{h\zeta\pi_0(X)}$ to $B^2 \check{Z}(Y)_{h\zeta\pi_0(X)}$ lives in $H^3(BX; V)$ as the fibre of $e_{h\pi_0(X)}$ is $B^2 V$ for some rational vector space V [7, Proposition 3.2]. Since $\pi_3(BX) = \pi_2(X) = 0$ [6], [13, Corollary 5.6], there exists a 4-connected map $BX \to B$ to a 2-stage Postnikov tower B with fundamental group $\pi_1(B) \cong \pi_0(X)$ and $\pi_2(B) \cong \pi_1(X)$. Hence $H^3(BX; V) \cong H^3(B; V)$ and as $H^*(\pi_1(X), 2; V)$ is a rational vector space, $H^3(B; V) \cong H^0(\pi_0(X); H^3(\pi_1(X), 2; V))$ by the Serre spectral sequence with local coefficients. Universal Coefficients and [15, Theorem V.7.8] asserting that $H_3(\pi_1(X), 2) = 0$ imply that the coefficient group $H^3(\pi_1(X), 2; V) = 0$. We conclude that the obstruction group $H^3(BX; V)$ vanishes. This shows that $\text{Ext}_\zeta(X, \check{Z})$ maps onto $\text{Ext}_\zeta(X, Z)$.

The obstruction to lifting a vertical homotopy to $B^2 \check{Z}_{h\zeta\pi_0(X)}$ lives in $H^2(BX; V) \cong H^2(B; V) \cong H^0(\pi_0(X); \text{Hom}(\pi_1(X), V))$ which vanishes if the fundamental group $\pi_1(X)$ is finite. This shows that the map in the lemma is injective provided X_0 is semisimple. $\square$

For a p-compact torus T of rank one [8, 6.3], $\text{Ext}(T, \check{T}) \cong [BT, B\check{T}] \cong \check{T}$ while $\text{Ext}(T, T) \cong [BT, B^2 T] = 0$ so the map e_* of Lemma 4.3 is not injective in case $X = T = Z$.

There exists a version of Lemma 4.3 allowing the fibres to be arbitrary, not just abelian, p-compact toral groups.

We now know that in case the identity component X_0 is semisimple and $\text{Ext}_\rho(X, Y) \neq \emptyset$, there are bijections

$$\text{Ext}_\rho(X, Y) \xrightarrow[\cong]{\Delta(-, G)} \text{Ext}_{Z\rho}(X, Z(Y)) \xleftarrow{\cong} \text{Ext}_{Z\rho}(X, \check{Z}(Y))$$

where the right hand group is isomorphic to the cohomology group $H^2_{Z\rho}(BX; \check{Z}(Y))$. If X_0 is even simply connected, there are bijections

$$
\begin{array}{ccc}
\text{Ext}_\rho(X, Y) & & \text{Ext}_{Z\rho}(\pi_0(X), \check{Z}(Y)) \\[4pt]
\cong \Big\uparrow \pi_0^* & & \Big\downarrow \cong \\[4pt]
\text{Ext}_\rho(\pi_0(X), Y) & \xrightarrow[\cong]{\Delta(-, G)} & \text{Ext}_{Z\rho}(\pi_0(X), Z(Y))
\end{array}
$$

where the upper right corner group is isomorphic to the cohomology group $H^2_{Z\rho}(\pi_0(X); \check{Z}(Y))$. Note, however, that the bijection $\Delta(-, G)$, depending on the choice of the extension G, is noncanonical.

Now follows an alternative description of the Theorem 3.4 difference $\Delta(H, G)$ between two short exact sequences $Y \to G \to X$ and $Y \to H \to X$ in $\mathrm{Ext}_\rho(X, Y)$.

PROPOSITION 4.4. *There exists a homotopy equivalence*

$$\Lambda \colon B\Delta(H, G) \to \mathrm{fibmap}(BG, BH)_{B1}^{B1} \quad \textit{over } BX.$$

Proof. Let $Bk = Bk_{\mathrm{hOut}(Y)} \colon BP(Y)_{\mathrm{hOut}(Y)} \to B^2 Z(Y)_{\mathrm{hOut}(Y)}$ denote the projection map of and $\lambda \colon W \to \mathrm{map}(I, BP(Y)_{\mathrm{hOut}(Y)})$ a connection [15, p. 29] for the universal fibration (2); i.e. λ assigns to any element of

$$W = \{(x, u) \in BP(Y)_{\mathrm{hOut}(Y)} \times \mathrm{map}(I, B^2 Z(Y)_{\mathrm{hOut}(Y)}) \mid Bk(x) = u(0)\}$$

a path $\lambda(x, u)$ in $BP(Y)_{\mathrm{hOut}(Y)}$ starting at $\lambda(x, u)(0) = x$ and lying over $Bk(\lambda(x, u)) = u$.

The fibres over any $b \in BX$ of $BG \to BX$, $BH \to BX$ are the fibres $Bk^{-1}(G(b))$, $Bk^{-1}(H(b))$ and the fibre of $B\Delta(G, H) \to BX$ is the space of vertical paths u in $B^2 Z(Y)_{\mathrm{hOut}(Y)}$ from $G(b)$ to $H(b)$. Define

$$\Lambda \colon B\Delta(H, G) \to \mathrm{fibmap}(BG, BH)_{B1}^{B1}$$

as the map over BX taking u to the map $\lambda(-, u)(1) : Bk^{-1}(G(b)) \to Bk^{-1}(H(b))$. The restriction of Λ to the fibre over the basepoint (where the classifying maps G and H have the same value) is the monodromy $\Omega B^2 Z(Y) \to \mathrm{map}(BY, BY)_{B1}$ for the universal fibration (1), hence a homotopy equivalence.   □

Thus also

$$\mathrm{fibmap}(BG, -)_{B1}^{B1} \colon \mathrm{Ext}_\rho(X, Y) \to \mathrm{Ext}_{Z\rho}(X, Z(Y))$$

is a bijection.

The evaluation map $B\mu \colon BY \times \mathrm{map}(BY, BY)_{B1} \to BY$ is a left action $\mu \colon Y \times Z(Y) \to Y$ of $Z(Y)$ on Y. Using the alternative description of Proposition 4.4 of the difference $\Delta(H, G)$ it is immediate that μ extends to a morphism

$$
\begin{array}{ccccc}
Y \times Z(Y) & \longrightarrow & \Delta^*(G \times \Delta(H, G)) & \longrightarrow & X \\
{\scriptstyle \mu}\downarrow & & \downarrow & & \| \\
Y & \longrightarrow & H & \longrightarrow & X
\end{array}
$$

of extensions. This property characterizes $\Delta(H, G)$ as an operator on $\mathrm{Ext}_\rho(X, Y)$.

COROLLARY 4.5. *Let $Z(Y) \to \Delta \to X$ be a short exact sequence representing an element $\Delta \in \mathrm{Ext}_{Z\rho}(X, Z(Y))$. Then $G + \Delta = H$ in $\mathrm{Ext}_\rho(X, Y)$ if and only if the action μ extends*

$$
\begin{array}{ccccc}
Y \times Z(Y) & \longrightarrow & \Delta^*(G \times \Delta) & \longrightarrow & X \\
{\scriptstyle \mu}\downarrow & & \downarrow & & \| \\
Y & \longrightarrow & H & \longrightarrow & X
\end{array}
$$

to a morphism over X.

Proof. The fibrewise adjoint of such a fibre map is an equivalence between Δ and $\Delta(H,G) = \mathrm{fibmap}(BG, BH)^{B1}_{B1}$. □

Corollary 4.5 concludes this section.

5. Rational automorphisms of non-connected p-compact groups

The purpose of this section is to investigate the monoid of rational automorphisms of not necessarily connected p-compact groups.

Let $Y \to G \to \pi$ be a short exact sequence of p-compact groups, where Y is connected and π is a finite p-group, representing an element $G \in \mathrm{Ext}_\rho(\pi, Y)$, $\rho\colon \pi \to \mathrm{Out}(Y)$ being the monodromy. (According to the remarks after Lemma 4.3, $\mathrm{Ext}_\rho(\pi, Y)$ is in bijection with the cohomology group $H^2_{Z\rho}(\pi; \check{Z}(Y))$.)

The pull back diagram

$$
\begin{array}{ccc}
\mathrm{map}_{\mathrm{End}(\pi)}(BG, BG) & \longrightarrow & \mathrm{map}(BG, BG) \\
\downarrow & & \downarrow \\
\mathrm{End}(\pi) & \longrightarrow & \mathrm{map}(BG, B\pi)
\end{array}
$$

where the bottom map takes $\chi\colon \pi \to \pi$ to $BG \to B\pi \xrightarrow{B\chi} B\pi$, serves as definition of the space in the upper left corner. Thus $\mathrm{map}_{\mathrm{End}(\pi)}(BG, BG)$ consists of self-maps of BG over maps $B\chi\colon B\pi \to B\pi$ induced from endomorphisms of the group π.

Recall that $\mathrm{End}(G) = [BG, *; BG]$ denotes the monoid of based homotopy classes of based self-maps of BG.

LEMMA 5.1. $\pi_0 \, \mathrm{map}_{\mathrm{End}(\pi)}(BG, BG) \cong \mathrm{End}(G)$.

Proof. The above pull back diagram of free mapping spaces also has a based version

$$
\begin{array}{ccc}
\mathrm{map}_{\mathrm{End}(\pi)}(BG, *; BG) & \longrightarrow & \mathrm{map}_*(BG, BG) \\
\downarrow & & \downarrow \\
\mathrm{End}(\pi) & \xrightarrow{\ \simeq\ } & \mathrm{map}_*(BG, B\pi)
\end{array}
$$

defining the space in the upper left corner. Thus $\mathrm{map}_{\mathrm{End}(\pi)}(BG, *; BG)$ is the space of based self-maps of BG over maps $B\chi\colon B\pi \to B\pi$ induced from endomorphisms of the group π. Note that in this based version, the horizontal maps are homotopy equivalences [12, Lemma 2.2]. Thus we have monoid homomorphisms

$$
\pi_0(\mathrm{map}_{\mathrm{End}(\pi)}(BG, BG)) \leftarrow \pi_0(\mathrm{map}_{\mathrm{End}(\pi)}(BG, *; BG)) \xrightarrow{\cong} \mathrm{End}(G)
$$

where the right hand arrow actually is an isomorphism. Also the left hand arrow is an isomorphism for, as the fibre BY is simply connected, two vertically homotopic fibre maps are also based vertically homotopic. □

The submonoid $\varepsilon_{\mathbb{Q}}(G) \subseteq \operatorname{End}(G) = \pi_0 \operatorname{map}_{\operatorname{End}(\pi)}(BG, BG)$ of rational automorphisms of G thus consists of all vertical homotopy classes of fibre maps of the form $(f, B\chi)$ where $f\colon BG \to BG$ restricts to a rational automorphism $f|BY\colon BY \to BY$ on the fibre BY. Consider the monoid homomorphism

$$\lambda\colon \varepsilon_{\mathbb{Q}}(G) \to \operatorname{End}(\pi) \times \varepsilon_{\mathbb{Q}}(Y)$$

that to the fibre self-map $(f, B\chi)$ associates the pair consisting of $\chi \in \operatorname{End}(\pi)$ and the restriction $f|BY \in \varepsilon_{\mathbb{Q}}(Y)$.

Let $\operatorname{End}_{\mathbb{Q}}(\pi, Y) \subseteq \operatorname{End}(\pi) \times \varepsilon_{\mathbb{Q}}(Y)$ denote the submonoid consisting of pairs (χ, g) where g is χ-equivariant. $\operatorname{End}_{\mathbb{Q}}(\pi, Y)$ contains the submonoid $\operatorname{End}_{\mathbb{Q}}(\pi, Y)_G$ of elements (χ, g) for which $\chi^*(G) = g_*(G)$ in $\operatorname{Ext}_{\rho\chi}(\pi, Y)$.

THEOREM 5.2. *Suppose that the connected component Y of G is completely reducible. Then there exists a short exact sequence*

$$0 \to H^1_{Z\rho}(\pi; \check{Z}(Y)) \to \varepsilon_{\mathbb{Q}}(G) \xrightarrow{\lambda} \operatorname{End}_{\mathbb{Q}}(\pi, Y)_G \to 1$$

of monoids. For any pair $(\chi, g) \in \operatorname{End}_{\mathbb{Q}}(\pi, Y)_G$, there is a bijection between the inverse image $\lambda^{-1}(\chi, g)$ and the cohomology group $H^1_{Z\rho\chi}(\pi; \check{Z}(Y))$.

Proof. The image of λ equals $\varepsilon_{\mathbb{Q}}(\pi, Y)_G$ by Theorem 3.9. The kernel of λ is the group of vertical homotopy classes of maps over $B\pi$ with restriction to BY homotopic to the identity. As a set, $\ker(\lambda)$ is in bijection with the vertical homotopy classes of sections, $\pi_0((BZ(Y))^{h\pi})$, of the fibration

$$\operatorname{map}(BY, BY)_{B1} \to BZ(Y)_{h\pi} \to B\pi \qquad (19)$$

with total space $BZ(Y)_{h\pi} = \operatorname{fibmap}(BG, BG)^{B1}_{B1}$. If $f, g\colon BG \to BG$ are maps over $B\pi$ representing elements of $\ker(\lambda)$, let s_f, s_g denote the corresponding sections of fibration (19). Note that $s_{fg} = \underline{f} \circ s_g$ where $\underline{f}$ denotes the self-map over $B\pi$ of $\operatorname{fibmap}(BG, BG)^{B1}_{B1}$ given by composition with f. By Lemma 4.3, fibration (19) has a fibrewise discrete approximation

$$B\check{Z}(Y) \to B\check{Z}(Y)_{h\pi} \to B\pi. \qquad (20)$$

and the fibre of the fibrewise discrete approximation $B\check{Z}(Y)_{h\pi} \to BZ(Y)_{h\pi}$ is BV for some rational vector space V. The vanishing of the cohomology groups $H^2(\pi; V)$ and $H^1(\pi; V)$ shows that in the situation

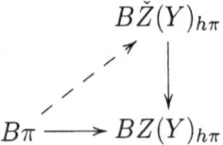

there are no obstructions to lifting maps or (vertical) homotopies. Thus fibrewise completion induces a bijection $\pi_0((B\check{Z}(Y))^{h\pi}) \cong \pi_0((BZ(Y))^{h\pi})$ and we may view the section s_f of (19) as a section of the fibrewise discrete approximation (20). Furthermore, since $B\check{Z}(Y)_{h\pi}$ is the classifying space of a p-discrete toral group so that $H^1(B\check{Z}(Y)_{h\pi}; V) = 0 = H^2(B\check{Z}(Y)_{h\pi}; V)$, similar considerations applied to the situation

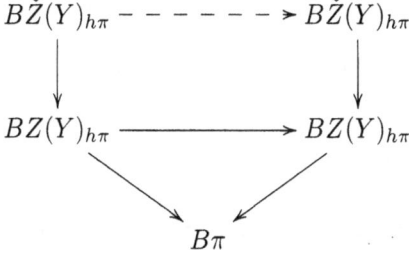

show that there is a bijection between vertical homotopy classes of self-maps over $B\pi$ of $BZ(Y)_{h\pi}$ and vertical homotopy classes of self-maps over $B\pi$ of $B\check{Z}(Y)_{h\pi}$. In particular, we may view the self-map $\underline{f}$ of $BZ(Y)_{h\pi}$ as a self-map of $B\check{Z}(Y)_{h\pi}$.

Associate to f the primary difference [15, p. 299] $\delta^1(s_f, s_1) \in H^1_{Z_\rho}(\pi; \check{Z}(Y))$ between the sections (of fibration (20)) corresponding to f and to the identity map. Then

$$\delta^1(s_{fg}, s_1) = \delta^1(s_{fg}, s_f) + \delta^1(s_f, s_1)$$
$$= \delta^1(\underline{f} \circ s_g, \underline{f} \circ s_1) + \delta^1(s_f, s_1)$$
$$= (\underline{f})_* \delta^1(s_g, s_1) + \delta^1(s_f, s_1)$$
$$= \delta^1(s_g, s_1) + \delta^1(s_f, s_1)$$

because $\underline{f}$ is homotopic to the identity map on the fibre $B\check{Z}(Y)$. This computation shows that the bijection $\ker(\lambda) \longrightarrow H^1_{Z_\rho}(\pi; \check{Z}(Y))$: $f \longrightarrow \delta^1(s_f, s_1)$ is a group homomorphism.

For an arbitrary pair $(\chi, g) \in \mathrm{End}_{\mathbb{Q}}(\pi, Y)_G$, the inverse image $\lambda^{-1}(\chi, g)$ is in bijection with the vertical homotopy classes of sections of the fibration fibmap$(BG, BG)^{Bg}_{B\chi} \longrightarrow B\pi$, or, equivalently (see the proof of Theorem 3.9), the vertical homotopy classes of sections of the fibration fibmap$(B(g_*G), B(\chi^*G))^{B1}_{B1} \longrightarrow B\pi$. Taking primary differences, as above, with respect to some fixed section provides a (noncanonical) bijection $\lambda^{-1}(\chi, g) \longrightarrow H^1_{Z_{\rho\chi}}(\pi; \check{Z}(Y))$. $\qquad\square$

More explicitly, the elements of $\lambda^{-1}(\chi, g)$ are represented by maps of the form

$$BG \longrightarrow B(g_*G) \longrightarrow B(\chi^*G) \longrightarrow BG$$

where the outer maps are fixed as the canonical ones [(16), (18)] and the middle arrow varies over all fibre homotopy equivalences over $B\pi$ and under the homotopy class of the identity map of BY.

The *space* of self-maps of BG over $B\chi$ and homotopic to Bg on the fibre BY is homotopy equivalent to the section space of fibmap$(B(g_*G), B(\chi^*G))^{B1}_{B1} \longrightarrow B\pi$, i.e. to the homotopy fixed point space

$$\mathrm{map}(BY, BY)^{h\pi}_{B1} \simeq BZ(Y)^{h\pi},$$

which, by obstruction theory, is a disjoint union of classifying spaces of p-compact toral groups.

Let $\mathrm{Aut}(\pi, Y)$ and $\mathrm{Aut}(\pi, Y)_G$ denote the subgroups of invertible elements of $\mathrm{End}_{\mathbb{Q}}(\pi, Y)$ and $\mathrm{End}_{\mathbb{Q}}(\pi, Y)_G$, respectively. The monoid short exact sequence of Theorem 5.2 restricts to a short exact sequence

$$0 \to H^1_{\check{Z}\rho}(\pi; \check{Z}(Y)) \to \mathrm{Aut}(G) \xrightarrow{\lambda} \mathrm{Aut}(\pi, Y)_G \to 1$$

of groups. The equivalence class of this group extension is unknown but it is perhaps worth noting that a somewhat similar group extension is determined by a differential in a Lyndon-Hochschild-Serre spectral sequence [12].

COROLLARY 5.3. *Suppose that Y is connected, completely reducible, and centerfree p-compact group. Then* $\mathrm{Ext}_\rho(\pi, Y) = \{Y \to G \to \pi\}$, $\varepsilon_{\mathbb{Q}}(G) = \mathrm{End}_{\mathbb{Q}}(\pi, Y)$, *and* $\mathrm{Aut}(G) = \mathrm{Aut}(\pi, Y)$.

Proof. Apply Lemma 3.3 and Theorem 5.2.   □

Finally, a couple of examples to illustrate the use of Theorem 5.2.

EXAMPLE 5.4. (1) The 2-compact group $\mathrm{SO}(2n + 1)^\wedge_2$, $n \geq 2$, is centerfree. Hence $\mathrm{Ext}_\rho(\pi, \mathrm{SO}(2n + 1)^\wedge_2)$ contains [Corollary 5.3] exactly one element G with $\varepsilon_{\mathbb{Q}}(G) = \mathrm{End}_{\mathbb{Q}}(\pi, \mathrm{SO}(2n + 1)^\wedge_2)$ and $\mathrm{Aut}(G) = \mathrm{Aut}(\pi, \mathrm{SO}(2n + 1)^\wedge_2)$ for any given homotopy action $\rho\colon \pi \to \mathrm{Out}(\mathrm{SO}(2n + 1)^\wedge_2)$.
(2) The center of the 2-compact group $\mathrm{SO}(2n)^\wedge_2$, $n > 4$, is cyclic of order 2 so the affine group $\mathrm{Ext}_\rho(\pi, \mathrm{SO}(2n)^\wedge_2)$ of equivalence classes of short exact sequences of 2-compact groups

$$\mathrm{SO}(2n)^\wedge_2 \to G \to \pi$$

realizing a fixed homotopy action $\rho\colon \pi \to \mathrm{Out}(\mathrm{SO}(2n)^\wedge_2)$ has $H^2(\pi; \mathbb{Z}/2)$ as group of operators. Assume that the homotopy action ρ is injective. Since $\varepsilon_{\mathbb{Q}}(\mathrm{SO}(2n)^\wedge_2) = \mathrm{Aut}(\mathrm{SO}(2n)^\wedge_2)$ [11, Theorem 5.6] is abelian [9], it follows that $\mathrm{End}_{\mathbb{Q}}(\pi, \mathrm{SO}(2n)^\wedge_2) = \mathrm{Aut}(\pi, \mathrm{SO}(2n)^\wedge_2) = \mathrm{Aut}(\mathrm{SO}(2n)^\wedge_2)$ consists of all automorphisms. For any (equivariant) automorphism g of $\mathrm{SO}(2n)^\wedge_2$, $Z(g)$ is the identity map, so [Lemma 3.8] g_* fixes G if and only if it fixes all elements of $\mathrm{Ext}_\rho(\pi, \mathrm{SO}(2n)^\wedge_2)$. The short exact sequence of Theorem 5.2 implies that $\varepsilon_{\mathbb{Q}}(G) = \mathrm{Aut}(G)$.

References

[1] G. Allaud, *On the classification of fiber spaces*, Math. Z. **92** (1966), 110–125.

[2] P. Booth, P. Heath, and R. Piccinini, *Characterizing universal fibrations*, Algebraic Topology. Proceedings, Vancouver 1977. Lecture Notes in Mathematics, vol. 673 (Berlin-Heidelberg-New York) (P. Hoffman, R. Piccinini, and D. Sjerve, eds.), Springer-Verlag, 1978, pp. 168–184.

[3] _____, *Fibre preserving maps and functional spaces*, Algebraic Topology. Proceedings, Vancouver 1977. Lecture Notes in Mathematics, vol. 673 (Berlin-Heidelberg-New York) (P. Hoffman, R. Piccinini, and D. Sjerve, eds.), Springer-Verlag, 1978, pp. 158–167.

[4] N. Bourbaki, *Algèbre, Chp. 2*, 3rd ed., Hermann, Paris, 1962.

[5] A.K. Bousfield and D.M. Kan, *Homotopy limits, completions and localizations*, 2nd ed., Lecture Notes in Mathematics, vol. 304, Springer-Verlag, Berlin–Heidelberg–New York–London–Paris–Tokyo, 1987.

[6] W. Browder, *Torsion in H-spaces*, Ann. of Math. (2) **74** (1961), 24–51.

[7] W.G. Dwyer and C.W. Wilkerson, *The center of a p-compact group*, Preprint, 1993.

[8] _____, *Homotopy fixed point methods for Lie groups and finite loop spaces*, Ann. of Math. (2) **139** (1994), 395–442.

[9] S. Jackowski, J. McClure, and R. Oliver, *Self homotopy equivalences of BG*, Preprint.

[10] J.M. Møller, *Completely reducible p-compact groups*, Proceedings of the Čech Centennial Homotopy Theory Conference (to appear).

[11] _____, *Rational isomorphisms of p-compact groups*, Topology (to appear).

[12] _____, *Self-homotopy equivalences of group cohomology spaces*, J. Pure Appl. Algebra **73** (1991), 23–37.

[13] J.M. Møller and D. Notbohm, *Centers and finite coverings of finite loop spaces*, J. reine angew. Math. **456** (1994), 99–133.

[14] D. Notbohm, *On the functor 'Classifying Space' for compact Lie groups*, Preprint, 1991.

[15] G. W. Whitehead, *Elements of homotopy theory*, Graduate Texts in Mathematics, vol. 61, Springer-Verlag, New York-Heidelberg-Berlin, 1978.

Matematisk Institut, Universitetsparken 5, DK–2100 København Ø
moller@math.ku.dk

Progress in Mathematics, Vol. 136
© 1996 Birkhäuser Verlag Basel/Switzerland

A version of an E^2 closed model category structure

ALEXANDER NOFECH

Introduction

Let $X_\bullet$ be a pointed simplicial set. Its constant bisimplicial extension $cX_{\bullet\bullet}$ can be thought of as a resolution of $X_\bullet$ by the zero spheres in the following sense: 1) by "flipping the axes", $cX_{\bullet\bullet}$ can be seen as a simplicial object having in each degree a wedge of zero spheres; 2) the diagonal $\Delta(cX_{\bullet\bullet})$ is isomorphic to $X_\bullet$. Replacing S^0 with some pointed simplicial set $A_\bullet$ one can ask what is the best possible approximation of $X_\bullet$ by a diagonal of a simplicial set having in all of its degrees wedges of $A_\bullet$ and its suspensions, and what is the relation of this approximation to $CW_A X_\bullet$, the $A_\bullet$-colocalization of $X_\bullet$. In this article we construct a closed model category structure in which such a resolution can be obtained by factoring the unique map from the basepoint into the given bisimplicial object as a cofibration followed by an acyclic fibration. These resolutions can be used to construct spectral sequences whose E^2-term depends on $A_\bullet$-homotopy of $X_\bullet$ and ordinary homotopy of $A_\bullet$ and which converge to the ordinary homotopy of the best possible approximation of $X_\bullet$ by homotopy colimits of diagrams whose objects are coproducts of $A_\bullet$ and its suspensions.

Another application of the resolution of this type is to construct spectral sequences in homology, which are those of André for the category of models being coproducts of $A_\bullet$ and its suspensions.

The properties of this closed model category structure are summarized in the following

1.6. MAIN THEOREM. *Let $X_\bullet$ be fibrant, and let the following be a CM4I factoring in the closed model category structure of Lemma 1.5. (a cofibration followed by an acyclic fibration):*

$$* \xrightarrow{Cof} P_{\bullet\bullet} \xrightarrow{\sim Fib} X_\bullet \underset{\sim}{\otimes} (pt).$$

Then

1) Each horizontal level $P_{\bullet,j}$ of $P_{\bullet\bullet}$ is a coproduct of copies of $\Sigma^i A_\bullet$ $(i \geq 0)$, with degeneracies being isomorphisms of summands of $P_{\bullet,j}$ onto summands of $P_{\bullet,k}$ with $k \geq j$.

2) For each map $\varphi : A_\bullet \to P_{\bullet,0}$ and maps induced by degeneracies of $P_{\bullet\bullet}$ considered as basepoints, there are isomorphisms:

$$\pi^v_* \pi^h_* (P_{\bullet\bullet}; A_\bullet)_\varphi \cong \pi^v_* \pi^h_* (X_\bullet \underset{\sim}{\otimes} (pt)_\bullet; A_\bullet)_\varphi$$

where the superscripts v, h stand for vertical and horizontal, the horizontal homotopy groups are "with coefficients in $A_\bullet$", and the vertical groups of the vertically constant object on the right are concentrated in degree zero.

The categories considered will be those of bisimplicial objects over the category of either pointed sets $(Sets_*)$ or, with a view to future applications, a category $\mathcal{C}$ of pointed universal algebras [BS]. An example is a category $RAlg$ of commutative supplemented graded algebras over a commutative ring with unit R, with the zero object being R in dimension zero.

For simplicity we will denote simplicial or bisimplicial categories with same subscripts as objects, as for example $P_{\bullet\bullet} \in (RAlg)_{\bullet\bullet}$.

This closed model category structure differs from the one of [DKS] in that the localizing simplicial object needs not be a cogroup object and in that it is close to the Reedy structure in the sense that the choice of S^0 (or constant FS^0 where F is the free functor for algebras) will return the Reedy structure (see Lemma 1.5).

One can use the derived functors of abelianization in this structure to define what can be called "$A_\bullet$-homology and $A_\bullet$-cohomology" that reduce to the usual ones for the above mentioned choices of $A_\bullet$. Of course in these theories there will in general be fewer admissible pairs or triples than in the usual ones. This and other applications will be considered elsewhere.

NOTATION. Let K, L be simplicial sets. Following [BF] we denote by $K \underset{\sim}{\times} L$ their external product, a bisimplicial set obtained as the product of the functors K and L. Similarly one defines an external tensor product $X_\bullet \underset{\sim}{\otimes} K$ of a simplicial object $X_\bullet$ over $\mathcal{C}$ with a simplicial set K, by taking coproducts of K_n copies of $X_\bullet$ in degree n. Clearly as in the case of external product the composition of the external tensor product with the bisimplicial diagonal gives the ordinary tensor product $X_\bullet \otimes K$ as defined in [Q, Part II, 1.8].

One can also define a tensor product of a bisimplicial object $X_{\bullet\bullet}$ over $\mathcal{C}$ with a bisimplicial set $Y_{\bullet\bullet}$: it has a coproduct of $Y_{k,l}$ copies of $X_{k,l}$ in dimension (k,l) with obvious bisimplicial operations. It is easy to see that $X_\bullet \underset{\sim}{\otimes} K \simeq cX_{\bullet\bullet} \otimes ((pt) \underset{\sim}{\times} K)$.

We will also consider external products and external tensor products of morphisms, defined as usual as the pushout in the commutative square formed by products of the given morphisms with identities of their domains and ranges.

For convenience we call the first bisimplicial index horizontal and the second vertical. Using the bisimplicial tensor product one defines the bisimplicial Hom-sets in the following way: the set of morphisms of either bisimplicial sets or bisimplicial algebras is given a bisimplicial structure by taking

$$\underline{\underline{Hom}}_{m,n}(X_{\bullet\bullet}, Y_{\bullet\bullet}) = Hom(X_{\bullet\bullet} \otimes (\Delta[m] \underset{\sim}{\times} \Delta[n]), Y_{\bullet\bullet})$$

where $\times$ is the external product and where a tensor product with a bisimplicial set is defined as above.

1 Construction of a resolution

Let A be a cofibrant object in a closed simplicial model category — for example, a retract of a pointed CW-complex in the category of pointed spaces. It was shown in [N, Th. 2.1.] that if some conditions related to "smallness" of objects are satisfied, there exists a closed model category in which fibrations remain the same, the weak equivalences, called A-equivalences, are those maps $\varphi : X \to Y$ that induce equivalences of simplicial sets

$$Hom_{\mathcal{C}}(A, \varphi_f) : Hom_{\mathcal{C}}(A, X_f) \simeq Hom_{\mathcal{C}}(A, Y_f)$$

where the subscript $_{-f}$ denotes fibrant approximation, and the cofibrations (determined by the left lifting property) are the retracts of A-cellular maps.

For convenience we include the full statement of this theorem:

DEFINITION. *Let $\mathcal{C}$ be a pointed closed simplicial model category and a A a cofibrant object of $\mathcal{C}$. We call $\mathcal{C}^A$ an A-cellular closed model category structure if the weak equivalences, fibrations and cofibrations in $\mathcal{C}^A$ are defined as follows ($f \nearrow g$ means the existence of a lifting for all commutative squares from f to g):*

$$W_{\mathcal{C}^A} = \{\varphi|\ Hom(A, \varphi_f) \in W_{\mathcal{C}}\}$$
$$Fib_{\mathcal{C}^A} = Fib_{\mathcal{C}}$$
$$Cof_{\mathcal{C}^A} = \{j|\ j \nearrow (W_{\mathcal{C}^A} \cap Fib_{\mathcal{C}^A}\}$$

We will call the cofibrations and the weak equivalences in $\mathcal{C}^A$ respectively A-cofibrations and A-equivalences. It follows from the definitions that weak equivalences in $\mathcal{C}$ are A-equivalences for any cofibrant A, and one can show that an A-trivial A-cofibration is an ordinary trivial cofibration.

THEOREM [N, 2.1]. *Let $\mathcal{C}$ be a pointed closed simplicial model category with arbitrary colimits and with a set $\{t_j\}$ of generators of trivial cofibrations. Suppose that $\mathcal{C}$ is proper (or if it is not then all its objects are fibrant). Let A be a cofibrant, s-definite object. Then there exists an A-cellular closed model category structure $\mathcal{C}^A$, as in the preceding definition, which admits functorial factorizations.*

1.1. THE REEDY STRUCTURE. In order to apply the Th. 2.1. of [N] to $\mathcal{C}_{\bullet\bullet}$ one needs an initial closed simplicial model category to be modified according to a chosen cofibrant object. For $\mathcal{C} = (Sets_*)$ it will be the Reedy structure described in [BF, Appendix B], with equivalences being degreewise (for convenience we fix this as equivalences on each of the vertical terms of a bisimplicial set, that is, $\varphi_{m,\bullet}$ is an equivalence for each $m \geq 0$). An important feature of this structure is that cofibrations are exactly the injective maps. Actually one only needs to specify a set of generators of cofibrations [N] to determine the acyclic fibrations by the RLP and these generators can be chosen as

$$(i[m] \underset{\sim}{\times} i[n])_+ \quad \text{where} \quad i[k] : \overset{\bullet}{\Delta}[k] \to \Delta[k]$$

are standard inclusions, and where the subscript "+" means a disjoint basepoint.

This implies that acyclic fibrations are transpose invariant, i.e. the same in both Reedy structures.

There exists a similar structure on $\mathcal{C}_{\bullet\bullet}$ when $\mathcal{C}$ is a category of pointed universal algebras. In this structure $F(i[m] \times i[n])$ will be the generators of cofibrations and $F(i[l] \times i[n,k])$ the generators of trivial cofibrations $(i[n,k] : V[n,k] \to \Delta[n]$ are the standard simplicial inclusions). The acyclic fibrations in this structure too are transpose invariant. We will refer to it also as the Reedy structure.

In the example of commutative R-algebras there are functors R, U, T, I, where R is the free R-module functor, U is forgetful, T is the symmetric algebra functor and I is the augmentation ideal. TR and UI are an adjoint pair which relates the Reedy structure on bisimplicial sets to the corresponding structure on bisimplicial R-algebras exactly as above.

Note that for bisimplicial sets a constantly extended S^0 (either way) is a constant bisimplicial set with one non-basepoint element in each bidegree and hence there is a natural isomorphism:

$$\underline{Hom}(S^0_\bullet \otimes (pt)_\bullet, X_{\bullet\bullet}) \cong X_{\bullet\bullet}.$$

The same is true for algebras with S^0 replaced by the free algebra FS^0.

1.2. REMARK. The simplicial structure of the Hom-sets which is used in [N, Th. 2.1.] is that of [BF] and not the one of the Introduction.

1.3. SUSPENSIONS. For cofibrant algebras one can use the simplicial structure to define suspensions. In the case of $(RAlg)_\bullet$, a general description of suspensions of simplicial supplemented algebras is given in [G, 3], but for a cofibrant simplicial algebra $A_\bullet$ one can use the simplicial structure of [Q] and define

$$\Sigma^n A_\bullet = (A_\bullet \otimes \Delta[n])/I(A_\bullet \otimes \overset{\bullet}{\Delta}[n])$$

where I is the augmentation ideal.

1.4. LEMMA. Let $A_\bullet$ be cofibrant. Then the following two cofibrant bisimplicial objects define the same closed model category structure on $\mathcal{C}_{\bullet\bullet}$:
(1) $\underset{i\geq 0}{\vee}(A_\bullet \otimes \Delta[i]) \otimes (pt)_\bullet$
(2) $\underset{i\geq 0}{\vee} \Sigma^i A_\bullet \otimes (pt)_\bullet$

Proof. Note that the equivalences of (1) are those bisimplicial maps $f_{\bullet\bullet}$ for which the following is an equivalence:

$$\underset{i\geq 0}{\Pi} Hom_i(A_\bullet, f_{\bullet\bullet})$$

where the simplicial structure is given by the second (vertical) index.

In order to compare the two consider cofiber sequences for each $i \geq 1$:

$$A_\bullet \otimes \overset{\bullet}{\Delta}[i] \to A_\bullet \otimes \Delta[i] \to \Sigma^i A_\bullet$$

that induce long exact sequences in homotopy similarly for simplicial sets and simplicial algebras, and use induction and the Five Lemma. Since both objects include $A_\bullet$ as a direct summand, there is no difficulty caused by the absence of a group structure on the set of homotopy classes of morphisms from $A_\bullet$.

1.5. LEMMA. *For the choice of the localizing object as in Lemma 1.4. the equivalences are bisimplicial maps that induce Reedy equivalences on*

$$\underline{Hom}(A_\bullet \underset{\sim}{\otimes} (pt)_\bullet, \_)$$

The choice $A_\bullet = S^0$ for simplicial sets and $A_\bullet = cFS^0$ for simplicial algebras leaves the original Reedy structure without change.

Proof. Note that

$$\underline{Hom}_{i,j}(A_\bullet \underset{\sim}{\otimes} (pt)_\bullet, X_{\bullet\bullet}) \simeq Hom(A_\bullet \otimes \Delta[i], X_{\bullet,j}).$$

From the description of equivalences in Lemma 1.4(1) it follows that they are equivalences on each vertical term of bisimplicial Hom-sets.

The second statement follows from the fact that S^0-equivalences (or FS^0-equivalences) are just equivalences on the vertical terms of bisimplicial objects.

1.6. MAIN THEOREM. *Let $X_\bullet$ be fibrant, and let the following be a CM4I factoring in the closed model category structure of Lemma 1.5. (a cofibration followed by an acyclic fibration):*

$$* \xrightarrow{Cof} P_{\bullet\bullet} \xrightarrow{\sim Fib} X_\bullet \underset{\sim}{\otimes} (pt).$$

Then

1)   *Each horizontal level $P_{\bullet,j}$ of $P_{\bullet\bullet}$ is a coproduct of copies of $\Sigma^i A_\bullet$ ($i \geq 0$), with degeneracies being isomorphisms of summands of $P_{\bullet,j}$ onto summands of $P_{\bullet,k}$ with $k \geq j$;*

2)   *For each map $\varphi : A_\bullet \to P_{\bullet,0}$ and maps induced by degeneracies of $P_{\bullet\bullet}$ considered as basepoints, there are isomorphisms:*

$$\pi_*^v \pi_*^h(P_{\bullet\bullet}; A_\bullet)_\varphi \cong \pi_*^v \pi_*^h(X_\bullet \underset{\sim}{\otimes} (pt)_\bullet; A_\bullet)_\varphi$$

where the superscripts v, h stand for vertical and horizontal, the horizontal homotopy groups are "with coefficients in $A_\bullet$", and the vertical groups of the vertically constant object on the right are concentrated in degree zero.

Proof. 1) This follows from the use of the vertical simplicial structure of [BF] in the small object argument construction of the factoring. It is enough to consider just one of the summands of the localizing object for any $i \geq 0$. An elementary step of the construction of [N, Th. 2.1.] attaches a copy of $\Sigma^i A_\bullet$ as a summand to $P_{\bullet,j}$ so that the face maps represent the attaching maps, and this structure is preserved by colimits at limit steps of the small object argument.

2) The acyclic (in the sense of Lemma 1.5.) fibration $P_{\bullet\bullet} \to X_{\bullet} \otimes (pt)$ induces a vertical Reedy acyclic fibration on $\underline{Hom}(A_{\bullet}, \_)$. Since acyclic fibrations are transposition invariant, the last map is a Reedy equivalence on horizontal terms, which implies the "first horizontal, then vertical" E^2-equivalence. Since $X_{\bullet} \otimes (pt)$ is fibrant, any two resolutions are equivalent cofibrant fibrant objects in the new structure and hence also equivalent in the initial structure [N] from which it follows that the E^2-term in reverse order given by the resolution is also well defined.

2 Interpretation of the homotopy colimit of the simplicial resolution

The most straightforward interpretation is as a map terminal up to homotopy from homotopy colimits of simplicial diagrams of the same type as the resolution $P_{\bullet\bullet}$ of Theorem 1.6. Alternatively, one can define an Artin-Mazur cocompletion dual to the one described in [BK, Ch. III, §8, Ch. XI, §10] choosing coproducts of $A_{\bullet}$ and its suspensions as the category of models and use simplicial replacement to show that the two definitions coincide. If $\hat{A}(X)$ denotes this cocompletion, there is a natural map:

$$i_X : \hat{A}(X) \to CW_A X.$$

One can say that X is $A_{\bullet}$-good if the map $\hat{A}(X) \to X$ is an A-equivalence. It is an easy consequence of the definitions that a fibrant X is $A_{\bullet}$-good if and only if the map i_X is an (ordinary) equivalence.

This construction was studied in [BT1] using different techniques and conditions were found for spaces to be M-good for certain M in [BT1, BT2]. A category of M-Π-algebras generalizing Π-algebras was defined in [BT1] and its structure for some M was determined.

Let the category of models $\mathcal{M}^A$ have as objects coproducts (possibly infinite) of $A_{\bullet}$ and its suspensions and let $\mathcal{C} - \Pi^A Alg$ be a corresponding generalization of $\mathcal{C} - \Pi Alg$. Let $\bar{\pi}_* : \mathcal{C} - \Pi^A \mapsto \mathcal{C} - \Pi$ be the *extension* of the ordinary homotopy functor which is well defined on free $\mathcal{C} - \Pi^A$-algebras by its values on the models, and let $\mathbf{L}_*$ stand for left derived.

2.1. A SPECTRAL SEQUENCE IN HOMOTOPY. *Applying the functor $\bar{\pi}_*$ to the resolution $P_{\bullet\bullet}$ gives the following spectral sequence:*

$$E^2_{s,t} = (\mathbf{L}_s \bar{\pi}_t) \pi^A_* X_{\bullet} \Longrightarrow \pi_{s+t} \hat{A}(X_{\bullet})$$

in which the 0-th row of the E^2-term can be identified with $\bar{\pi}_* \pi^A_* X_{\bullet}$ (compare [BS, Th. 4.2.]).

2.2. A SPECTRAL SEQUENCE IN HOMOLOGY. *A functor $\bar{h}_*$ is defined similarly to the one before except that after taking the models representing a free $\mathcal{C} - \Pi^A$-algebra one applies an abelianization functor. The resulting extension can be thought of as a zero homology and we will denote*

$$h^A_0(X_{\bullet}; h_* A_{\bullet}) = \bar{h}_* \pi^A_* X_{\bullet}.$$

Then the resulting spectral sequence is

$$E_{s,t}^2 = (\mathbf{L}_s \bar{h}_0) \pi_*^A X_\bullet \implies h_{s+t} \hat{A}(X)$$

and it can be rewritten as

$$E_{s,t}^2 = h_s^A(X_\bullet; h_t(A_\bullet)) \implies h_{s+t} \hat{A}(X_\bullet)$$

thus coinciding with a spectral sequence of André for the given choice of models.

Acknowledgement. The author is grateful to Jeff Smith for many conversations during the work on this paper.

References

[BS]   D. Blanc and C. Stover, *A generalized Grothendieck spectral sequence*, London Math. Soc. Lect. Notes Ser. No. 175 (1992), 145–161.

[BT1] D. Blanc and R. D. Thompson, *A suspension spectral sequence for v_n-periodic homotopy groups*, preprint (1992).

[BT2] D. Blanc and R. D. Thompson, *M-equivalences and homotopy colimits*, preprint (1993).

[BF]   A. K. Bousfield and E. M. Friedlander, *Homotopy theory of Γ-spaces, spectra and bisimplicial sets*, Lecture Notes in Math. **658** (1978), Springer, 80–130.

[BK]   A. K. Bousfield and D. M. Kan, *Homotopy limits, Completions and Localizations*, Lecture Notes in Math. **304** (1972), Springer.

[DF]   E. Dror Farjoun, *Localizations, fibrations and conic structures*, preprint (1991).

[DKS] W. G. Dwyer, D. M. Kan and C. R. Stover, *An E^2 model category structure for pointed simplicial spaces*, J. Pure Appl. Alg. **90 (2)** (1993), 137–152.

[G]   P. G. Goerss, *A Hilton-Milnor theorem for categories of simplicial algebras*, Amer. J. Math. **111** (1989), 927–971.

[N]   A. Nofech, *A-cellular homotopy theories*, preprint (1994).

[Q]   D. G. Quillen, *Homotopical Algebra*, Lecture Notes in Math. **43** (1967), Springer.

[R]   C. L. Reedy, *Homotopy category of model categories*, preprint.

Alexander Nofech, University of Rochester, Rochester, NY 14620, USA

Progress in Mathematics, Vol. 136
© 1996 Birkhäuser Verlag Basel/Switzerland

p-adic Lattices of Pseudo Reflection Groups

D. Notbohm

ABSTRACT. Let U be a vector space over the p–adic rationals, and let $W \longrightarrow Gl(U)$ be faithful representation of a finite group such that W is generated by pseudo reflections. For odd primes we study the p–adic W–lattices of this representation and achieve a complete classification. Examples of such situations are given by the Weyl group acting on the 1-dimesional homology of the maximal torus of a connected compact Lie group, or of the so called p–compact groups, a homotopy theoretic generalisation of compact Lie groups. The associated lattices are an important algebraic invariant in the study of these geometric object.

Introduction

Let U be a finite dimensional vector space over the p–adic rationals $\mathbb{Q}_p^\wedge$. An element $1 \neq \sigma \in Gl(U)$ is called a *pseudo reflection* if σ has finite order and if the kernel of $\sigma - id_U$ has codimension 1. The element σ is called a *honest reflection* or a *reflection* if σ has order 2. Because we are working in characteristic 0, the order of σ divides $p - 1$ and the linear transformation $\rho(\sigma)$ is diagonalzable. i.e. U has a basis of eigenvectors with respect to σ.

A *pseudo reflection group* is a couple $W = (W, \rho)$, where W is a finite group and $\rho : W \to Gl(U)$ a faithful representation such that the image $\rho(W)$ is generated by pseudo reflections. To emphasize the representation we also denote the pseudo reflection group by $W \to Gl(U)$.

A W–*lattice* of $W \to Gl(U)$ is a lattice $L \subset U$ of U of maximal rank fixed under the action of W, i.e. L is a $\mathbb{Z}_p^\wedge[W]$-module and $L \otimes \mathbb{Q} \cong U$ as vector spaces.

In this work we are concerned with the classification of all p–adic W–lattices $L \subset U$ of a given finite pseudo reflection group $W \to Gl(U)$. For odd primes we will achieve a complete classification. Our motivation to study this question comes from homotopy theory. The Weyl group W_G of a connected compact Lie group G acting on the tangent space of the maximal torus T of G or on the 1-dimensional homology $H_1(T; \mathbb{Z})$ provides an example of a honest reflection group and also of an integral lattice. This action is an important algebraic invariant in the study of connected compact Lie groups. In [5], Dwyer and Wilkerson gave the notion of p–compact groups, which is the homotopy theoretic generalisation of the notion of compact Lie groups. In their work, pseudo reflection groups occured in the same manner as honest reflection groups

1991 *Mathematics Subject Classification.* 20C11, 06B15, 55R35.
Key words and phrases. reflection group, pseudo reflection group, lattice, p–adic representations, p–compact groups.

for connected compact Lie groups, namely as Weyl groups acting on a 'maximal torus'. These p–compact groups provide examples of pseudo reflection groups and associated p–adic lattices. Besides these geometric and homotopy theoretic aspects we believe that the study of p–adic W–lattices has interest from it's own.

We will use the following notation and definitions in this paper.

1.1 NOTATION, DEFINITIONS AND REMARKS.

Let $W \to GL(U)$ be a pseudo reflection group.

1.1.1   The representation $W \to Gl(U)$ is called *fixed–point free* if the fixed–point set $U^W = 0$ is trivial. A lattice $L \subset U$ is called *fixed–point free* if $L^W = 0$.

1.1.2   The pseudo reflection group $W \to Gl(U)$ is called *irreducible* if the representation is irreducible. By [4], this is equivalent to the fact that the associated complex representation $U \otimes_{\mathbb{Q}_p^\wedge} \mathbb{C}$ is irreducible.

1.1.3   For every W–lattice L, we have a short exact sequence

$$0 \to L \to L_\mathbb{Q} \to L_\mathbb{Q}/L =: T_{L,\infty} \to 0$$

of W–modules. The quotient $T_{L,\infty} \cong (\mathbb{Z}/p^\infty)^n \subset (S^1)^n$ is called a p–discrete torus and can be considered as a subgroup of a torus whose dimension equals the rank of L. Completing the classifying spaces of $T_{L,\infty}$ and passing to 2-dimensional homology establishes an isomorphism $H_2((BT_{L,\infty})_p^\wedge; \mathbb{Z}_p^\wedge) \cong L$ of W–modules.

1.1.4   For a W–lattice $L \subset U$, the fixed–point set $Z(L) := (T_{L,\infty})^W$ is called the *center* of L. There is an associated W–lattice PL given by the kernel of the composition

$$L \otimes \mathbb{Q} \to T_{L,\infty} \to T_{L,\infty}/Z(L) = T_{PL,\infty} \ .$$

and a W–equivariant map $L \to PL$.

The lattice L is called *centerfree* if $Z(L) = 0$.

1.1.5   For a W–lattice $L \subset U$, the *covariants* L_W are given by the quotient L/SL, where $SL \subset L$ is the lattice generated by all elements of the form $l - w(l)$ with $l \in L$ and $w \in W$.

The lattice L is called *simply connected* if $L_W = 0$.

1.1.6   A monomorphism $L \to M$ of W–lattices is called a W–*trivial restriction* or W–*trivial extension* if W acts trivially on the quotient M/L and if M/L is finite.

1.1.7   Two W–lattices $L_1, L_2 \subset U$ are called isomorphic, if L_1 and L_2 are isomorphic as $\mathbb{Z}_p^\wedge[W]$–modules. That is that the two associated integral representations $\rho_1, \rho_2 : W \to GL((\mathbb{Z}_p^\wedge)^n)$ are conjugated.

Most of these notions are motivated by an analogy to connected compact Lie groups. Let W_G denote the Weyl group and T_G the maximal torus of a connected compact Lie group G. Then, the action of W_G on $U_G := H_2(BT_G; \mathbb{Z}_p^\wedge) \otimes \mathbb{Q}$ represents W as a finite reflection group, and $L_G := H_2(BT_G; \mathbb{Z}_p^\wedge) \subset U_G$ is a *p*–adic lattice. For odd primes, the center $Z(L_G)$ of L_G is a *p*–discrete approximation of the center of G [9]. Because the fundamental group $\pi_1(G)$ of G is isomorphic to the quotient of $\pi_1(T)$ by the translations of the extended Weyl group, one can also show that, for odd primes, $(L_G)_{W_G}$ is a *p*–discrete approximation of $\pi_1(G)$, i.e. $(L_G)_{W_G} \otimes \mathbb{Z}_p^\wedge \cong (L_G \otimes \mathbb{Z}_p^\wedge)_{W_G} \cong \pi_1(G) \otimes \mathbb{Z}_p^\wedge$.

By D_{2n} we denote the dihedral group of order $2n$.

1.2 THEOREM. *Let p be an odd prime. Let $W \to Gl(U)$ be a finite fixed–point free pseudo reflection group, and let $p \neq 3$ or $W \neq D_{12}$. Then the following holds:*

(1) *There exists a centerfree lattice $P \subset U$, unique up to isomorphism.*
(2) *There exists a simply connected lattice $S \subset U$, unique up to isomorphism.*

1.3 THEOREM. *Let $p = 3$, $W = D_{12}$ and $W \to GL(U)$ be the irreducible representation of W as pseudo reflection group. Then, up to isomorphism, there exist two lattices $L_1, L_2 \subset U$, which are both centerfree and simply connected.*

A detailed description of these two lattices is given in Section 4. It also turns out that, for the two associated two dimensional integral representationa $\rho_1, \rho_2 : D_{12} \to GL(\mathbb{Z}_p^{\wedge 2})$, there exists an automorphism $\alpha : D_{12} \to D_{12}$ such that ρ_1 and $\rho_2 \alpha$ are conjugate.

Next we describe centerfree and simply connected lattices of fixed–point free pseudo reflection groups. For a finite pseudo reflection group $W \to Gl(U)$, there exist splittings $U \cong U^W \oplus U_1 \oplus \cdots \oplus U_n$ of U as W–modules and $W \cong W_1 \times \cdots \times W_n$ of W such that W_i acts on U_i as an irreducible pseudo reflection group and trivially on every U_j for $j \neq i$ (e.g. see [10] or [4]).

1.4 THEOREM. *Let p be an odd prime. Let $W \to Gl(U)$ be a finite fixed–point free pseudo reflection group, and let $W \cong \prod_i W_i$ and $U \cong \bigoplus_i U_i$ be the associated splittings into irreducible pseudo reflection groups. Then the following holds:*

(1) *Every centerfree lattice $P \subset U$ splits into a direct sum $P \cong \bigoplus_i P_i$ of centerfree lattices $P_i \subset U_i$.*
(2) *Every simply connected lattice $S \subset U$ splits into a direct sum $S \cong \bigoplus_i S_i$ of simply connected lattices $S_i \subset U_i$.*

The last three theorems show that centerfree and simply connected lattices are "almost unique up to isomorphism". The only indeterminancy can come from the two different lattices of D_{12}.

1.5 THEOREM. *Let $W \to GL(U)$ be a finite fixed–point free pseudo reflection group. Then the following holds:*

(1) *A lattice $P \subset U$ is centerfree if and only if every W–trivial restriction $L \to P$ factors over $L \to PL \to P$.*
(2) *A lattice $S \subset U$ is simply connected if and only if every W-trivial restriction $S \to L$ factors over $S \to SL \to L$.*

Finally we consider the general case of a pseudo reflection group $W \rightarrow Gl(U)$. Let $U \cong U^W \oplus U'$ be the splitting into the fixed–point set and the fixed–point free factor U'. Because W acts trivially on U^W, all lattices of U^W are isomorphic as W–modules. We choose a lattice of U^W and denote it by Z.

1.6 THEOREM. *Let p be an odd prime. Let $W \rightarrow Gl(U)$ be a pseudo reflection group, and let L be a W–lattice. Then the following holds:*

    (1) *There exist a simply connected W–lattice S and a W–trivial restriction $Z \oplus S \rightarrow L$ with quotient $L/(Z \oplus S) \cong (L/L^W)_W$.*

    (2) *There exist a centerfree W–lattice P and a W–trivial restriction $L \rightarrow Z \oplus P$.*

Lattices, which are centerfree and simply connected, do not allow non trivial W-restrictions. Also they only show up as direct summands as the next theorem shows.

1.7 THEOREM. *Let $W \rightarrow Gl(U)$ be a finite pseudo reflection group. Let $L \rightarrow Z \oplus P$ be the W–trivial restriction of Theorem 1.6. Let $P \cong P_1 \oplus P_2$ be a splitting into centerfree lattices such that P_1 is also simply connected. Then, the lattice P_1 is a direct summand of L.*

The following corollary of all of the above results is obvious and classifies all lattices of finite pseudo reflection groups. This classification result is an analogue of the classification of connected compact Lie groups, where irreducible pseudo reflection groups play the same role as simple connected compact Lie groups and trivial representation the role of tori.

1.8 COROLLARY. *Let p be an odd prime. Then, every W–lattice is a W–trivial extension of a trivial W–lattice and a simply connected W–lattice, and every simply connected W–lattice is a direct sum of simply connected lattices of irreducible pseudo reflection groups.*

Our main theorems are statements about odd primes. This comes simply from the following lemma, which plays a little but important role in several proofs.

1.9 LEMMA. *Let p be an odd prime, let W be a pseudo reflection group and let M be a $\mathbb{Z}_p^\wedge$–module with trivial W–action. Then, we have $H_1(W; M) = H^1(W; M) = 0$.*

Proof. As a pseudo reflection group for an odd prime, W is generated by elements of order coprime to p. By the Hurewicz theorem, the first homology group $H_1(W, \mathbb{Z})$ is isomorphic to the abelianization of W, which is a finite abelian group of order coprime to p. Universal coefficient theorems imply the statement.   □

The paper is organized as follows: In Section 2, we discuss centerfree lattices and prove the first parts of Theorem 1.2, Theorem 1.4 and of Theorem 1.5. In Section 3 we study simply connected lattices and prove the second part of these theorems. Section 4 is devoted to the study of the dihedral group D_{12}

at the prime 3. The last section contains an analysis of general finite pseudo reflection groups and the proof of Theorem 1.6 and Theorem 1.7.

Although the nature of this paper is mostly algebraic, sometimes we deal with completed topological spaces. Completion is always meant in the sense of Bousfield and Kan [2].

Independently of us, Dwyer and Wilkerson got also proofs for some of the main results of this work [6].

Finally a warning: We are only dealing with odd primes, i.e. p always denotes an odd prime.

2 Centerfree lattices

As mentioned in the introduction $W \to Gl(U)$ is a finite pseudo reflection group.

2.1 LEMMA.

(1) For a W–lattice L, the center $Z(L)$ is a finite abelian p–group if and only if L is fixed–point free.
(2) If L is fixed–point free, then we have $Z(L) \cong H^1(W; L)$.

Proof. Taking fixed–points in the short exact sequence

$$0 \to L \to L_{\mathbb{Q}} \to T_{L,\infty} \to 0$$

gives rise to an exact sequence

$$0 \to L^W \to L_{\mathbb{Q}}^W \to (T_{L,\infty})^W \to H^1(W; L) \to H^1(W; L_{\mathbb{Q}}) = 0 .$$

The first two terms vanishes if L is fixed–point free. Otherwise the quotient $L_{\mathbb{Q}}^W / L^W$ is a p–discrete torus. In particular, the quotient is not finite. Because $H^1(W; L)$ is always finite, both parts follow. □

For a W–lattice L we denote by $L/p := L \otimes_{\mathbb{Z}_p^\wedge} \mathbb{F}_p$ the associated $\mathbb{F}_p[W]$–module.

2.2 LEMMA. A W–lattice P is centerfree if and only if $(P/p)^W = 0$.

Proof. The multiplication $\mu_p : P \to P$ by p establishes a short exact sequence

$$0 \to P \xrightarrow{\mu_p} P \to P/p \to 0 .$$

Passing to fixed–points gives an exact sequence

$$0 \to P^W \xrightarrow{\mu_p} P^W \to (P/p)^W \to H^1(W; P) \xrightarrow{\mu_p} H^1(W; P) .$$

If $(P/p)^W = 0$ then $\mu_p : P^W \to P^W$ and $\mu_p : H^1(W; P) \xrightarrow{\mu_p} H^1(W; P)$ are isomorphisms ($H^1(W; P)$ is a finite group). From the first isomorphism follows that P is fixed–point free. Because $H^1(W; P)$ is a finite abelian p–group, the second isomorphism implies that $H^1(W; P) = 0$. The other direction follows from Lemma 2.1. □

The next statement shows that, for every W–lattice L, the associated W–lattice PL is centerfree.

2.3 PROPOSITION. *Let L be a fixed–point free W lattice.*

(1) *There exists an exact sequence*

$$0 \to L \to PL \to Z(L) \to 0 ,$$

and hence, L is a W–trivial restriction of PL.

(2) *The lattice PL is centerfree and $H^1(W; PL) = 0$.*

Proof. By definition of the lattice PL, there exists a commutative diagram

$$
\begin{array}{ccccccccc}
0 & \longrightarrow & L & \longrightarrow & L \otimes \mathbb{Q} & \longrightarrow & T_{L,\infty} & \longrightarrow & 0 \\
& & \downarrow & & \cong \downarrow & & \downarrow & & \\
0 & \longrightarrow & PL & \longrightarrow & PL \otimes \mathbb{Q} & \longrightarrow & T_{PL,\infty} & \longrightarrow & 0 ,
\end{array}
$$

where the middle arrow is an isomorphism, where the left arrow is a monomorphism and where the right arrow is a epimorphism. Hence, the cokernel of the left arrow and the kernel of the right arrow, given by $Z(L)$ are isomorphic. This establishes the desired exact sequence of part (1).

Taking fixed–points in the exact sequence of (1) gives rise to the exact sequence

$$0 = PL^W \to Z(L)^W = Z(L) \to H^1(W; L) \to H^1(W; PL) \to H^1(W; Z(L)) .$$

By Lemma 2.1, the second arrow is an isomorphism. By Lemma 1.9, the last term vanishes. Hence, we have $H^1(W; PL) = 0$. Again by Lemma 2.1, the lattice PL is centerfree. □

The following technical proposition is the key for the proof of Theorem 1.2.

2.4 PROPOSITION. *Let $p \neq 3$ or $W \neq D_{12}$. Let $W \to Gl(U)$ be an irreducible pseudo reflection group, and let P be a centerfree lattice. If there exists an exact sequence*

$$0 \to V_0 \to P/p \to V_1 \to 0$$

of $\mathbb{F}_p[W]$–modules, such that $V_1^W = 0$, then either $V_0 = 0$ or $V_1 = 0$.

Proof. Let assume that V_0 and V_1 are nontrivial vector spaces. We choose a basis for V_0 and extend it to a basis of P/p. Then every element $w \in W$ can be represented by a matrix of the form

$$
\begin{pmatrix}
A_w & C_w \\
0 & B_w
\end{pmatrix}
$$

where A_w decribes the action of w on V_0, B_w the action on V_1 and $C_w : V_1 \to V_0$ the twisting, i.e. the failure to be a direct product. This description establishes a homomorphism $\phi : W \to Gl(V_0) \times Gl(V_1)$ given by $\phi(w) := (A_w, B_w)$. Let W_i be the image of W in the factor $Gl(V_i)$. That is we have a homomorphism

$\phi : W \longrightarrow W_0 \times W_1$. Because V_0 and V_1 have no non trivial fixed–point, both groups W_0 and W_1 are non trivial. For V_0 this follows from Lemma 2.2.

The kernel K of ϕ consists of those elements which are desribed by a matrix of the form $\begin{pmatrix} id & C \\ 0 & id \end{pmatrix}$. Therefore, every element of the kernel has order p and the kernel is an elementary abelian p–group and a normal subgroup of W.

Now let $\sigma \in W$ be a p–adic pseudo reflection. The matrix

$$\sigma - id = \begin{pmatrix} A_\sigma - id & C_\sigma \\ 0 & B_\sigma - id \end{pmatrix}$$

has rank 1. That is that all columns and all rows are multiple of one column or one row. We have $A_\sigma - id \neq 0$ if and only if $B_\sigma = id$. The equivalence follows from the fact that the order of σ is coprime to p. Therefore, W_0 and W_1 are generated by p-adic reflections. Let (w_0, w_1) be an element of $W_0 \times W_1$. We can assume that w_0 is the image of a product of p–adic reflections which are mapped onto the identity in W_1, and similiar for w_1. This shows that ϕ is an epimorphism.

The above considerations show that W allows a short exact sequence

$(*)$ $\qquad\qquad\qquad 1 \rightarrow K \rightarrow W \rightarrow W_0 \times W_1 \rightarrow 1 .$

where W_0 and W_1 are nontrivial groups, generated by elements coming from pseudo reflections in W, and where $K \subset W$ is an elementary abelian normal subgroup. For abbreviation, we say that W has the property $(*)$.

We want to show that either W_0 or W_1 is the trivial group. This would imply that either $V_0 = 0$ or $V_1 = 0$. The proof of this conclusion splits into two part, the nonmodular case, i.e. $(|W|, p) = 1$, and the modular case. For the modular case we use the classification of the irreducible pseudo reflection groups by Clark and Ewing [4]. We also use their numbering of the different cases.

First let $(|W|, p) = 1$. Then $K = 0$ and $W \cong W_0 \times W_1$ splits into a product of pseudo reflection groups. Because the representation $W \rightarrow Gl(U)$ is irreducible, this implies that either W_0 or W_1 is the trivial group.

Now let p divide $|W|$. We are considering separately the case where K is central in W and the case in which it is not. If $K \subset W$ is a central subgroup, then every element of K establishes a W–equivariant automorphism $U \rightarrow U$. The W–representation U is irreducible if and only if $U \otimes_{\mathbb{Q}_p^\wedge} \mathbb{C}$ is irreducible [4]. Hence, the induced map $f \otimes \mathbb{C}$ is a multiple, in particular a p–adic multiple of the identity. That is to say that there exists a homomorphism $K \rightarrow \mathbb{Z}_p^{\wedge *} \cong \mathbb{Z}/p - 1 \times \mathbb{Z}_p^\wedge$. Because W is finite and because W acts faithfully on U, this homomorphism is injective, and the kernel K is trivial. We can proceed as in the nonmodular case.

If $K \rightarrow W$ is not central, then there exists a pseudo reflection $\sigma \in W$ acting nontrivially on K. Because the order of σ is coprime to p, the representation K of the group $< \sigma >$, generated by σ, splits into 1-dimesional irreducible summands. Let $K' \subset K$ be one of the summands with a nontrivial action of σ and let $x \in K'$ be a generator. The subgroup $D :=< \sigma, x\sigma x^{-1} >=< \sigma, x\sigma >=< \sigma, x >$ of W, generated by two pseudo reflections, fits into a short exact sequence

$$1 \to K' \to D \to< \sigma >\to 1 .$$

The order $m = |\sigma|$ of σ is coprime to p. Therefore, the sequence splits and $D \cong \mathbb{Z}/p \rtimes \mathbb{Z}/m$ acts on U as a pseudo reflection group. As a $\mathbb{Z}/p$–module, $U \cong \bigoplus_i U_i$ splits into a direct sum of irreducible $\mathbb{Z}/p$–modules which are permuted by $\mathbb{Z}/m$. Each factor is either 1–dimensional with trivial $\mathbb{Z}/p$–action ($\mathbb{Q}_p^\wedge$ contains no p–th root of unity) or isomorphic to $U' \cong (\mathbb{Q}_p^\wedge)^{p-1}$ where we consider U' as the kernel of the map $(\mathbb{Q}_p^\wedge)^p \to \mathbb{Q}_p^\wedge$ given by summing up the coordinates and where $\mathbb{Z}/p$ acts via cyclic permutation on $(\mathbb{Q}_p^\wedge)^p$. The factors with trivial $\mathbb{Z}/p$–action does not lead to a faithful representation of D. Every factor isomorphic to U' is fixed under the action of $\mathbb{Z}/m$, and $\mathbb{Z}/m$ acts on U' via permutation associated to the action on $\mathbb{Z}/p$ considered as a set. Therefore, U' represents D as a pseudo reflection group if and only if $m = 2$. That is to say that $D \cong D_{2p}$ is a dihedral group. By the classification list of irreducible pseudo reflection groups [4] the only modular cases are given by D_6 and D_{12}. Hence, we have $p = 3$ and $D \cong D_6$.

By the above arguments it is only left to consider modular cases for $p = 3$. We will finish the proof by a case by case checking following the list of [4]. We only have to discuss the numbers 1, 2a, 2b, 12, 28, 35, 36 and 37.

CASES NUMBER 1 AND 2A. In this case $\Sigma_n \subset W \subset \mathbb{Z}/l \wr \Sigma_n$ where l divides $p - 1$. In particular, the subgroup $K \subset W$ is a normal subgroup of Σ_n as well as of $A_n \subset \Sigma_n$. Here, A_n denotes the group of permutations of positive sign. For $n \geq 5$, the group A_n is simple. For $n = 4$, we have $A_4 \cong (\mathbb{Z}/2 \times \mathbb{Z}/2) \rtimes \mathbb{Z}/3$. Therefore, in these cases there exists no normal elementary abelian 3–subgroup, and we can proceed as in the nonmodular case.

Now let $n = 3$. If $W \neq \Sigma_3$, then there also exists no normal elementary abelian subgroup. If $W = \Sigma_3$, then the representation $W \to Gl(U)$ is desribed by the matrices $\sigma = \begin{pmatrix} 0 & 1 \\ 1 & 0 \end{pmatrix}$ and $\tau = \begin{pmatrix} 1 & 0 \\ -1 & -1 \end{pmatrix}$, which are reflections and generate Σ_3. For the obviuos associated lattice L, a straightforward calculation shows the existence of a short exact sequence $0 \to \mathbb{Z}/3 \to L/3 \to det/3 \to 0$ of W–modules. Here, W acts trivially on $\mathbb{Z}/3$ and det is the 1–dimensional representation given by the sign of the permutation. Every other lattice $L' \subset U$ has mod-3 the same composition factors. In particular, one of the modules V_0 or V_1 is isomorphic to the trivial representation and the other to $det/3$. This contradicts the fact that $V_0^{\Sigma_3} = 0 = V_1^{\Sigma_3}$.

For later purpose we note the following observation: For $n \geq 5$, the above argument shows that, if W has the property $(*)$, every pseudo reflection representation of W splits into two summands, where both factors carry a nontrivial W–action.

CASE NUMBER 2B. In this case, we have $W = D_6$ or $W = D_{12}$ and U is 2–dimensional. The first case we already discussed and the second is excluded by the assumptions.

CASE NUMBER 12. In this case we have $dim_{\mathbb{Q}_p^\wedge} U = 2$ and $W = Gl(2, \mathbb{F}_3)$. There exists a lattice $L \subset U$ such that the action of W on $L/3 \cong \mathbb{Z}/3 \oplus \mathbb{Z}/3$ is isomorphic to the standard action, which gives an irreducible representation (details may be found in [1] or [10]). Hence, every other lattice $L' \subset U$ gives mod–p also an irreducible representation, which proves the statement in this case.

CASE NUMBER 28. In this case, we have $W = W_{F_4} \cong ((\mathbb{Z}/2)^3 \rtimes \Sigma_4) \rtimes \Sigma_3$. The last isomorphism may be found in [7, p. 45]. A straight forward calculations shows that $K = 0$. We can proceed as in the nonmodular case.

CASE NUMBER 35, 36, 37. In this case we have $W = W_{E_6}$, $W = W_{E_7}$ or $W = W_{E_8}$. We describe two maximal subgroups of maximal rank for each of these connected compact Lie groups.

G	H'	H''
E_6	$S^1 \times_{\mathbb{Z}/2} Spin(10)$	$SU(2) \times_{\mathbb{Z}/2} SU(6)$
E_7	$S^1 \times_{\mathbb{Z}/2} Spin(12)$	$S^1 \times_{\mathbb{Z}/3} E_6$
E_8	$SSpin(16)$	$SU(2) \times_{\mathbb{Z}/2} E_7$

A list of all maximal subgroups of maximal rank may be found in [8]. This establishes subgroups of W as follows:

W	W'	W''
W_{E_6}	$W_{H'} \cong (\mathbb{Z}/2)^5 \rtimes \Sigma_5$	$W_{SU(6)} \cong \Sigma_6$
W_{E_7}	$W_{H'} \cong (\mathbb{Z}/2)^5 \rtimes \Sigma_6$	W_{E_6}
W_{E_8}	$W_{H'} \cong (\mathbb{Z}/2)^7 \rtimes \Sigma_8$	W_{E_7}

In all cases, the two groups W' and W'' generate W. This follows because $H' \subset G$ is maximal of maximal rank. Moreover, the intersection $W' \cap W''$ is nonempty. We want to show that there exists no epimorphism $W \to W_0 \times W_1$ as in (*) with kernel given by an elementary abelian p–group.

Let us look at the case $W = W_{E_6}$. By the observation at the end of cases number 1 and 2a, if W' has the property (*), the W'–module U splits into a direct sum of nontrivial W'–modules. The same is true for W''. But by the choice of the groups, both belong to case 2a with $n \geq 5$, we only can split of a trivial summand of U considered as a W' or W''–module. Therefore, W' as well as W'' have not the property (*), and an epimorphism $W_{E_6} \to W_0 \times W_1$ maps W' and W'' only into one factor. Because $W' \cap W''$ is nonempty, both are mapped into the same factor, let us say into W_0. Because W_{E_6} is generated by W' and W'', the group W_{E_6} is only mapped into W_0, too. Hence, W_1 is trivial. This proves the statement in this case. In particular, this argument also shows that there exists no epimorphism of the form (*) with kernel given by an elementary abelian p–group.

For W_{E_7} and W_{E_8}, we can argue analogously using the result for W_{E_6} or W_{E_7}. This finishes the discusion of all possible cases and the proof of the statement.   □

REMARK. The last proposition as well as the proof originates in a discussion with C. Broto and J. Aguadé on a similar question.

2.5 LEMMA. *Let $P \to L$ be a monomorphism between W-lattices of U. If P is centerfree, then we have $(L/P)^W = 0$.*

Proof. Because P is centerfree, it is also fixed–point free (lemma 2.1). Hence $U \cong P \otimes \mathbb{Q}$ as well as every lattice of U is fixed–point free. The short exact sequence $P \to L \to L/P$ gives rise to an exact sequence $L^W = 0 \to (L/P)^W \to H^1(W; P) = 0$. Thus, the quotient L/P has no fixed–points. $\qquad\square$

PROOF OF THEOREM 1.2 (1). Let P and Q be two centerfree lattices of an irreducible pseudo reflection group $W \to Gl(U)$. Then, for r big enough, the lattice $p^r P := \{p^r v : v \in P\}$ is a sublattice of Q. Because $p^r P$ and P are isomorphic W–lattices, there exists a W–equivariant monomorphism $\alpha : P \to Q$. Moreover, by choosing a minimal r, we can assume that $rk(Q/P) < rk(Q) = rk(P)$. Here, $rk(M)$ denotes the rank of a module, which we define to be the dimension of M/p over $\mathbb{F}_p$. Otherwise we have $P \subset pQ := \{px : x \in Q\}$ and $p^{r-1}P \subset Q$. Because P is centerfree we know that $(Q/P)^W = 0$ (Lemma 2.5). The monomorphism $P \xrightarrow{\alpha} Q$ is rationally an isomorphism, and the quotient Q/P is finite.

Applying the functor $\otimes \mathbb{F}_p$ yields an exact sequence

$$0 \to Tor(Q/P, \mathbb{F}_p) \to P/p \xrightarrow{\bar{\alpha}} Q/p \to Q/P \otimes \mathbb{F}_p \to 0$$

of W–modules. Let $V_0 := Tor(Q/P; \mathbb{F}_p)$ and let $V_1 := Im(\bar{\alpha})$ be the image of $\bar{\alpha}$ which is isomorphic to the kernel of $Q/p \to Q/P \otimes \mathbb{F}_p$. Because P and Q are centerfree we have $V_0^W = 0 = V_1^W$ (Lemma 2.2). Applying Proposition 2.4 (U is irreducible) shows that either V_0 or V_1 are trivial vector spaces. If $V_1 = 0$ then $rk(Q/P) = rk(Tor(Q/P; \mathbb{F}_p)) = rk(P)$, which is a contradiction. Thus, $V_0 = 0$ and $Q/P = 0$. That is to say that $\alpha : P \to Q$ is an isomorphism. This proves the statement for irreducible pseudo reflection groups.

Next we consider the case of a reducible fixed–point free pseudo reflection group W, i.e. $W \cong W_1 \times W_2$ splits into a nontrivial product of pseudo reflection groups. Moreover, $U \cong U_1 \times U_2$ also splits into a direct sum where $U_1 = U^{W_2}$ and $U_2 = U^{W_1}$. An application of the following proposition reduces the proof in this case to the case of irreducible pseudo reflection groups and finishes therefore the proof of Theorem 1.2 (1). $\qquad\square$

2.6 PROPOSITION. *Let $W \to Gl(U)$ be a reducible fixed–point free pseudo reflection group, and let P be a centerfree W–lattice of $U = U_1 \oplus U_2$. Then, the following holds:*

(1) *The fixed–point set P^{W_1} is centerfree with respect to the W_2–action.*
(2) *We have $P \cong P^{W_1} \oplus P^{W_2}$ as W–modules.*

Proof. The quotient P/P^{W_1} is torsion free. Hence, the sequence of W–modules

$$0 \to P^{W_1}/p \to P/p \to (P/p)/(P^{W_1}/p) \to 0$$

is short exact. Taking fixed–points yields an exact sequence

$$0 \to (P^{W_1}/p)^W \cong (P^{W_1}/p)^{W_2} \to (P/p)^W = 0 \;.$$

The last fixed–point set vanishes because P is centerfree and because of Lemma 2.2. Again by Lemma 2.2, the fixed–point set P^{W_1} is centerfree with respect to the W_2–action.

Applying the functor $\otimes \mathbb{Q}$ establishes an exact sequence

$$0 \to P^{W_1} \otimes \mathbb{Q} \to P \otimes \mathbb{Q} \to (P/P^{W_1}) \otimes \mathbb{Q} \to 0 \;.$$

Because $P^{W_1} \otimes \mathbb{Q} \cong (P \otimes \mathbb{Q})^{W_1}$, this sequence splits and shows that $(P/P^{W_1}) \otimes \mathbb{Q}$ as well as P/P^{W_1} are trivial W_2–module. Taking W_2–fixed–points establishes the exact sequence

$$0 = P^W = (P^{W_1})^{W_2} \to P^{W_2} \to (P/P^{W_1})^{W_2} = P/P^{W_1} \to H^1(W_2; P^{W_1}) = 0 \;.$$

The last identity follows from Lemma 2.1 since P^{W_1} is W_2–centerfree. This implies that the middle arrow is an isomorphism, and that $P^{W_1} \oplus P^{W_2} \to P$ is an isomorphism of W–modules. $\qquad\square$

PROOF OF THEOREM 1.4 (1). Let $W \to Gl(U)$ be a reducible pseudo reflection group. Using an induction over the number of irreducible summands of U, the statement follows from Proposition 2.6. $\qquad\square$

PROOF OF THEOREM 1.5 (1). Passing to associated centerfree lattices is a functor. If P is centerfree, every W–trivial restriction $L \to P$ establishes a W–trivial restriction $PL \to PP = P$. This establishes the desired factorisation.

To prove the other direction we consider the identity $id : P \to P$. By assumption this factors over $P \to PP \to P$. Hence, the second arrow is an epimorphism and therefore, as a map of torsionfree $\mathbb{Z}_p^\wedge$–modules, an isomorphism. This shows that $PP \cong P$ and that P is centerfree. $\qquad\square$

3 Simply connected lattices

Again, $W \to Gl(U)$ denotes a finite pseudo reflection group. The situation for simply connected lattices is somehow dual to the case of centerfree lattices (see Proposition 5.1 and Corollary 5.2).

3.1 LEMMA.

(1) For a W–lattice L, the group L_W of covariants is finite if and only if L is fixed–point free.
(2) If L is fixed–point free, then we have $L_W \cong H_1(W, T_{L,\infty})$.

Proof. Passing to covariants and using the fact that $L_W \cong H_0(W, L)$, the short exact sequence

$$0 \to L \to L \otimes \mathbb{Q} := L_{\mathbb{Q}} \to T_{L,\infty} \to 0$$

gives rise to the exact sequence

$$0 = H_1(W; L_{\mathbb{Q}}) \to H_1(W; T_{L,\infty}) \to L_W \to (L_{\mathbb{Q}})_W \to (T_{L,\infty})_W \to 0 .$$

We can split $L_{\mathbb{Q}} \cong U_1 \oplus U_2$ into a direct sum of a fixed–point free W–module U_1 and summand with trivial W–operation. Because every exact sequence of W–modules over $\mathbb{Q}_p^{\wedge}$ splits, we have $(U_1)_W = 0$ and $(L_{\mathbb{Q}})_W \cong L_{\mathbb{Q}}^W$. The homology group $H_1(W; T_{L,\infty})$ is finite. Thus, L_W is finite if and only if L is fixed–point free. The second part is obvious. $\square$

In the introduction, for a W–lattice L, we defined SL to be the kernel of $L \to L_W$.

3.2 PROPOSITION. *Let L be a fixed–point free W–lattice.*

(1) *There exists an exact sequence*

$$0 \to SL \to L \to L_W \to 0 ,$$

   and L is a W–trivial extension of SL.
(2) *The lattice SL is simply connected.*

Proof. The first part follows from definition of L_W and SL and from Lemma 3.1. Passing to covariants, the short exact sequence of (1) establishes the exact sequence

$$H_1(W, L_W) \to SL_W \to L_W \to L_W \to 0 .$$

The first term vanishes (Lemma 1.9) and the second last arrow is an isomorphism. $\square$

The next results connects simply connected and centerfree lattices.

3.3 PROPOSITION. *Let S be a simply connected W–lattice. Let $P := PS$ be the associated centerfree lattice. Then, we have $SP \cong S$ and $Z(S) \cong P_W$.*

Proof. By construction there exists a short exact sequence

$$0 \to S \to P \xrightarrow{q_S} Z(S) \to 0 .$$

Because $Z(S)$ is a trivial W–module, the map q_S factors over the covariants P_W. This establishes a commutative diagram of short exact sequences

$$
\begin{array}{ccccccccc}
0 & \longrightarrow & SP & \longrightarrow & P & \longrightarrow & P_W & \longrightarrow & 0 \\
  &   & \downarrow &   & \| &   & \downarrow &   &   \\
0 & \longrightarrow & S & \longrightarrow & P & \longrightarrow & Z(S) & \longrightarrow & 0
\end{array}
$$

where the cokernel S/SP of the monomorphism $SP \to S$ is a W–submodule of P_W. Therefore, the quotient S/SP is a module with trivial W–action, and the epimorphism $S \to S/SP$ factors over $S_W = 0$. This shows that all vertical arrows are isomorphisms. $\square$

We finish this section with proofs of Part (2) of theorems 1.2, 1.4 and 1.5.

PROOF OF THEOREM 1.2 (2). The existence of a simply connected lattice follows from Proposition 3.2. Let S and S' be two simply connected lattices. Let P and P' be the associated centerfree lattices. By Theorem 1.2 (1), we know that $P \cong P'$, and by Proposition 3.3 follows that $S \cong SP \cong SP' \cong S'$. $\square$

PROOF OF THEOREM 1.4 (2). Let $S \subset U$ be a simply connected lattice and let $P := PS \subset U$ be the associated centerfree lattice. By Theorem 1.4 (1), we have a splitting $P \cong \bigoplus_i P_i$ of P into centerfree lattices $P_i \subset U_i$ of the irreducible pseudo reflection groups $W_i \to Gl(U_i)$. The lattices $S_i := SP_i \subset U_i$ are simply connected. The sequence $S \cong SP \cong \bigoplus_i SP_i = \bigoplus_i S_i$ proves the statement. The first isomorphism follows from Proposition 3.3, and the second from Proposition 2.6 and Proposition 3.2. $\square$

PROOF OF THEOREM 1.5 (2). Passing to the associated simply connected lattices is a functor. If S is simply connected, every W–trivial restriction $S \to L$ establishes a W–trivial restriction $S = SS \to SL$. This is desired factorisation.

To prove the other direction we consider the identity $id : S \to S$. By assumption, it factors over $S \to SS \to S$. The second arrow is an isomorphism, because it is an epimorphism and because all modules are torsion free. This shows that S is simply connected. $\square$

4 The case $p = 3$ and $W = D_{12}$

The 2–dimensional homolgy $S := H_2(BT_{SU(3)}; \mathbb{Z}_p^\wedge)$ of the classifying space of the maximal torus of $SU(3)$ gives a representation of Σ_3 as pseudo reflection group. The action can be represented as in case no. 2a of the proof of Proposition 2.4. A straightforward calculation shows that S is simply connected and that $Z(S) \cong \mathbb{Z}/3$. Hence, by Proposition 3.2 and Theorem 1.2, there exist two Σ_3–lattices of $U := S \otimes \mathbb{Q}$, namely S and $P := PS$. They fit into a W–trivial restriction

$$0 \to S \to P \to \mathbb{Z}/3 \to 0 .$$

The action of Σ_3 on U can be extended to an action of $D_{12} \cong \Sigma_3 \times \mathbb{Z}/2$ by saying that the subgroup $\mathbb{Z}/2$ acts via multiplication by -1 or trivially. Because the centralizer of $\Sigma_3 \subset Gl(U)$ is given by $\{id, -id\}$, these are the only possible extensions. The first represents D_{12} as a pseudo reflection group, the second does not. Let $D_{12} \to Gl(U)$ be the representation of D_{12} as pseudo reflection group and let S and P also denote the D_{12}–lattices of U.

4.1 PROPOSITION. *The D_{12}–lattices S and P are both simply connected and centerfree, but not isomorphic.*

Proof. Because the representation $D_{12} \to GL(U)$ contains multiplication by -1, we have $S_{D_{12}} = 0 = P_{D_{12}}$ and $Z(S) = 0 = Z(P)$. The two lattices are not isomorphic because they are nonisomorphic as Σ_3–modules. $\square$

PROOF OF THEOREM 1.3. We already constructed two non isomorphic lattices. Let $L \subset U$ another lattice. Then, as Σ_3–lattice L has to be isomorphic to either S or P. Because for both lattices there exists a unique extension to a D_{12}–lattice, representing D_{12} as a pseudo reflection group, we also know that L is isomorphic to S or P as D_{12}–lattice.   □

REMARK. Let $\rho_S, \rho_P : D_{12} \to Gl(\mathbb{Z}_p^{\wedge 2})$ denote the representations associated with S and P. Here we identify both lattices with $\mathbb{Z}_p^{\wedge 2}$. There exists an automorphism $\alpha : D_{12} \to D_{12}$ which maps every reflection $s \in D_{12}$ on $-id \circ s$. Using α we can construct new representations given by $\rho_S \alpha, \rho_P \alpha : D_{12} \to Gl(\mathbb{Z}_p^{\wedge 2})$ with associated lattices S_α and P_α. Rationally, the representations $\rho_S \alpha$ and ρ_s are isomorphic, because restriction represents Σ_3 in both cases as pseudo reflection group, because there is only one representation with this property and because there exists a unique extension to D_{12} as a pseudo reflection group. Hence, S_α and P_α also describe lattices of $U := S \otimes \mathbb{Q}$.

The lattices P_α and P are not isomorphic. Otherwise, we have $S_\alpha \cong S$, because $\alpha\alpha = id : D_{12} \to D_{12}$. Moreover, there exists a short exact sequence $0 \to S \to P \to (\mathbb{Z}/3)_\alpha \to 0$. Taking invariants with respect to the action of Σ_3 shows that $H^1(\Sigma_3; S) \cong H^1(\Sigma_3; P) = 0$. The last equation follows from Lemma 2.2. The same lemma would also show that S is a centerfree Σ_3–lattice.

Because there exist only two lattices of U, these considerations show that S and P_α are isomorphic as well as P and S_α.

5 General lattices

Before we discuss the case of general finite pseudo reflection groups, i.e. before we prove the theorems 1.6 and 1.7, we need some informations about the dual representations. Let $W \to Gl(U)$ be a pseudo reflection group. We consider U as a left $\mathbb{Q}_p^\wedge[W]$–module. The set $Hom_{\mathbb{Q}_p^\wedge}(U, \mathbb{Q}_p^\wedge)$ becomes a left $\mathbb{Q}_p^\wedge[W]$–module by defining $w(x^*) := x^* w^{-1}$ for $x^* \in U^*$ and $w \in W$. The vector space U^* again represents W as a pseudo reflection group. For a W–lattice $L \subset U$ we define $L^* := Hom_{\mathbb{Z}_p^\wedge}(L; \mathbb{Z}_p^\wedge)$ which becomes analogously as above a left $\mathbb{Z}_p^\wedge[W]$–module. Because $L^* \otimes \mathbb{Q} \cong (L \otimes \mathbb{Q})^*$ as $\mathbb{Q}_p^\wedge[W]$–modules, the lattice L^* is a lattice of U^*.

5.1 PROPOSITION. *Let $W \to Gl(U)$ be a fixed–point free finite pseudo reflection group. Then, for every lattice $L \subset U$ we have $(SL)^* \cong P(L^*)$ and $(PL)^* \cong S(L^*)$.*

Proof. Dualizing a W–trivial restriction

$$0 \to L \to M \to M/L =: Q \to 0$$

gives a W–trivial restriction

$$0 \to M^* \to L^* \to Ext_{\mathbb{Z}_p^\wedge}(Q, \mathbb{Z}_p^\wedge) \to 0 \ .$$

Because Q carries the trivial W–action, we have $Ext_{\mathbb{Z}_p^\wedge}(Q, \mathbb{Z}_p^\wedge) \cong Q$ as W–modules.

Let $L \subset U$ be a lattice. Dualizing the W–trivial restriction

$$0 \to (SL)^* \to P((SL)^*) \to P((SL)^*)/(SL)^* \cong Z((SL)^*) \to 0$$

gives

$$0 \to (P((SL)^*))^* \to SL \to Z((SL)^*) \to 0 .$$

The equivalence in the top row follows from Proposition 2.3. Because SL is simply connected, taking covariants show that $Z((SL)^*)_W = 0$. But, as a center of a lattice, $Z((SL)^*)$ is a module with trivial W–action. This shows that $Z((SL)^*) = 0$ and that $(SL)^*$ is centerfree. Dualizing the W–trivial restriction $SL \to L$ gives the W–trivial restriction $L^* \to (SL)^*$. Applying the construction P establishes the W–trivial restriction $\alpha : P(L^*) \to (SL)^*$. Since $P(L^*)$ is centerfree, the map α is a W–trivial restriction and since $H^1(W, P(L^*)) = 0$ (Proposition 2.3 (2)), applying fixed–points, the cokernel of α is trivial. Hence, α is an isomorphism.

The other equation is proved analogously, but dual. □

The proof of the following is obvious.

5.2 COROLLARY. *Let $W \to Gl(U)$ be a finite fixed–point free pseudo reflection group.*

    (1) *A lattice $P \subset U$ is centerfree if and only if $P^* \subset U^*$ is simply connected.*

    (2) *A lattice $S \subset U$ is simply connected if and only if $S^* \subset U^*$ is centerfree.*

PROOF OF THEOREM 1.6. Let $L \subset U$ be a W–lattice. The quotient $L/L^W =: \overline{L}$ is a fixed–point free W–lattice of U', where $U \cong U^W \oplus U'$ splits into the direct sum of the fixed–points U^W and a fixed–point free part U'. Let $S := S(\overline{L}) \subset U'$ be the associated simply connected lattice. Using pullbacks the W–trivial restriction $S \to \overline{L}$ establishes a commutative diagram of short exact sequences

$$
\begin{array}{ccccccccc}
0 & \longrightarrow & L^W \cong Z & \longrightarrow & L' & \longrightarrow & S & \longrightarrow & 0 \\
& & \| & & \downarrow & & \downarrow & & \\
0 & \longrightarrow & L^W \cong Z & \longrightarrow & L & \longrightarrow & \overline{L} & \longrightarrow & 0 .
\end{array}
$$

The top row describes an element of the group $Ext_{\mathbb{Z}_p^\wedge[W]}(S, Z)$ of extensions. We have the following sequence of isomorphisms:

$$
\begin{aligned}
Ext_{\mathbb{Z}_p^\wedge[W]}(S, Z) &\cong H^1(W; Hom_{\mathbb{Z}_p^\wedge}(S, Z)) \\
&\cong H^1(W, S^* \otimes Z) \\
&\cong H^1(W; S^*) \otimes Z \\
&= 0 .
\end{aligned}
$$

The first identity follows, because S and Z are free modules over $\mathbb{Z}_p^\wedge$ [3, III; 2.2], the second from the isomorphism between the coefficients, the third because W acts trivially on Z and because of Lemma 1.9, and the last because S^* is centerfree (Corollary 5.2 and Lemma 2.1). That is to say that $L' \cong Z \oplus S$. Moreover, we have an isomorphism $L/(Z \oplus S) \cong \overline{L}/S \cong \overline{L}_W$ which shows that $Z \oplus S \to L$ is a W-trivial restriction. This proves part (1).

For the second statement we dualize the above argument. There exists a W-trivial restriction $Z^* \oplus P^* \to L^*$ where $P^* \subset U^*$ is a simply connected lattice (Corollary 5.2 and part (1)). Dualizing again gives a short exact sequence

$$0 \to L \to Z \oplus P \to Ext(L^*/(Z^* \oplus P^*); \mathbb{Z}_p^\wedge) \to 0 ,$$

which shows that the first arrow is a W-trivial restriction.   □

REMARK.   Using Proposition 5.1 and Corollary 5.2 one can easily prove the second parts of the theorems 1.2, 1.4 and 1.5 as a consequence of the first parts. The idea is the same as in the proof of Theorem 1.6.

PROOF OF THEOREM 1.7.   Let $0 \to L \to P \oplus Z \to P \oplus Z/L =: Q \to 0$ be the W-trivial restriction of Theorem 1.6. Let $W \cong W_1 \times W_2$ and $P \cong P_1 \oplus P_2$ be a splitting into centerfree W_i-lattices. We also assume that P_1 is simply connected. The composition $P_1 \to P \oplus Z \to Q$ factors over $(P_1)_{W_1}$ and is therefore trivial. Hence, the inclusion $P_1 \to P \oplus Z$ lifts to L which shows that P_1 is a direct summand of L.   □

References

[1]   J. Aguadé, *Constructing modular classifying spaces*, Israel J. Math. **66** (1989), 23–40.

[2]   A.K. Bousfield and D.M. Kan, Homotopy limits, completions and localizations, Lecture Notes in Mathematics 304, Springer-Verlag, Berlin-Heidelberg-New York, 1972.

[3]   K.S. Brown, Cohomology of groups, Springer Verlag, Berlin–Heidelberg–New York, 1982.

[4]   A. Clark and J. Ewing, *The realization of polynomial algebras as cohomology rings*, Pacific J. Math. **50** (1974), 425–434.

[5]   W.G. Dwyer and C.W. Wilkerson, *Homotopy fixed–point methods for Lie groups and finite loop spaces*, Preprint.

[6]   W.G. Dwyer and C.W. Wilkerson, *Product splittings of p–compact groups*, Preprint.

[7]   J.E. Humphreys, Reflection groups and Coxeter groups, Cambridge University Press, 1990.

[8]   S. Jackowski, J.McClure, and R.Oliver, *Homotopy classification of self maps of BG via G actions II*, Ann. Math. **135** (1992), 227–270.

[9]   J. Møller and D. Notbohm, *Finite loop spaces with maximal tori*, Preprint.

[10] G.C. Shephard and J.A. Todd, *Finite unitary reflection groups*, Canadian J. Math. **6** (1954), 274–304.

Mathematisches Institut, Bunsenstr. 3-5, D–3400 Göttingen, Germany,
e-mail: notbohm@cfgauss.uni-math.gwdg.de.

Progress in Mathematics, Vol. 136
© 1996 Birkhäuser Verlag Basel/Switzerland

On the cohomology of configuration spaces

ERICH OSSA

1. Introduction

The aim of this note is to show how previous combinatorial calculations in the computation of the cohomology of configuration spaces can be considerably simplified by more conceptual arguments involving some representation theory. Since I first lectured on these results some other accounts have been given ([CT93, Str93]), partly overlapping with this. Nevertheless. it seemed still worthwhile to publish a full account of these considerations.

If Y is any space, we denote by

$$\tilde{C}_m(Y) = \{(p_1,\ldots,p_m) \in Y^m \mid p_i \neq p_j \text{ for } i \neq j\}$$

the space of m-tuples of pairwise different points in Y. The symmetric group S_m acts on $\tilde{C}_m(Y)$ in the obvious way; the quotient space

$$C_m(Y) = \tilde{C}_m(Y)/S_m$$

is the configuration space of m-element subsets of Y. The importance of these spaces lies in their relationship to the theory of iterated loop spaces ([BV68, May72, Seg73, CLM76]).

In the following we shall describe a calculation of the mod p cohomology of the configuration space $C_p(\mathbb{R}^n)$, based on the analysis of the S_p-action on the integral cohomology of $\tilde{C}_p(\mathbb{R}^n)$. This is one essential step in the general program of computing the cohomology of all $C_m(\mathbb{R}^n)$. This program was carried through by Fred Cohen in [CLM76]. In particular, all the cohomology results are entirely due to him (at least at odd primes); our only contribution consists in supplying more appropriate proofs.

2. The S_m action

We write H^* for cohomology with $\mathbb{Z}$ coefficients and recall first the computation of the cohomology of $\tilde{C}_m(\mathbb{R}^n)$.

Let $(q_i)_{i\in\mathbb{N}}$ be a fixed sequence of distinct points in $\mathbb{R}^n$ and put $Q_m = \{q_1,\ldots,q_m\}$. We use

$$Q_{m,l} = (q_{l+1},\ldots,q_{l+m}) \in \tilde{C}_m(\mathbb{R}^n - Q_l)$$

as the standard base point of the space $\tilde{C}_m(\mathbb{R}^n - Q_l)$.

For $k < m$ we have a projection $\pi : \tilde{C}_m(\mathbb{R}^n - Q_l) \to \tilde{C}_k(\mathbb{R}^n - Q_l)$ given by $\pi(p_1, \ldots, p_m) = (p_1, \ldots, p_k)$. It was shown by Fadell and Neuwirth [FN62] that π is actually a locally trivial fibre bundle. Obviously the fibre $\pi^{-1}Q_{k,l} \subset \tilde{C}_m(\mathbb{R}^n - Q_l)$ is $\cong \tilde{C}_{m-k}(\mathbb{R}^n - Q_{k+l})$. We note in passing that these fibrations have sections, defined by adding a fixed configuration of $m - k$ points at a spot far outside the varying configuration of k points.

Now, for $1 \leq i, j \leq m$, $i \neq j$, define $\pi_{ij} : \tilde{C}_m(\mathbb{R}^n) \to \tilde{C}_2(\mathbb{R}^n)$ by $\pi_{ij}(p_1, \ldots, p_m) = (p_i, p_j)$. There is an obvious S_2-equivariant homotopy equivalence $S^{n-1} \to \tilde{C}_2(\mathbb{R}^n)$. Denote by $A \in H^{n-1}(S^{n-1}; \mathbb{Z})$ the standard generator and let

$$A_{i,j} = \pi_{i,j}^*(A) \in H^{n-1}(\tilde{C}_m(\mathbb{R}^n); \mathbb{Z}) .$$

Clearly, one has $A_{j,i} = (-1)^n A_{i,j}$.

Note too, that under restriction to $\tilde{C}_{m-k}(\mathbb{R}^n - Q_k) \cong \pi^{-1}(Q_k) \subset \tilde{C}_m(\mathbb{R}^n)$ the classes $A_{i,j}$ with $1 \leq i, j \leq k$ map to zero (since then the map $\pi_{i,j}$ is constant on this space).

PROPOSITION 2.1: $H^*(\tilde{C}_{m-k}(\mathbb{R}^n - Q_k); \mathbb{Z})$ is a free abelian group with generators

$$A_{i_1 j_1} A_{i_2 j_2} \ldots A_{i_s j_s}$$

where

$$k < j_1 < j_2 < \ldots < j_s \leq m \quad \text{and} \quad i_\nu < j_\nu \quad \text{for} \quad \nu = 1, \ldots, s .$$

Proof: Let $\rho : \tilde{C}_{m-k}(\mathbb{R}^n - Q_k) \to \mathbb{R}^n - Q_k$ be defined by $\rho(p_1, \ldots, p_{m-k}) = p_1$. Obviously, $\mathbb{R}^n - Q_k$ is homotopy equivalent to a wedge of k spheres S_i^{n-1}, and ρ^* maps the generator of $H^{n-1}(S_i^{n-1})$ to $A_{i,k+1}$. The fibre of ρ is $\tilde{C}_{m-k-1}(\mathbb{R}^n - Q_{k+1})$; by induction its cohomology is as stated in the proposition. In particular, the inclusion of the fibre is an epimorphism in cohomology. It follows that the spectral sequence collapses (and also that in the case $n = 2$ the cohomology of the fibre is fixed under the fundamental group of the base).   □

We state again the most important case $k = 0$. For $n = 2$ this result and the proposition below were first obtained by Arnold [Arn69].

COROLLARY 2.2: $H^*(\tilde{C}_m(\mathbb{R}^n); \mathbb{Z})$ is the free abelian group with generators

$$A_{i_1 j_1} A_{i_2 j_2} \ldots A_{i_s j_s}$$

where

$$1 < j_1 < \ldots < j_s \leq m \quad \text{and} \quad 1 \leq i_\nu < j_\nu \quad \text{for} \quad \nu = 1, \ldots, m .$$

In particular, the Poincaré series of $H^*(\tilde{C}_m(\mathbb{R}^n); \mathbb{Z})$ is

$$\prod_{1 < j \leq m} (1 + (j - 1)t^{n-1})$$

and the total rank is $m!$.

The multiplicative structure of $H^*(\widetilde{C}_m(\mathbb{R}^n); \mathbb{Z})$ is completely determined by the commutativity relation

$$A_{kl}\, A_{ij} = (-1)^{n-1} A_{ij}\, A_{kl}$$

once one knows how to express $A_{ik}\, A_{jk}$ for $i < j < k$ in terms of the basis of corollary 2.2. Here we have

PROPOSITION 2.3: $H^*(\widetilde{C}_m(\mathbb{R}^n); \mathbb{Z})$ *is multiplicatively generated by the*

$$A_{i,j} \in H^{n-1}(\widetilde{C}_m(\mathbb{R}^n); \mathbb{Z}), \quad 1 \leq i < j \leq m \;,$$

subject to the only relations

$$A_{ik}\, A_{jk} = A_{ij}\, A_{jk} - A_{ij}\, A_{ik} \quad for \quad i < j < k \;.$$

Proof: It suffices to prove this relation for $(i, j, k) = (1, 2, 3)$ since then it can be pulled back by the projection $p_{i,j,k} : \widetilde{C}_m(\mathbb{R}^n) \to \widetilde{C}_3(\mathbb{R}^n)$ to give the general formula.

Now there must be a relation of the form

$$A_{13}\, A_{23} = \lambda A_{12}\, A_{23} + \mu A_{12}\, A_{13} \;,$$

and in $H^*(\widetilde{C}_3(\mathbb{R}^n))$ this must be compatible with the action of the symmetric group S_3. Applying the transpositions $(1, 2)$ and $(1, 3)$ one obtains $\lambda = -\mu = 1$ by an elementary calculation. $\qquad\square$

We now begin to study $H^*\widetilde{C}_m(\mathbb{R}^n)$ as a module over the symmetric group. We first calculate the character of this representation (which determines its rational type). Let sign_m denote the sign representation $\mathrm{sign}_m : S_m \to \mathrm{Aut}(\mathbb{Z})$.

PROPOSITION 2.4: *As a representation of S_m there is an isomorphism*

$$H^*(\widetilde{C}_m(\mathbb{R}^n); \mathbb{Q}) \cong \mathrm{Ind}_{S_2}^{S_m}(\mathbb{Q}) \oplus \mathrm{Ind}_{S_2}^{S_m}(\mathrm{sign}_2^{\otimes n} \otimes \mathbb{Q}) \;.$$

Proof: Let us first suppose that n is odd. Then $H^*(\widetilde{C}_m(\mathbb{R}^n))$ is entirely concentrated in even degrees. It is easily seen that $\widetilde{C}_m(\mathbb{R}^n)$ is S_m-equivariantly homotopy equivalent to a finite S_m-CW-complex (for instance by using only configurations $(p_1, \ldots, p_m)$ satisfying $\delta \leq \|p_i - p_j\| \leq \Delta$ and $\|p_i\| \leq \Delta$ for suitable δ and Δ, but see [Str93] for a canonical choice). Hence we may apply the Lefschetz fixed point theorem to $\sigma \in S_m$. Write

$$\tau_i(\sigma) = \mathrm{trace}\left(\sigma_* : H^{i(n-1)}(\widetilde{C}_m(\mathbb{R}^n)) \to H^{i(n-1)}(\widetilde{C}_m(\mathbb{R}^n))\right) \;.$$

Since S_m acts freely on $\widetilde{C}_m(\mathbb{R}^n)$, we have

$$\sum_{i=0}^{n-1} (-1)^{(n-1)i} \tau_i(\sigma) = \begin{cases} 0 & , \quad \sigma \neq 1 \\ m! & , \quad \sigma = 1 \;. \end{cases}$$

For odd n this proves that the character χ of the representation of S_m on $H^*(\widetilde{C}_m(\mathbb{R}^n); \mathbb{Q})$ is precisely the character of the regular representation of S_m.

If n is even, we use the map $c : \mathbb{R}^n \to \mathbb{R}^n$ given by $c(x_1, \ldots, x_n) = (x_1, \ldots, x_{n-1}, -x_n)$. This induces a map $c : \widetilde{C}_m(\mathbb{R}^n) \to \widetilde{C}_m(\mathbb{R}^n)$. Let σ be an element of S_m different from the identity. If $c \circ \sigma : \widetilde{C}_m(\mathbb{R}^n) \to \widetilde{C}_m(\mathbb{R}^n)$ has a fixed point $(p_1, \ldots, p_m)$, we must have $p_{\sigma(i)} = c(p_i)$. It follows that σ is of order 2. If σ is the product of k disjoint transpositions, the fixed point set of $\sigma \circ c$ is obviously homotopy equivalent to the disjoint union of 2^k copies of $\widetilde{C}_k(H^n) \times \widetilde{C}_{m-2k}(\mathbb{R}^{n-1})$ where $H^n = \{(x_1, \ldots, x_n) \in \mathbb{R}^n \mid x_n > 0\}$ is the upper half space. Since $\widetilde{C}_k(H^n) \cong \widetilde{C}_k(\mathbb{R}^n)$, and since n is even, the Euler characteristic of $\widetilde{C}_k(H^n)$ is zero for $k > 1$. Thus we obtain

$$\sum_i \tau_i(\sigma) = \sum_i (-1)^i \tau_i(c \circ \sigma) = \begin{cases} m! & \text{for } \sigma = 1, \\ 2(m-2)! & \text{if } \sigma \text{ is a transposition,} \\ 0 & \text{otherwise.} \end{cases}$$

This is precisely the character of the representation in the proposition. $\square$

As a corollary we obtain the fixed point set of S_m on $H^*(\widetilde{C}_m(\mathbb{R}^n)) \otimes \text{sign}^\varepsilon$.

COROLLARY 2.5: *Let μ_n^+ denote the multiplicity of the trivial representation of S_m and μ_n^- the multiplicity of the sign representation of S_m in $H^*(\widetilde{C}_m(\mathbb{R}^n); \mathbb{Q})$. Then*

$$\mu_n^+ = \begin{cases} 2 & \text{for even } n, \\ 1 & \text{for odd } n, \end{cases} \quad \text{and} \quad \mu_n^- = \begin{cases} 0 & \text{for even } n, \\ 1 & \text{for odd } n. \end{cases}$$

Proof: Let χ be the character of S_m acting on $H^*(\widetilde{C}_m(\mathbb{R}^n); \mathbb{Q})$. Then

$$m! \, \mu_n^+ = \sum_{\sigma \in S_n} \chi(\sigma) \quad \text{and} \quad m! \, \mu_n^- = \sum_{\sigma \in S_n} \text{sign}(\sigma) \cdot \chi(\sigma) \, .$$

The above formulae for χ yield the result. $\square$

We shall see later (in proposition 2.8) where these representations occur.

However, we need to study also the representation of S_m on the mod p cohomology of $\widetilde{C}_m(\mathbb{R}^n)$. For this we first derive more information about the action on integral cohomology.

Proposition 2.1 suggests that we order the set of A_{ij} (with $i < j$) by the lexicographic order from the right. Consider now a graph Γ with vertex set $\{1, \ldots, m\}$. Then we assign to Γ the element

$$A(\Gamma) = \prod_{(i,j) \in \Gamma} A_{i,j} \in H^*(\widetilde{C}_m(\mathbb{R}^n); \mathbb{Z})$$

where the product is over all $i < j$ such that (i, j) is an edge of Γ, and where the product is taken with respect to the above ordering of generators. For instance, if $i < j < k$, we may express the relation of proposition 2.3 in the form

$$A(\overset{\bullet}{\underset{\bullet}{\overset{\bullet}{|}}}) = A(\overset{\bullet}{\underset{\bullet}{\overset{\bullet}{|}}}) - A(\overset{\bullet}{\underset{\bullet}{\overset{\bullet}{|}}}) \, .$$

LEMMA 2.6: *Let Γ be a graph which contains a cycle. Then $A(\Gamma) = 0$.*

Proof: We proceed by induction on the length of the cycle. If Γ contains a cycle (i, j, k), then $A(\Gamma)$ contains $A_{ij} A_{ik} A_{jk} = A_{ij} A_{ij} (A_{jk} - A_{ik}) = 0$. Otherwise Γ contains edges (i, k) and (j, k) with $i < j < k$ such that there is a path in $\Gamma - \{k\}$ connecting i and j. Let l be the length of this path so that the length of the cycle is $z = l + 2$. Applying the relation of proposition 2.3 we obtain $A(\Gamma) = A(\Gamma_1) - A(\Gamma_2)$ where Γ_1 and Γ_2 both contain a cycle of length $l+1 < z$.
□

We see that $H^*(\widetilde{C}_m(\mathbb{R}^n); \mathbb{Z})$ is spanned by the monomials $A(\Gamma)$ where Γ is a forest, that is, a disjoint union of trees. For any forest Γ denote by $P(\Gamma) \subset \mathcal{P}\{1, \ldots, m\}$ the set of connected components of Γ. Let $S(\Gamma) \subset S_m$ be the subgroup of those $\sigma \in S_m$ which map $P(\Gamma)$ into itself:

$$\sigma \in S(\Gamma) \quad \text{iff} \quad \{\sigma(M) \mid M \in P(\Gamma)\} = P(\Gamma) .$$

Finally, let $\pi(\Gamma) = (m_1, \ldots, m_r)$ with $m_1 \geq \ldots \geq m_r$ be the family of cardinalities $\#M$ for $M \in P(\Gamma)$. Then $\pi(\Gamma)$ is a partition of m whose length is $l(\Gamma) = \#P(\Gamma)$. Denote by $V(\Gamma) \subset H^*(\widetilde{C}_m(\mathbb{R}^n); \mathbb{Z})$ the span of those monomials $A(\Gamma')$ such that $\pi(\Gamma') = \pi(\Gamma)$ and by $W(\Gamma)$ the span of those monomials $A(\Gamma'')$ such that $P(\Gamma'') = P(\Gamma)$.

Of great importance is the representation of S_m on the top dimensional cohomology $H^{(m-1)(n-1)}(\widetilde{C}_m(\mathbb{R}^n); \mathbb{Z})$; we shall call it simply the "top representation" and denote it by T_m, suppressing the dependence on n.

PROPOSITION 2.7:

1. *Let Γ be a forest on $\{1, \ldots, m\}$ with $l(\Gamma) = r$. Then*

$$V(\Gamma) \subset H^{(m-r)(n-1)}(\widetilde{C}_m(\mathbb{R}^n); \mathbb{Z})$$

   is a S_m-invariant submodule.

2. *For each partition ϖ of m let Γ_ϖ be a forest on $\{1, \ldots, m\}$ with $\pi(\Gamma_\varpi) = \varpi$. Then*

$$H^*(\widetilde{C}_m(\mathbb{R}^n); \mathbb{Z}) = \bigoplus_\varpi V(\Gamma_\varpi) .$$

3. *There is an isomorphism of representations of S_m*

$$V(\Gamma) = \text{Ind}_{S(\Gamma)}^{S_m} W(\Gamma) .$$

   Moreover, if $\pi(\Gamma) = (m_1, \ldots, m_r)$, the representation $W(\Gamma)$ of $S(\Gamma)$ is isomorphic to the standard representation of $S(\Gamma)$ on $T_{m_1} \otimes T_{m_2} \otimes \ldots T_{m_r}$.

Proof: It is sufficient to note that the components of Γ are preserved under the relations of proposition 2.3, that is, if

$$A(\Gamma) = \sum n_i A(\Gamma_i) ,$$

with $n_i \in \mathbb{Z}$ and where the Γ_i are basic forests, then

$$n_i \neq 0 \Rightarrow P(\Gamma_i) = P(\Gamma) \ .$$

The rest of the proposition is then clear.   □

We can now clarify the occurrence of the sign-representation and of the trivial representation in $H^*(\widetilde{C}_m(\mathbb{R}^n); \mathbb{Q})$. Of course $H^0(\widetilde{C}_m(\mathbb{R}^n); \mathbb{Q}) \cong \mathbb{Q}$ is the trivial representation.

PROPOSITION 2.8:

1. *Assume that n is even. Then the trivial representation of S_m is contained in $H^{n-1}(\widetilde{C}_m(\mathbb{R}^n); \mathbb{Q})$.*

2. *Assume that n is odd. Let l be the biggest integer such that $2l \leq m$. Then the sign representation of S_m is contained in $H^{l(n-1)}(\widetilde{C}_m(\mathbb{R}^n); \mathbb{Q})$.*

Proof: Assume first that n is even. $U := H^{n-1}(\widetilde{C}_m(\mathbb{R}^n); \mathbb{Q})$ has the $\mathbb{Q}$-basis $\{A_{i,j} \,|\, 1 \leq i < j \leq m\}$. Obviously, the element $\sum_{i<j} A_{i,j}$ is a nonzero fixed vector for the action of S_m. Alternatively, let ϖ be the partition $(2, 1, \ldots, 1)$ of m. Then $S(\Gamma_\varpi) \cong S_2 \times S_{m-2}$ acts trivially on the one dimensional space $W(\Gamma_\varpi)$, and $H^{n-1}(\widetilde{C}_m(\mathbb{R}^n); \mathbb{Q}) \cong \mathrm{Ind}_{S(\Gamma_\varpi)}^{S_m} \mathbb{Q}$.

Now assume that n is odd. Note that the cohomology ring $H^*(\widetilde{C}_m(\mathbb{R}^n))$ is concentrated in even dimensions and hence is strictly commutative. Let ϖ be the partition $(2, 2, \ldots, 2)$ (if m is even) respectively $(2, 2, \ldots, 2, 1)$ (if m is odd). $W(\Gamma_\varpi)$ has basis $A_{12}A_{34} \ldots A_{2l-1, 2l}$ (for the obvious choice of Γ_ϖ), and $S(\Gamma_\varpi)$ acts by the sign-representation. Thus, by Frobenius reciprocity, the sign-representation of S_m occurs in $\mathrm{Ind}_{S(\Gamma_\varpi)}^{S_m} W(\Gamma_\varpi) \subset H^{l(n-1)}(\widetilde{C}_m(\mathbb{R}^n))$.   □

The result above will be sufficient for the computation of $H^*(C_m(\mathbb{R}^n))$. Note also that for a complete understanding of $H^*(\widetilde{C}_m(\mathbb{R}^n))$ as S_m-module it is now sufficient to study the representation T_m. We offer the following observation which is not too difficult to prove:

PROPOSITION 2.9: *There is an isomorphism of representations of $S_{m-1} \subset S_m$*

$$\mathrm{Res}_{S_{m-1}}^{S_m} T_m \cong \mathbb{Z}[S_{m-1}] \ .$$

T_m is generated as module over $\mathbb{Z}[S_{m-1}]$ by $w_m = A_{12} A_{23} \ldots A_{m-1, m}$.

Proof: Let L_m be the span of all A_{im} for $i < m$. Then L_m is invariant under S_{m-1}, and as a representation of S_{m-1} it is isomorphic to

$$\mathrm{Ind}_{S_{m-2}}^{S_{m-1}}(\mathrm{sign}_{m-2}^{\otimes n})$$

where sign_{m-2} is the sign representation of S_{m-2}. Obviously, $\mathrm{Res}_{S_{m-1}}^{S_m} T_m \cong T_{m-1} \otimes L_m$. Thus, by Frobenius reciprocity,

$$
\begin{aligned}
\mathrm{Res}_{S_{m-1}}^{S_m} T_m &\cong T_{m-1} \otimes L_m \cong T_{m-1} \otimes \mathrm{Ind}_{S_{m-2}}^{S_{m-1}}(\mathrm{sign}_{m-2}^{\otimes n}) \\
&\cong \mathrm{Ind}_{S_{m-2}}^{S_{m-1}}(\mathrm{Res}_{S_{m_2}}^{S_{m-1}} T_{m-1}) \otimes \mathrm{sign}_{m-1}^{\otimes n}
\end{aligned}
$$

By induction

$$\operatorname{Ind}_{S_{m-2}}^{S_{m-1}} \operatorname{Res}_{S_{m-2}}^{S_{m-1}} T_{m-1} \cong \operatorname{Ind}_{S_{m-2}}^{S_{m-1}} \mathbb{Z}[S_{m-2}]$$
$$\cong \mathbb{Z}[S_{m-1}]$$

Since $\mathbb{Z}[S_{m-1}] \otimes \operatorname{sign}_{m-1} \cong \mathbb{Z}[S_{m-1}]$, the claim follows. $\quad\square$

The representation T_m occurs in various contexts. It is actually isomorphic to the tensor product of the Lie representation of S_m with $\operatorname{sign}_m^{\otimes(n-1)}$. This was certainly known to F. Cohen (see [CLM76, Thm.12.3, p.302] and [CT93, p.95] or [CT78, p.115]). A detailed proof has also been given by John Rognes in [Rog93, §13].

3. The cohomology of $C_p(\mathbb{R}^n)$

We turn our attention now to the mod p cohomology of $C_p(\mathbb{R}^n)$. In fact we compute $H^*(C_p(\mathbb{R}^n); \mathbb{F}_p \otimes \operatorname{sign}_p^\varepsilon)$ for $\varepsilon = 0$ and 1 (which is needed for further applications). From the p-fold covering $\widetilde{C}_p(\mathbb{R}^n) \to C_p(\mathbb{R}^n)$ we obtain a spectral sequence

$$E_2^{s,t}(\varepsilon) = H^s(S_p; H^t(\widetilde{C}_p(\mathbb{R}^n); \mathbb{F}_p) \otimes \operatorname{sign}_p^\varepsilon) \Rightarrow H^{s+t}(C_p(\mathbb{R}^n); \mathbb{F}_p \otimes \operatorname{sign}_p^\varepsilon) .$$

Recall that $H^*(S_p; \mathbb{F}_p) \cong \mathbb{F}_p[\alpha] \otimes \Lambda_p(\beta)$, with the polynomial generator $\alpha \in H^{2p-2}$ and the exterior generator $\beta \in H^{2p-3}$. Moreover, $H^*(S_p; \mathbb{F}_p \otimes \operatorname{sign}_p)$ as module over $H^*(S_p; \mathbb{F}_p)$ is generated by $\alpha' \in H^{p-1}(S_p; \mathbb{F}_p \otimes \operatorname{sign}_p)$ and $\beta' \in H^{p-2}(S_p; \mathbb{F}_p \otimes \operatorname{sign}_p)$ subject to the only relations $\beta\beta' = 0$ and $\beta\alpha' = \alpha\beta'$.

We observe first that proposition 2.7 implies the following:

PROPOSITION 3.1: *For $s > 0$ and $t \notin \{0, (p-1)(n-1)\}$ the group $E_2^{s,t}(\varepsilon)$ vanishes.*

Proof: In these cases the representation of S_p on $H^t(\widetilde{C}_p(\mathbb{R}^n); \mathbb{F}_p) \otimes \operatorname{sign}_p^\varepsilon$ is induced from a subgroup of S_p whose order is prime to p. By Shapiro's Lemma the cohomology has to vanish in positive dimensions. $\quad\square$

Thus, except for differentials starting on the vertical axis $s = 0$, the only possible differential in the spectral sequence is $d_{(p-1)(n-1)+1}$. The following was observed by F. Cohen [CLM76]:

PROPOSITION 3.2: *For $s > 0$ the differential*

$$d_{(p-1)(n-1)+1} : E_2^{s,(p-1)(n-1)}(\varepsilon) \to E_2^{s+(p-1)(n-1)+1,0}(\varepsilon)$$

is an isomorphism.

Proof: The cohomology of S_p in positive dimensions is periodic with period $2p-2$. The multiplicativity of the spectral sequence implies that the differentials commute with this periodicity. But since $C_p(\mathbb{R}^n)$ is finite dimensional, $E_\infty^{s,t}$

vanishes in big enough dimensions. Since $E_2^{0,t}$ vanishes for $t > (p-1)(n-1)$, differentials starting on the vertical axis $s = 0$ cannot interfere with differentials starting in the range $s > 0$.   □

For $p > 3$ it follows in particular that $E_2^{1,t}(\varepsilon)$ vanishes for all t. Thus it follows from the Bockstein exact sequence (alias the universal coefficient theorem) that $E_2^{0,t}(\varepsilon)$, the fixed point set of S_p on $H^t(\widetilde{C}_p(\mathbb{R}^n); \mathbb{F}_p) \otimes \mathrm{sign}_p^\varepsilon$, is a quotient of the corresponding fixed point set on integral cohomology. By corollary 2.5 and proposition 2.8 we can conclude

PROPOSITION 3.3: *Assume $p > 3$.*

1. *Let n be even. Then $E_2^{0,t}(\varepsilon)$ vanishes for $t \notin \{0, n-1\}$. One has $E_2^{0,0}(0) = E_2^{0,n-1}(0) = \mathbb{F}_p$ and $E_2^{0,0}(1) = E_2^{0,n-1}(1) = 0$.*

2. *Let n be odd and let $(p-1)(n-1) = 2k$. Then $E_2^{0,t}(\varepsilon)$ vanishes for $t \notin \{0, k\}$. One has $E_2^{0,0}(0) = \mathbb{F}_p$ and $E_2^{0,k}(1) = \mathbb{F}_p$, but $E_2^{0,0}(1) = E_2^{0,k}(0) = 0$.*

We have now a clear picture of the spectral sequence for $p > 3$ and can compute the $\mathbb{F}_p$-cohomology of $C_p(\mathbb{R}^n)$:

PROPOSITION 3.4: *Assume that p is > 3 and let $(p-1)(n-1) = 2k$. Then $H^t(C_p(\mathbb{R}^n); \mathbb{F}_p \otimes \mathrm{sign}_p^\varepsilon)$ vanishes in dimensions $t > 2k$. In dimensions $t \leq 2k$ the map*

$$H^t(BS_p; \mathbb{F}_p \otimes \mathrm{sign}_p^\varepsilon) \to H^t(C_p(\mathbb{R}^n); \mathbb{F}_p \otimes \mathrm{sign}_p^\varepsilon)$$

is injective. It is an isomorphism in this range if $\varepsilon \not\equiv n \pmod 2$; otherwise its cokernel is isomorphic to $\mathbb{F}_p$, concentrated in dimension $n-1$ for $n \equiv \varepsilon \equiv 0 \pmod 2$ respectively in dimension k for $n \equiv \varepsilon \equiv 1 \pmod 2$.

The same statement is true for $p = 3$. The argument, however, is slightly different: the groups $H^0(S_p; H^{2n-2}(\widetilde{C}_p(\mathbb{R}^n); \mathbb{F}_p) \otimes \mathrm{sign}_p^\varepsilon)$ are isomorphic to $\mathbb{F}_p$. Since multiplication by $\alpha \in H^{2p-2}(S_p; \mathbb{F}_p)$ is non trivial on these groups, the same must hold for the differential d_{2n-1}. We leave the (easy) details to the interested reader.

References

[Arn69]   V. Arnold: *The Cohomology Ring of the Colored Braid Groups*, Edinburgh Math. Notes 5 (1969), 138–140

[BV68]   J. M. Boardman, R. Vogt: *Homotopy-everything H-spaces*, Bull. AMS 74 (1968), 1117–1122

[CLM76]   F. R. Cohen, T. Lada, J. P. May: *The homology of iterated loop spaces*, Lecture Notes in Math. 533 (1976)

[Coh85]   F. R. Cohen: *Artin's braid groups and classical homotopy theory*, Contemp. Math. 44 (1985), 207–220

[Coh88]   F. R. Cohen: *Artin's braid groups, classical homotopy theory, and sundry other curiosities*, Contemp. Math. 78 (1988), 167–206

[CMT78]   F. R. Cohen, J. P. May, L. R. Taylor: *Splitting of certain spaces CX*, Proc. Camb. Phil. Soc. 84 (1978), 465–496

[CT78]   F. R. Cohen, L. Taylor: *Computations of Gelfand-Fuks cohomology, the cohomology of function spaces and the cohomology of configuration spaces*, Lecture Notes in Math. 657 (1978), 106–143

[CT93]   F. R. Cohen, L. Taylor: *On the representation theory associated to the cohomology of configuration spaces*, Contemp. Math. 146 (1993), 167–206

[FN62]   E. Fadell, L. Neuwirth: *Configuration spaces*, Math. Scand. 10 (1962), 119–126

[Fuk70]   D. B. Fuks: *Cohomologies of the group cos* mod 2, Functional Analysis Appl. 4 (1970), 143–151

[Leh86]   G. I. Lehrer: *On the action of the symmetric group on the cohomology of the complement of its reflecting hyperplanes*, J. Algebra 104 (1986), 410–424

[Leh87]   G. I. Lehrer: *On the Poincaré series associated with Coxeter group actions on complements of hyperplanes*, J. Lond. Math. Soc. 36 (1987), 275–294

[Leh88]   G. I. Lehrer: *A survey of Hecke algebras and the Artin braid groups*, Contemp. Math. 78 (1988), 365–383

[May72]   J. P. May: *The geometry of iterated loop spaces*, Springer Lecture Notes 271 (1972)

[Rog93]   J. Rognes: *The rank filtration in algebraic K-theory*, Topology 31 (1993), 813–845

[Seg73]   G. Segal: *Configuration spaces and iterated loop spaces*, Invent. Math. 21 (1973), 213–221

[Str93]   N. P. Strickland: *Geometry and topology of configuration spaces*, Preprint MIT 1993

Erich Ossa, Bergische Universität, Gesamthochschule Wuppertal
Fachbereich 7 - Mathematik, Gaußstrasse 20, 42097 Wuppertal, Germany

Progress in Mathematics, Vol. 136
© 1996 Birkhäuser Verlag Basel/Switzerland

On isomorphism classes of locally unitary groups

AKIMOU OSSE

Herrn F. EHRLER gewidmet, in Erinnerung eines Briefes des Sommers 1983.

1. Introduction

Since the solution of the celebrated Hilbert's Fifth Problem, one knows that a locally euclidean topological group is a Lie group. In accordance with Hilbert's original idea of getting rid of the differentiability hypothesis in Lie theory, one can look for a homotopy analogue of this result. Unfortunately the most straightforward idea is false: Hilton, Roitberg and Stasheff have exhibited a topological group having the homotopy type of a compact manifold and which is not homotopy equivalent to any Lie group (see [Sta69]). Nevertheless, the problem of finding a homotopy characterisation of Lie groups remains open and has been intensively investigated for the last three decades. Obviously, the first step of such a programme is a better understanding of the homotopy properties of Lie groups. One has to restrict this study to the class of compact connected Lie groups: the Iwasawa decomposition tells us that any connected Lie group has the homotopy type of a compact one.

In the sixties, P. Baum and W. Browder have shown that two compact, connected and simple Lie groups are homotopy equivalent if and only if they are isomorphic as Lie groups. In the same direction, H. Scheerer has discovered the following property:

> THEOREM([Sch68]) If two compact connected Lie groups have the same homotopy type, then they have the same Lie algebra.

This last result justifies the topologists' interest in the following problem:

> Given a compact connected Lie group G, what are the properties of the set $\Lambda(G)$ of isomorphism classes of compact connected Lie groups having the same Lie algebra as G?

The first systematic study of these sets has been made by P. Baum (see [Bau67]); he proved there that $\Lambda(G)$ is always finite and explored its properties for some unitary groups. More recently, D. Notbohm and L. Smith have been led to the set $\Lambda(G)$ by their study of the topological genus of the classifying space of the group G. In [NS91, §4] and for a fixed positive integer n, they gave some nice characterisations of the compact connected Lie groups having the same Lie algebra as the unitary group $\mathbf{U}(n)$; these groups are the *locally unitary groups* of the title. In fact, the results of [NS91, §4] answer, in a particular case, a question of G. Mislin (see [Smi91]):

If two compact connected Lie groups have classifying spaces of the
same homotopy type, can one conclude that they are isomorphic as
Lie groups?

General proofs of an affirmative solution to that question are now available
([Not92, Oss92, Oss94]). In other words, the classifying space of a compact
connected Lie group contains all the information needed to reconstruct the
group. This result confirms that the best strategy for the programme above
should consist in finding a characterisation of classifying spaces of compact
Lie groups. We refer to [Rec71, AW80, NS91, DMW92] for some of the most
exciting results of this approach.

The present paper is devoted to a detailed study of the two problems above
in the particular case of the locally unitary groups. Even if their answers are
known, we believe that a further analyis of these groups is worth while. Indeed
these compact connected Lie groups have been an important test case in the
solution of Mislin's problem, and such a study is expected to shed some light
on the general pattern. We will use the good knowledge of the unitary groups
-they are probably the most studied Lie groups- to give as much information
as possible on all the locally unitary groups.

Our main result is a new proof of the classification theorem of the locally
unitary groups given in the last section of [NS91]. While Notbohm-Smith's
proof uses a result of W. Plesken on integral representations of the symmetric
groups, our approach is entirely based on the theory of λ-rings. This is not a
surprise if one observes that the most striking part of their theorem says that
the locally unitary groups are distinguished by the K-theory of their classifying
spaces. The interested reader may refer to [Oss94] for a possible extension of
Notbohm-Smith's result to all compact connected Lie groups.

The paper is organised as follows. After some recollections in section 2,
the following one is devoted to the analysis of the finite covers of the unitary
groups. For this particular class of locally unitary groups, we exhibit particularly simple realisations; these are used in section 4 for some representation
rings computations. The treatement of the general case relies on some results
of P. Baum, which are recalled in section 5. We compute the representation
rings of the locally unitary groups in section 6. This last section also contains
the announced proof of Notbohm-Smith's theorem.

Notations and conventions

- If G is a compact Lie group, BG denotes a classifying space of G and
  $\pi_1(G)$ is its fundamental group. The representable K-theory ring of BG,
  defined by $K^0(BG) := [BG, \mathbb{Z} \times BU]$, will simply be denoted $K(BG)$.

- The symbol $\cong$ stands for isomorphism of Lie groups, while $\simeq$ means
  homotopy equivalence.

- $\mathbb{N}, \mathbb{Z}, \mathbb{C}, \mathbb{S}^1$ denotes respectively the sets of the non-negative integers, the
  integers, the complex numbers and the complex numbers of norm 1.

- $Diag(X_1, \cdots, X_k)$ denotes the matrix with the blocks $X_1, \cdots, X_k$ on the diagonal and zero elsewhere. In particular, $\mathbb{1} = Diag(1, 1, \ldots, 1)$ is just the identity matrix.

- n will always denote an integer greater or equal to 2.

2. On the theory of λ-rings

The best introduction to the subject is the now classical paper of Atiyah-Tall ([AT69]). For the purpose of this paper, we briefly recall that a λ-*ring* R is a commutative ring with unit, together with a sequence $\{\lambda^k : R \longrightarrow R; \quad k \in \mathbb{N}\}$ of maps satisfying a list of axioms which are abstracted from the properties of exterior powers of vector spaces. Following the usual notation, we will write $\lambda_t(x) = \sum_{k \geq 0} \lambda^k(x) \cdot t^k$, for all $x \in R$; then for example, one of the axioms translates into

$$\lambda_t(x + y) = \lambda_t(x) \cdot \lambda_t(y) \qquad \forall x, y \in R \, .$$

Let x be an element of a λ-ring R. If the formal series $\lambda_t(x)$ is a polynomial of degree k in the variable t, we say that x is of λ-*dimension* k and we note $\lambda\text{-dim}(x) = k$. For later use in section 6., we recall here a consequence of the axioms mentionned above: if $x \in R$ is of λ-dimension 1, then

$$\forall y \in R, \forall k \geq 0 \quad \lambda^k(x \cdot y) = x^k \cdot \lambda^k(y) \, .$$

Let R and S be λ-rings. A ring homomorphism $f : R \longrightarrow S$ is a λ-*morphism* if $f(\lambda^k(x)) = \lambda^k(f(x))$ for all $x \in R$ and for all positive integers k. A λ-*isomorphism* is a λ-morphism which is also a ring isomorphism. We shall write $R \overset{\lambda}{\cong} S$ to express that the rings R and S are λ-isomorphic.

If R is a λ-ring, the *Adams operations* are maps $\Psi^k : R \longrightarrow R$ defined by the formal power series:

$$\Psi_{-t}(x) = \sum_{k \geq 0} \Psi^k(x) \cdot (-t)^k = -t \cdot (\frac{d}{dt} \lambda_t(x))/\lambda_t(x) \qquad \forall x \in R \, .$$

The Ψ^k are λ-morphisms and satisfy the identity $\Psi^k \circ \Psi^l = \Psi^{kl}$.

Let us now introduce the λ-rings which intervene in the present paper. For a fixed compact Lie group G, consider the free abelian group generated by the isomorphism classes of complex irreducible representations of G. With the multiplication induced by the tensor product of representations, $R(G)$ becomes a commutative ring with unit: it is called the *(complex) representation ring* of G. There's a λ-structure on $R(G)$, given by the map induced by the exterior powers of representations of the group G. It is not difficult to show that the correspondence $G \to R(G)$ is functorial.

EXAMPLE. If G is the r-dimensional torus $T^r = \mathbb{S}^1 \times \cdots \times \mathbb{S}^1$, then $R(G)$ is the Laurent polynomial ring $\mathbb{Z}[\alpha_1^{\pm 1}, \ldots, \alpha_r^{\pm 1}]$, with

$$\alpha_i^{\pm 1} : T^r \longrightarrow \mathbb{S}^1, \ (z_1, \ldots, z_r) \longmapsto z_i^{\pm 1} \qquad \text{for} \quad i = 1, \ldots, r \, .$$

The λ-series and the Adams operations are given by

$$\lambda_t(\alpha_i^{\pm 1}) = 1 + \alpha_i^{\pm 1} \cdot t, \qquad \Psi^k(\alpha_i^{\pm 1}) = \alpha_i^{\pm k} \qquad i = 1, 2, \ldots, r, \quad k = 1, 2, \ldots$$

The representation ring of a compact connected Lie group is always an integral domain; its behaviour with respect to finite covers is described by

PROPOSITION 2.1 *Let* $p : G \longrightarrow H$ *be a finite cover of compact connected Lie groups (i.e a surjective morphism with a finite kernel). The induced morphism* $p^* : R(H) \longrightarrow R(G)$ *is injective, so that* $R(H)$ *can be identified with a* λ-*subring of* $R(G)$. *With this identification, the fraction field* $\mathrm{Frac}(G)$ *of* $R(G)$ *is of degree* $|Ker(p)|$ *over the field* $\mathrm{Frac}(H)$.

Proof. The first part is straightforward and the second follows from proposition 8.5 (page 235) of [BD85]. $\qquad\qquad\qquad\qquad\qquad\qquad\qquad\qquad\qquad\qquad\qquad\quad$ $\square$

Our main computational tool will be the following result of H. Weyl:

THEOREM 1 *([Ada69, Theorem 6.20])* G *is a compact connected Lie group,* T *a maximal torus of* G *and* $W = N(T)/T$ *the Weyl group with respect to* T. *The finite group* W *acts (by* λ-*automorphisms) on* $R(T)$ *and the ring of the invariants* $R(T)^W$ *is* λ-*isomorphic to* $R(G)$.

By combining this theorem and Proposition 7.4 (page 105) of [BD85], one proves

PROPOSITION 2.2 *Let* G *be a compact connected Lie group of rank* r. *For every integer* $k > 0$, *the Adams operation* $\Psi^k : R(G) \longrightarrow R(G)$ *is injective. Moreover, if* $\mathrm{Frac}(G)$ *denotes the fraction field of* $R(G)$ *and* $\mathrm{Frac}_k(G)$ *the fraction field of* $\Psi^k(R(G))$, *the degree of the field extension* $\mathrm{Frac}_k(G) \subset \mathrm{Frac}(G)$ *is* k^r.

3. The unitary group U(n) and its finite covers

Recall that the n-th unitary group is defined by

$$\mathbf{U}(n) = \{A \in \mathbf{GL}_n(\mathbb{C}); \quad A \cdot A^* = \mathbb{1}\},$$

where $A^* = \bar{A}^t$ denotes the conjugate transpose of the matrix A. The *special unitary group* $\mathbf{SU}(n)$ is the closed subgroup of $\mathbf{U}(n)$ formed by the matrices of determinant 1. The standard maximal torus of $\mathbf{U}(n)$ is the group T of diagonal matrices in $\mathbf{U}(n)$. And the Weyl group $W(\mathbf{U}(n), T)$ is the symmetric group Σ_n, acting on T by permutation of the diagonal coefficients. From the canonical identification of T with the product $\mathbb{S}^1 \times \ldots \times \mathbb{S}^1$, it is easily shown that $R(T) = \mathbb{Z}[\alpha_1^{\pm 1}, \ldots, \alpha_n^{\pm 1}]$, where the Weyl group Σ_n acts by permutation of the indices of the α_j. The fundamental theorem of the theory of symmetric functions and theorem 1 yield the following description of the representation ring of $\mathbf{U}(n)$:

$$R(\mathbf{U}(n)) = \mathbb{Z}[\lambda_1, \ldots, \lambda_{n-1}, \lambda_n, \lambda_n^{-1}],$$

where $\lambda_1, \ldots, \lambda_n$ are the elementary symmetric functions in the variables α_1, ..., α_n; its λ-structure is completely determined by the formula

$$\lambda_t(\lambda_1) = 1 + \sum_{j=1}^{n} \lambda_j \cdot t^j .$$

Let m be a positive integer. As $m\mathbb{Z}$ is the only subgroup of index m in $\mathbb{Z} = \pi_1(\mathbf{U}(n))$, there is exactly one compact connected Lie group $\mathbf{U}_m(n)$ which is a m-fold cover of $\mathbf{U}(n)$. The groups $\mathbf{U}_m(n)$ are different as groups over $\mathbf{U}(n)$; but they may be isomorphic as (abstract) Lie groups. In this section, we will present a more concrete construction of the $\mathbf{U}_m(n)$ and use it to compute their representation rings. These informations will be crucial in section 4 for the determination of the (Lie) isomorphism classes of these groups.

PROPOSITION 3.1 *Set $d = gcd(m, n)$ and consider the morphism of Lie groups*

$$\varphi_m : \mathbb{S}^1 \times \mathbf{SU}(n) \longrightarrow \mathbf{U}(n+1), \quad (z; A) \mapsto Diag(z^{\frac{n}{d}}; \ z^{\frac{m}{d}} \cdot A) .$$

The image of φ_m is isomorphic to the Lie group $\mathbf{U}_m(n)$; in particular, one has $\mathbf{U}_m(n) \cong (\mathbb{S}^1 \times \mathbf{SU}(n))/E_m$ with

$$E_m = \{(\exp(\frac{2\pi\mathbf{i}}{n} \, d\,j); \ \exp(-\frac{2\pi\mathbf{i}}{n} \, m\,j) \cdot \mathbb{1}); \quad j = 0, 1, \ldots, \frac{n}{d} - 1\} .$$

The covering map $p_m : \mathbf{U}_m(n) = \varphi_m(\mathbb{S}^1 \times \mathbf{SU}(n)) \longrightarrow \mathbf{U}(n)$ is just the restriction of the projection map $pr_2 : \mathbb{S}^1 \times \mathbf{SU}(n) \longrightarrow \mathbf{U}(n), \quad (z, B) \longmapsto B$.

Proof. Because of the uniqueness of $\mathbf{U}_m(n)$ (as a group over $\mathbf{U}(n)$), the proposition follows from the fact that $Ker(p_m) = \{(\zeta; \mathbb{1}); \ \zeta^m = 1\} \cong \mathbb{Z}/m$. $\qquad\square$

REMARK. Our description of $\mathbf{U}_m(n)$ shows that its centre is the subgroup

$$Z_m(n) = \{Diag(z^{\frac{n}{d}}, z^{\frac{m}{d}} \cdot \theta, \ldots, z^{\frac{m}{d}} \cdot \theta); \quad \theta, z \in \mathbb{S}^1 \quad \text{with} \quad \theta^n = 1\}$$

of the unitary group $\mathbf{U}(n+1)$. The Lie group $Z_m(n)$ is abelian and compact, but not necessarily connected. In fact, one has the decomposition $Z_m(n) = Z_m(n)^0 \times F_m(n)$ where

$$Z_m(n)^0 = \{Diag(z^{\frac{n}{d}}, z^{\frac{m}{d}}, \ldots, z^{\frac{m}{d}}); \quad z \in \mathbb{S}^1\} \subset Z_m(n) ,$$

$$F_m(n) = \{Diag(\theta^{-u \cdot \frac{m}{d}}, \theta^{v \cdot \frac{n}{d}}, \ldots, \theta^{v \cdot \frac{n}{d}}); \quad \theta^n = 1\} \subset Z_m(n) .$$

and the integers u, v satisfy the relation $u \cdot \frac{m}{d} + v \cdot \frac{n}{d} = 1$. As the connected Lie group $Z_m(n)^0$ is isomorphic to the circle group and $F_m(n)$ to $\mathbb{Z}/d$, the centre $Z_m(n)$ is connected if and only if the integers m and n are coprime.

Recall that $T = \{Diag(z_1, z_2, \ldots, z_n); \ z_j \in \mathbb{S}^1\}$ is the standard maximal torus of $\mathbf{U}(n)$ and let $\bar{T} = \{A \in T; \ det(A) = 1\}$ be the corresponding one for $\mathbf{SU}(n)$. Then the group T_m, defined by

$$T_m = p_m(\mathbb{S}^1 \times \bar{T}) = \{Diag(z_0, z_1, \ldots, z_n) \in \mathbf{U}(n+1); \ \prod_{j=1}^{n} z_j = z_0^m\} ,$$

is a maximal torus of the compact connected Lie group $\mathbf{U}_m(n)$. One sees easily that the restriction of the covering map p_m to this torus is given by

$$p_{m|T_m} : T_m \longrightarrow T, \quad Diag(z_0^{\frac{n}{d}}, z_0^{\frac{m}{d}} \cdot z_1, \ldots, z_0^{\frac{m}{d}} \cdot z_n) \longmapsto Diag(z_0^{\frac{m}{d}} \cdot z_1, \ldots, z_0^{\frac{m}{d}} \cdot z_n) .$$

By applying the representation ring functor to this restriction, we obtain an injective morphism of rings

$$(p_{m|T_m})^* : R(T) = \mathbb{Z}[\alpha_1^{\pm 1}, \ldots, \alpha_n^{\pm 1}] \hookrightarrow R(T_m) = \mathbb{Z}[\gamma_0^{\pm 1}, \ldots, \gamma_n^{\pm 1}]/(\gamma_0^m - \prod_{j=1}^{n} \gamma_j) ,$$

which sends α_j on γ_j for $j = 1, \ldots, n$.
We use the injection $(p_{m|T_m})^*$ to identify $R(T)$ with the subring $\mathbb{Z}[\gamma_1^{\pm 1}, \ldots, \gamma_n^{\pm 1}]$ of $R(T_m)$, so that

$$R(T_m) \cong \mathbb{Z}[\alpha_1^{\pm 1}, \ldots, \alpha_n^{\pm 1}, \tau_m]/(\tau_m^m - \alpha_1 \cdot \alpha_2 \cdots \alpha_n) .$$

By naturality, the Weyl group $W(\mathbf{U}_m(n), T_m) = \Sigma_n$ acts on $R(T_m)$ by permuting the indices of the α_j. It is then easy to see that the ring of invariants of this action, which is the representation ring of $\mathbf{U}_m(n)$ by theorem 1, is

$$R(T_m)^{\Sigma_n} \cong \mathbb{Z}[\lambda_1, \ldots, \lambda_{n-1}, \lambda_n^{1/m}, \lambda_n^{-1/m}] ,$$

where $\lambda_1, \ldots, \lambda_n$ are the elementary symmetric functions in the variables α_1, $\ldots, \alpha_n$ and $\lambda_n^{1/m}$ is the element τ_m of $R(T_m)$. If one observes that this description also gives λ-structure of $R(\mathbf{U}_m(n))$, then we have completed the proof of

PROPOSITION 3.2 *Given a positive integer m, the representation ring of the m-fold cover $\mathbf{U}_m(n)$ of the unitary group $\mathbf{U}(n)$ is the subring*

$$\mathbf{R}_m = \mathbb{Z}[\lambda_1, \lambda_2, \ldots, \lambda_{n-1}, \tau_m^{\pm 1}]$$

of the Laurent polynomial ring $R(T_m)$. Its λ-structure is completely determined by the two formulae:

$$\lambda_t(\lambda_1) = 1 + \sum_{j=1}^{n-1} \lambda_j \cdot t^j + \tau_m^m \cdot t^n, \qquad \lambda_t(\tau_m^{\pm 1}) = 1 + \tau_m^{\pm 1} \cdot t .$$

Before ending this section, we must make some comments on the groups $\mathbf{U}_m(n)$ and our description of their representations rings.

1. Taking $m = n$ in the result above, one gets

$$\mathbf{R}_n = R(\mathbb{S}^1 \times \mathbf{SU}(n)) = \mathbb{Z}[\lambda_1, \lambda_2, \ldots, \lambda_{n-1}, \tau_n^{\pm 1}] \quad with \; \tau_n^n = \lambda^n(\lambda_1) .$$

Observe that the usual presentation of this ring is

$$R(\mathbb{S}^1 \times \mathbf{SU}(n)) = R(\mathbb{S}^1) \otimes R(\mathbf{SU}(n)) = \mathbb{Z}[t, t^{-1}] \otimes \mathbb{Z}[\Lambda_1, \Lambda_2, \ldots, \Lambda_{n-1}] ,$$

where Λ_1 is the standard representation of $\mathbf{SU}(n)$ and $\Lambda_j = \lambda^j(\Lambda_1)$. To pass from this presentation to the first one, it suffices to map t on τ_n and Λ_j on $\tau_n^{-j} \cdot \lambda_j$ *for* $j = 1, \ldots, n - 1$.

2. Setting $\bar{m} = \frac{m\,n}{d}$, one can use the homomorphism induced by the covering map $\varphi_m : \mathbb{S}^1 \times \mathbf{SU}(n) \longrightarrow \mathbf{U}_m(n)$ to identify $\mathbf{R}_m = R(\mathbf{U}_m(n))$ with the subring of $\mathbf{R}_{\bar{m}}$ generated by the elements $\lambda_1, \lambda_2, \ldots, \lambda_{n-1}, \tau_{\bar{m}}^{\pm\frac{n}{d}}$. Since the groups $\mathbf{U}_n(n)$ and $\mathbf{U}_{\frac{m\,n}{d}}(n)$ are isomorphic to $\mathbb{S}^1 \times \mathbf{SU}(n)$, one can conclude that their representation rings are λ-isomorphic.

3. The morphism φ_m provides an inclusion of $\mathbf{U}_m(n)$ in $\mathbb{S}^1 \times \mathbf{U}(n)$ and gives rise to the exact sequence of Lie groups

$$1 \longrightarrow \mathbf{U}_m(n) \xrightarrow{\;i_m\;} \mathbb{S}^1 \times \mathbf{U}(n) \xrightarrow{\;j_m\;} \mathbb{S}^1 \longrightarrow 1 \, ,$$

where $j_m : \mathbb{S}^1 \times \mathbf{U}(n) \longrightarrow \mathbb{S}^1$; $(z; M) \longmapsto z^{-m} \cdot det(M)$. From the Gysin sequence of the associated fibration

$$\mathbb{S}^1 \longrightarrow B\mathbf{U}_m(n) \longrightarrow B\mathbb{S}^1 \times B\mathbf{U}(n) \, ,$$

one computes that the integral cohomology ring of the classifying space $B\mathbf{U}_m(n)$ is given by

$$H^*(B\mathbf{U}_m(n); \mathbb{Z}) = \mathbb{Z}[\bar{x}; \bar{c}_1, \ldots, \bar{c}_n]/(\bar{c}_1 - m\,\bar{x}) \, ;$$

$\bar{c}_1, \ldots, \bar{c}_n$ are the images of the universal Chern classes and $\bar{x}$ the image of a generator of $H^*(B\mathbb{S}^1; \mathbb{Z})$.

4.   Some properties of the λ-rings $\mathbf{R}_m$

For the purpose of the present paper, the most important property of the λ-ring $\mathbf{R}_m$ is recorded in the following

PROPOSITION 4.1 *Let m be a fixed positive integer and x an element of $\mathbf{R}_m$. If λ-$dim(x) = n$, then*

   i) *either $x = \tau_m^l \cdot \Psi^k(\lambda_i)$,*

   ii) *or $x = \sum_{j=1}^n \tau_m^{l_j}$;*

with $l, l_1, \ldots, l_n \in \mathbb{Z}$, $i \in \{1, n-1\}$ and $k \in \mathbb{N} \setminus \{0\}$.

Proof. Because of the λ-embedding

$$\mathbf{R}_m \hookrightarrow R(T_m) = \mathbb{Z}[\alpha_1^{\pm 1}, \ldots, \alpha_n^{\pm 1}, (\alpha_1 \cdot \alpha_2 \cdots \alpha_n)^{1/m}] \, ,$$

one can write $x = \tau_m^l \cdot \sum_{j=1}^n a_j$ where $l \in \mathbb{Z}$ and the a_j are monomials in $\alpha_1, \ldots, \alpha_n$. If all the a_j are invariant under the action of the symmetric group Σ_n, then $x = \sum_{j=1}^n \tau_m^{l_j}$ with $l_j \in \mathbb{Z}$ and this is case ii).

We now assume that there is a monomial $a_i = \alpha_1^{i_1} \cdot \alpha_2^{i_2} \cdots \alpha_n^{i_n}$ $(i_j \in \mathbb{N})$ which is not invariant. As λ-$dim(x) = n$, the order of the orbit of this a_i, under the action of Σ_n, is less or equal to n. Without loss of generality, we can assume that a_i has the form

$$a_i = (\alpha_1 \cdots \alpha_{k_1})^{s_1} \cdot (\alpha_{k_1+1} \cdots \alpha_{k_1+k_2})^{s_2} \cdots (\alpha_{k_1+\cdots+k_{r-1}+1} \cdots \alpha_n)^{s_r}$$

where

- $s_1, \ldots, s_r \in \mathbb{N}$ and $s_p \neq s_q$ whenever $\forall p \neq q$;

- $k_p \geq 1$ for all p and $k_1 + \cdots + k_r = n$.

In such a situation, the stabiliser of a_i, under the action of Σ_n, is the subgroup $\Sigma_{k_1} \times \cdots \times \Sigma_{k_r}$. Therefore, the order of the orbit of a_i is the multinomial coefficient

$$N = \frac{n!}{k_1! \cdots k_r!} \, .$$

Note that $r \geq 2$, since a_i is not invariant by hypothesis. We also claim that there is a k_j equal to 1. Otherwise, all the k_j are greater or equal to 2. Using, if necessary, an adequate permutation, we can suppose $n - k_1 \geq k_1$. For $j = k_1 + 1, \ldots, n$, the transpositions $(1\ j)$ and $(2\ j)$ give $2 \cdot (n - k_1)$ different elements of the orbit of a_i. As

$$2 \cdot (n - k_1) = (n - k_1) + (n - k_1) \geq (n - k_1) + k_1 = n \, ,$$

we can add a_i to these elements to obtain $n+1$ distinct monomials in our orbit. But this is impossible since N is less or equal to n.

Choose an index j with $k_j = 1$ and write $N = n \cdot \frac{(n-1)!}{k_1! \cdots k_r!}$. The factor $\tilde{N} = \frac{(n-1)!}{k_1! \cdots k_r!}$ is a positive integer because $k_1 + \cdots + k_r - k_j = n - 1$. Moreover, $\tilde{N}$ is equal to 1 if and only if there is another k_l equal to $n - 1$. This implies that $r = 2$, $N = n$ and we may take $a_i = \alpha_1^{s_1} \cdot (\alpha_2 \cdots \alpha_n)^{s_2}$ with $s_1 \neq s_2$. If $s_1 > s_2$, we obtain $x = \tau_m^{l+m \cdot s_2} \cdot \Psi^{(s_1 - s_2)}(\lambda_1)$; and otherwise, x is equal to $\tau_m^{l+m \cdot s_1} \cdot \Psi^{(s_2 - s_1)}(\lambda_{n-1})$.   □

REMARK. The preceding proof also shows that any $x \in \mathbf{R}_m$ of λ-dimension less than n must be a sum of powers of τ_m. In particular, the multiplicative set $\mathbf{R}_m(1) = \{x \in \mathbf{R}_m \setminus \{0\}; \ \lambda_t(x) = 1 + x \cdot t\}$ is equal to the set $\{\tau_m^l; \ l \in \mathbb{Z}\}$; so that every element of $\mathbf{R}_m(1)$ is invertible.

We are now ready for the main result of the first part of this paper.

THEOREM 2 *Let m, m' be positive integers. The following are equivalent:*

i) $BU_m(n) \simeq BU_{m'}(n)$

ii) $K(BU_m(n)) \overset{\lambda}{\cong} K(BU_{m'}(n))$

iii) $R(\mathbf{U}_m(n)) \overset{\lambda}{\cong} R(\mathbf{U}_{m'}(n))$

iv) $m \equiv \pm m' \pmod{n}$

v) $\mathbf{U}_m(n) \cong \mathbf{U}_{m'}(n)$.

Proof. i) $\Rightarrow$ ii) : Clear.

ii) $\Rightarrow$ iii) : This a consequence of [AM76, corollary 1.12].

$iii) \Rightarrow iv)$: Let $\varphi : \mathbf{R}_m \longrightarrow \mathbf{R}_{m'}$ be a λ-isomorphism. By proposition 4.1, one must have

$$\varphi(\tau_m) = \tau_{m'}^\varepsilon \quad \text{and} \quad \varphi(\lambda_1) = \tau_{m'}^l \cdot \Psi^k(\lambda_i) \,,$$

with $\varepsilon = \pm 1, l \in \mathbb{Z}, k \in \mathbb{N}\backslash\{0\}$ and $i \in \{1, n-1\}$. This means that the ring $\mathbf{R}_{m'} = \varphi(\mathbf{R}_m)$ is generated by $\Psi^k(\mathbf{R}_{m'})$ and $\tau_{m'}$. Since $\Psi^k(\tau_{m'}) = \tau_{m'}^k$ is an element of $\Psi^k(\mathbf{R}_{m'})$, one can infer that the degree of the field extension

$$\text{Frac}(\Psi^k(\mathbf{R}_{m'})) \subset \text{Frac}(\varphi(\mathbf{R}_m)) = \text{Frac}(\mathbf{R}_{m'})$$

is equal to k. But proposition 2.2 says that this degree must be k^n, hence $k^n = k$. As n is bigger than 2, it follows that k is equal to 1. On the one hand, we have

$$\lambda^n(\varphi(\lambda_1)) = \varphi(\lambda^n(\lambda_1)) = \varphi(\tau_m^m) = \tau_{m'}^{\varepsilon \cdot m} \,,$$

and on the other hand (recall that $\tau_{m'}^l$ is of λ-dimension 1)

$$\lambda^n(\varphi(\lambda_1)) = \lambda^n(\tau_{m'}^l \cdot \lambda_i) = \tau_{m'}^{n\,l} \cdot \lambda^n(\lambda_i) = \tau_{m'}^{n\,l+i\,m'}$$

Observe that the second equality of the sequence above follows from the axioms of λ-rings. It suffices now to remember that $i \in \{1, n-1\}$, so that $\varepsilon m \equiv \pm m' \pmod{n}$.

$iv) \Rightarrow v)$: By hypothesis, there exists $\epsilon = \pm 1$ and $l \in \mathbb{Z}$ such that $m + \epsilon \cdot m' = l \cdot n$. This relation implies $d = pgcd(m, n) = pgcd(m', n) = d'$. We can then check that the Lie group isomorphism

$$F : \mathbb{S}^1 \times \mathbf{SU}(n) \longrightarrow \mathbb{S}^1 \times \mathbf{SU}(n), \quad (z, A) \longmapsto (z^\epsilon, A)$$

sends the kernel of the covering map φ_m onto the kernel of $\varphi_{m'}$. So F induces an isomorphism between $\mathbf{U}_m(n)$ and $\mathbf{U}_{m'}(n)$.

$v) \Rightarrow i)$: Clear. $\square$

5. The other locally unitary groups

The aim of this part is to give as much information as possible on all the compact connected Lie groups having the same Lie algebra as $\mathbf{U}(n)$. We recall some results of [Bau67] and use them to generalise the results of the preceding section. We present realisations of the locally unitary groups and compute their representation rings.

Given a compact connected Lie group G, there exists a finite cover $p : T^r \times G_0 \longrightarrow G$, where G_0 is a 1-connected compact Lie group and T^r is a r-dimensional torus with $r \geq 0$. The group $\tilde{G} = T^r \times G_0$ is unique up to isomorphism but the projection p may not, as we have seen for $G = \mathbf{U}(n)$ in section 3.

To deal with the local isomorphism classes of the group G, P. Baum introduced the following notion:

DEFINITION 5.1 *A special subgroup* (K, φ) *of* $\tilde{G} = T^r \times G_0$ *is of the form*

$$(K, \varphi) = \{(\varphi(g), g); \quad g \in K\} \subset \tilde{G}$$

where K *is a subgroup of the center of* G_0 *and* φ *is a homomorphism of* K *into* T^r.

THEOREM 3 *(Proposition 2 and Corollary 6 of* [Bau67]*)*

1. *If* H *is a compact connected Lie group locally isomorphic to* G, *then there is a special subgroup* (K, φ) *of* $\tilde{G}$ *such that* H *is isomorphic to the quotient* $\tilde{G}/(K, \varphi)$. *We then say that* $\tilde{G}/(K, \varphi)$ *is a special form of the group* H.

2. *Let* (K_1, φ_1) *and* (K_2, φ_2) *be special subgroups of* $\tilde{G}$. *The group* $\tilde{G}/(K_1, \varphi_1)$ *covers* (respectively *is isomorphic to)* $\tilde{G}/(K_2, \varphi_2)$ *if and only if there exists an automorphism* α *of* G_0 *and a finite cover* $\beta : T^r \longrightarrow T^r$ *(respectively an automorphism* β *of* T^r*) such that* α *maps* K_1 *into* (respectively *onto)* K_2 *and the diagram*

$$
\begin{array}{ccc}
K_1 & \xrightarrow{\varphi_1} & T^r \\
\alpha \downarrow & & \downarrow \beta \\
K_2 & \xrightarrow{\varphi_2} & T^r
\end{array}
$$

commutes.

In our case, the group $\mathbf{SU}(2)$ admits only inner automorphisms and for $n > 2$, $\mathbf{SU}(n)$ has only one non-trivial outer automorphism; it is given by the complex conjugation $(a_{ij}) \longmapsto (\bar{a}_{ij})$. These facts imply the following consequence of theorem 3:

PROPOSITION 5.1   *i) Any compact connected Lie group which is locally isomorphic to* $\mathbf{U}(n)$ *is of the form*

$$\mathbf{FU}_{s,k}(n) = \mathbb{S}^1 \times \mathbf{SU}(n)/N_{s,k}$$

with $s|n$, $k \in \mathbb{N} \setminus \{0\}$ *and where*

$$N_{s,k} = \{(\exp(-\frac{2\pi\mathbf{i}}{s} k\, j); \exp(\frac{2\pi\mathbf{i}}{s} j) \cdot \mathbb{1}); \quad j = 0, \ldots, s-1\}.$$

ii) For s, s' *dividing* n *and* $k, k' \in \mathbb{N}$, *the groups* $\mathbf{FU}_{s,k}(n)$ *and* $\mathbf{FU}_{s',k'}(n)$ *are isomorphic if and only if* $s = s'$ *and* $k' \equiv \pm k \pmod{s}$.
In other words, the isomorphism class of a group $\mathbf{FU}_{s,k}(n)$ *is completely determined by the divisor* s *of* n *and the class of the positive integer* k *in the quotient-set* $(\mathbb{Z}/s)/\{\pm 1\}$.

iii) *For s, s' dividing n and $k, k' \in \mathbb{N} \setminus \{0\}$, $\mathbf{FU}_{s,k}(n)$ covers $\mathbf{FU}_{s',k'}(n)$ if and only if $s|s'$ and there exists an integer r such that $k' \equiv \pm r \cdot k \pmod{s}$. In particular, the "surprising result" of [Bau67], namely groups which covers each other without being isomorphic, arises for $\mathbf{FU}_{s,k}(n)$ and $\mathbf{FU}_{s,k'}(n)$ whenever $r \not\equiv \pm 1 \bmod(n)$ and r invertible modulo s.*

COROLLARY 1 *If $\Lambda(\mathbf{U}(n))$ denotes the set of isomorphism classes of the compact connected Lie groups locally isomorphic to $\mathbf{U}(n)$, then*

$$card(\Lambda(\mathbf{U}(n))) = \sum_{s|n}([\frac{s}{2}] + 1)$$

For some values of n, P. Baum has also studied the transitive relation on $\Lambda(\mathbf{U}(n))$ given by:

For every $[H_1], [H_2] \in \Lambda(\mathbf{U}(n))$, $[H_1] \geq [H_2]$ if and only if there exists a surjective Lie group morphism $H_1 \longrightarrow H_2$ with finite kernel.'

Consider the set $L(n) = \{(s, \alpha); \quad s|n$ and $\alpha \in (\mathbb{Z}/s)/\{\pm 1\}\}$, equipped with the relation:

For every $(s, \alpha), (s', \alpha') \in L(n)$, $(s, \alpha) \geq (s', \alpha')$ if and only if $s|s'$ and there exists an integer r such that $\alpha' \equiv \pm r \cdot \alpha \pmod{s}$.

By elementary number theory, the relation on $L(n)$ can also be given in the more tractable form:

$$(s, \alpha) \geq (s', \alpha') \qquad \text{if and only if} \qquad \left\{ \begin{array}{c} s|s' \\ gcd(s, \alpha) | gcd(s', \alpha') \end{array} \right.$$

Using the terminology of [Bau67, Definitions 6 & 7], proposition 3.1 says that there is an isomorphism of *quasi-ordered sets* between $\Lambda(\mathbf{U}(n))$ and $L(n)$. These observations may be used to write down a computer program which draws the graph of the quasi-ordered set $\Lambda(\mathbf{U}(n))$.

In order to get the special form of the finite covers of $\mathbf{U}(n)$, we choose a positive integer m and set $d = gcd(m, n)$. In section 3, we have shown that the Lie group $\mathbf{U}_m(n)$ can be realised as a quotient $\mathbb{S}^1 \times \mathbf{SU}(n)/E_m$ where

$$E_m = \{(\exp(\frac{2\pi i}{n} d\,j); \exp(-\frac{2\pi i}{n} m\,j) \cdot \mathbb{1}); \quad j = 0, 1, \ldots, \frac{n}{d} - 1\}.$$

By Bezout's theorem, there exist $u, v \in \mathbb{Z}$ such that $u \cdot \frac{m}{d} + v \cdot \frac{n}{d} = 1$. Using the group automorphism

$$E_m \longrightarrow E_m, \qquad x \longmapsto x^u,$$

one can write

$$E_m = \{(\exp(-\frac{2\pi i d}{n} u\,j); \exp(\frac{2\pi i d}{n} j) \cdot \mathbb{1}); \quad j = 0, \ldots, \frac{n}{d} - 1\}.$$

So E_m is a special group in the sense of Baum and the special form we are looking for is:

$$\mathbf{U}_m(n) = \mathbf{FU}_{\frac{n}{d},u}(n) \ .$$

Conversely, if s is a divisor of n and k is invertible modulo s, then

$$\mathbf{FU}_{s,k}(n) = \mathbf{U}_{\frac{n\,\tilde{m}}{s}}(n) \ ,$$

where $\tilde{m}$ is a positive integer such that $k \cdot \tilde{m} \equiv 1 \pmod{s}$. It follows that $\mathbf{FU}_{s,k}(n)$ is a finite cover of $\mathbf{U}(n)$ if and only if the integers s and k are coprime.

In fact, the finite covers $\mathbf{U}_m(n)$ play a central rôle among the locally unitary groups. Note first that the group $\mathbb{S}^1 \times \mathbf{SU}(n)$ covers all these groups. If s is a divisor of n and k is a positive integer, set $r = gcd(s, k)$. Then proposition 5.1.iii) implies that the group $\mathbf{FU}_{\frac{s}{r},\frac{k}{r}}(n)$ is a finite cover of $\mathbf{FU}_{s,k}(n)$. Since the integers $\frac{s}{r}$ and $\frac{k}{r}$ are coprime, the group $\mathbf{FU}_{\frac{s}{r},\frac{k}{r}}(n)$ is also a finite cover of $\mathbf{U}(n)$. In other words, any locally unitary group which is non-isomorphic to $\mathbb{S}^1 \times \mathbf{SU}(n)$ is covered by a group $\mathbf{U}_m(n)$ for some $m \not\equiv 0 \pmod n$.

REMARK. In [NS91], Notbohm and Smith have observed that $\pi_1(\mathbf{FU}_{s,k}(n)) = \mathbb{Z} \oplus (\mathbb{Z}/r)$, where $r = gcd(s, k)$. Therefore $\mathbf{FU}_{s,k}(n)$ is isomorphic to a finite cover of $\mathbf{U}(n)$ if and only if its fundamental group is $\mathbb{Z}$.

6. The representation rings of the $\mathbf{FU}_{s,k}(n)$

In section 3, the existence of explicit embeddings of the groups $\mathbf{U}_m(n)$ in the unitary group $\mathbf{U}(n+1)$ made all the computations straightforward. It is not difficult to show that, except for the group $\mathbf{U}(n)$ itself, these embeddings are optimal. For all the locally unitary groups, we are only able to offer the following

PROPOSITION 6.1 *Given a locally unitary group* $\mathbf{FU}_{s,k}(n)$, *there is an injective Lie group morphism*

$$\varrho_{s,k} : \mathbf{FU}_{s,k}(n) \hookrightarrow \mathbf{U}(N) \ ,$$

where $N = 1 + n^2 + \binom{n}{r}$ *and* $r = gcd(s, k)$.

Proof. Let $\mathbf{FU}_{s,k}(n)$ be a locally unitary group, set $r = gcd(s, k)$ and choose integers μ, ν such that $\mu \cdot \frac{k}{r} + \nu \cdot \frac{s}{r} = 1$. We have already seen that the group $\mathbf{FU}_{\frac{s}{r},\frac{k}{r}}(n)$ is a finite cover of $\mathbf{FU}_{s,k}(n)$. A covering map $\bar{\varphi}$ is induced by the Lie group morphism

$$\varphi : \mathbb{S}^1 \times \mathbf{SU}(n) \longrightarrow \mathbb{S}^1 \times \mathbf{SU}(n); \quad (z; A) \longmapsto (z^r; A) \ .$$

If $\mathbf{FU}_{\frac{s}{r},\frac{k}{r}}(n)$ is identified with the image of the morphism

$$\varphi_\mu : \mathbb{S}^1 \times \mathbf{SU}(n) \longrightarrow \mathbb{S}^1 \times \mathbf{U}(n); \quad (z; A) \longmapsto (z^{\frac{s}{r}}; z^\mu \cdot A)$$

constructed in proposition 3.1, then the kernel of the covering map $\bar{\varphi}$ is the group

$$\Gamma_{s,k} = \{(\exp(-\frac{2\pi i}{r}(\frac{k}{r}j-\frac{s}{r}l)); \exp(\frac{2\pi i}{r}(\nu j+\mu l))\cdot \mathbb{1}); \begin{array}{ccc} j & = & 0,1,\ldots,\text{s-1} \\ l & = & 0,1,\ldots,\text{r-1} \end{array} \}$$

By using the equality $\mu \cdot \frac{k}{r} + \nu \cdot \frac{s}{r} = 1$, one checks that

$$\Gamma_{s,k} = \{(\exp(-\frac{2\pi i}{r}j); \exp(\frac{2\pi i}{r}l)\cdot \mathbb{1}); \quad j,l = 0,1,\ldots,r-1\} \cong \mathbb{Z}/r \oplus \mathbb{Z}/r$$

It follows that $\mathbf{FU}_{s,k}(n)$ is isomorphic to a subgroup of the direct product $\mathbb{S}^1 \times (\mathbf{U}(n)/\mathbb{Z}/r)$ and it suffices to find an embedding of the group $\mathbf{FU}_{n,r}(n) = \mathbf{U}(n)/\mathbb{Z}/r$ in some unitary group.

To get such an embedding, consider the standard complex representation V of the group $\mathbf{U}(n)$ on the vector space $V = \mathbb{C}^n$. Denote the dual representation of V by V^* and its r-th exterior power by $\Lambda^r V$. If $\{e_1,\ldots,e_n\}$ is the standard basis of V, its dual $\{\varepsilon_1,\ldots,\varepsilon_n\}$ is a basis of V^* and the vectors $e_{i_1} \wedge \ldots \wedge e_{i_r}$ $(1 \le i_1 < \cdots < i_r \le n)$ form a basis of $\Lambda^r V$. Under the action of a matrix $Diag(z_1,\ldots,z_n) \in \mathbf{U}(n)$, the basis vector $v_i \otimes \varepsilon_j$ of $V \otimes V^*$ is multiplied by the complex number $z_i z_j^{-1}$. This implies that the kernel of the representation $\mathbf{U}(n) \longrightarrow \mathbf{GL}(V \otimes V^*)$ is the group of the scalar matrices, that is the centre of $\mathbf{U}(n)$. As the reader should have noticed, this representation is nothing but the adjoint representation of the group $\mathbf{U}(n)$. On the basis vector $e_{i_1} \wedge \ldots \wedge e_{i_r}$ $(1 \le i_1 < \cdots < i_r \le n)$ of $\Lambda^r V$, the matrix $Diag(z_1,\ldots,z_n)$ acts like the multiplication by the number $z_{i_1} \cdots z_{i_r}$.

Thus the $\mathbf{U}(n)$-space $W = V \otimes V^* \oplus \Lambda^r V$ gives a homomorphism

$$\varrho : \mathbf{U}(n) \longrightarrow \mathbf{U}(n^2 + \binom{n}{r})$$

with kernel $\mathbb{Z}/r = \{\theta \cdot \mathbb{1}; \quad \theta^r = 1\} \subset \mathbf{U}(n)$. For the promised $\varrho_{s,k}$, it suffices to take the composite

$$\mathbf{FU}_{s,k}(n) \cong \mathbf{FU}_{\frac{s}{r},\frac{k}{r}}(n)/\Gamma_{s,k} \hookrightarrow \mathbb{S}^1 \times (\mathbf{U}(n)/\mathbb{Z}/r) \xrightarrow{Id \times \bar{\varrho}} \mathbb{S}^1 \times Im(\varrho) \subset \mathbb{S}^1 \times \mathbf{U}(n^2 + \binom{n}{k}),$$

where $\bar{\varrho}$ is the canonical isomorphism between $\mathbf{U}(n)/\mathbb{Z}/r$ and the image of ϱ. $\qquad \square$

QUESTION: How can one improve the bound $N = 1 + n^2 + \binom{n}{r}$ of the preceding proposition? More generally, if a compact Lie group G is given, what is the minimal integer N_0 such that G embeds into the unitary group $\mathbf{U}(N_0)$? Note that Peter-Weyl's theorem assures the existence of such an integer; it also shows that the question is not new.

The degree of the embedding $\varrho_{s,k}$ in proposition above is not optimal; but it is sufficient for our next goal, namely the computation of the representation rings of the locally unitary groups. We shall freely use the notations of section 3 and continue with those of the preceding proof. Let a group $\mathbf{FU}_{s,k}(n)$ be given.

Set $r = gcd(s, k)$ and choose $\mu, \nu \in \mathbb{Z}$ such that $\mu > 0$ and $\mu \cdot \frac{k}{r} + \nu \cdot \frac{s}{r} = 1$. From the considerations of section 5, we can take

$$\mathbf{FU}_{\frac{s}{r}, \frac{k}{r}}(n) = \mathbf{U}_{\frac{n}{s}\mu r}(n) .$$

Now recall from proposition 3.1 that there is a covering map

$$\varphi_\mu : \mathbf{FU}_{1,\mu}(n) = \mathbb{S}^1 \times \mathbf{SU}(n) \longrightarrow \mathbf{FU}_{\frac{s}{r}, \frac{k}{r}}(n) = \mathbf{U}_{\frac{n}{s}\mu r}(n) .$$

If we compose it with the covering map $\bar{\varphi}$ of the above proof, we get a new covering map

$$\tilde{\varphi} = \bar{\varphi} \circ \varphi_\mu : \mathbf{FU}_{1,\mu}(n) = \mathbb{S}^1 \times \mathbf{SU}(n) \longrightarrow \mathbf{FU}_{s,k}(n) \subset \mathbb{S}^1 \times (\mathbf{U}(n)/\mathbb{Z}/r) .$$

The image of the standard maximal torus $T_{1,\mu} \subset \mathbf{FU}_{1,\mu}(n)$ by the morphism $\tilde{\varphi}$ is a maximal torus $T_{s,k}$ in the group $\mathbf{FU}_{s,k}(n)$. If we identify $\mathbf{FU}_{s,k}(n)$ to $Im(\varrho_{s,k}) \subset \mathbf{U}(N)$, the inclusion $T_{s,k} \subset T(N)$ induces a *surjective* morphism $R(T(N)) \longrightarrow R(T_{s,k})$. The restriction $\hat{\varphi}$ of the composite $\varrho_{s,k} \circ \tilde{\varphi}$ to the maximal torus $T_{1,\mu}$ can be explicitly computed. This (straightforward, but tedious) computation shows that the ring $R(T_{s,k})$, when identified with its image by φ_μ^*, coincides with the subring of

$$R(T_{1,\mu}) = \mathbb{Z}[\alpha_1^{\pm 1}, \dots, \alpha_n^{\pm 1}, \tau_\mu]/(\tau_\mu^{n \cdot \mu} - \alpha_1 \alpha_2 \cdots \alpha_n)$$

generated by the monomials

$$\tau_\mu^{\pm s}, \quad \alpha_i \alpha_j^{-1} \ (i, j = 1, 2, \dots, n), \quad (\alpha_{i_1} \cdots \alpha_{i_r})^{\pm 1} \ (1 \le i_1 < \cdots < i_r \le n) .$$

In the particular case $s = n$ and $k = r$, we obtain that the representation ring of the maximal torus $T_{n,r}$ of $\mathbf{FU}_{n,r}(n) = \mathbf{U}(n)/\mathbb{Z}/r$ is the subring of $R(T_{1,\mu})$ generated by the monomials

$$(\alpha_1 \alpha_2 \cdots \alpha_n)^{\pm 1}, \quad \alpha_i \alpha_j^{-1} (i, j = 1, 2, \dots, n), \quad (\alpha_{i_1} \cdots \alpha_{i_r})^{\pm 1} (1 \le i_1 < \cdots < i_r \le n).$$

From the presentation of the two rings $R(T_{s,k})$ and $R(T_{n,r})$, we can deduce the following decomposition:

$$R(T_{s,k}) = \bigoplus_{j=0}^{\frac{n}{s}\mu - 1} R(T_{n,r}) \cdot \tau_\mu^{s j} .$$

As τ_μ is visibly invariant under the Weyl group action, we have proved

PROPOSITION 6.2 *Let* $\mathbf{FU}_{s,k}(n)$ *be a locally unitary group. Set* $r = pgcd(s, k)$ *and choose an integer* $\mu > 0$ *such that* $\mu \frac{k}{r} \equiv 1 \pmod{\frac{s}{r}}$. *If* $\mathbf{R}_{s,k} := R(\mathbf{FU}_{s,k}(n))$ *and* $\mathbf{R}_{n,r} := R(\mathbf{FU}_{n,r}(n))$ *are identified to* λ-*subrings of* $\mathbf{R}_{1,\mu} := R(\mathbf{FU}_{1,\mu}(n))$, *then*

$$\mathbf{R}_{s,k} = \bigoplus_{j=0}^{\frac{n}{s}\mu - 1} \mathbf{R}_{n,r} \cdot \tau_\mu^{s j} \subset \mathbf{R}_{\frac{s}{r}, \frac{k}{r}} = \bigoplus_{j=0}^{\frac{n}{s}\mu r - 1} \mathbf{R}_{n,1} \cdot \tau_\mu^{\frac{s}{r} j} .$$

In particular, the multiplicative group $\mathbf{R}_{s,k}(1)$ *is isomorphic to the infinite cyclic group generated by* τ_μ^s.

This proposition reduces our problem to the computation of the representation rings of the quotient-groups $\mathbf{U}(n)/\mathbb{Z}/r$. A straightforward generalisation of the discussion in [Ada69, §3.68] shows that $R(\mathbf{U}(n)/\mathbb{Z}/r)$ is λ-isomorphic to the subring of $R(\mathbf{U}(n))$ generated by $\lambda_n^{\pm 1}$ and the monomials $\lambda_1^{a_1} \cdot \lambda_2^{a_2} \cdots \lambda_{n-1}^{a_{n-1}}$ where the non-negative integers $a_1, \ldots, a_{n-1}$ satisfy the diophantine congruence

$$a_1 + 2\, a_2 + \ldots + (n-1)\, a_{n-1} \equiv 0 \pmod r .$$

This equation doesn't seem to be easily solvable (see [Oss94] for a discussion and some references). However, it can be used to check that

$$\Psi^r(R(\mathbf{U}(n))) = \mathbb{Z}[\Psi^r(\lambda_1), \Psi^r(\lambda_2), \ldots, \Psi^r(\lambda_{n-1}), \Psi^r(\lambda_n^{\pm 1})]$$

is a (proper, in general) λ-subring of $R(\mathbf{U}(n)/\mathbb{Z}/r)$.

The following assertion is an immediate consequence of both proposition 6.2 and proposition 4.1:

COROLLARY 2 *Let* $x \in \mathbf{R}_{s,k} \subset \mathbf{R}_{1,\mu}$ *be of* λ-*dimension* n; *then*

i) *either* $x = \sum_{j=1}^{n} \tau_\mu^{s\, l_j}$ *with* $l_1, \ldots, l_n \in \mathbb{Z}$;

ii) *or* $x = \tau_\mu^{s\, l} \cdot \Psi^{r\, k}(\lambda_i)$ *with* $l \in \mathbb{Z}$, $k \in \mathbb{N} \setminus \{0\}$ *et* $i \in \{1, n-1\}$.

We are now ready to present our proof of Notbohm-Smith's result:

THEOREM 4 (NOTBOHM & SMITH) *Fix an integer* n *greater or equal to 3. Let* s, s' *be divisors of* n *and let* k, k' *be positive integers. The following assertions are equivalent:*

i) $\mathbf{FU}_{s,k}(n) \cong \mathbf{FU}_{s',k'}(n)$;

ii) $\mathbf{BFU}_{s,k}(n) \simeq \mathbf{BFU}_{s',k'}(n)$;

iii) $K(\mathbf{BFU}_{s,k}(n)) \overset{\lambda}{\cong} K(\mathbf{BFU}_{s',k'}(n))$;

iv) $R(\mathbf{FU}_{s,k}(n)) \overset{\lambda}{\cong} R(\mathbf{FU}_{s',k'}(n))$;

v) $s = s'$ *and* $k \equiv \pm k' \pmod s$.

Proof. The two implications $i) \implies ii) \implies iii)$ are obvious. The part $iii) \implies iv)$ is a consequence of [AM76, Corollary 1.12] and the equivalence of $i)$ and $v)$ is provided by proposition 5.1. Hence, we are done if we can prove that the fourth assertion implies the fifth one. For this purpose, we set $r = pgcd(s, k)$, $r' = pgcd(s', k')$ and choose positive integers μ, μ' such that $\mu \frac{k}{r} \equiv 1 \pmod{\frac{s}{r}}$, $\mu' \frac{k'}{r'} \equiv 1 \pmod{\frac{s'}{r'}}$. We will use the presentation of the rings $\mathbf{R}_{s,k}$, $\mathbf{R}_{s',k'}$ given in proposition 6.2.

Let $\varphi : \mathbf{R}_{s,k} \longrightarrow \mathbf{R}_{s',k'}$ be a λ-isomorphism; then $\varphi(\tau_\mu^s) = \tau_\mu^{\epsilon\,s'}$ with $\epsilon = \pm 1$. By corollary 2, there exist $l' \in \mathbb{Z}$, $m' \in \mathbb{N} \setminus \{0\}$ and $i \in \{1, n-1\}$ such that

$$(*) \qquad \varphi(\Psi^r(\lambda_1)) = \tau_{\mu'}^{s'\,l'} \cdot \Psi^{r'\,m'}(\lambda_i) .$$

Let A_r (resp. $A'_{r'\,m'}$) denote the subring of $\mathbf{R}_{s,k}$ (resp. $\mathbf{R}_{s',k'}$) generated by the elements $\Psi^r(\lambda_1), \ldots, \Psi^r(\lambda_{n-1}), \tau_\mu^{\pm s}$ (resp. $\Psi^{r'\,m'}(\lambda_1), \ldots, \Psi^{r'\,m'}(\lambda_{n-1}), \tau_{\mu'}^{\pm s'}$). As in the proof of theorem 2, we pass to the fraction fields of all these rings and use the propositions in section 2 to compute the degrees of some of the field extensions. These degrees are recorded in the diagrams:

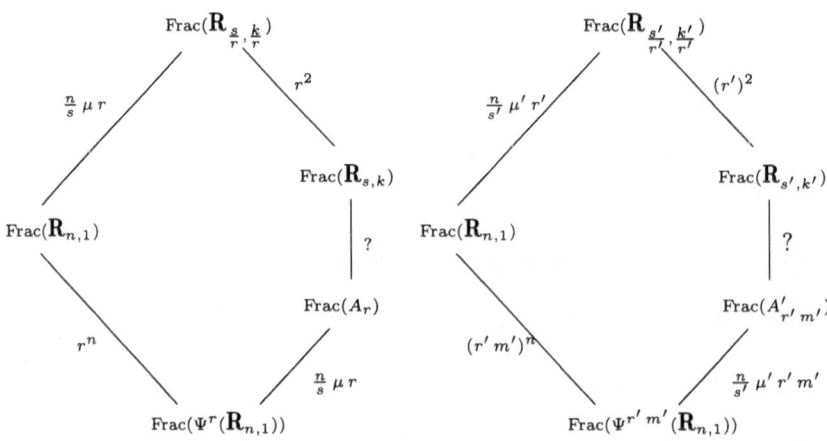

Since φ is a ring isomorphism, we write

$$\begin{aligned} r^{n-2} &= \deg(\operatorname{Frac}(A_r) \subset \operatorname{Frac}(\mathbf{R}_{s,k})) \\ &= \deg(\operatorname{Frac}(A'_{r'\,m'}) \subset \operatorname{Frac}(\mathbf{R}_{s',k'})) \\ &= (r')^{n-2} \cdot (m')^{n-1} . \end{aligned}$$

Reasonning in the same way for φ^{-1} gives us a positive integer m such that $(r')^{n-2} = r^{n-2} \cdot m^{n-1}$. The combination of these two relations leads to $r^{n-2} = (m\,m')^{n-1} \cdot r^{n-2}$, so that $m = m' = 1$. We invoke here the hypothesis $n \geq 3$ to infer that we must also have $r = r'$. With this last equality, we are now able to extend φ to a λ-isomorphism $\tilde{\varphi} : \mathbf{R}_{\frac{s}{r}, \frac{k}{r}} \longrightarrow \mathbf{R}_{\frac{s'}{r'}, \frac{k'}{r'}}$, just by setting

$$\tilde{\varphi}(\tau_\mu^{\frac{s}{r}}) = \tau_\mu^{\epsilon\,\frac{s'}{r'}}; \qquad \tilde{\varphi}(\lambda_1) = \tau_{\mu'}^{\frac{s'}{r'}\,l'} \cdot \lambda_i .$$

By theorem 2, the existence of the λ-isomorphism $\tilde{\varphi}$ implies the following congruence:

$$(**) \qquad \frac{n}{s}\,\mu\,r \equiv \pm \frac{n}{s'}\,\mu'\,r' \pmod{n} .$$

Moreover, the proof of the implication $iv) \Longrightarrow v)$ of the same theorem provides a λ-isomorphism

$$\sigma : \mathbf{R}_{1,\mu} \longrightarrow \mathbf{R}_{1,\mu'} \qquad \text{such that} \quad \sigma(\mathbf{R}_{\frac{s}{r}, \frac{k}{r}}) = \mathbf{R}_{\frac{s'}{r'}, \frac{k'}{r'}} .$$

We go back once more to the degree of fraction fields extensions to get

$$
\begin{aligned}
\frac{s}{r} &= \deg(\mathrm{Frac}(\mathbf{R}_{\frac{s}{r},\frac{k}{r}}) \subset \mathrm{Frac}(\mathbf{R}_{1,\mu})) \\
&= \deg(\mathrm{Frac}(\mathbf{R}_{\frac{s'}{r'},\frac{k'}{r'}}) \subset \mathrm{Frac}(\mathbf{R}_{1,\mu'})) \\
&= \frac{s'}{r'} ;
\end{aligned}
$$

therefore $s = s'$ and the congruence $(**)$ reduces to

$$
\mu \equiv \pm\mu' \ (\mathrm{mod}\ \frac{s}{r}) .
$$

To complete the proof, we just have to multiply both sides of this last congruence by $\frac{k}{r}\frac{k'}{r'}$ and chase the denominators. $\qquad\square$

REMARK. The result is false for $n = 2$; it suffices to consider the groups $\mathbb{S}^1 \times$ $\mathbf{SO}(3)$ and $\mathbb{S}^1 \times \mathbf{SU}(2)$. So the condition $n \geq 3$ should also be added to the hypothesis of theorem 4.5 of [NS91].

Acknowledgements. I came upon the theme of this paper for the first time in 1989, during my master thesis; I was then looking at the representation rings of the finite covers of the groups $\mathbf{U}(n)$. My interest in the subject was renewed after the talks given by L. Smith and D. Notbohm in 1991, at the Göttingen Symposium. So the present paper should be considered as an account and a detailed complement to [NS91, §4]. Its material is taken from the third chapter of my PhD thesis ([Oss94]). I would like to express my sincere gratitude to my advisor, Prof. U. Suter, for his guidance and his encouragements. I'm also indebted to my friends A. Jeanneret and D. Lines for many fruitful discussions. Finally many thanks are due to G. Mislin and to the referee for many valuable suggestions.

References

[Ada69] J.F. ADAMS. Lectures on Lie groups, *W.A. Benjamin Inc.* (1969).

[AM76] J.F. ADAMS & Z. MAHMUD. Maps Between Classifying Spaces, *Invent. Math.* 35 (1976), 1–41.

[AW80] J.F. ADAMS & C. WILKERSON. Finite H-spaces and Algebras over the Steenrod Algebra, *Ann. Math.* 111 (1980), 95–143.

[AT69] M.F. ATIYAH & D.O. TALL. Groups Representations, λ-Rings and The J-homomorphism, *Topology* 8 (1969), 213–297.

[Bau67] P.F. BAUM. Local Isomorphism of Compact Connected Lie Groups, *Pac. J. Math.* 22, (1967), 197–204.

[BD85] Th. BRÖCKER & T. tom DIECK. Representations of Compact Lie Groups, *Springer* (1985).

[DMW92] W.G. DWYER, H.R. MILLER & C.W. WILKERSON. Homotopical Uniqueness of classifying Spaces, *Topology* 31 (1992), 29–45.

[Not92] D. NOTBOHM. On the Functor 'Classifying Space' for Compact Lie Groups, *preprint* (1992).

[NS91]   D. NOTBOHM & L. SMITH. Fake Lie Groups and Maximal Tori III,
         Math. Ann. 290 (1991), 629–642.

[Oss92] A. OSSE. Sur le Classifiant d'un Groupe de Lie Compact Connexe, Comptes-
         Rendus Acad. Sci. Paris 315 Serie I (1992), 833–838.

[Oss94] A. OSSE. Anneaux des représentations et espaces classifiants des groupes de
         Lie, Thèse, Université de Neuchâtel (1994).

[Rec71] D.L. RECTOR. Subgroups of Finite Dimensional Topological Groups,
         J. Pure and Applied Alg. 1 (1971), 253–273.

[Sch68] H. SCHEERER. Homotopieäquivalente Kompakte Liesche Gruppen, Topol-
         ogy 7 (1968)

[Smi91] L. SMITH (ed.). Classifying Spaces of Compact Lie Groups and Finite Loop
         Spaces; Problem Session, Mathematica Gottingensis Heft 41 (Sept 1991).

[Sta69] J. STASHEFF. Manifolds of the Homotopy type of (non-Lie) Groups, Bull.
         AMS 75 (1969), 998–1000.

Institut de Mathématiques, Rue Emile Argand 11, CH-2007 Neuchâtel (SUISSE)
Current Address: Ohio State University, Columbus, Ohio.

Progress in Mathematics, Vol. 136
© 1996 Birkhäuser Verlag Basel/Switzerland

Localization and genus in group theory and homotopy theory

GEORGE PESCHKE[1]

The following is an expanded version of the talk I gave at the 1994 Barcelona Conference on Algebraic Topology. We consider the

Fundamental problem. Let $\mathcal{C}$ be a category and let $\mathcal{L}$ be a family of localizing functors on $\mathcal{C}$. Given objects X and Y in $\mathcal{C}$, suppose that information on LX, LY and/or morphisms $LX \to LY$ is available for all of the functors $L \in \mathcal{L}$. – *Deduce information about X, Y and/or morphisms $X \to Y$.*

We respond to this problem by

(1) presenting a category theoretical framework which is suitable to discuss such local $\xrightarrow{\text{info}}$ global processes;

(2) implementing these ideas for localizations of groups and spaces at sets of primes P.

To support the development in (1) we combine the works of Adams [1], Casacuberta-Peschke-Pfenniger [7] and chapter I of the present exposition. Thus we arrive at the following pivotal point. The objects obtained by applying the functors in $\mathcal{L}$ to $X \in \mathcal{C}$ can always be organized (by intrinsic (!) category theoretical structure) into a unique commutative diagram. Moreover, this process is functorial and is left adjoint to the inverse limit functor (if it exists) restricted to such diagrams. Losely speaking this means that the "production of local data" is adjoint to "the process of assembling local data to global data".

The development in (2) employs (1) and constitutes a continuation of earlier work of Sullivan [16], [17], Bousfield-Kan [5] and Hilton-Mislin-Roitberg [10] in the sense that local $\xrightarrow{\text{info}}$ global processes familiar from these sources are here extended in two directions. Firstly, we show that these classical local $\xrightarrow{\text{info}}$ global processes admit a much stronger formulation even within the traditional environment of simply connected or nilpotent spaces. Secondly, the class of spaces for which these processes are valid is enlarged well beyond nilpotent spaces. For this purpose we use the P-localization of spaces based on the algebro-geometric property that the loop space of a P-local space be a P-local group

[1]supported by NSERC of Canada

up to homotopy. This idea is suggested by [11, 1.7 and 1.8]. The corresponding
P-localization theory is developed in [6].

I Local $\xrightarrow{\text{info}}$ global processes

I.1 Local diagrams and the genus concept

Recall that a localizing functor L on a category $\mathcal{C}$ involves

(i) *L-localizing maps*   For each $X \in \mathcal{C}$ there is a natural arrow $\varepsilon : X \to LX$. Moreover, $\varepsilon : LX \to LLX$ is an isomorphism.

(ii) *L-local objects* $\{Y \in \mathcal{C} : Y \xrightarrow{\varepsilon}_{\cong} LY\}$.

(iii) *universal property*

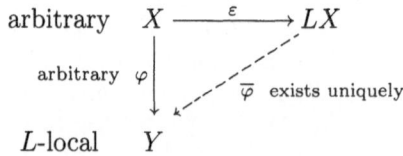

Now let $\mathcal{L} = \{L_\lambda\}_{\lambda \in \Lambda}$ be a family of localizing functors on $\mathcal{C}$. The collection of
the corresponding classes of local objects is partially ordered by inclusion, and
this partial ordering is paralleled by natural transformations of the functors
L_λ. Indeed, the universal property of localizing functors gives a natural trans-
formation $\tau : L_{\lambda_1} \to L_\lambda$, whenever $\{L_{\lambda_1}\text{-local objects}\} \supset \{L_\lambda\text{-local objects}\}$.
Further τ is uniquely determined by the requirement that the diagram of nat-
ural transformations below commutes.

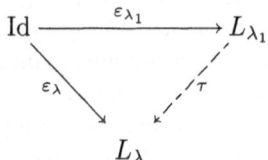

To design an appropriate environment for local $\xrightarrow{\text{info}}$ global processes we
regard $\mathcal{L} = \{L_\lambda\}_{\lambda \in \Lambda}$ as a small category, using the natural transformations
above as morphisms. Thus we organize L_λ-local data, $\lambda \in \Lambda$, into a diagram
modeled on $\mathcal{L}$.

I.1.1 DEFINITION   An $\mathcal{L}$-diagram $\underline{Y}$ in $\mathcal{C}$ consists of

(i) an L_λ-local object Y_λ, for each $\lambda \in \Lambda$;
(ii) an L_λ-localizing morphism $Y_{\lambda_1} \to Y_\lambda$, if

$$\{L_{\lambda_1}\text{-local objects}\} \quad \supset \quad \{L_\lambda\text{-local objects}\}.$$

If in search of examples, the reader might at this point want to skip ahead to I.2.5 and I.2.6.

The collection of all $\mathcal{L}$-local data forms the category $\mathcal{L}$-diag($\mathcal{C}$) whose objects are $\mathcal{L}$-diagrams and whose morphisms are diagram morphisms.

I.1.2 REMARK "Evaluation" of $\mathcal{L}$ on $\mathcal{C}$ is a covariant functor

$$\mathcal{L} : \mathcal{C} \longrightarrow \mathcal{L}\text{-diag}(\mathcal{C}), \quad X \longmapsto \mathcal{L}(X). \qquad \square$$

The ideal situation for local$\xrightarrow{\text{info}}$global processes occurs when $\mathcal{L}$ is an equivalence of categories. However, normally there will exist $\mathcal{L}$-diagrams $\underline{Y}$ which cannot be obtained by $\mathcal{L}$-localizing an object of $\mathcal{C}$. This makes desirable a "realizability criterion" which distinguishes those $\mathcal{L}$-diagrams $\underline{Y}$ with $\underline{Y} \cong \mathcal{L}(X)$, for some $X \in \mathcal{C}$, from the remaining ones. On the other hand, there may exist $\mathcal{L}$-diagrams $\underline{Z}$ for which $\underline{Z} \cong \mathcal{L}(X)$, for several objects $X \in \mathcal{C}$. This phenomenon gives rise to the following

I.1.3 DEFINITION The $\mathcal{L}$-genus of an object $\underline{Y} \in \mathcal{L}$-diag($\mathcal{C}$) consists of all isomorphism classes $[X] \subset \mathcal{C}$ with $\mathcal{L}(X) \cong \underline{Y}$.

I.1.4 DEFINITION Two objects X and Y of $\mathcal{C}$ are $\mathcal{L}$-**generically isomorphic** or **of the same $\mathcal{L}$-genus** if $\mathcal{L}(X) \cong \mathcal{L}(Y)$.

I.1.5 REMARK If $\mathcal{C}$ has coproducts it is possible [12, 1.5] to define an additive structure on the class of all $\mathcal{L}$-genera of $\mathcal{C}$ by setting

$$[X]_{\mathcal{L}} + [Y]_{\mathcal{L}} := [X \sqcup Y]_{\mathcal{L}}. \qquad \square$$

I.1.6 REMARK A number of classical concepts are genera in our sense. For example, given a prime p, let L_p denote p-localization on $\mathcal{N}_f$, the category of finitely generated nilpotent groups. The category of p-localizing functors $\mathcal{L} := \{L_2, L_3, L_5, \dots\}$ is discrete. The $\mathcal{L}$-genus of $G \in \mathcal{N}_f$ is known as "the" genus of G. This example, together with several others, is considered in [12, section 1]. $\qquad \square$

I.2 The role of inverse limits

As in section I.1, $\mathcal{C}$ denotes a category and $\mathcal{L}$ a family of localizing functors. We regard $\mathcal{L}$ as small category using the natural transformations coming from the partial ordering of classes of local objects by inclusion. A systematic local$\xrightarrow{\text{info}}$global process benefits from the existence of inverse limits in $\mathcal{C}$ for diagrams modeled on $\mathcal{L}$.

I.2.1 PROPOSITION If $\mathcal{C}$ has inverse limits, then [12, 1.8]

$$\mathcal{C} \underset{\underleftarrow{\lim}}{\overset{\mathcal{L}}{\rightleftarrows}} \mathcal{L}\text{-diag}(\mathcal{C})$$

is a pair of adjoint functors; i.e. for each $X \in \mathcal{C}$ and $\underline{Y} \in \mathcal{L}$-diag($\mathcal{C}$) there is a natural isomorphism of morphism sets

$$\phi : \mathcal{C}(X, \varprojlim \underline{Y}) \xrightarrow{\cong} \mathcal{L}\text{-diag}(\mathcal{C})(\mathcal{L}(X), \underline{Y}). \qquad \square$$

I.2.2 COROLLARY   In the previous proposition suppose $\underline{Y} = \mathcal{L}(Z)$ and assume that the canonical map

$$c : Z \xrightarrow{\;\cong\;} \varprojlim \mathcal{L}(Z)$$

is an isomorphism. Then

$$\phi : \mathcal{C}(X, Z) \xrightarrow{\;\cong\;} \mathcal{L}\text{-diag}(\mathcal{C})(\mathcal{L}(X), \mathcal{L}(Z)). \qquad \square$$

I.2.3 REMARK   If $\mathcal{C}$ has functorial weak inverse limits for the diagrams modeled on $\mathcal{L}$, then the map ϕ in I.2.1 is still onto but will, in general, fail to be 1–1. $\qquad \square$

Due to the significance of the objects $\varprojlim \mathcal{L}(Z)$ we introduce the following terminology.

I.2.4 DEFINITION   Suppose $\mathcal{C}$ has (functorial weak) inverse limits for the diagrams modeled on $\mathcal{L}$. The $\mathcal{L}$-completion of $Z \in \mathcal{C}$ is

$$Z_{\mathcal{L}}^{\wedge} := \varprojlim \mathcal{L}(Z).$$

We offer two familiar examples of $\mathcal{L}$-completions.

I.2.5 EXAMPLE   For $c \geq 1$, let $N_c : \mathcal{G} \to \mathcal{G}$ be the functor which kills commutators of order $\geq c$; i.e. N_1 is abelianization. Thus there is a natural tower

$$\mathcal{N} := N_1 \leftarrow N_2 \leftarrow \cdots \leftarrow N_c \leftarrow \cdots$$

of localizing functors and natural transformations. The $\mathcal{N}$-completion $G_{\mathcal{N}}^{\wedge}$ of a group G is known as the nilpotent completion of G. $\qquad \square$

I.2.6 EXAMPLE   Given a prime p and an integer $n \geq 0$, there is a homology theory $E(n)_*$ whose coefficients are

$$E(n)_* \cong \mathbb{Z}_{(p)}[v_1, \ldots, v_n, v_n^{-1}],$$

where $\dim(v_i) = 2p^i - 2$. Localization with respect to $E(n)_*$ in the sense of Bousfield [3] is conventionally denoted by L_n. The functors L_n form a tower

$$\mathcal{L} := L_0 \leftarrow L_1 \leftarrow \cdots \leftarrow L_n \leftarrow \cdots .$$

Its value on a space (spectrum) X is called the chromatic tower of X. In keeping with our terminology, $X_{\mathcal{L}}^{\wedge}$ could be called the chromatic completion of X. If the telescope conjecture were true, it would follow that the canonical map $X \to X_{\mathcal{L}}^{\wedge}$ is a homotopy equivalence [14]. $\qquad \square$

II   Local-global principles for P-localization of groups and spaces

We now interpret the general ideas for local$\xrightarrow{\text{info}}$global processes in chapter I within the context of localization at sets of primes P in the category of groups $\mathcal{G}$ or in CW_*, the homotopy category of based connected CW-complexes. For the reader's convenience we collect some relevant facts about P-localization in $\mathcal{G}$ and CW_*.

Throughout let P be a set of primes, and let P' denote the (multiplicative closure of) the complimentary set of primes. The P-local objects in $\mathcal{G}$ and CW_*, together with their conceptual relationship, are displayed in the table below.

category	$\mathcal{G}$	$\subset$ {loop spaces} $\underset{\Omega}{\overset{B}{\underset{\longleftarrow}{\longrightarrow}}}$	CW_*
typical object	G	ΩX	X
P-local condition	$g \mapsto g^n, n \in P'$, bijection	$\omega \mapsto \omega^n, n \in P'$, self homotopy equivalence	ΩX is a P-local group up to homotopy

Note that the n-th power map ρ_n on G is a bijection if and only if ρ_n is a self homotopy equivalence of G (regarded as a discrete space). This change in point of view allows us to extend the, a priori purely algebraic, notion of "G is a P-local group" to the geometric-algebraic notion of "G is a P-local group up to homotopy".

II.0.1 THEOREM [15] (compare [12, section 0])   There exists a P-localizing functor $\mathcal{G} \to \mathcal{G}$, $G \to G_P = L_P G$, whose local objects are as described above. □

II.0.2 THEOREM [6] (compare [1], [4] and [8])   For each set of primes P there exists a P-localizing functor $L_P : CW_* \to CW_*$, $X \to X_P = L_P X$, whose local objects are as described above. □

II.0.3 THEOREM   The P-localizing functor $L_P : CW_* \to CW_*$ from II.0.2 has the following properties.

(i) On nilpotent spaces it agrees with the P-localizing functor described in [5], [10], [16], [17].

(ii) L_P induces P-localization of fundamental groups.

(iii) L_P is a homology localization with appropriate twisted coefficients. In particular, Bousfield's $H_*(\bullet; \mathbb{Z}_P)$-localization [3] of X factors uniquely through $X \to X_P$. □

II.1 Local global principles for P-localization of groups

For a prime p, let $L_p : \mathcal{G} \to \mathcal{G}$, $G \to G_{(p)}$ denote localization at p. Further note that, for every group G, localization at the empty set yields the group $L_\emptyset G = G_\emptyset$ in which every element has unique n-th roots, for every $n \geq 1$. Throughout this section we fix the diagram

$$\mathcal{L} := \left\{ \begin{array}{ccccc} L_2 & L_3 & L_5 & \cdots \\ & \searrow \downarrow \swarrow & & \\ & L_\emptyset & & \end{array} \right\}$$

of P-localizing functors on $\mathcal{G}$.

   Guided by the insight gained from corollary I.2.2 we are looking for groups G which map isomorphically into their $\mathcal{L}$-completion. For this purpose it is necessary to look first for groups with a well behaved map into their $\mathcal{L}$-completion. Such a search is necessary because there are groups, like the infinite alternating group, which become trivial as soon as one inverts any given prime. In this case there is not enough local information left to work with.

II.1.1 INFORMATION [12]   There is a certain subcategory $\mathcal{G}'$ of $\mathcal{G}$ satisfying

   (i) all nilpotent groups are in $\mathcal{G}'$;
   (ii) all $L_\emptyset$-local groups are in $\mathcal{G}'$;
   (iii) $\mathcal{G}'$ is closed under directed colimits;
   (iv) for each set of primes P, L_P restricts to an exact functor on $\mathcal{G}'$. □

   The category $\mathcal{G}'$ contains more groups than the ones visible from I.1.1. Our interest in $\mathcal{G}'$ comes from the following

II.1.2 THEOREM [12, 4.1]   If $G \in \mathcal{G}'$, then

   (i) the canonical map $c : G \to G_{\mathcal{L}}^\wedge$ is a monomorphism;
   (ii) c is an isomorphism if and only if G has torsion at no more than finitely many primes. □

   Applications of II.1.2 include the following three results.

II.1.3 THEOREM [12, section 6]   An epimorphism or a monomorphism $\varphi : G \to H$ between two groups in $\mathcal{G}'$ has a section if and only if $\mathcal{L}(\varphi) : \mathcal{L}(G) \to \mathcal{L}(H)$ has a section. □

II.1.4 PROPOSITION   Let G be a group for which $c : G \to G_{\mathcal{L}}^\wedge$ is an isomorphism. Then there is a natural isomorphism

$$\varphi : \mathrm{Aut}(G) \longrightarrow \varprojlim \{\mathrm{Aut}(G_{(p)}) \to \mathrm{Aut}(G_\emptyset)\}. \qquad \square$$

II.1.5 REMARK   Theorem II.1.2 can be used to compute $\mathrm{Hom}(\Pi T_p, H)$, where T_p is p-torsion and $H \to H_{\mathcal{L}}^\wedge$ is an isomorphism; see [12, 5.3]. □

II.2 The $\mathcal{L}$-genus in $\mathcal{G}$

Let $\mathcal{L} = \{L_p \to L_\emptyset\}$ be the diagram of localizing functors on the category of groups $\mathcal{G}$ defined in II.1. At present our knowledge of the $\mathcal{L}$-genus of a group G consists firstly of the study of groups which become trivial after localization at any given prime [2] and, secondly, of the proposition below.

II.2.1 PROPOSITION [12, 7.2]   Let $\underline{G} = \{r_p : G_p \to G_0\}$ be an $\mathcal{L}$-diagram in $\mathcal{G}'$. Set $K_p := \ker(r_p)$ and assume that the inclusion of $I := \cap\mathrm{im}(r_p)$ into G_0 induces isomorphisms $L_p I \to \mathrm{im}(r_p)$. Then the following hold.

(i) If an arbitrary group Γ fits into the commutative diagram below, then $\mathcal{L}(\Gamma) \cong \underline{G}$.

(ii) Any group $\Gamma \in \mathcal{G}'$ which belongs to the $\mathcal{L}$- genus of $\underline{G}$ is an extension of I by the restricted product $\Pi^w K_p$

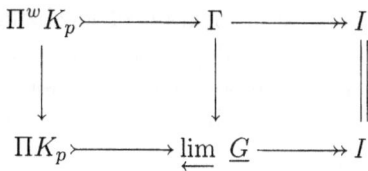

If we replace $\mathcal{G}$ in II.2.1 by $\mathcal{AB}$, the category of abelian groups, we obtain a complete description of the $\mathcal{L}$-genus of an abelian group in terms of invariants from homological algebra.

II.2.2 THEOREM [13, 3.3]   Let $\underline{A} = \{r_p : A_p \to A_0\}$ be an $\mathcal{L}$-diagram in $\mathcal{AB}$. Let $T := \oplus T[A_p]$ be the direct sum of the torsion subgroups of the groups A_p and let $\Pi := \Pi T[A_p]$. If the inclusion of $I := \mathrm{im}(r_p)$ in A_0 induces an isomorphism $I_\emptyset \to A_0$ then the following hold.

(i) $\varprojlim \underline{A}$ fits into the short exact sequence

(λ) $\qquad\qquad 0 \longrightarrow \Pi \rightarrowtail \varprojlim \underline{A} \longrightarrow I \longrightarrow 0.$

(ii) The inclusion $\tau : T \rightarrowtail \Pi$ induces a map $\tau_* : \mathrm{Ext}(I,T) \to \mathrm{Ext}(I,\Pi)$ and the elements of the $\mathcal{L}$-genus of $\underline{A}$ are in bijective correspondence with $(\tau_*)^{-1}\{\lambda\}$ modulo the action of the subgroup of $\mathrm{Aut}(I) \times \mathrm{Aut}(T)$ on $\mathrm{Ext}(I,T)$ which stabilizes λ.

II.3 Local global principles for spaces

In this section $L_p : CW_* \to CW_*$, $X \to X_{(p)}$, denotes localization of a space at the prime p. $L_\emptyset$ denotes rationalization; see section II.0.

We consider local$\xrightarrow{\mathrm{info}}$global processes in CW_* with respect to the diagram

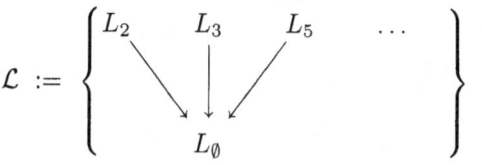

In analogy with section II.1 we begin by describing a large class of spaces for which P-localization is well behaved.

II.3.1 INFORMATION Let CW'_* be the full subcategory of those spaces in CW_* whose fundamental group belongs to $\mathcal{G}'$ and such that L_P induces P-localization of homotopy groups, for each set of primes P. Then CW'_* satisfies

(i) all nilpotent spaces are in CW'_*;
(ii) all $L_\emptyset$-local spaces are in CW'_*;
(iii) CW'_* is closed under directed colimits of cellular maps.   □

Next we describe the relationship between a space X and its $\mathcal{L}$-completion $X_{\mathcal{L}}^{\wedge} := \text{ho}\varprojlim \mathcal{L}(X)$; compare section I.2. In this description we take advantage of the following

II.3.2 REMARK Regard $\mathcal{L}$ as a diagram of localizing functors on abelian groups. The associated functor, sending an abelian group to its $\mathcal{L}$-completion, is left exact, hence has derived functors denoted $\frac{n}{\mathcal{L}}$, for $n \geq 0$. Now let A be an abelian group. The functors $\frac{n}{\mathcal{L}}$ have the following properties.

(i) $A_{\mathcal{L}}^0 \cong A_{\mathcal{L}}^{\wedge} = \varprojlim \mathcal{L}(A)$;
(ii) $A_{\mathcal{L}}^n = 0$ if $n \geq 2$;
(iii) $A_{\mathcal{L}}^1$ is a $\mathbb{Q}$-module;
(iv) if A is torsion, then $A_{\mathcal{L}}^1$ is trivial.   □

II.3.3 THEOREM Let $X \in CW'_*$. If each homotopy group of X has torsion at only finitely many primes, then the following hold.

(i) For each $n \geq 1$, there is a short exact sequence

$$0 \longrightarrow (\pi_{n+1}X)_{\mathcal{L}}^1 \rightarrowtail \pi_n X_{\mathcal{L}}^{\wedge} \longrightarrow \pi_n X \longrightarrow 1.$$

(ii) The homotopy long exact sequence of the fibration $F \hookrightarrow X \xrightarrow{c} X_{\mathcal{L}}^{\wedge}$ breaks up into short exact sequences

$$0 \longrightarrow \pi_n X \dashleftarrow\!\!\!\!\!\!\dashrightarrow \pi_n X_{\mathcal{L}}^{\wedge} \longrightarrow \pi_{n-1}F \longrightarrow 0.$$

which split naturally as indicated for all $n \geq 2$.
(iii) F has rational homotopy groups.   □

II.3.4 COROLLARY Suppose $X \in CW'_*$ has homotopy groups which are torsion and such that each homotopy group has torsion at only finitely many primes. Then X is homotopy equivalent to the product of its P-localizations; $X \simeq \Pi L_p X$.   □

Turning to local$\xrightarrow{\text{info}}$global processes for maps, we have the following theorem. Part of its proof has been obtained in interaction with Shiu Wong.

II.3.5 THEOREM For $X \in CW'_*$ and finite connected $W \in CW_*$ the map

$$\mathcal{L} : [W, X]_* \longrightarrow [\mathcal{L}(W), \mathcal{L}(X)]_*$$

is a bijection if the two conditions below are satisfied

- (i) each of the homotopy groups of X has torsion at only finitely many primes;
- (ii) • X is an H-space or W is a suspension or
  - • the higher homotopy groups of X have finitely generated torsion free quotients.   □

II.3.6 REMARK   At least one of the hypotheses in II.3.5.ii is needed to make the current proof of II.3.5 work. These hypotheses are not based on phenomenological evidence. Hence it is conceivable that the hypotheses (ii) can be considerably weakened or omitted altogether.   □

The following proposition allows us to approach the "lifting problem" in fibrations by working a prime at a time. More precisely,

II.3.7 PROPOSITION   Let $f : X \to Y$ be a fibration between spaces in CW'_*, and let $W \in CW_*$ be finite and connected. If X and Y satisfy conditions (i) and (ii) of II.3.5 then a lifting problem

$$
\begin{array}{ccc}
 & & (X, x_0) \\
 & \overset{\widetilde{\varphi}}{\nearrow} & \downarrow f \\
(W, w_0) & \underset{\varphi}{\longrightarrow} & (Y, y_0)
\end{array}
$$

has a solution if and only if the associated lifting problem

$$
\begin{array}{ccc}
 & & \mathcal{L}(X) \\
 & \overset{\widetilde{\mathcal{L}(\varphi)}}{\nearrow} & \downarrow \mathcal{L}(f) \\
\mathcal{L}(W) & \underset{\mathcal{L}(\varphi)}{\longrightarrow} & \mathcal{L}(Y)
\end{array}
$$

has a solution.   □

III Arithmetic squares

III.1.1 DEFINITION   An arithmetic square in $\mathcal{G}$ or in CW_* is a (homotopy) pullback diagram

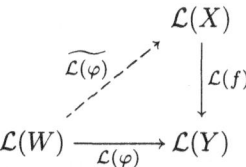

$$
\begin{array}{ccc}
X & \overset{e}{\longrightarrow} & \Pi X_{(p)} \\
\downarrow & & \downarrow \\
X_\emptyset & \underset{e_\emptyset}{\longrightarrow} & (\Pi X_{(p)})_\emptyset
\end{array}
$$

The p-th coordinate map of e is p-localization of X. The bottom map is obtained by rationalizing e.

III.1.2 REMARK   Diagrams of the type $X_\emptyset \to (\Pi X_{(p)})_\emptyset \leftarrow \Pi X_{(p)}$ are qualitatively different from the $\mathcal{L}$-diagrams used in chapters I and II. This is so because the functor sending X to $\Pi X_{(p)}$ is not idempotent, hence is not a localizing functor. Still arithmetic squares provide a very effective local$\xrightarrow{\text{info}}$ global process; see III.1.3 and III.1.4 below.

The link between X and $\mathrm{ho}\varprojlim\{X_\emptyset \to (\Pi X_{(p)})_\emptyset \leftarrow \Pi X_{(p)}\}$ is tighter than the link between X and $\mathrm{ho}\varprojlim\{X_{(p)} \to X_\emptyset\}$. In fact, a diagram of the type $X_\emptyset \to (\Pi X_{(p)})_\emptyset \leftarrow \Pi X_{(p)}$ determines an $\mathcal{L}$-diagram $\{X_{(p)} \to X_\emptyset\}$ together with "higher coherence information"; see [13, section 5]. $\square$

For groups we have

III.1.3 THEOREM   (compare [9, 3.1])   If $G \in \mathcal{G}'$, then

$$
\begin{array}{ccc}
G & \xrightarrow{\ e\ } & \Pi G_{(p)} \\
\downarrow & & \downarrow \\
G_\emptyset & \xrightarrow{\ e_\emptyset\ } & (\Pi G_{(p)})_\emptyset
\end{array}
$$

is an arithmetic square in the category of groups. $\square$

For spaces we have

III.1.4 THEOREM   (compare [9, 3.3])   If $X \in CW'_*$, then X is homotopy equivalent to the basepoint connected component of $\overline{X}$,

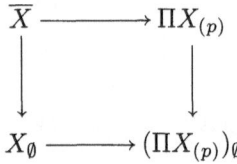

where $\overline{X}$ is defined as the homotopy pullback displayed above. $\square$

Acknowledgements.   This article was written while I was visiting the Centre de Recerca Matemàtica of Barcelona in the summer of 1994. It is a pleasure to express my gratitude for the hospitality extended to me during this time.

Bibliography

1. J. F. Adams, *Localisation and Completion*, Lecture Notes University of Chicago, 1975.

2. A. J. Berrick, C. Casacuberta, *Groups and Spaces with all Localizations Trivial*, 1990 Barcelona Conference on Algebraic Topology; Springer Lect. N. **1509** (1992), 20–29.

3. A. K. Bousfield, *The localization of spaces with respect to homology*, Topology **14** (1975), 133–150.

4. A. K. Bousfield, *Constructions of factorization systems in categories*, J. Pure Appl. Algebra **9** (1977), 207–220.

5. A.K. Bousfield, D.M. Kan, *Homotopy Limits, Completions and Localizations*, Springer-Verlag, Berlin/New York, 1987 (second corrected printing).

6. C. Casacuberta, G. Peschke, *Localizing with respect to self maps of the circle*, Trans. AMS **339**, 117–140.

7. C. Casacuberta, G. Peschke, M. Pfenniger, *On Orthogonal Pairs in Categories and Localisation, Adams Memorial Symposium on Algebraic Topology*, (N. Ray, G. Walker editors); London Math. Soc. Lect. N. Ser. 175, Cambridge UP, 1992, pp. 211–223.

8. E. Dror Farjoun, *Homotopy Localization and v_1-periodic Spaces, Proceedings 1990 Barcelona Conference on Algebraic Topology*, Springer-Verlag, 1992, pp. 104–113.

9. P. J. Hilton, G. Mislin, *Bicartesian Squares of Nilpotent Groups*, Comment. Math. Helv. **50** (1975), 477–491.

10. P. J. Hilton, G. Mislin, J. Roitberg, *Localization of Nilpotent Groups and Spaces*, North-Holland, Amsterdam, 1975.

11. G. Peschke, *Localizing groups with action*, Publ. Matem. **33** (1989), 227–234.

12. G. Peschke, *Localization and Genus in Group Theory*, Trans. AMS, to appear.

13. G. Peschke, P. Symonds, *Various local global principles for abelian groups*, to appear in Publ. Matem.

14. D. C. Ravenel, *Localization with respect to certain periodic homology theories*, Amer. J. Math. **106** (1984), 351–414.

15. P. Ribenboim, *Torsion et localisation de groupes arbitraires*, Lecture Notes in Math. **740** (1978), 444–456.

16. D. Sullivan, *Geometric Topology I and II*, MIT Notes (1970).

17. D. Sullivan, *Genetics of homotopy theory and the Adams conjecture*, Ann. of Math. **100** (1974), 1–79.

Department of Mathematics, University of Alberta
Edmonton, Canada T6G 2G1
e-mail: gepe@jazz.math.ualberta.ca

Progress in Mathematics, Vol. 136
© 1996 Birkhäuser Verlag Basel/Switzerland

On the connectivity of posets in the mapping class group

Yining Xia

ABSTRACT. Let $\Delta_p(\Gamma)$ denote the poset of elementary abelian p-subgroups of Γ, and let Γ_g denote the mapping class group of genus g. We give a complete description of the connectivity of the simplicial complex $|\Delta_p(\Gamma_g)|/\Gamma_g$ for any prime p and integer g.

Introduction

Let Γ be a group and let $\Delta_p(\Gamma)$ be the set of non-trivial elementary abelian p-subgroups of Γ ordered by inclusion, where p is a prime. It is well-known that one can associate a simplicial complex $|\Delta_p(\Gamma)|$ to the poset $\Delta_p(\Gamma)$. That Γ acts by conjugation on $\Delta_p(\Gamma)$ induces an action on $|\Delta_p(\Gamma)|$. The quotient simplicial complex $|\Delta_p(\Gamma)|/\Gamma$ plays a very essential role in studying group cohomology $H^*(\Gamma)$. One can see this point from some fundamental works by Brown [B], Quillen [Q], Webb [W] and some others.

The mapping class group, Γ_g, is defined to be the group of path components of orientation-preserving diffeomorphisms of the oriented closed surface S_g of genus g. The study of the mapping class group is currently a central topic in topology, algebraic geometry and conformal field theory. Our aim in this note is to study connectivity of the simplicial complex $|\Delta_p(\Gamma_g)/\Gamma_g|$ (denoted as Δ_p/Γ_g for short). A deeper question is to describe the homotopy type of the complex Δ_p/Γ_g systematically. One would naively expect to understand at least high dimensional cohomology $H^*(\Gamma_g)$ at prime p via the fibration of the Borel construction

$$\{\Gamma_\sigma\} \longrightarrow E\Gamma_g \times_{\Gamma_g} \Delta_p \longrightarrow \Delta_p/\Gamma_g$$

where Γ_σ denotes as the stabilizer of a cell $\sigma \subset \Delta_p/\Gamma_g$, which can be described as a finite index subgroup of Γ_h^n, a mapping class group of lower genus h with n punctures (see [X]).

Our main result (Theorem 2.1) says that Δ_p/Γ_g is generically disconnected for $p \geq 5$ as long as Γ_g contains p torsion, although Δ_2/Γ_g is always connected and Δ_3/Γ_g is connected if and only if g is 1 mod(3). One should note at this point that the classical Sylow theorem implies that, for a finite group G, the complex Δ_p/G is always connected for any prime p (the empty set is connected).

The rest of this note is divided into two sections. In section 1, we state definitions and some facts concerning the poset of elementary abelian p-subgroups of a group and Nielsen's fixed point data theory of the mapping class group. In section 2, we discuss the connectivity of Δ_p/Γ_g in detail and prove our main result.

1 The poset and fixed point data theory of Γ_g

Given a poset (a partially ordered set) S, an ordered simplicial complex $|S|$ is constructed as follows. The vertices of $|S|$ are the elements of S, and the n-simplices of $|S|$ are the linearly ordered finite subsets $s_0 < \cdots < s_n$ of S. A particular interesting poset associated to a group Γ is the poset of non-trivial elementary abelian p-subgroups of Γ ordered by natural inclusions. This poset $\Delta_p(\Gamma)$ (in fact Δ_p/Γ) is closely related to cohomology of Γ at p via Brown's spectral sequence (see [B]). It also contains the same amount of information as Quillen's category which is used by Quillen to prove some fundamental results of group cohomology theory (see [Q]).

Recall that (see [GMX]), in the general case of an oriented surface with orientation preserving action of a finite group G, the isotropy groups are all cyclic, and the set $\delta(G)$ of fixed point data is given as a set (with multiplicities) of conjugacy classes in G of generators for these isotropy groups G_x, one for each singular orbit. If G is abelian, $\delta(G)$ becomes an unordered set of non-trivial, not necessary distinct elements of G with product the identity. Thus, for any finite abelian subgroup $G \subset \Gamma_g$ $(g > 1)$, we define the fixed point data of G as

$$\delta(G) = \langle g, G \,|\, g_1, \dots, g_q \rangle$$

where g is the genus of the surface S_g, q the number of singular orbits of the G action on S_g and g_i a generator of $stab_G(x_i)$ which acts on the tangent space at the i-th singular point x_i by rotation through $2\pi/|stab_G(x_i)|$. There is a natural action of $Aut(G)$ on the set of fixed point data given by

$$\alpha(\delta(G)) = \langle g, G \,|\, \alpha(g_1), \dots, \alpha(g_q) \rangle$$

for any $\alpha \in Aut(G)$.

It is well-known that a conjugacy class of finite abelian subgroup $G \subset \Gamma_g$ $(g > 1)$ is determined by two algebraic invariants of G: The fixed point data $\delta(G)$ of G up to the $Aut(G)$ action and the free cobordism class of the induced free action of $G/H(\delta(G))$ (for details see [E]), where $H(\delta(G))$ is a well-defined subgroup of G generated by all elements of $\delta(G)$, S_g the closed oriented surface of genus g. We also think $G \subset \Gamma_g$ as a lift $G \subset Diff^+(S_g)$, i.e., an effective action of G on S_g.

When G is a finite cyclic group $\mathbb{Z}/n$, the result above by Edmonds [E] specializes to give the result of Nielsen [N] that an action of $\mathbb{Z}/n$ on S_g up to conjugacy is determined by only its fixed point data up to automorphism actions.

When $H \subset G$ are a pair of elementary abelian p-subgroups of Γ_g, the fixed point data of H can be read from the fixed point data of G.

Let us assume that $\delta(G) = \langle g, G \,|\, g_1, \ldots, g_q \rangle$ and $g_{i_1}, \ldots, g_{i_m} \in H$. Then,

$$\delta(H) = \langle g, H \,|\, g_{i_1}, \ldots, g_{i_1}, \ldots, g_{i_m}, \ldots, g_{i_m} \rangle$$

where each of $i_1, \ldots, i_m$ repeats $|G/H|$ times.

The simplicial complex Δ_p/Γ_g is finite by a classical result which says that Γ_g contains only finite many conjugacy classes of elementary abelian p-subgroups. Our results of connectivity of Δ_p/Γ_g in the next section directly simplify the calculation of $H^*(\Gamma_g)$ at prime p by calculating some pieces of cohomology of certain subgroups of Γ_g at prime p (see [B]).

2 Connectivity of Δ_p/Γ_g

The main result of this section is, for $g > 2$,

THEOREM 2.1.

(A) Δ_2/Γ_g is connected.

(B) Δ_3/Γ_g is connected for $g \equiv 1 \bmod(3)$.

(C) Δ_3/Γ_g is disconnected for $g \equiv 0$ or $2 \bmod(3)$.

(D) If $g = kp + 1$ $(0 < k < (p-3)/2)$ or $g = kp$ $(0 < k < (p-1)/2)$ for an odd prime $p > 3$, then Δ_p/Γ_g is connected.

(E) If $g = kp + 1$ $(k \geq (p-3)/2)$ or $g = kp$ $(k \geq (p-1)/2)$ for an odd prime $p > 3$, then Δ_p/Γ_g is disconnected.

(F) If g is not 0 or $1 \bmod(p)$ for an odd prime $p > 3$, and Γ_g contains a subgroup of order p, then Δ_p/Γ_g is disconnected.

We remark for completeness here that Δ_p/Γ_1 and Δ_p/Γ_2 are connected for any prime p.

We prove Theorem 2.1 by showing that (1) the complex Δ_p/Γ_g is a single point under the assumption of (D); (2) the complex Δ_p/Γ_g is discrete and more than two points under the assumptions of (C) and (F); (3) any vertex of the complex Δ_p/Γ_g is connected to a fixed vertex under the assumptions of (A) and (B); (4) the complex Δ_p/Γ_g has an isolated point and contains at least two points under the assumption of (E).

LEMMA 2.2. If g is not $1 \bmod(p)$, then Γ_g does not contain $\mathbb{Z}/p \times \mathbb{Z}/p$.

Proof. If Γ_g contains $\mathbb{Z}/p \times \mathbb{Z}/p$, then the Riemann-Hurwitz formula implies $2g - 2 = p^2(2h - 2) + p(p-1)n$. This is a contradiction.

The following lemma is well-known.

LEMMA 2.3. There is an action of $\mathbb{Z}/p$ on S_g $(g > 1)$ if and only if the Riemann-Hurwitz formula $2g - 2 = p(2h - 2) + (p-1)n$ has a pair of non-negative solutions (h, n) $(n \neq 1)$.

LEMMA 2.4. *If $g = kp+1$ ($0 < k < (p-3)/2$) or $g = kp$ ($0 < k < (p-1)/2$) for an odd prime $p > 3$, then Γ_g contains only one conjugacy class of elementary abelian p subgroup $\mathbb{Z}/p$.*

Proof. If $g = kp + 1$ and $k < (p - 3)/2$, then the Riemann-Hurwitz formula $2g - 2 = p(2h - 2) + (p - 1)n$ forces that there is only free actions of $\mathbb{Z}/p$ on S_g and no $\mathbb{Z}/p \times \mathbb{Z}/p$ action on S_g. The fixed point data of free actions are the same. So there is only one conjugacy class of $\mathbb{Z}/p \subset \Gamma_g$.

If $g = kp$ and $k < (p - 1)/2$, then the Riemann-Hurwitz formula forces that there are only $\mathbb{Z}/p$ actions of S_g with two fixed points and no $\mathbb{Z}/p \times \mathbb{Z}/p$ action on S_g. By listing all such possible fixed point data, one sees that there is only one conjugacy class of $\mathbb{Z}/p \subset \Gamma_g$.

LEMMA 2.5. *Assume that $\pi \subset \mathbb{Z}/p \times \mathbb{Z}/p \subset \Gamma_g$ is a subgroup of order p. Then, the fixed point data*

$$\delta(\pi) = \langle g, \pi \mid g_1, \ldots, g_q \rangle$$

satisfies the condition that, for any non-trivial element $x \in \pi$, the number of $g_i = x$ must be a multiple of p.

Lemma 2.5 follows above from the definition of the fixed point data.

Proof. (for Theorem 2.1) Recall that vertices of the simplicial complex Δ_p/Γ are represented by all conjugacy classes of elementary abelian p-subgroups of Γ. Let v_0 be the vertex of Δ_2/Γ_g represented by the conjugacy class of $\mathbb{Z}/2 = \langle x \rangle$, where x acts on S_g with the minimal number of fixed points among all involutions of S_g. Such a x exists and is unique up to conjugacy since the number of fixed points determines the fixed point data in the case $p = 2$, thus, determines the conjugacy class of $\mathbb{Z}/2 \subset \Gamma_g$. To reach the conclusion of (A), we only need to show that every vertex represented by a $\mathbb{Z}/2 \subset \Gamma_g$ is connected to v_0 via some vertex represented by a $\mathbb{Z}/2 \times \mathbb{Z}/2$. Note that, of course, every vertex of Δ_2/Γ_g must connect to a vertex represented by a $\mathbb{Z}/2 \subset \Gamma_g$. Applying the Riemann-Hurwitz formula $2g - 2 = 2(2h - 2) + n$ to a covering map $m : S_g \longrightarrow S_g/\mathbb{Z}/2 = S_h$, one obtains that the minimal number of fixed points of $\mathbb{Z}/2$ action is 0 or 2 depending whether g is odd or even.

Case A1: g is odd. v_0 is represented by a $\mathbb{Z}/2$ generated by a free involution x of S_g. The number of fixed points of any $\mathbb{Z}/2$ action on S_g is a multiple of 4. Let v_k be the vertex represented by a $\mathbb{Z}/2$ generated by an involution y acting on S_g with $4k$ fixed points ($1 \leq k \leq (g + 1)/2$). We prove by construction that there exists a $\mathbb{Z}/2 \times \mathbb{Z}/2 \subset \Gamma_g$ which contains both x and y up to conjugacy. We divide two subcases now.

Subcase (1a): $g - 1 - 2k \equiv 0 \bmod(4)$. In this case, one takes $h = (g - 2k + 3)/4$ and $n = 2k$ as a solution for the Riemann-Hurwitz formula $2g - 2 = 4(2h - 2) + 2n$ associated to a $\mathbb{Z}/2 \times \mathbb{Z}/2$ covering $S_g \longrightarrow S_h$. Thus, the fact that $h > 0$ (since $2k \leq g + 1$) implies that there is a surjective map $f : \pi_1(S_h - \{x_1, \ldots, x_{2k}\}) \longrightarrow \mathbb{Z}/2 \times \mathbb{Z}/2$ such that $f(x_i) = a$ ($1 \leq i \leq 2k$), where a and b are generators of $\mathbb{Z}/2 \times \mathbb{Z}/2$. Such a map gives a $\mathbb{Z}/2 \times \mathbb{Z}/2$ action on S_g with the fixed point data $\langle g, \mathbb{Z}/2 \times \mathbb{Z}/2 \mid a, \ldots, a \rangle$, a repeats $2k$ times. Therefore, $a \in \mathbb{Z}/2 \times \mathbb{Z}/2$ acts on S_g with $4k$ fixed points and $b \in \mathbb{Z}/2 \times \mathbb{Z}/2$ acts on S_g freely. This shows that a is conjugate to y and b is conjugate to x.

Subcase (1b): $g - 1 - 2k \equiv 2 \mod(4)$. In this case, one takes $h = (g - 2k + 1)/4$ and $n = 2k + 2$ as a solution for the Riemann-Hurwitz formula $2g - 2 = 4(2h - 2) + 2n$. Then, there is a surjective map $f : \pi_1(S_h - \{x_1, \ldots, x_{2k+2}\}) \longrightarrow \mathbb{Z}/2 \times \mathbb{Z}/2$ such that $f(x_i) = a$ $(1 \leq i \leq 2k)$ and $f(x_{2k+1}) = f(x_{2k+2}) = b$. Such a map gives a $\mathbb{Z}/2 \times \mathbb{Z}/2$ action on S_g with the fixed point data $\langle g, \mathbb{Z}/2 \times \mathbb{Z}/2 \mid a, \ldots, a, b, b \rangle$, where again a repeats $2k$ times. Thus, $a \in \mathbb{Z}/2 \times \mathbb{Z}/2$ acts on S_g with $4k$ fixed points and $ab \in \mathbb{Z}/2 \times \mathbb{Z}/2$ acts on S_g freely. This again tells a is conjugate to y and ab is conjugate to x.

Case A2: g is even. v_0 is represented by a $\mathbb{Z}/2$ generated by an involution x of S_g with two fixed points. In this case, the number of fixed points of any $\mathbb{Z}/2$ action on S_g is 2 modulo 4. Let v_k be the vertex represented by a $\mathbb{Z}/2$ generated by an involution y of S_g with $4k + 2$ $(1 \leq k \leq g/2)$ fixed points. We find a $\mathbb{Z}/2 \times \mathbb{Z}/2 \subset \Gamma_g$ which contains both x and y up to conjugacy. Again, we divide into two subcases as follows.

Subcase (2a): $g - 2k \equiv 0 \mod(4)$. In this case, one takes $h = (g - 2k)/4$ and $n = 2k + 3$ as a solution for the Riemann-Hurwitz formula $2g - 2 = 4(2h - 2) + 2n$. Thus, one finds a $\mathbb{Z}/2 \times \mathbb{Z}/2$ action on S_g with the fixed point data $\langle a, \ldots, a, b, ab \rangle$, where a repeats $2k + 1$ times, by constructing a surjective map $f : \pi_1(S_h - \{x_1, \ldots, x_{2k+3}\}) \longrightarrow \mathbb{Z}/2 \times \mathbb{Z}/2$ with $f(x_i) = a$ $(1 \leq i \leq 2k+1)$, $f(x_{2k+2}) = b$ and $f(x_{2k+3}) = ab$. Again, we see that a is conjugate to y and b is conjugate to x in Γ_g.

Subcase (2b): $g - 2k \equiv 2 \mod(4)$. In this case, one takes $h = (g - 2k - 2)/4$ and $n = 2k + 5$ as a solution for the Riemann-Hurwitz formula $2g - 2 = 4(2h - 2) + 2n$. Then, one sees that a $\mathbb{Z}/2 \times \mathbb{Z}/2$ action on S_g with the fixed point data $\langle a, \ldots, a, b, ab, ab, ab \rangle$, where a repeats $2k + 1$ times, by constructing a surjective map $f : \pi_1(S_h - \{x_1, \ldots, x_{2k+5}\}) \longrightarrow \mathbb{Z}/2 \times \mathbb{Z}/2$ with $f(x_i) = a$ $(1 \leq i \leq 2k+1)$, $f(x_{2k+2}) = b$ and $f(x_{2k+i}) = ab$ $(3 \leq i \leq 5)$. So, a is conjugate to y and b is conjugate to x. We finish the proof of A.

The idea for the proof of part B is the same as the one above for part A. Notice that there is a free action of $\mathbb{Z}/3 = \langle y \rangle$ on S_g when $g = 3k + 1$ $(k > 0)$. Let v_0 be the vertex of Δ_3/Γ_g represented by the $\mathbb{Z}/3$ generated by this free action. Also, notice that the number of fixed points of any $\mathbb{Z}/3 = \langle x \rangle$ action on S_{3k+1} must be a multiple of 3 and the fixed point data of $\langle x \rangle$ is in the form $\langle x, \ldots, x, x^2, \ldots, x^2 \rangle$, where x repeats $3i$ and x^2 repeats $3j$ times $(i + j \leq k + 1)$. Let $v_{i,j}$ be the vertex of Δ_3/Γ_g represented by the $\mathbb{Z}/3$ above. we need to find a $\mathbb{Z}/3 \times \mathbb{Z}/3$ action on S_g which contains both x and y (of course, $\langle x \rangle$ and $\langle y \rangle$ are different conjugacy classes) up to conjugacy. It is easy to see from the Riemann-Hurwitz formula that the number of branch points of a $\mathbb{Z}/3 \times \mathbb{Z}/3$ action on S_{3k+1} is $k + 3 - 3h$ $(h = 0, 1, \ldots, [k/3] + 1)$. Now we divide the proof into three cases.

Case B1: $k - i - j \equiv 0 \mod(3)$. Take $h = (k - i - j)/3$ and $n = i + j + 3$ as a solution for the Riemann-Hurwitz equation $2g - 2 = 9(2h - 2) + 6n$. one is easy to see a $\mathbb{Z}/3 \times \mathbb{Z}/3 = \langle a, b \rangle$ action on S_g, by constructing a surjective map from

$$\pi_1(S_h - \{x_1, \ldots, x_{i+j+3}\}) \longrightarrow \mathbb{Z}/3 \times \mathbb{Z}/3$$

with the fixed point data

$$\langle g, \mathbb{Z}/3 \times \mathbb{Z}/3 \,|\, a, \ldots, a, a^2, \ldots, a^2, g_1, g_2, g_3 \rangle$$

where a repeats i and a^2 repeats j times, $g_t \in \mathbb{Z}/3 \times \mathbb{Z}/3$ $(1 \le t \le 3)$ is determined by $i + 2j$. Namely, if $i + 2j \equiv 0 \bmod(3)$, take $g_1 = g_2 = g_3 = ab$, then $\langle a \rangle$ is conjugate to $\langle x \rangle$ and $\langle b \rangle$ is conjugate to $\langle y \rangle$; if $i + 2j \equiv 1 \bmod(3)$, take $g_1 = ab$, $g_2 = g_3 = a^2 b$, then $\langle a \rangle$ is conjugate to $\langle x \rangle$ and $\langle b \rangle$ is conjugate to $\langle x \rangle$; if $i + 2j \equiv 2 \bmod(3)$, take $g_1 = g_2 = ab$, $g_3 = a^2 b$, then, again, $\langle a \rangle$ is conjugate to $\langle x \rangle$ and $\langle b \rangle$ is conjugate to $\langle x \rangle$.

Case B2: $k-i-j+1 \equiv 0 \bmod(3)$. Take $h = (k-i-j+1)/3$ and $n = i+j+2$ as a solution for the Riemann-Hurwitz equation $2g - 2 = 9(2h - 2) + 6n$, one is easy to obtain a $\mathbb{Z}/3 \times \mathbb{Z}/3 = \langle a, b \rangle$ action on S_g with the fixed point data

$$\langle g, \mathbb{Z}/3 \times \mathbb{Z}/3, \,|\, a, \ldots, a, a^2, \ldots, a^2, g_1, g_2 \rangle$$

where a repeats i and a^2 repeats j times, g_1 and g_2 are determined by $i + 2j$. Namely, if $i+2j \equiv 0 \bmod(3)$, take $g_1 = ab$ and $g_2 = a^2 b^2$; if $i+2j \equiv 1 \bmod(3)$, take $g_1 = ab$ and $g_2 = ab^2$; and if $i + 2j \equiv 2 \bmod(3)$, take $g_1 = a^2 b$ and $g_2 = a^2 b^2$. This amounts to say that $\langle a \rangle$ is conjugate to $\langle x \rangle$ and $\langle b \rangle$ is conjugate to $\langle y \rangle$.

Case B3: $k-i-j-1 \equiv 0 \bmod(3)$. Take $h = (k-i-j-1)/3$ and $n = i+j+2$ as a solution for the Riemann-Hurwitz equation $2g - 2 = 9(2h - 2) + 6n$, one can construct a $\mathbb{Z}/3 \times \mathbb{Z}/3 = \langle a, b \rangle$ action on S_g with the fixed point data

$$\langle g, \mathbb{Z}/3 \times \mathbb{Z}/3, \,|\, a, \ldots, a, a^2, \ldots, a^2, g_1, g_2, g_3, g_4 \rangle$$

where a repeats i and a^2 repeats j times, g_t $(1 \le t \le 4)$ again depends on $i+2j$, i.e., if $i + 2j \equiv 0 \bmod(3)$, takes $g_1 = g_2 = ab$ and $g_3 = g_4 = a^2 b^2$; if $i + 2j \equiv 1 \bmod(3)$, takes $g_1 = g_2 = ab$, $g_3 = ab^2$ and $g_4 = a^2 b^2$; if $i+2j \equiv 2 \bmod(3)$, takes $g_1 = g_2 = ab$ and $g_3 = g_4 = ab^2$. This finishes the proof of part B.

In order to prove part C, we find two different conjugacy classes of $\mathbb{Z}/3$ acting on S_g under the assumption of C.

Case C1: $g \equiv 0 \bmod(3)$. There are at least two solutions $n = 2$, $h = g/3$ and $n = 5$, $h = g/3-1$ for the Riemann-Hurwitz equation $2g-2 = 3(2h-2)+2n$. Thus, there are two $\mathbb{Z}/3$ actions on S_g with different fixed point data.

Case C2: $g \equiv 2 \bmod(3)$. There are at least two solutions $n = 4$, $h = (g-2)/3$ and $n = 7$, $h = (g-5)/3$ for the Riemann-Hurwitz equation $2g - 2 = 3(2h - 2) + 2n$. Again, this says that there are two different conjugacy classes subgroups of order 3 in Γ_g.

Lemma 2.2 implies that Δ_3/Γ_g is disconnected if $g = 0$ or $2 \bmod(3)$.

The proof of the part D follows from Lemma 2.4.

To prove parts E and F of this theorem, it is sufficient to find an isolated vertex of Δ_p/Γ_g under the assumptions of E and F. One needs to find a $\mathbb{Z}/p \subset \Gamma_g$ which is not a subgroup of any $\mathbb{Z}/p \times \mathbb{Z}/p \subset \Gamma_g$.

By Lemma 2.5, we construct a $\mathbb{Z}/p$ action with the fixed point data

$$\delta(\mathbb{Z}/p) = \langle g, \mathbb{Z}/p \mid g_1, \ldots, g_q \rangle$$

in that, for a $x \in \mathbb{Z}/p$, the number of $g_i = x$ is not a multiple of p.

Case E1: $g = kp + 1$ ($p > 3$ and $k \geq (p-3)/2$). One sees that there is a $\mathbb{Z}/p$ action on S_g with p fixed points and the orbit space $S_{k-(p-3)/2}$ from the Riemann-Hurwitz equation $2g - 2 = p(2h - 2) + (p-1)n$. Thus, one can construct a $\mathbb{Z}/p$ action on S_g with the fixed point data

$$\delta(\mathbb{Z}/p) = \langle g, \mathbb{Z}/p \mid a, \ldots, a, a^3, a^{p-1} \rangle$$

where a, a generator of the $\mathbb{Z}/p$, repeated $p - 2$ times. This $\mathbb{Z}/p$ (called π) is not included as a subgroup of any $\mathbb{Z}/p \times \mathbb{Z}/p \subset \Gamma_g$ by Lemma 2.5. Of course, there is a $\mathbb{Z}/p$ free action on S_g by taking $h = k + 1$.

Case E2: $g = kp$ ($p > 2$ and $k \geq (p-1)/2$). One can see at least two $\mathbb{Z}/p$'s acting on S_g, the one with 2 fixed points and the other one with $2 + p$ fixed points. Furthermore, we may assume the fixed point data of one $\mathbb{Z}/p = \pi$ is

$$\langle g, \pi \mid a, \ldots, a, a^{p-1} \rangle.$$

This one is obviously not included in any $\mathbb{Z}/p \times \mathbb{Z}/p$ by Lemma 2.5. So, the vertex of Δ_p/Γ_g represented by π is an isolated point in either one of these two cases.

Case F: If $g > 2$ is not zero or one modulo (p) ($p > 2$), Lemma 2.2 says that Δ_p/Γ_g is discrete. We prove that there are at least two conjugacy classes of order p subgroups in Γ_g if there is an order p subgroup of Γ_g. if there is one in this case. To this end, we find two distinct conjugacy classes of $\mathbb{Z}/p \subset \Gamma_g$ with both of them having n fixed points ($n \in \{3, 4, \ldots, p-1, p+1\}$). Let $0 < \beta < p$ be the unique integer satisfying $n - 1 + \beta \equiv 0 \bmod(p)$ and let $0 < \bar{\beta} < p$ be the unique integer satsfying $n - 3 + \bar{\beta} \equiv 0 \bmod(p)$. Note that $\beta \neq 1$ or $p - 1$ since $n \neq 0$ or 2 $\bmod(p)$ and $\bar{\beta} \neq 1$. We choose two $\mathbb{Z}/p$'s called π_1 and π_2 with

$$\delta(\pi_1) = \langle g, \pi_1 \mid a, \ldots, a, a^\beta \rangle$$

where a a generator repeated $n - 1$ times and

$$\delta(\pi_2) = \langle g, \pi_2 \mid a, \ldots, a, a^{p-1}, a^{\bar{\beta}} \rangle$$

where a repeats $n - 2$ times. π_1 is not conjugate to π_2 since their fixed point data are not the same. Now we finish the proof of theorem 2.1.

The author would like to thank the referee for his or her valuable suggestions. The author would also like to thank the organizers of BCAT for providing such a lively and interesting environment.

References

[B]   K. S. Brown, Cohomology of Groups, Graduate Texts in Math., vol. 87, Springer Verlag, New York Heidelberg Berlin, 1982.

[E]   A. Edmonds, *Surface symmetry I*, Michigan Math. J. **29** (1982), 171–183.

[GMX] H. H. Glover, G. Mislin and Y. Xia, *On the Yagita invariant of mapping class groups*, Topology **33** (1994), 557–574.

[N]   J. Nielsen, *Die Struktur periodischer Transformationen von Flächen*, Danske Vid. Selsk. Mat.-Fys. Medd. **15** (1937), 1–77.

[Q]   D. Quillen, *The spectrum of an equivariant cohomology ring*, Ann. Math. **94** (1971), 549–602.

[W]   P. Webb, *A local method in group cohomology*, Comment. Math. Helvetici **62** (1987), 135–167.

[X]   Y. Xia, *On the cohomology of* Γ_p, to appear, Trans. AMS.

Northern Illinois University, DeKalb, Illinois 60115

Progress in Mathematics

Edited by:

H. Bass
Columbia University
New York
10027
U.S.A.

J. Oesterlé
Dépt. de Mathématiques
Université de Paris VI
4, Place Jussieu
75230 Paris Cedex 05, France

A. Weinstein
Dept. of Mathematics
University of CaliforniaNY
Berkeley, CA 94720
U.S.A.

Progress in Mathematics is a series of books intended for professional mathematicians and scientists, encompassing all areas of pure mathematics. This distinguished series, which began in 1979, includes authored monographs, and edited collections of papers on important research developments as well as expositions of particular subject areas.

We encourage preparation of manuscripts in such form of TeX for delivery in camera-ready copy which leads to rapid publication, or in electronic form for interfacing with laser printers or typesetters.

Proposals should be sent directly to the editors or to: Birkhäuser Boston, 675 Massachusetts Avenue, Cambridge, MA 02139, U.S.A.

BIRKHÄUSER • MATHEMATICS

LM – Lectures in Mathematics

The Ball and Some

R.-P. Holzapfel, Humboldt-U

1995. 168 pages. Softcover
ISBN 3-7643-2835-5

ious twelfth Hilbert problem calls for holomorphic functions in several variables with properties analogous to the exponential function and the elliptic modular function with a view to the explicit construction of (Hilbert) class fields by means of special values. The lecture notes present those functions living on the two-dimensional complex unit ball. In the course of their construction, the reader is introduced to work with complex multiplication, moduli fields, moduli space of curves, surface uniformizations, Gauss-Manin connection, Jacobian varieties, Torelli's theorem, Picard modular forms, Theta functions, class fields and transcendental values in an effective manner.

Please order through your bookseller or:
Birkhäuser Verlag AG
P.O. Box 133
CH-4010 Basel / Switzerland
FAX: ++41 / 61 / 271 76 66
e-mail: 100010.2310@compuserve.com

For orders originating in the USA or Canada:
Birkhäuser
333 Meadowlands Parkway
Secaucus, NJ 07094-2491
USA

BIRKHÄUSER BASEL • BOSTON • BERLIN

BIRKHÄUSER • *MATHEMATICS*

I.M. Gelfand, Rutgers University, New Brunswick, NJ, USA /
M.M. Kapranov, Northwestern Univ., Evanston, IL, USA /
A. Zelevinsky, Northeastern University, Boston, MA, USA

DISCRIMINANTS, RESULTANTS, AND MULTIDIMENSIONAL DETERMINANTS

MTA
Mathematics: Theory and Applications

1994. 532 pages. Hardcover
ISBN 3-7643-3660-9

This book presents a systematic study of discriminants and re-sultants in the general context of projective algebraic geometry. It will be of interest to graduate students and researchers in algebra and algebraic, discrete, and computational geometry. It will also appeal to an audience of combinatorialists, Lie theo-rists, and differential geometers.

CONTENTS: *PART I. General Discriminants and Resultants: PART II. A-Discriminants and A-Resultants   PART III. Classical Discriminants and Resultants   Appendix A. Determinants of Complexes   Appendix B. A. Cayley: On the Theory of Elimination*

Please order through your bookseller or:
Birkhäuser Verlag AG
P.O. Box 133
CH-4010 Basel / Switzerland
FAX: ++41 / 61 / 271 76 66
e-mail: 100010.2310@compuserve.com

For orders originating in the USA or Canada:
Birkhäuser
333 Meadowlands Parkway
Secaucus, NJ 07094-2491
USA

BIRKHÄUSER BASEL • BOSTON • BERLIN